Springer Series on Naval Architecture, Marine Engineering, Shipbuilding and Shipping

Volume 8

Series Editor

Nikolas I. Xiros, University of New Orleans, New Orleans, LA, USA

The Naval Architecture, Marine Engineering, Shipbuilding and Shipping (NAMESS) series publishes state-of-art research and applications in the fields of design, construction, maintenance and operation of marine vessels and structures. The series publishes monographs, edited books, as well as selected PhD theses and conference proceedings focusing on all theoretical and technical aspects of naval architecture (including naval hydrodynamics, ship design, shipbuilding, shipyards, traditional and non-motorized vessels), marine engineering (including ship propulsion, electric power shipboard, ancillary machinery, marine engines and gas turbines, control systems, unmanned surface and underwater marine vehicles) and shipping (including transport logistics, route-planning as well as legislative and economical aspects).

The books of the series are submitted for indexing to Web of Science.

More information about this series at http://www.springer.com/series/10523

Ievgen Bilousov · Mykola Bulgakov · Volodymyr Savchuk

Modern Marine Internal Combustion Engines

A Technical and Historical Overview

Ievgen Bilousov
Kherson State Maritime Academy
Kherson, Ukraine

Mykola Bulgakov
Kherson State Maritime Academy
Kherson, Ukraine

Volodymyr Savchuk
Kherson State Maritime Academy
Kherson, Ukraine

ISSN 2194-8445 ISSN 2194-8453 (electronic)
Springer Series on Naval Architecture, Marine Engineering, Shipbuilding and Shipping
ISBN 978-3-030-49751-4 ISBN 978-3-030-49749-1 (eBook)
https://doi.org/10.1007/978-3-030-49749-1

This Springer imprint is published by the registered company Springer Nature Switzerland AG
The registered company address is: Gewerbestrasse 11, 6330 Cham, Switzerland

Acknowledgments

The authors express their gratitude to the reviewers: Doctor of Technical Sciences, Professor **Andrei Marchenko** from the National Technical University «Kharkiv Polytechnic Institute», (Kharkiv, Ukraine) and Doctor of Technical Sciences, Professor **Boris Timoshevsky** from the National University of Shipbuilding (Nikolaev, Ukraine) for a number of valuable comments, made during the preparation of the manuscript. The authors also thank the entire staff of the **Department of Operation of Ship Power Plants** of the Kherson State Maritime Academy for their help in collecting and preparing materials, as well as for the support, provided in the process of writing this book.

Introduction

Marine internal combustion engines have firmly taken the leading positions in the world merchant fleet as the main and auxiliary sources of energy that provide both the movement of the vessel and the production of electrical and other types of energy used to support technological processes and the crew and passengers life. Due to the peculiarities of the workflow organization, when the fuel combusted directly in the working cylinders, internal combustion engines today have the highest efficiency of converting thermal energy into mechanical work. In addition, the use of heavy fuels for marine diesel engines, produced as residual products of oil refining, has significantly reduced the cost of energy produced. The introduction of new technologies associated with the use of electronic control systems, as well as the transition to the gas and gas-diesel cycles, has successfully solved the issues of improving the environmental performance of marine engines. In operation, internal combustion engines are relatively simple and technologically advanced, have a long resource, often commensurate with the life of the vessel where they are installed.

All of the above has led to the fact that today 98% of the vessels of the world merchant fleet are equipped with internal combustion engines. Ship engine-building is an advanced sector of the global economy, which is developing rapidly, trying to respond to market demands as quickly as possible.

Currently, marine internal combustion engines produce several dozen manufacturers that have their own approaches to solving various problems, related to improving fuel efficiency, environmental friendliness, reliability and reducing operating costs. Summarizing the accumulated experience in solving these problems can be useful to specialists, engaged in the field of design and in the field of engine operation, as well as students and post-graduate students, studying in maritime educational institutions. The book is designed for this very audience. Authors collected information about more than a hundred vessel engines of leading world manufacturers. The information in the book is accompanied by detail drawings that give an idea of the constructive solutions, which a particular manufacturer has applied. The characteristics of the engine allow us to estimate the level of its technical capabilities and the degree of perfection.

It should be noted that the volume and structure of information, provided by various manufacturers, can be very different, therefore the information, given in the book for different engines, can differ in the set of parameters and the fullness of presentation. The information can be used only for preliminary acquaintance with the main characteristics of engines. For more detailed information, please refer to the special literature or directly to the manufacturer.

All of the default engine specifications are given for standard operating conditions in tropical conditions at sea level, according to IACS M28 (1978), which include:

the air pressure at the engine inlet is 0,1 MPa;
the temperature of the air entering the engine is 45 °C;
seawater temperature—32 °C;
relative air humidity at the engine inlet—60%;
the lowest calorific value of the fuel is 42700 kJ/kg.

Contents

Conditional Abbreviations

ACERT	Advanced Combustion Emission Reduction Technology
ADEM	Advanced diesel engine management (system)
AVL	Austrian engineering company
BS EN ISO	British, European and International Standards
CCNR	Central Commission for Navigation on the Rhine
ClassNK (NK)	Nippon Kaiji Kyokai
DF	Dual-fuel
DNV	Det Norske Veritas (classification society)
ECA	Emission Control Area
EEDI	Energy Efficiency Design Index
EU	European Union
EUI	Electronic Unit Injector
GD	Direct Injected Gas
GI	Gas Injector
HFO	Heavy fuel oil
IACS	International Association of Classification Societie's
IMO	International Maritime Organization
ISO	International Organization for Standardization
JIS	Japanese Industrial Standards
LNG	Liquefied Natural Gas
LPG	Liquefied petroleum gas
MARPOL	International Convention for the Prevention of Pollution from Ships
MDO	Marine Diesel Oil
MGO	Marine Gas Oil
SCR	Selective catalytic reduction
TDC	Top dead centre
UN	United Nations
WHR	Waste heat recovery

Chapter 1
Four-Stroke Marine Engines

The internal combustion engines operating in the four-stroke cycle are most widely used on all types vessels with different purpose. Mostly, the vessels use medium and high-speed engines of this type, however, a number of Asian manufacturers produce low-speed four-stroke engines which are capable to operate with direct power transmission to the propeller.

Four-stroke engines, compared with two-stroke low-speed engines, have smaller weight and dimensions. The specific cost of the unit power during the construction of such engines is much less. However, high rotational frequencies lead to an increased level of noise and vibrations during their operation. When using four-stroke diesel engines as the main ones, the use of gearboxes and elastic joints, which somewhat reduce the efficiency of the propulsion unit, is required to coordinate their rotational speed with the propeller rotational speed. Working cylinders of smaller diameter is compensated by an increase in their number, which leads to an increase in the complexity of servicing such engines. The latter circumstance is partly compensated by the smaller dimensions of the parts and, consequently, by their smaller mass.

To reduce the size and weight, as well as to increase the specific power of medium- and high-speed vessels' engines often perform as V-shaped.

Cylinders of four-stroke diesel engines have a simpler design, since they do not have scavenging ports, while cylinder heads are more complicated in design, as they contain camshafts with a drive mechanism, fuel injectors, a cylinder test valve, a valve for limit of maximum pressure, a starting valve and other units.

Medium-speed engines are used on the vessels of the merchant fleet as the main ones in the composition of diesel-geared or diesel-electric gears, as well as for the drive of vessel power plants.

High-speed engines are mainly used in the composition of the main and emergency diesel generators, as well as to drive other process equipment. On small displacement ships, or on ships of the technical fleet with high energy saturation, high-speed

I. Bilousov et al., *Modern Marine Internal Combustion Engines*, Springer Series on Naval Architecture, Marine Engineering, Shipbuilding and Shipping 8,
https://doi.org/10.1007/978-3-030-49749-1_1

engines, which are distinguished by smaller weight and size, can also be used as the main ones.

The widespread introduction of gas turbine supercharging on four-stroke engines, has significantly increased the level of forcing the workflow and brought the average effective pressure to 2.4–3.0 MPa. The solution of a complex of issues, related to the transfer of medium and some high-speed engines to heavy fuel, as well as a significant reduction in specific fuel consumption, allowed four-stroke engines to successfully compete with low-speed two-stroke engines on vessels with a displacement of up to 12,500 tons. As a result, the quantity of four-stroke engines used as the main ones in the structure of the world merchant fleet has now risen to about 25%.

The introduction of new types of fuel systems that provide fuel injection under high pressures, the optimization of mixing processes in the combustion chambers, new approaches to organizing the working process in medium and high-speed engines make them more environmentally friendly, especially in terms of nitrogen oxides NO_x in the exhaust gases. This circumstance makes these engines attractive for vessels of technical fleet, working, as a rule, in water areas with current restrictions on emissions of harmful substances, as well as for passenger and cargo ferries and cruise ships.

1.1 Anglo Belgian Corporation

Anglo Belgian Corporation (ABC) was founded in October 26, 1912 by a group of 9 investors, including brothers Marcel and Richard Drory. In addition, engineer Karen Garens, who owned a license to manufacture diesel engines, joined the company's founders. One year later, ABC began producing engines of 6, 8, 12, 16, 24 and 40 hp, as well as a 2-cylinder engine of 45 hp for marine applications. At the international exhibition in Ghent in 1913, ABC presented 3 engines (8, 16 and 40 hp). After the First World War, ABC signed a license agreement with the British engineering company Paxman Ricardo (London) and obtains the opportunity to produce engines with 1500 min^{-1}, which marked the second significant stage of the company development. In the years of the Second World War, in spite of declining in production, two prototypes of a medium-speed engine were developed—2 and 3-cylinder engines, as well as a new 4-stroke engine of simple action, called DU (Diesel Universal), which later became the basis for the creation of many models of ABC engines in the postwar years.

The engine type DU in its parameters fully meet the requirements of the post-war market and quickly gained popularity. Soon 5, 6 and 8 cylinder variants of this engine were created. Subsequently, ABC equipped its DU series engine with a turbocharging system, then releasing this engine under the DUS series. As a result, the engine power increased by 1.5 times compared with the base variant.

The next generation of ABC engines was created as a further development of the DU engine concept. The new engine received the designation DX in the version without turbocharging and DXS with turbocharging. Speed was increased from 600 to 750 min^{-1}. The next stage in the development of the ABC engine design was the equipping of turbo charged engines with charge air coolers. This modification received the designation DXC. The power of this series of engines was two times more, then the power of the basic version (DX series engines).

In 1973, ABC acquired a license to manufacture the high-speed engine PA4 from the French company Semt-Pielstick. At the same time, the company created a completely new type of high-speed engine, designated DZ. The new engine could work on heavy fuel with a viscosity of up to 380 cSt. The engine speed was 1000 min^{-1}.

From 1980 to 1985, the company launched the first DZC series engines. This modified engine brought the company success, reaching, in addition to DXC, 75% of the total trading. In addition to optimizing existing engines, both types of engines have been modified and adapted to work on gas. Due to significant engine improvements, the DZC type took a leading position in its market segment.

In 1997, ABC began the development of a V-shaped DZ engine. 12 and 16 cylinder engines have expanded the range of the proposed unit capacity to 5000 hp. The engine, produced at the plant in Gent under the designation VDZC (Fig. 1.1), meted all modern requirements, which largely contributed to the growth of sales by the company. In 2011, ABC invested in the development and production of a new engine, expanding the range of offered products. The engineering and design department of ABC, together with the Austrian engineering company AVL worked intensively on this project. The new DL36 series (Fig. 1.2) allows to expand the market for ABC in the segment of vessels with a large deadweight, such as coasting, offshore and military vessels, powerful tugs and ferries. With an 8DL36 engine (5200 kW) and with a continuation of the series in V-versions (unit capacity up to 4400 kW), ABC will be able to provide shipyards with optimal solutions for almost all types and sizes of vessels.

1.1.1 The Engine of the VDZC Series

The engine of the VDZC series (Fig. 1.1) is a medium- or high-speed (depending on the modification) four-stroke V-shaped engine with gas turbine supercharging and cooling of charge air, left or right-hand rotation. The fuel system is a mechanical type, with a direct fuel injection. The engine is used as a main for tugs, riverboats, locomotives and generator sets. All models of these engines can operate on both light and heavy fuel, as well as on biofuel of plant and animal origin, gas and biogas. It is also possible dual-fuel performance. Meets all requirements for emissions of exhaust gases in accordance with IMO Tier-2, CCNR-2 and EU3A.

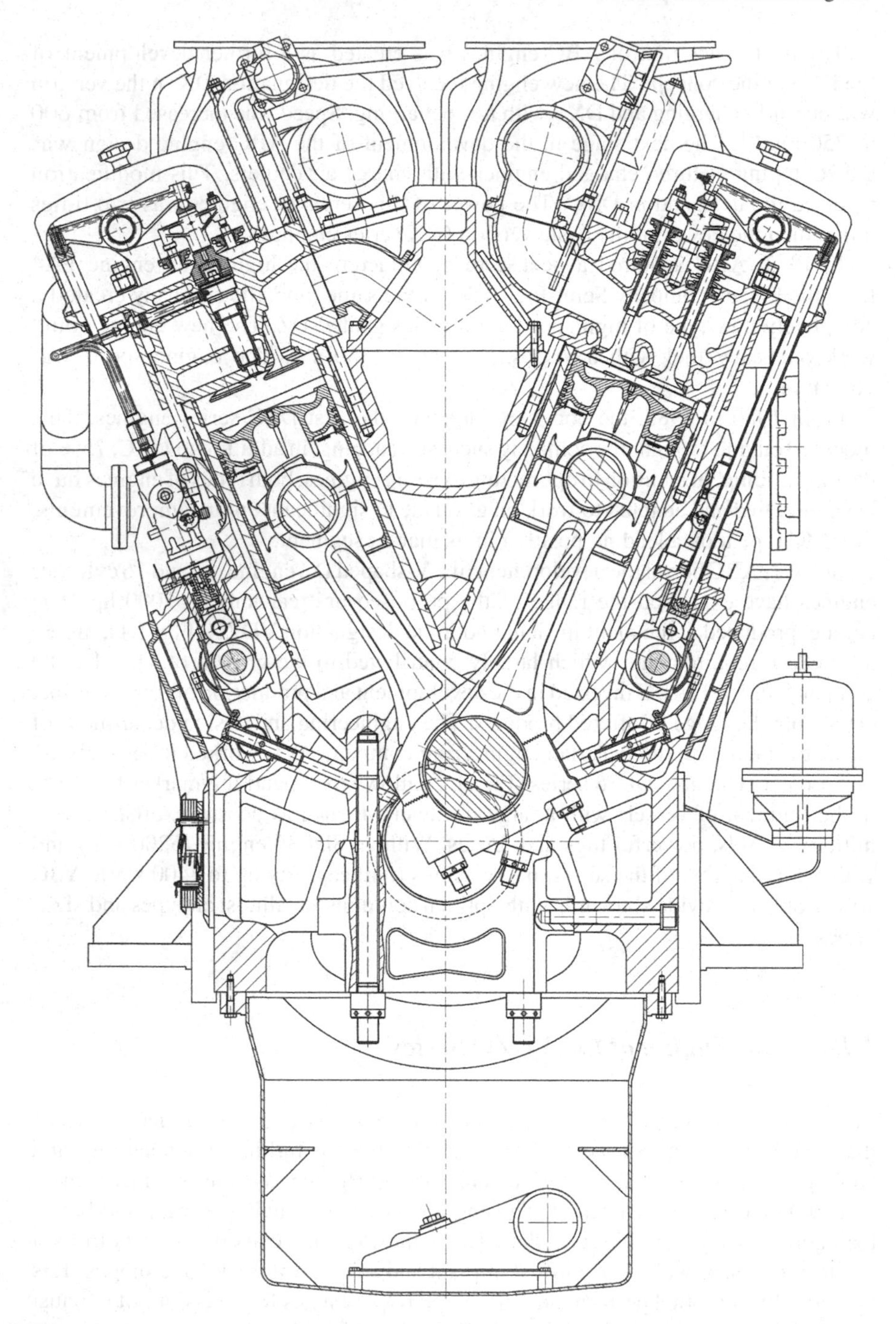

Fig. 1.1 Cross-section of the VDZC engine series [1]

Main technical characteristics of VDZC series engines

Parameter	Value
Number and cylinders arrangement	12, 16 V-shaped
Cylinder bore (mm)	256
Piston stroke (mm)	310
Cylinder capacity (dm^3)	
12 cylinders	191.5
16 cylinders	255.2
Rotation speed (min^{-1})	720–1000
Cylinder power at 1000 min^{-1} (kW)	234
Cylinder power at 720 min^{-1} (kW)	172
Compression ratio	12.1
Maximum cycle pressure (MPa)	15.0
Mean effective pressure at 1000 min^{-1} (MPa)	1.88
Brake specific fuel oil consumption (g/kWh)	205
Mean piston speed at 1000 min^{-1} (m/s)	10.3

Dimensions and weight engines series VDZC

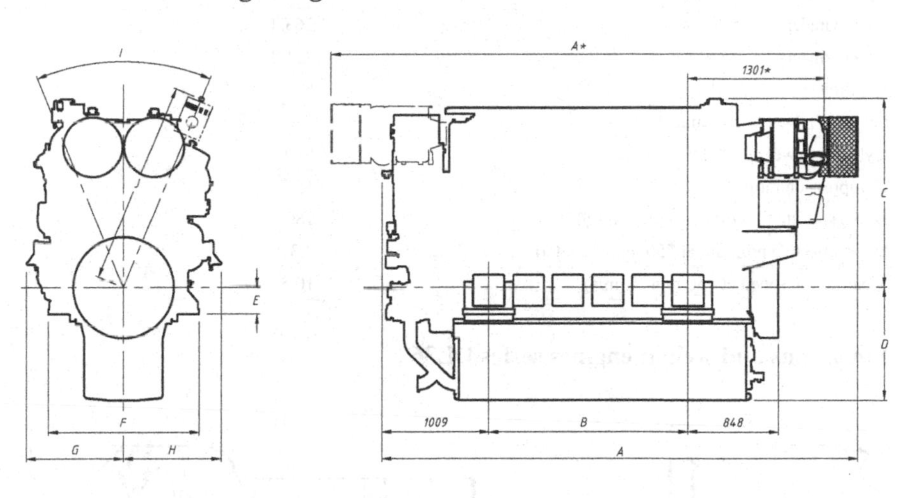

Engine version	A (mm)	A* (mm)	B (mm)	C (mm)	D (mm)	E (mm)
12DZC	4529	4686	1900	1780	1060	250
16DZC	5289	5446	2660	1780	1060	250
Engine version	F (mm)	G (mm)	H (mm)	I (°)	J (mm)	Weight (kg)
12DZC	1425	925	925	45	1950	18,000
16DZC	1425	925	925	45	1950	21,750

1.1.2 The Engine of the DL36 Series

The engine of the DL36 series (Fig. 1.2) is a medium-speed four-stroke in-line engine of left or right rotation with two-stage gas turbine supercharging and cooling of the charge air after each stage. Fuel system accumulator (Common Rail) or mechanical type, with direct fuel injection. The engine is used as the main one for tugboats, river and sea vessels, and stationary generator sets. All models of these engines can operate on both light and heavy fuel, as well as on biofuel of plant and animal origin, gas and biogas. It is also possible dual-fuel performance. Meets all the requirements for the content of harmful emissions in the exhaust gases in accordance with IMO Tier-3 (when Common Rail fuel system is installed). All engines are certified for conformity with ISO 9001:2008.

Main technical parameters of DL36 series engines

Parameter	Value
Number and cylinders arrangement	6, 8 in-line
Cylinder bore (mm)	365
Piston stroke (mm)	420
Cylinder capacity (dm^3)	
6 cylinders	263.4
8 cylinders	351.2
Rotation speed (min^{-1})	600–750
Cylinder power at 750 min^{-1} (kW)	650
Cylinder power at 600 min^{-1} (kW)	520
Compression ratio	15.5
Brake specific fuel oil consumption (g/kWh)	180
Mean effective pressure at 750 min^{-1} (MPa)	2.39
Mean piston speed at 750 min^{-1} (m/s)	10.5

Dimensions and weight engines series DL36

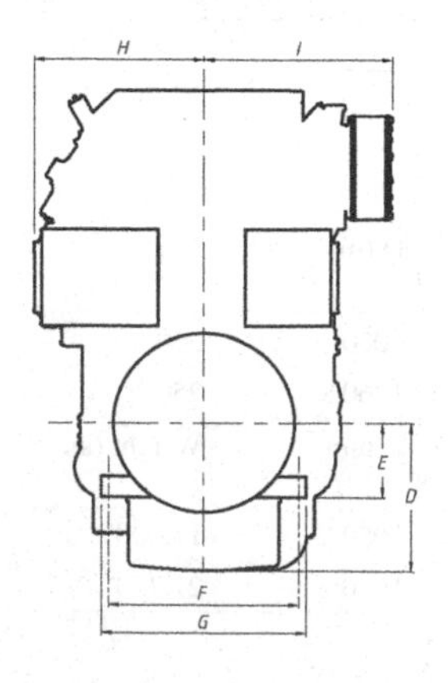

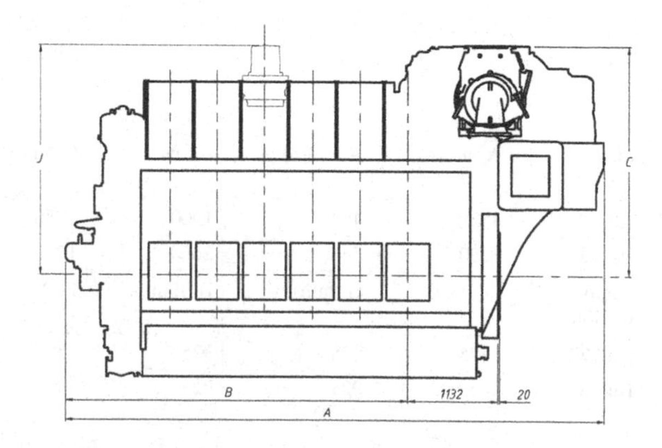

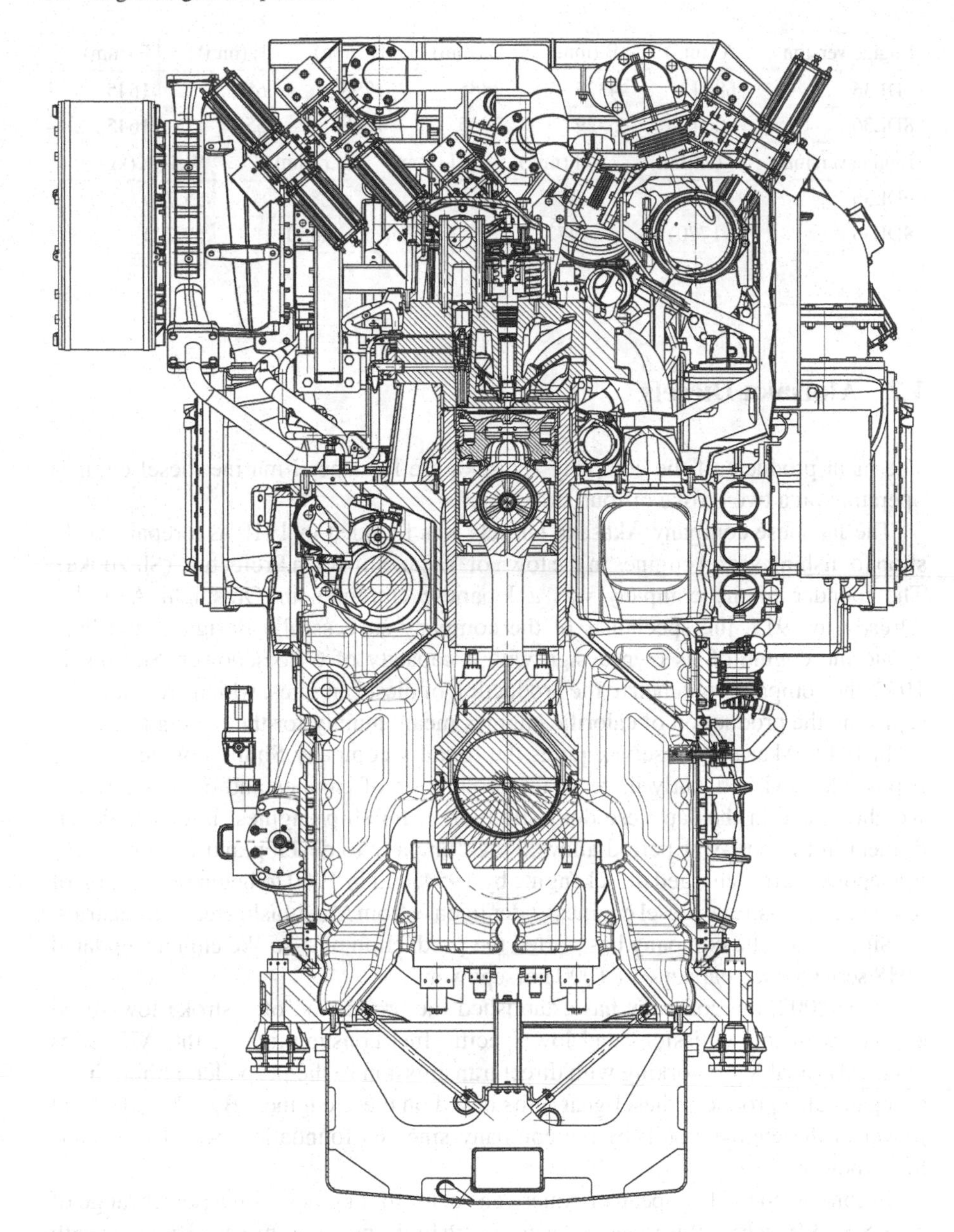

Fig. 1.2 Cross-section of the engine series DL36 [2]

Engine version	A (mm)	B (mm)	C (mm)	D (mm)	E (mm)	F (mm)
6DL36	6804	4329	2885	1298	650	1645
8DL36	8004	5529	2885	1298	650	1645

Engine version	G (mm)	H (mm)	I (mm)	J (mm)	Weight (kg)
6DL36	1770	1477	1626	2850	55,700
8DL36	1770	1477	1626	2850	67,700

1.2 Akasaka Diesels

The main products of the company Akasaka are low-speed marine diesel engines operating on a two-stroke or four-stroke cycle.

The Japanese company Akasaka Diesels was founded in 1910 as a repair workshop for fishing vessel engines in the town of Yaizu, Shizuoka Prefecture (Shizuoka). The founder of the company was a Japanese entrepreneur Otoshichi Akasaka. Already in 1915, the specialists of the company successfully designed and built a calorific engine of its own design with a capacity of 6 horsepower. Starting in 1933, the company switched to the production of diesel engines, which are gradually replacing the production of calorific engines meant mainly for the fishing fleet.

In 1942, Akasaka Diesels acquires the foundry company Shunyo, which makes it possible to significantly increase the production of engines and occupy a rather weighty niche in the Japanese domestic market for ship engines. Intensive development in the post-war period allowed the company to develop and launch a 900-horsepower turbo-charged diesel engine by 1954, and in 1960 to begin production of low-speed two-stroke diesel engines under license from Mitsubishi Heavy Industries.

Since 1967, the company has started the production of 4-stroke engines updated UHS series with a capacity of 1000 horsepower.

Since 2002, the company has established the production of 4-stroke low-speed engines with low emissions and low specific fuel consumption of the AX series (Fig. 1.3) capable of working with direct transmission to the propeller, although the company also produces diesel-gear units based on these engines. By 2005, the total power of the engines, built by the company since its foundation, was 15 millions horse power.

In contrast to the European or American sector for engines with a power range of 500–8000 kW, where this range is covered with high and medium-speed engines with a fixed cylinder size, and the required power is reached by their number, Japanese low-speed engines are available in six-cylinder versions with changing cylinder diameters for achieve the required power output. Thus, over the past years, the Akasaka Diesels program covers six-cylinder engines with a range of cylinder diameters of 280, 310, 340, 370, 380, 410 and 450 mm. The stroke ratio of the piston to the cylinder diameter in these the engines is in the range of 1.95–2.2.

The advantages of this approach are:

- increased reliability and ease of maintenance, which is associated with a smaller number of components;
- reduction of noise and vibration;
- improved fuel consumption and lubricating oil;
- the best adaptability to work on heavy fuels.

The disadvantages of low-speed four-stroke engines can be attributed to a rather high rotation speed of the propeller in case of direct transmission; much weight and dimensions.

The power range, which is covered by engines manufactured by Akasaka, is 375–3300 kW. The largest four-stroke low-speed engine model A45S develops power 3309 kW at a rotational speed of 220 min^{-1}.

Currently, Akasaka Diesel products are mainly intended for the domestic market of Japan, but recently the company has undertaken a number of efforts to create a management system and service its engines in Southeast Asia and Oceania. As a result of these actions, shipyards in Indonesia, South Korea, the Philippines and Taiwan are increasingly showing interest in engines, manufactured by this company. Since 1996, all products of the company have been certified by the ISO 9001 NK quality system.

1.2.1 AX28 Series Engine

AX series engines are a further development of the A series engines that have worked well in the domestic market of Japan. The total number of engines produced series A, exceeded 800 units. As a result of the lead works, it was possible to reduce the weight of the engine, as well as to reduce fuel consumption compared with the A series engines. The AX series engines are a low-speed four-stroke engine of the new generation. The power range of the AX series of engines is in the range from 1150 to 2000 kW. They are used as main ones on vessels with deadweight from 2000 to 4000 tons, General Cargo class, Oil Tanker, Chemical Tanker, Tanker LPG, etc. The main features and benefits of the AX series of engines are as follows:

- reliability and durability, based on proven technical solutions;
- low rotational speed and high torque, which makes possible to transfer power to the propeller through a direct transmission, resulting in high propulsion efficiency, and reduced fuel consumption;
- the optimal configuration of the combustion chamber and optimal fuel system construction allow the engines to operate on heavy fuels from 180 to 240 cSt meeting the requirements for the NO_x content in the exhaust gases according to IMO Tier-III.

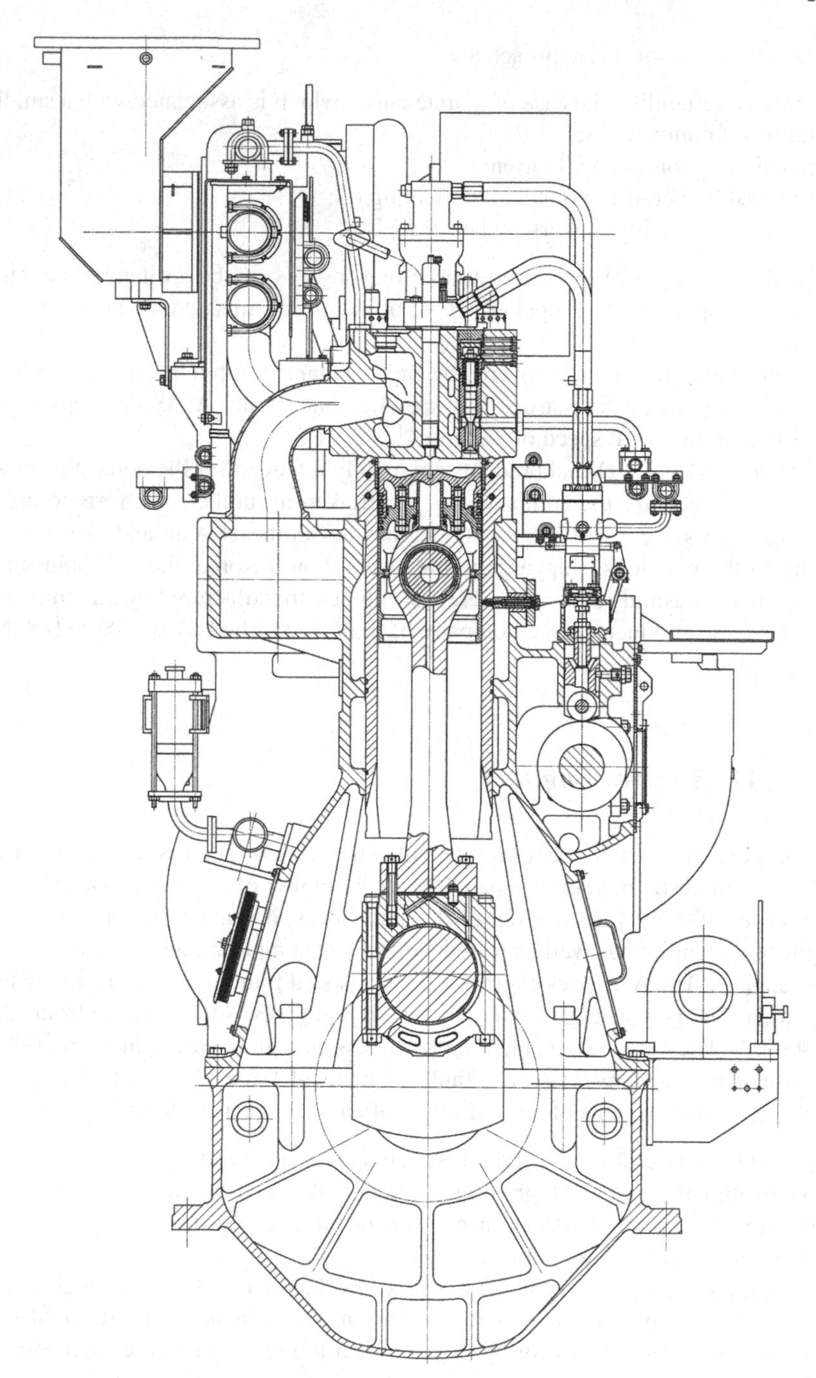

Fig. 1.3 Cross-section of the engine AKASAKA AX28 [3]

Main technical parameters of AX28 series engines

Parameter	Value
Number and cylinders arrangement	6 in-line
Cylinder bore (mm)	280
Piston stroke (mm)	600
Cylinder capacity (dm^3)	221.7
Rotation speed max (min^{-1})	310–320
Rotation speed min (min^{-1})	260
Cylinder power at 310 min^{-1} (kW)	196
Cylinder power at 320 min^{-1} (kW)	220.5
Maximum cycle pressure (MPa)	13.5
Brake specific fuel oil consumption (g/kWh)	184
Mean effective pressure at 320 min^{-1} (MPa)	2.0
Mean piston speed at 320 min^{-1} (m/s)	6.4

Dimensions and weight engines series AX28

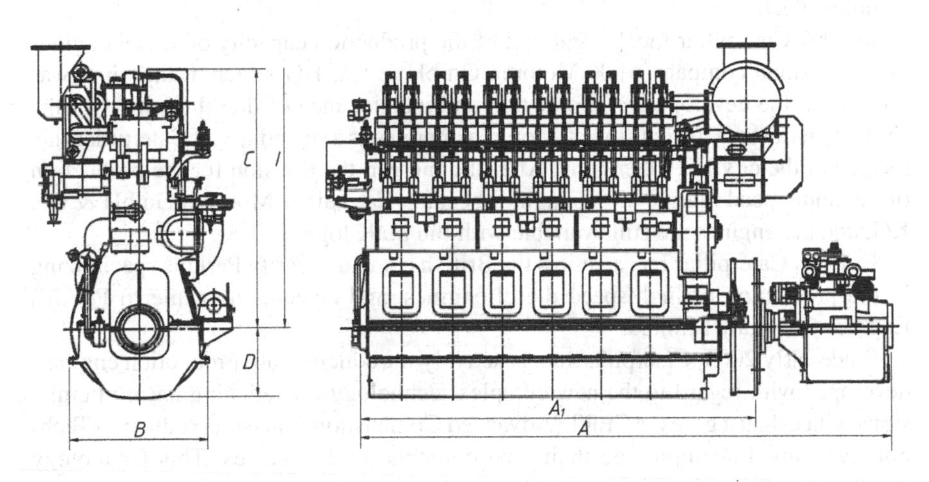

Engine version	A (mm)	A_1 (mm)	B (mm)	D (mm)	C (mm)	I (mm)	Weight (kg)
AX28	4882	3342	1360	720	2361	2830	26,300

1.3 Caterpillar Inc.

The predecessor of the company was the Stockton Wheel Company, founded in 1883 by the brothers Charles Henry Holt and Benjamin Holt in California. In 1886, the company has begun producing combine harvesters, and in 1904 produced its first commercially successful steam crawler tractor, which soon began to be sold under the brand name Caterpillar (from English caterpillar). In 1906, a subsidiary Aurora Engine Company headed by Pliny Holt, began production of gasoline engines, to replace the steam engines used previously on tractors. In 1931, under the leadership of Charles Holt, the first six-cylinder diesel engine was designed, which is widely used to drive agricultural and road equipment, produced by the company. In the same year, the company has been begun production of mobile diesel generator sets.

In 1938, Caterpillar produced the first series of engines specifically designed for use on small vessels. The series consisted of three models, two in-line six-cylinder D11000 and D13000 with 80 and 100 hp. respectively, and the eight-cylinder V-shaped D17000 with 135 hp. All engines had a rotational speed of 900 min^{-1} and were equipped with a reverse gearbox.

On December 2, 1982, the first production engine of the Model 3500 (Fig. 1.4) was assembled at the newly built Lafayette Engine Center in Lafayette, Indiana. Today, Lafayette Engine Center is the main manufacturing unit of Caterpillar Inc. for the production of diesel engines for marine, oil, electricity, locomotive and industrial applications. Currently, the plant produces engines of the three series 3500, 3600 (Figs. 1.5 and 1.6) and C280, which are widely used on the commercial vessels and technical fleet.

In 1997 Caterpillar Inc. bought part of the production capacity of vessel engines of the German company MaK Motoren GmbH & Co. KG which at that time was a leader in the development, production and maintenance of diesel engines for the shipping industry. Currently MaK Motoren operates as a subsidiary of Caterpillar Inc. located in the city Kiel, Germany. After the merger, the division for the production of medium-speed marine diesel engines became Caterpillar Motoren GmbH & Co. KG, and the engines are still available with the MaK logo.

In 1998, Caterpillar Inc. acquires the British company Varity Perkins specializing in the production of high-speed diesel engines, and changes its name to Perkins Engines Company Limited.

Since early 2003, Caterpillar Inc. is actively introducing into production engines, developed with regard to the new complex technology for reducing harmful emissions with exhaust gases, ACERT (Advanced Combustion Emission Reduction Technology), aimed at improving their environmental performances. This technology combines technical improvements of a number of main engine systems: air supply, fuel supply, electronic control, as well as additional cleaning systems. The electronic unit (ADEM) controls the operation of all engine systems to achieve a reduced level of harmful substances in the exhaust gases and to keep high fuel efficiency.

In 2010, Caterpillar Inc. bought the company Electro-Motive Diesel, Inc., a major American manufacturer of high-speed engines for locomotives and vessels.

In 2011 Caterpillar Inc. acquires the German company MWM GmbH specializing in the production of gas engines.

Currently, Caterpillar Inc. offers two brands of marine engines. Medium-speed MaK engines with a power capacity from 1020 to 16,000 kW and high-speed Caterpillar engines with a power capacity from 162 to 5420 kW, which are performed on vessels as main and auxiliary.

1.3.1 Caterpillar Engine 3500 Series

Caterpillar engine 3500 series (Fig 1.4). The engines 3500 series appeared on the market in 1981, at that time, the power covered a range of 507–1417 kW. The engines were manufactured in 8, 12 and 16 cylinder versions with a V-shaped arrangement of working cylinders. The number of working cylinders, according to the company's classification, are indicated after the designation of the series, so the 12-cylinder engine has the designation—3512. In 1995, the engines were modernized, in particular, the pump-injectors with a mechanical drive and control from the hydromechanical speed regulator were replaced by the injectors with an electronic control from the microprocessor unit, installed on the engine; compression ratio increased from 13.5 to 14; boost pressure increased and as a result the cylinder power capacity increased from 84 to 140 kW. This modification of the engine is available with the index "B".

Main technical parameters of engines series 3516

Parameter	Value
Number and cylinders arrangement	16 V-shaped
Cylinder bore (mm)	170
Piston stroke (mm)	190
Cylinder capacity (dm^3)	69
Rotation speed max (min^{-1})	1200
Rotation speed min (min^{-1})	450
Cylinder power at 1200 min^{-1} (kW)	84
Compression ratio	14
Air charging pressure (MPa)	0.254
Maximum cycle pressure (MPa)	13.5
Brake specific fuel oil consumption (g/kWh)	206
Mean effective pressure at 1200 min^{-1} (MPa)	2.025
Mean piston speed at 1200 min^{-1} (m/s)	6.8

Dimensions and weight engines series 3516

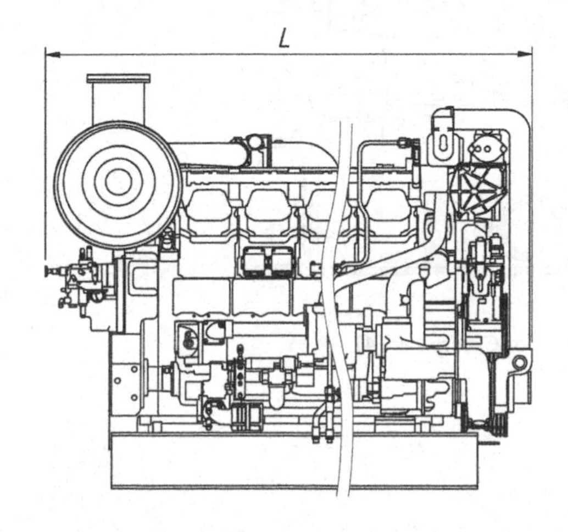

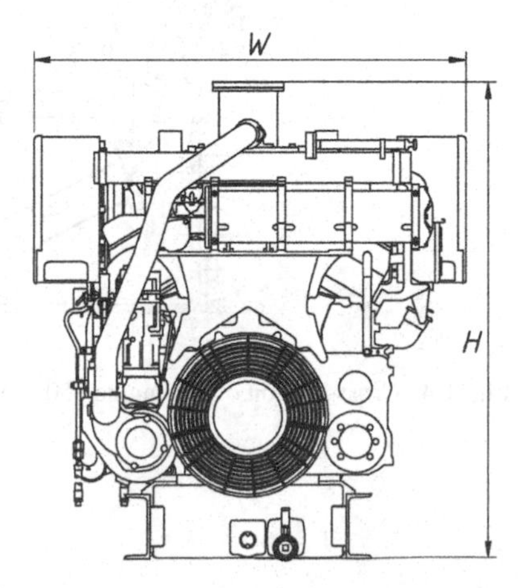

Engine version	W (mm) B (mm)	H (mm) C (mm)	L (mm)	Weight (kg)
3508	1702	1719.5	2136	4309
3512	1702	1719.5	2676	6078
3516	1702	1719.5	3366	7484

The growth of thermal and mechanical loads required the replacement of a one-piece piston with a composite head which is made of a steel machined forging, and the skirt is made of aluminum alloy. The piston head is cooled by oil, coming from the general circulation lubrication system to the nozzles, installed in the crankcase, from which a jet of oil is directed into the holes in the piston skirt leading to the inner

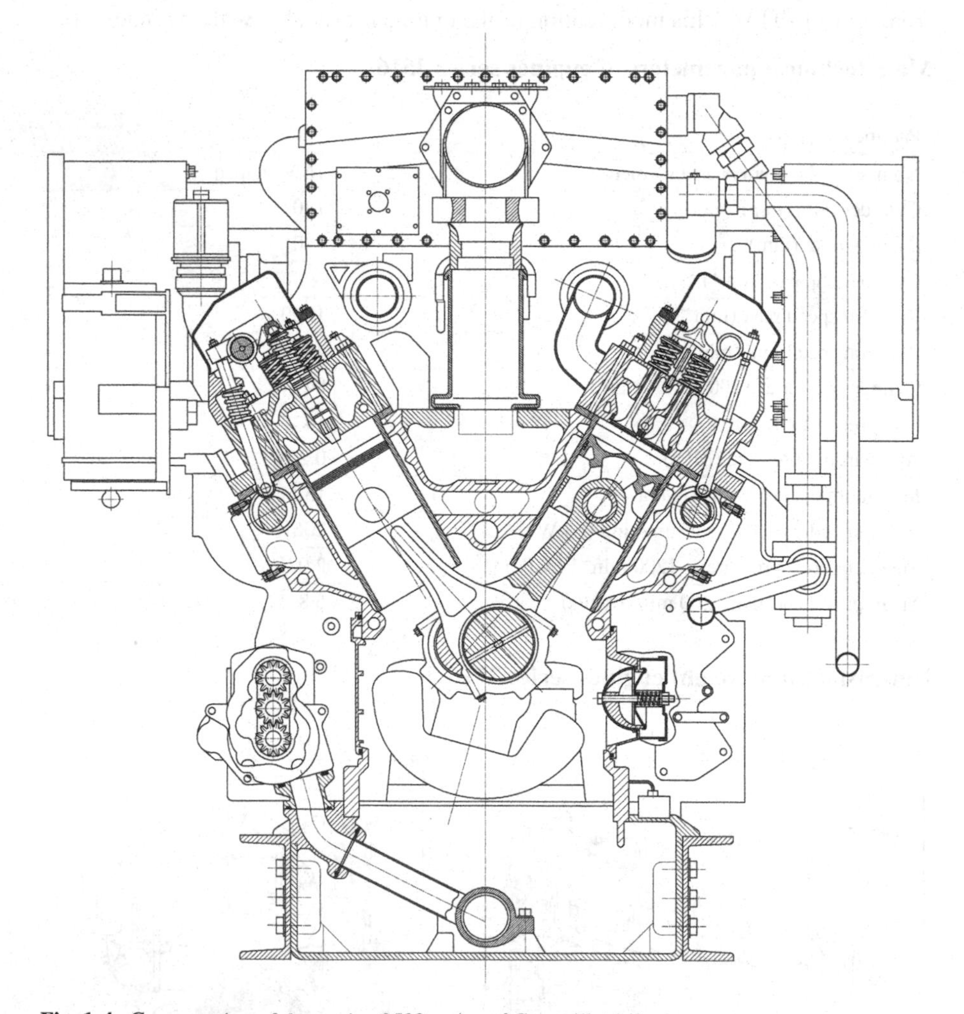

Fig. 1.4 Cross-section of the engine 3500 series of Caterpillar [4]

chambers of the head. Intensive cooling of the head allowed to raise the first piston ring, which significantly improved the environmental performance of the engine. The last engine upgrade included a boost in rotational speed, which was brought to 1800 rpm, while the cylinder engine power was increased to 198.5 kW. In this performance, the engine received the designation "C". In addition to the basic version of the engine, engines with increased power are produced, in which the piston stroke is increased to 215 mm. It should be noted that the company produces engines that can be optimized for various modes of operation as part of the propulsion installations of the vessel, and as a part of the drives of various units used on vessels.

1.3.2 *The Caterpillar Engine 3600 Series*

The Caterpillar engine 3600 series first appeared on the market in 1986. Series 3600 is a high-speed four-stroke diesel engines designed for: offshore industry, industrial energy, as well as for use in drilling rigs, to drive pumps, excavators and locomotives. They cover the power range from 1500 to 4500 kW and have a cylinder diameter of 280 and a piston stroke of 300 mm. Cylinder capacity is 300 kW at 1000 min^{-1} and 260 kW at 900 min^{-1}. This series of engines can operate on both light distillate fuels of the type MDO and heavy residual heavy fuels of the type HFO. The engines with the number of cylinders 6 and 8 are produced in-line version (Fig. 1.5), and 12 and 16 with a V-shaped arrangement of cylinders (Fig. 1.6). Fuel injection is performed with a mechanically driven pump-injector unit. Until 2003, the pump-injectors had a mechanical adjustment. Subsequently, the engines began to be equipped with a pump-nozzle with electronic regulation type EUI (Electronic Unit Injector) from the installed, in this case, microprocessor control unit, which is part of the ACERT system. The use of this system on the engine allows to effectively manage the combustion process. As a result, the engines of this series meet modern and future emission standards, without decreasing reliability, durability and efficiency.

Main technical parameters of engines series 3616

Parameter	Value
Number and cylinders arrangement	6, 8 in-line; 12, 16 V-shaped
Cylinder bore (mm)	280
Piston stroke (mm)	300
Cylinder capacity (dm^3)	296
Rotation speed max (min^{-1})	750, 800, 900, 1000
Rotation speed min (min^{-1})	350
Cylinder power at 1000 min^{-1} (kW)	307.5
Compression ratio	12.4
Air charging pressure (MPa)	0.195

(continued)

(continued)

Parameter	Value
Maximum cycle pressure (MPa)	16.2
Brake specific fuel oil consumption (g/kWh)	198.2
Mean effective pressure at 1000 min^{-1} (MPa)	1.998
Mean piston speed at 1000 min^{-1} (m/s)	10.0

Dimensions and weight engines series 3616

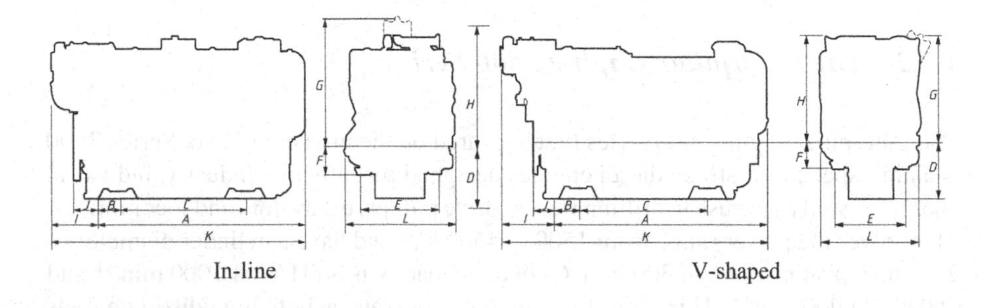

Engine version	A (mm)	B (mm)	C (mm)	D (mm)	E (mm)	F (mm)	G (mm)
3606	3261	265	2050	841	1120	450	2035
3608	4081	265	2870	841	1120	450	2035
3612	3657	300	2300	976	1120	450	1850
3616	4577	300	3220	976	1120	450	1850

Engine version	H (mm)	I (mm)	J (mm)	K (mm)	L (mm)	Weight (kg)
3606	1785	727	360	3988	1748	34,070
3608	1785	727	360	4808	1748	41,390
3612	2255	905	360	4562	1714	51,230
3616	2255	905	360	5482	1714	64,470

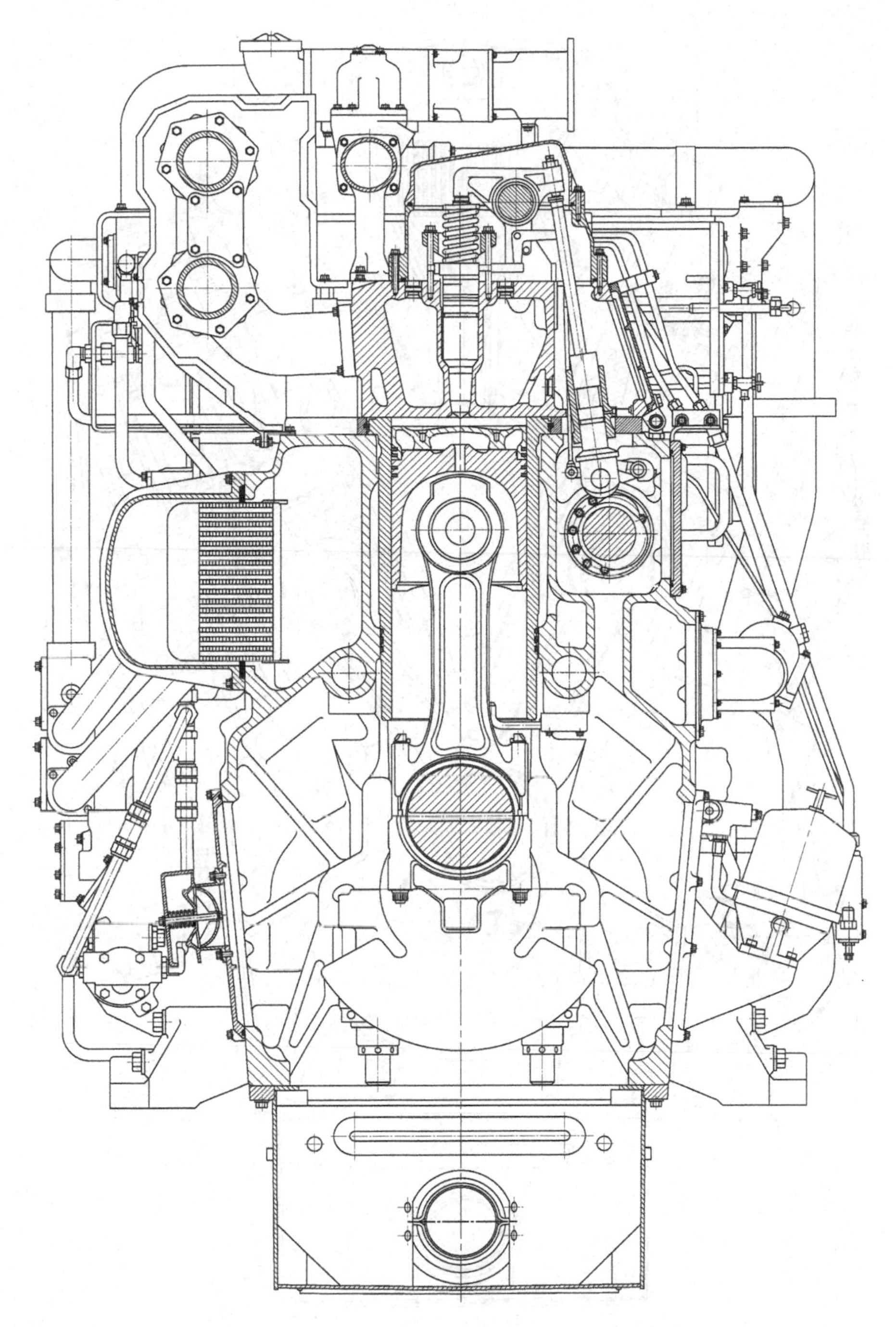

Fig. 1.5 Cross-section of the engine 3600 series of Caterpillar (in-line) [5]

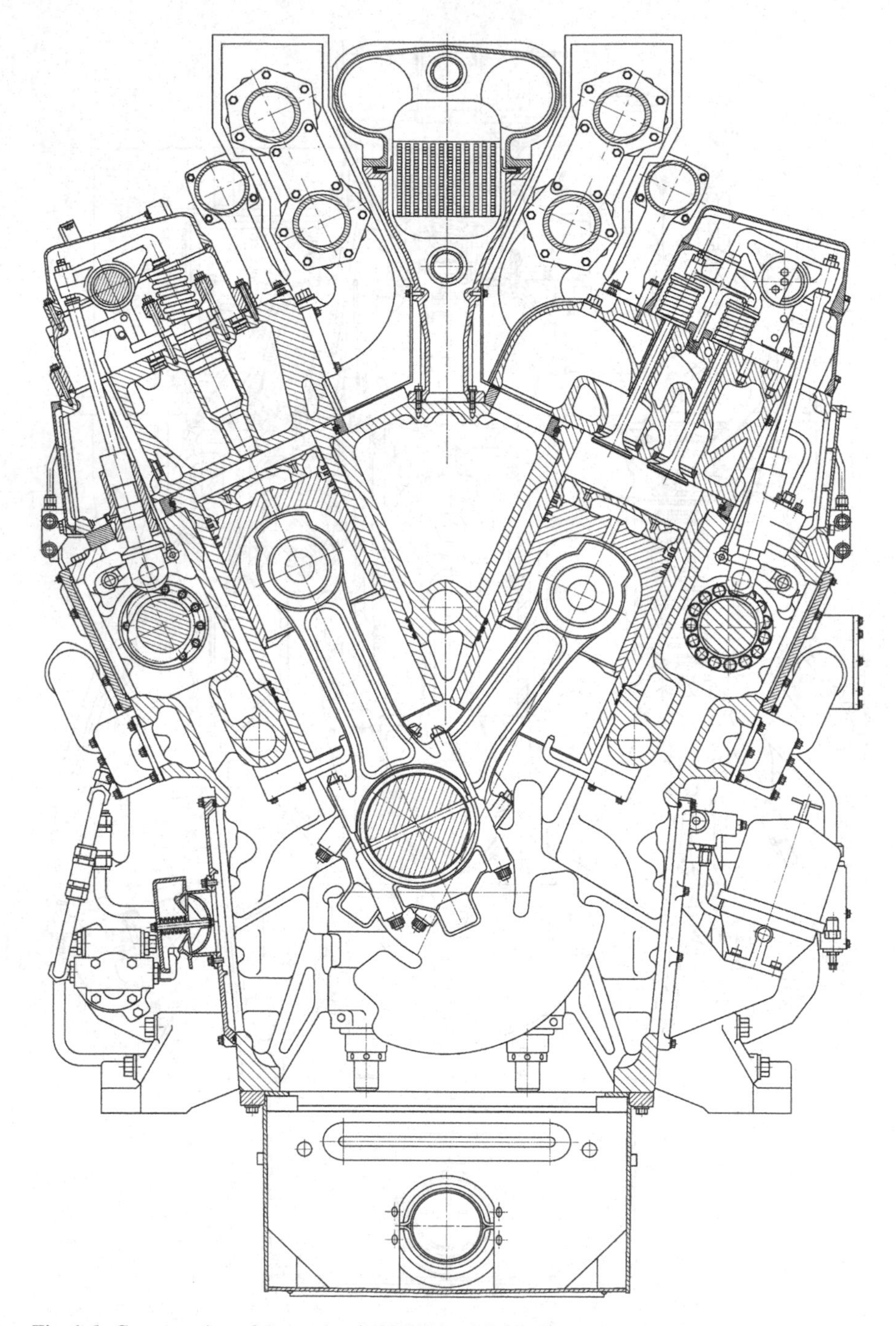

Fig. 1.6 Cross-section of the engine 3600 series of Caterpillar (V-shape) [5]

1.4 MaK Motoren GmbH & Co. KG

MaK Motoren GmbH & Co. KG is currently a fully owned subsidiary of Caterpillar Inc. The foundation of the company can be considered the year 1918, when, according to the Versailles Treaty, arms production in Germany was restricted. In this regard, defense factories, located in the Kiel, in the north of the country, began to look for other markets. On the basis of these enterprises, Kiel Deutsche Werke AG (DWK) was created, which started producing diesel engines for locomotives and vessels. During the Second World War, the company produced submarines and diesel locomotives. At the end of the Second World War, as a result of the Allied bombing, the main facilities of the enterprises in Kiel were destroyed, which led to the cessation of the company's activities. The revival of the company took place on May 25, 1948, when a limited liability company called Maschinenbau Kiel was organized. It included several factories of the former Deutsche Werke AG. Inheriting the production of marine diesel engines from Deutsche Werke, the company continued to produce them under this brand for 30 years. Products of this society were produced with the logo of the MAC.

In 1954, after a lengthy legal dispute with MAN, the logo was changed from MAK to MaK. In 1964, Maschinenbau Kiel joined the Friedrich Krupp AG group of companies headquartered in Essen and was renamed in Atlas-MaK Maschinenbau GmbH. After restructuring in 1971, the company presented its products on the market as Krupp-MaK Maschinenbau GmbH. During the 60–80s, the company launched the production of marine engines of its own design, which were supplied to the market with the MaK logo. In total, in the lineup were presented four models with diameters of the working cylinder from 240 to 580 mm, which covered the power range from 1000 to 10,000 kW. The engines were designed to work on heavy fuels, which at that time made them extremely attractive for the ship industry. The operating experience of the largest engines of this series, such as the M552C and M601C (Fig. 1.7) as the main ones, showed that they can quite compete in fuel efficiency and a number of other indicators with low-speed engines, and in terms of weight and dimensions significantly exceed them. Given this experience, in the early 90s, the company began developing a new engine line with improvements in fuel efficiency. The first engine of this series in 1992 is a diesel M20C, with a cylinder diameter of 200 and a piston stroke of 300 mm. In 1994, a new model appeared M32 (Fig. 1.9) with a cylinder diameter of 320 and a piston stroke of 480 mm. In 1996, the production of the M25 diesel engine (Fig. 1.8) was started with a cylinder diameter of 255 and a piston stroke of 400 mm. By 1990, difficult times have begun for the Krupp-MaK Maschinenbau group of companies, Germany significantly reduced the production of weapons, which they specialized in producing. In this regard, units for the production of military equipment were allocated into the new company MaK System GmbH, and the production of diesel locomotives was sold in 1992 to Siemens. In 1997, the manufacture of marine medium-speed marine engines was sold to Caterpillar Inc. The company was renamed to Caterpillar Motoren in 2000, and its production facilities in Kiel and Rostock eventually became part of Caterpillar Marine Power Systems, headquartered in Hamburg. Some MaK models are also manufactured at Caterpillar's factory in Guangdong, China. Gradually, the production of old models was discontinued, and the new line was supplemented in 1998 with the model M43 (Fig. 1.11 and 1.12).

All engines of this series are available in-line, 6, 7, 8, and 9-cylinder versions. Since 2002, the production of engines with a diameter of working cylinders of 320 and 430 mm in a V-shape with the number of cylinders 12 and 16 has been adjusted. Today, the engine power range covers 1000–16,200 kW. All engines are turbocharged and can run on diesel fuel type MDO and heavy with a viscosity of up to 700 cSt.

1.4.1 Engine M601 Series by MaK Motoren GmbH & Co. KG

The engine with the largest cylinder bore ever produced by a company with the Mac logo was first marketed in 1977. Currently, the production of engines of this series is terminated, but on ships it is still in use. The diameter of the cylinder is 580, and the piston stroke is 600 mm. The engines were produced in-line 6, 8 and 9-cylinder versions and covered the power range from 6000 to 11,250 kW. This series of engines can operate on both light distillate fuels of the type MDO and heavy residual heavy fuels of the type HFO. In this case, the letter "C" (MaK 601C) is added to the engine marking. The basis of the engine design is the foundation frame, cast from high-quality ductile-iron with a crankcase, installed on it. All parts of the engine framework are tightened with the help of steel anchor studs, which unload the parts of the entablature from tensile stresses, including those, caused by the pressure of gases in the working cylinders.

Main technical parameters of engines series M 601 C

Parameter	Value
Number and cylinders arrangement	6, 8, 9 in-line
Cylinder bore (mm)	580
Piston stroke (mm)	600
Cylinder capacity (dm^3)	951/1268/1426.5
Rotation speed max (min^{-1})	400/425
Rotation speed min (min^{-1})	350
Cylinder power at 425 min^{-1} (kW)	1250
Compression ratio	12.8
Air charging pressure (MPa)	0.213
Maximum cycle pressure (MPa)	14.5
Brake specific fuel oil consumption (g/kWh)	177.0
Mean effective pressure at 400/425 min^{-1} (MPa)	2.21/2.23
Mean piston speed at 400/425 min^{-1} (m/s)	8.0/8.5

Dimensions and weight engines series M 601 C

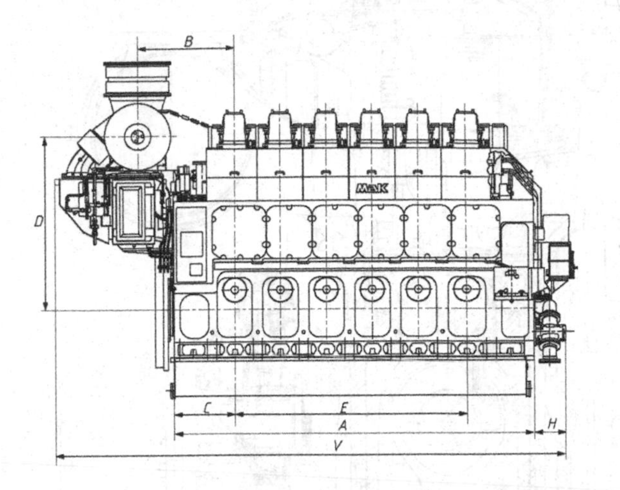

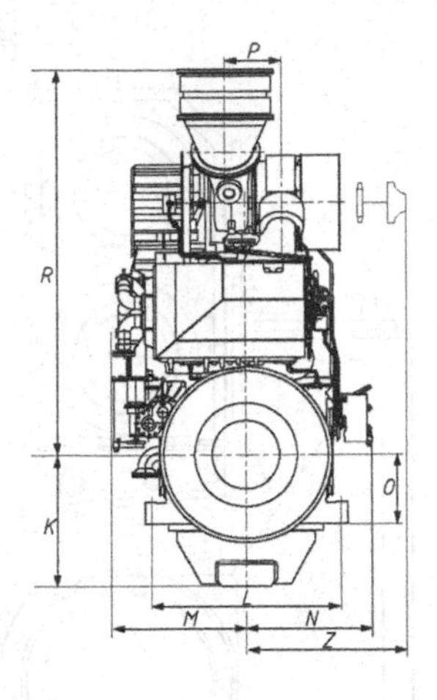

Engine version	A (mm)	B (mm)	C (mm)	D (mm)	E (mm)	H (mm)
6M601C	6880	1490	1490	2549	4300	640
8M601C	8600	1720	1490	2696	6020	832
9M601C	8600	1720	1490	2696	6020	832

Engine version	K (mm)	L (mm)	M (mm)	N (mm)	O (mm)	P (mm)
6M601C	1057	1940	1750	1423	600	450
8M601C	1057	1940	1750	1423	600	355
9M601C	1057	1940	1750	1423	600	355

Engine version	R (mm)	V (mm)	W (mm)	Z (mm)	Weight (kg)
6M601C	3349	8163	3173	1620	110,000
8M601C	3696	10,472	3173	2529	146,000
9M601C	3696	11,626	3173	2529	164,000

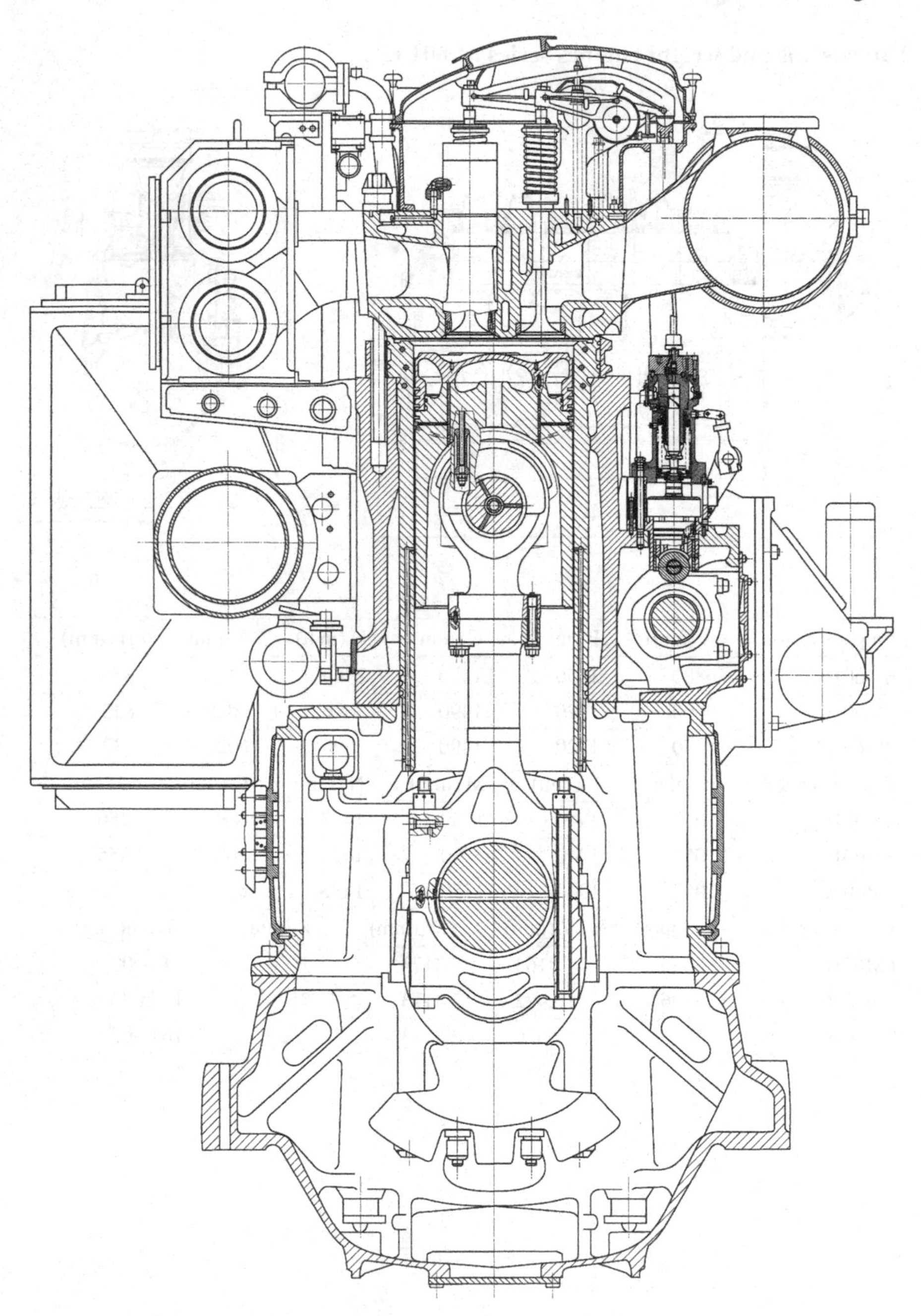

Fig. 1.7 Cross-section of the engine MaK M 601C series [6]

1.4.2 Engines of the M20C, M25C, M32C and M43C Series of MaK Motoren GmbH & Co. KG

The engines of this series are four-stroke diesel engines, irreversible, with pulsed or isobaric turbocharging, intermediate cooling of charging air with direct fuel injection. The engines with the number of cylinders 6, 7, 8 and 9 are produced in the in-line version, and with the number of cylinders 12, 16 in the V-shaped version. The latter ones are produced with a diameter of 320 and 430 mm working cylinders and are designated as VM32C and VM43C. The direction of rotation is clockwise, however, on the customer's request, it can be reversed. All engines of this series are adapted to work on heavy fuels like MDO with a viscosity up to 700 cSt.

These engines cover the power range from 1020 to 16,000 kW and are used on all types of vessels, including: ocean, offshore, cabotage, as well as vessels for inland waterways. Diesels are used both as main ones, with power transmission through a reduction gearbox to an adjustable pitch propeller, and as a part of ship power plants.

Engines in this series have an increased ratio of piston stroke to cylinder diameter, which for V-shaped models is 1.31 for VM32C and 1.42 for VM43C (Fig. 1.12). For in-line engines, the largest ratio is 1.57 for the M25C model, for the M20C and M32C models it is 1.5; and for the M43C model (Fig. 1.11), this ratio is 1.42. This solution made it possible to increase the height of the combustion chamber, providing more space for the organization of effective mixing. In 2000, MaK Motoren launched a program to develop, on the basis of existing engine models, new modifications with improvements in environmental performance. This program is called LEE (Low Emission Engine). Under this programme, a number of design improvements were developed, which made it possible to reduce exhaust gas at low load conditions and at transition regimes.

According LEE program, the engine can be equipped with an exhaust gas reduction system FCT (Flex Cam Technology) by changing the valve timing and advance angle of the fuel supply to the combustion chamber. The use of highly efficient turbocompressors with pressure ratios of 5 and above in combination with increased compression ratios allows organizing workflow according to the Miller cycle with internal cooling of the air charge in modes, close to maximum loads. This made it possible to significantly reduce the maximum temperature of the cycle and, as a result, reduce the content of nitrogen oxides (NO_x) in the engine exhaust gas by about 30% without reducing fuel efficiency. At medium and low loads, the efficiency of the engine for the Miller cycle is reduced. To maintain it at a target level, the FCT system allows to change the valve timing and fuel supplying, providing a high maximum cycle pressure. It allows to maintain high fuel efficiency and engine operation without visible smoking in the entire range of loads.

Since 2007, all engines manufactured by the company, equipped with the FCT system. In addition, the elements of the system are unified for the basic design of the engine, which allows upgrading engines already in operation by equipping them with this system.

The next stage in the development of the LEE program was the development of a flexible accumulator-type fuel supply system called CCR (Caterpillar Common Rail). This system is controlled by a microprocessor unit, based on the algorithms, embedded in it (the so-called injection maps), which allows optimizing the fuel supplying characteristics for each engine operating mode.

As a result of all listed innovation application, all engines of the series correspond to the requirements of IMO Code MARPOL 73/78, Chapter VI (NO_x emission limits), including when working on heavy fuels.

Main technical parameters of engines series M 25 C

Parameter	Value
Number and cylinders arrangement	6, 8, 9 in-line
Cylinder bore (mm)	255
Piston stroke (mm)	400
Cylinder capacity (dm^3)	20.4
Rotation speed max (min^{-1})	720/750
Rotation speed min (min^{-1})	250
Cylinder power in conformity with ISO 3046/1 at 720/750 min^{-1} (kW)	317/330
Compression ratio	16.5
Air charging pressure (MPa)	0.32
Compression pressure (MPa)	17.5
Pressure pump timing angle, deg to TDC	11.0
Injector angle, deg to TDC	8.0
Maximum cycle pressure (MPa)	20.5
Exhaust gas temperature at inlet of turbocharger (°C)	380
Exhaust gas temperature at outlet of turbocharger (°C)	320
Brake specific fuel oil consumption at $n = const$ (g/kWh)	
– at the load 100%	184.0
– at the load 85%	183.0
– at the load 75%	185.0
– at the load 50%	193.0
Brake specific air consumption at 20 °C (m^3/kWh)	5.8
Lubrication system oil consumption (g/kWh)	0.6
Mean effective pressure at 720/750 min^{-1} (MPa)	2.58/2.61
Mean piston speed at 720/750 min^{-1} (m/s)	9.6/10.0
Starting air pressure (MPa)	3.0
Turbocharger hardware model	HPR6000

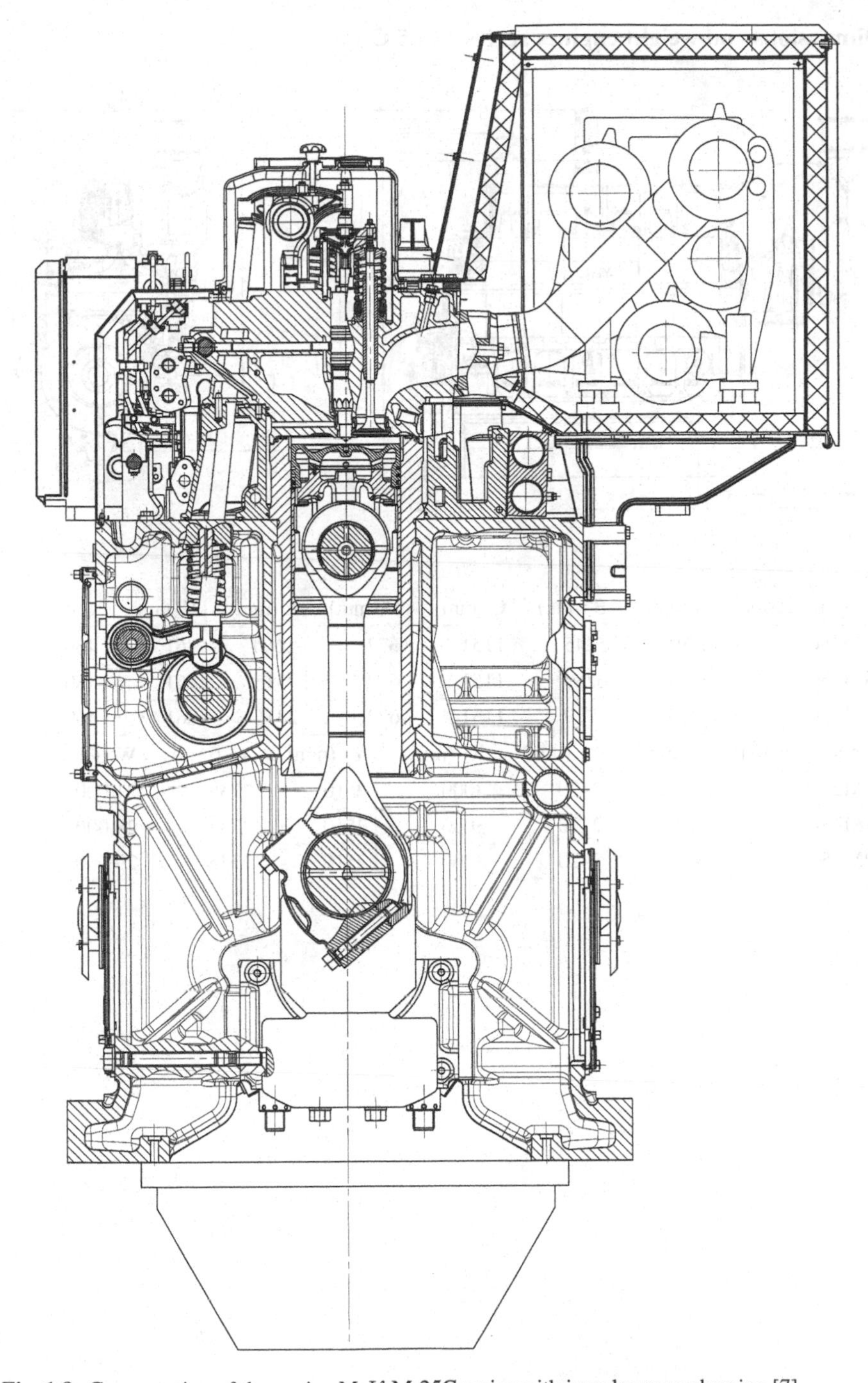

Fig. 1.8 Cross-section of the engine MaK M 25C series with impulse supercharging [7]

Dimensions and weight engines series M 25 C

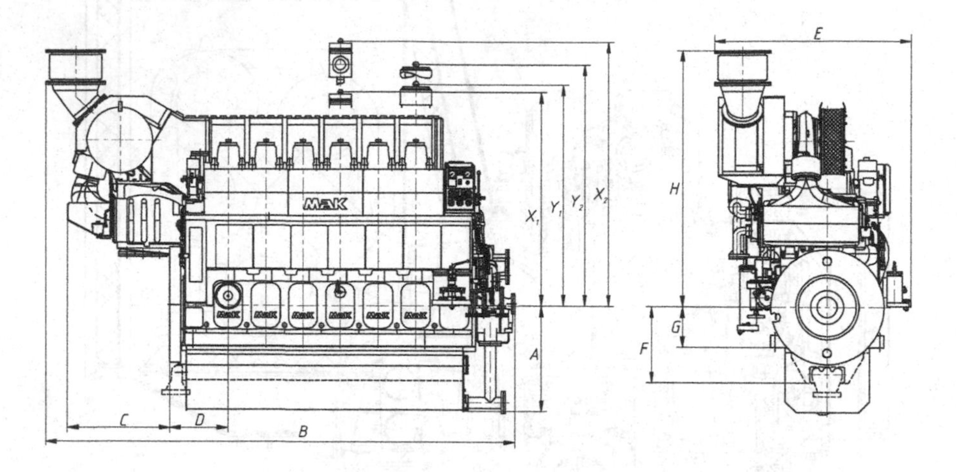

Engine version	A (mm)	B (mm)	C (mm)	D (mm)	E (mm)	F (mm)	G (mm)
6M25C	1191	5345	1151	672	2260	861	460
8M25C	1191	6289	1151	672	2315	861	460
9M25C	1191	6719	1151	672	2315	861	460

Engine version	H (mm)	X_1 (mm)	X_2 (mm)	Y_1 (mm)	Y_2 (mm)	Weight (kg)
6M25C	2906	2420	3000	2510	2735	21,000
8M25C	3052	2420	3000	2510	2735	28,000
9M25C	3052	2420	3000	2510	2735	29,000

Main technical parameters of engines series M 32 C

Parameter	Value
Number and cylinders arrangement	6, 8, 9 in-line
Cylinder bore (mm)	320
Piston stroke (mm)	460
Cylinder capacity (dm^3)	38.7
Rotation speed max (min^{-1})	600
Rotation speed min (min^{-1})	360
Cylinder power at 600 min^{-1} (kW)	500
Compression ratio	16.2
Air charging pressure (MPa)	0.34
Compression pressure (MPa)	16.0
Pressure pump timing angle, deg to TDC	6.5
Injector angle, deg to TDC	1.5
Maximum cycle pressure (MPa)	19.8
Exhaust gas temperature at inlet of turbocharger (°C)	430
Exhaust gas temperature at outlet of turbocharger (°C)	330
Brake specific fuel oil consumption at $n = const$ (g/kWh)	
– at the load 100%	178.0
– at the load 85%	177.0
– at the load 75%	181.0
– at the load 50%	190.0
Brake specific air consumption at 20 °C (m^3/kWh)	5.8
Lubrication system oil consumption (g/kWh)	0.6
Mean effective pressure at 600 min^{-1} (MPa)	2.59
Mean piston speed at 600 min^{-1} (m/s)	9.6
Starting air pressure (MPa)	3.0
Turbocharger hardware model	NA 357

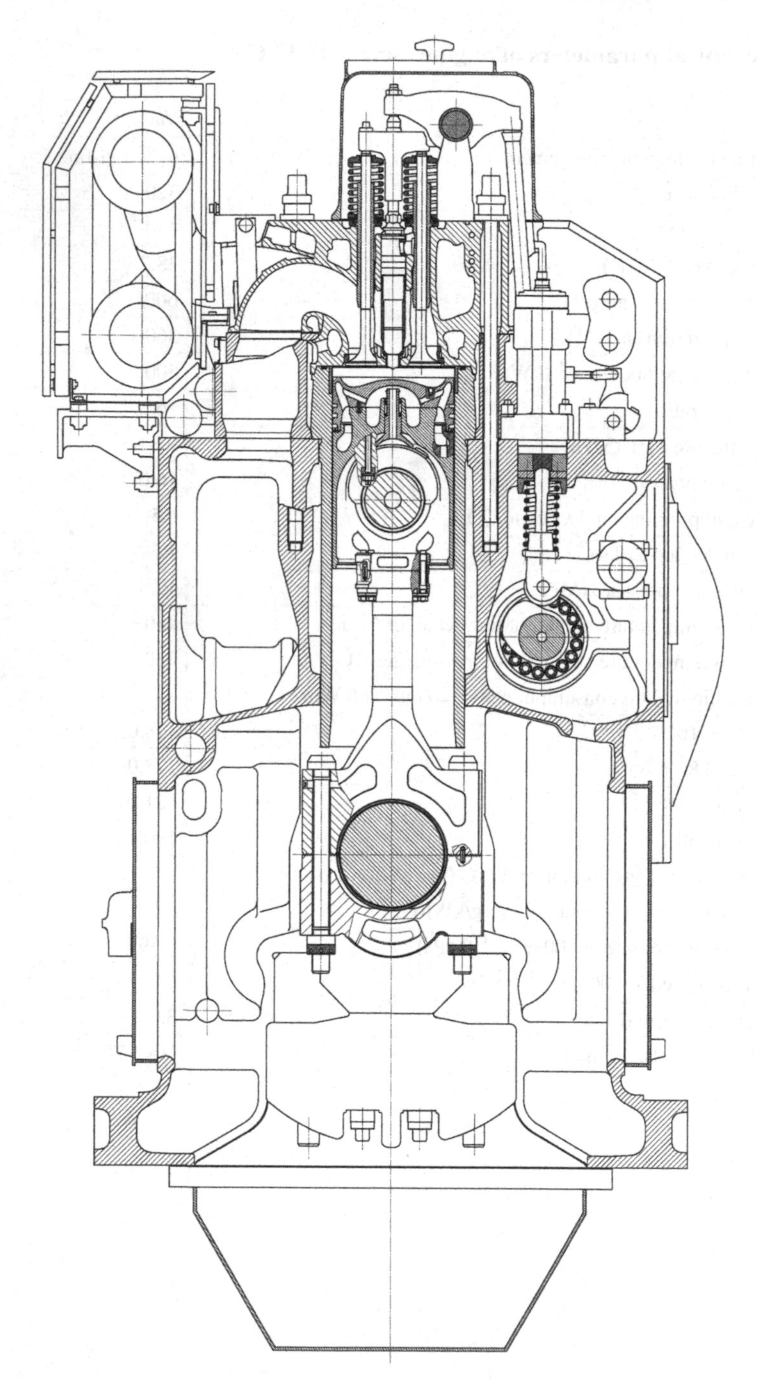

Fig. 1.9 Cross-section of the engine MaK M 32C series with impulse supercharging [8]

Dimensions and weight engines series M 32 C

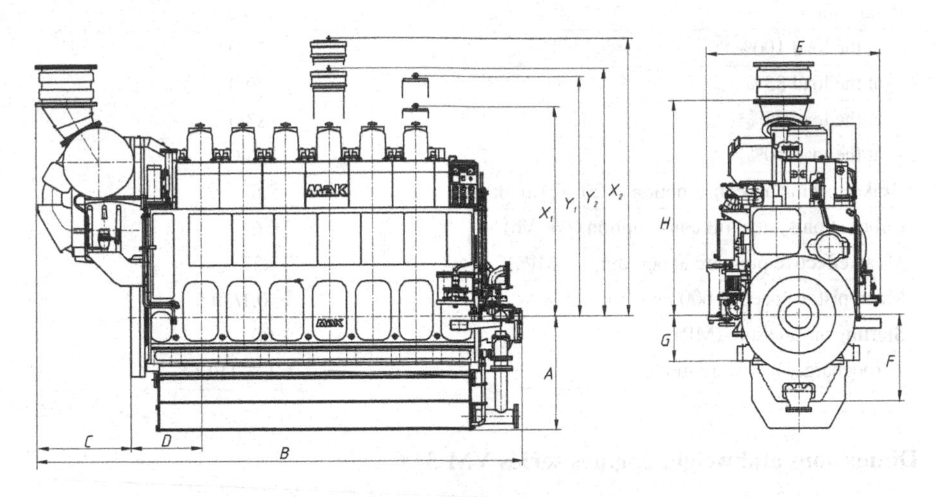

Engine version	A (mm)	B (mm)	C (mm)	D (mm)	E (mm)	F (mm)	G (mm)
6 M 32C	1387	5921	1137	852	2177	1052	550
8 M 32C	1387	7061	1185	852	2177	1052	550
9 M 32C	1387	7591	1185	852	2177	1052	550

Engine version	H (mm)	X_1 (mm)	X_2 (mm)	Y_1 (mm)	Y_2 (mm)	Weight (kg)
6 M 32C	3094	2570	2940	3040	3405	37,500
8 M 32C	3446	2570	2940	3040	3405	46,400
9 M 32C	3446	2570	2940	3040	3405	49,400

Main technical parameters of engines series VM 32 C

Parameter	Value
Number and cylinders arrangement	12, 16 V-shaped
Cylinder bore (mm)	320
Piston stroke (mm)	420
Cylinder capacity (dm^3)	33.8
Rotation speed max (min^{-1})	720/750
Rotation speed min (min^{-1})	250
Cylinder power at 720/750 min^{-1} (kW)	480/500
Compression ratio	16.4
Air charging pressure (MPa)	0.30
Compression pressure (MPa)	15.0
Maximum cycle pressure (MPa)	19.8
Exhaust gas temperature at inlet of turbocharger (°C)	430
Exhaust gas temperature at outlet of turbocharger (°C)	330
Brake specific fuel oil consumption at $n = const$ (g/kWh)	

(continued)

(continued)

Parameter	Value
– at the load 100%	179.0
– at the load 85%	179.0
– at the load 75%	183.0
– at the load 50%	195.0
Brake specific air consumption at 20 °C (m^3/kWh)	5.8
Lubrication system oil consumption (g/kWh)	0.6
Mean effective pressure at 600 min^{-1} (MPa)	2.37
Mean piston speed at 600 min^{-1} (m/s)	10.1/10.5
Starting air pressure (MPa)	3.0
Turbocharger hardware model	2 × TPL65

Dimensions and weight engines series VM 32 C

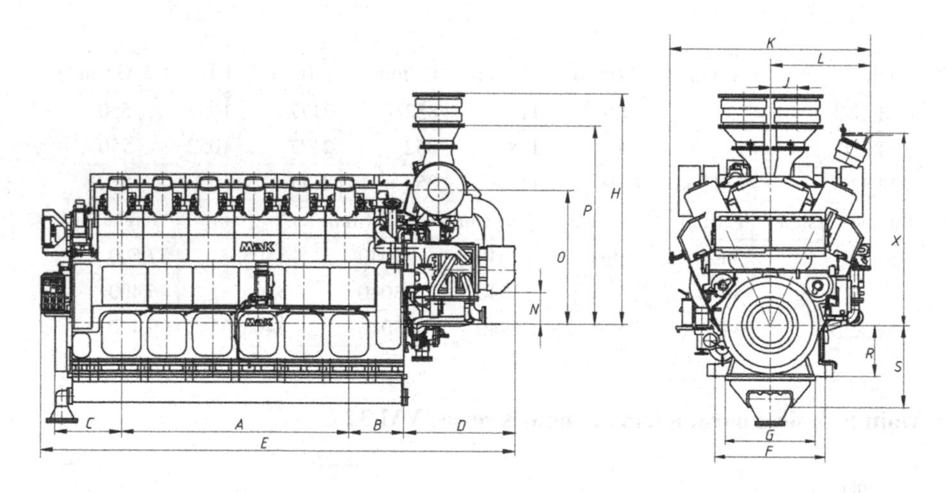

Engine version	A (mm)	B (mm)	C (mm)	D (mm)	E (mm)	F (mm)
12M32C	3375	807	949	1630	6963	1630
16M32C	4725	807	949	1630	8313	1630

Engine version	G (mm)	H (mm)	J (mm)	K (mm)	L (mm)	M (mm)
12M32C	1338	3395	396,5	2985	1485	1307
16M32C	1338	3350	553	2923	1488	1211

Engine version	N (mm)	O (mm)	P (mm)	R (mm)	S (mm)	Weight (kg)
12M32C	464	1968	2920	750	1205	64,400
16M32C	464	1899	2806	750	1205	81,600

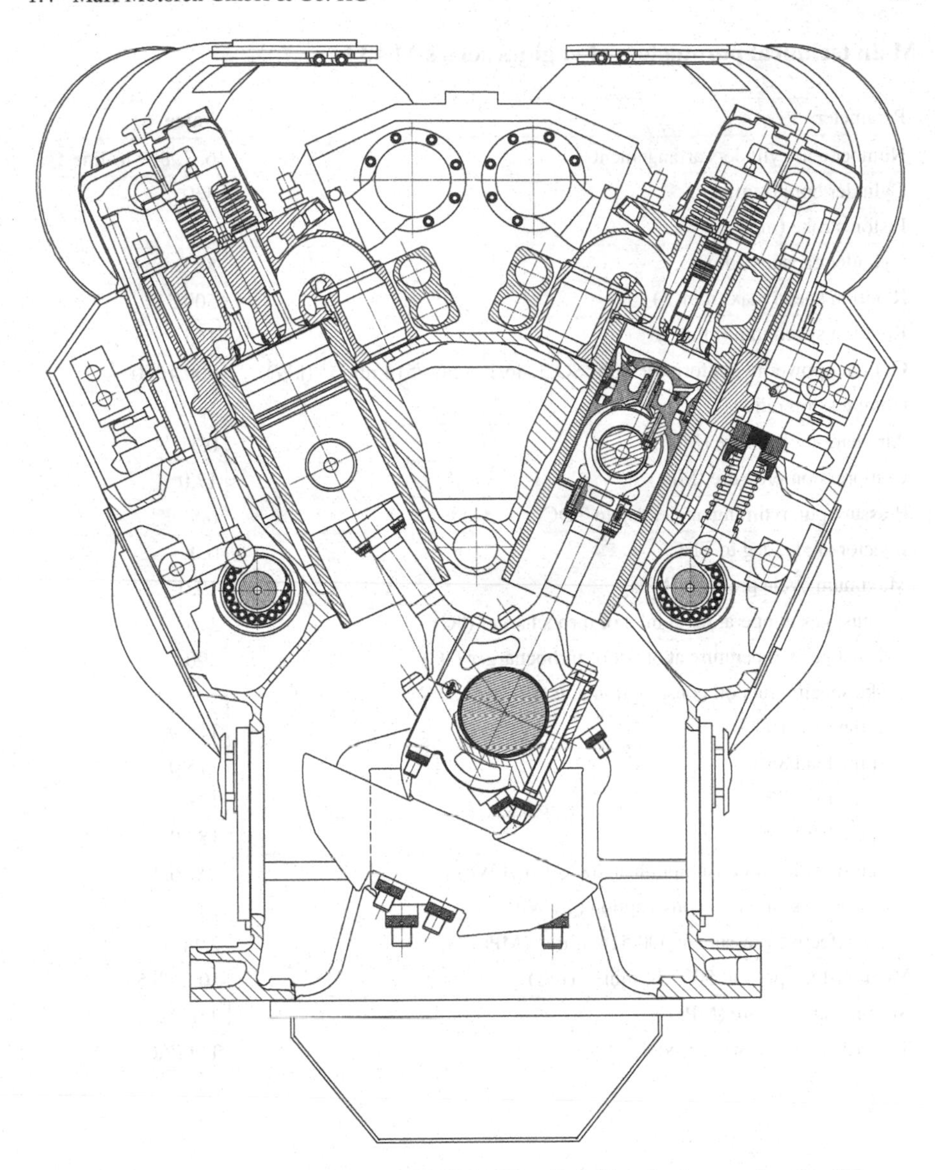

Fig. 1.10 Cross-section of the engine MaK VM 32C series with isobaric supercharging [9]

Main technical parameters of engines series M 43 C

Parameter	Value
Number and cylinders arrangement	6, 7, 8, 9 in-line
Cylinder bore (mm)	430
Piston stroke (mm)	610
Cylinder capacity (dm^3)	88.6
Rotation speed max (min^{-1})	500/514
Rotation speed min (min^{-1})	165
Cylinder power in conformity with ISO 3046/1 at 500/514 min^{-1} (kW)	900/1000
Compression ratio	16.2
Air charging pressure (MPa)	0.31
Compression pressure (MPa)	17.0
Pressure pump timing angle, deg to TDC	6.5
Injector angle, deg to TDC	0.0
Maximum cycle pressure (MPa)	19.0
Exhaust gas temperature at inlet of turbocharger (°C)	375
Exhaust gas temperature at outlet of turbocharger (°C)	290
Brake specific fuel oil consumption at $n = const$ (g/kWh)	
– at the load 100%	176.0
– at the load 85%	175.0
– at the load 75%	178.0
– at the load 50%	185.0
Brake specific air consumption at 20 °C (m^3/kWh)	5.9–6.25
Lubrication system oil consumption (g/kWh)	0.6
Mean effective pressure at 500/514 min^{-1} (MPa)	2.44
Mean piston speed at 500/514 min^{-1} (m/s)	10.2/10.5
Starting air pressure (MPa)	3.0
Turbocharger hardware model	TPL76C

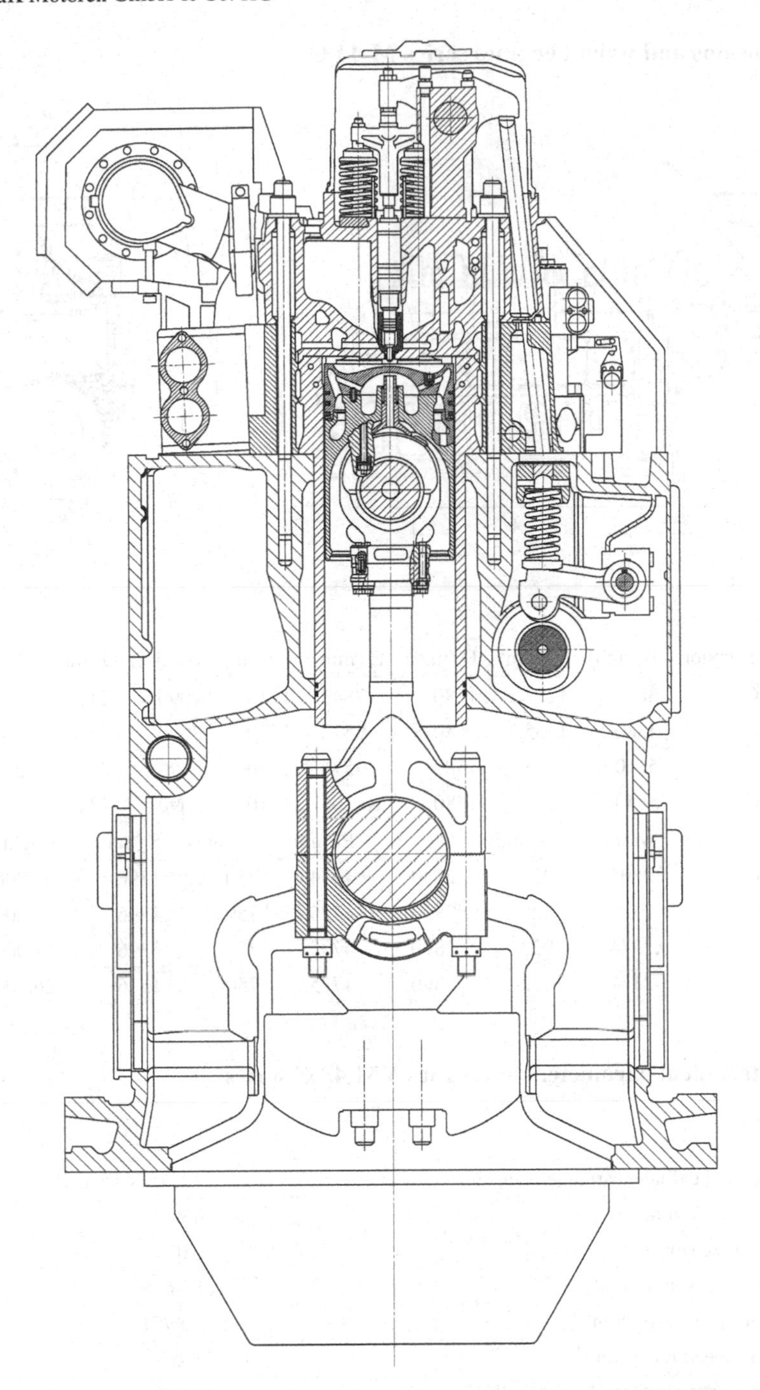

Fig. 1.11 Cross-section of the engine MaK M 43C series with isobaric supercharging [10]

Dimensions and weight engines series M 43 C

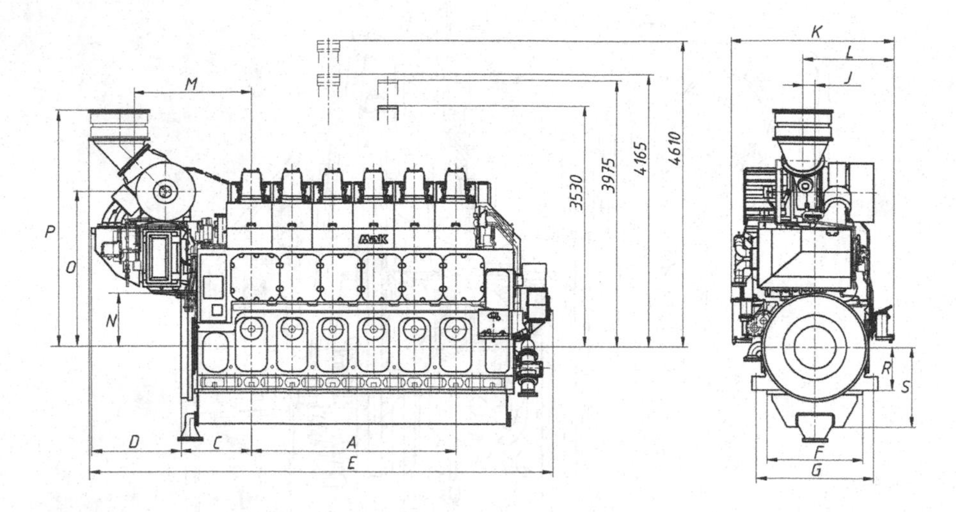

Engine version	A (mm)	C (mm)	D (mm)	E (mm)	F (mm)	G (mm)	J (mm)	K (mm)
6M43C	3650	1255	1580	7842	1710	2090	215	2890
7M43C	4380	1255	1580	8572	1710	2090	232	2890
8M43C	5110	1255	1580	9576	1710	2090	232	2890
9M43C	5840	1255	1580	10,306	1710	2090	232	2890

Engine version	M (mm)	N (mm)	O (mm)	P (mm)	R (mm)	S (mm)	Weight (kg)
6M43C	2505	922	2730	4434	750	1396	93,000
7M43C	2374	922	2890	4725	750	1396	106,000
8M43C	2374	922	2890	4725	750	1396	114,000
9M43C	2374	922	2890	4725	750	1396	126,000

Main technical parameters of engines VM 43 C series

Parameter	Value
Number and cylinders arrangement	12, 16 V-shaped
Cylinder bore (mm)	430
Piston stroke (mm)	610
Cylinder capacity (dm^3)	88.6
Rotation speed max (min^{-1})	500/514
Rotation speed min (min^{-1})	165
Cylinder output at 500/514 min^{-1} (kW)	1.000
Compression ratio	16.2

(continued)

(continued)

Parameter	Value
Air charging pressure (MPa)	0.34
Final compression pressure (MPa)	17.5
Maximum cycle pressure (MPa)	21.0
Exhaust gas temperature, before turbocharger (°C)	365
Exhaust gas temperature, after turbocharger (°C)	280
Brake-specific fuel consumption at $n = const$ (g/kWh)	
– at the load 100%	177.0
– at the load 85%	176.0
– at the load 75%	178.0
– at the load 50%	185.0
Effective specific air consumption at 20 °C (m^3/kWh)	6.1–6.2
Effective specific oil consumption (g/kWh)	0.6
Mean effective pressure at 500/514 min^{-1} (MPa)	2.71/2.64
Mean piston speed at 500/514 $мин^{-1}$ (m/s)	10.2/10.5
Starting air pressure (MPa)	3.0
Turbocharger hardware model	2 × TPL76C

Dimensions and weight engines series VM 43 C

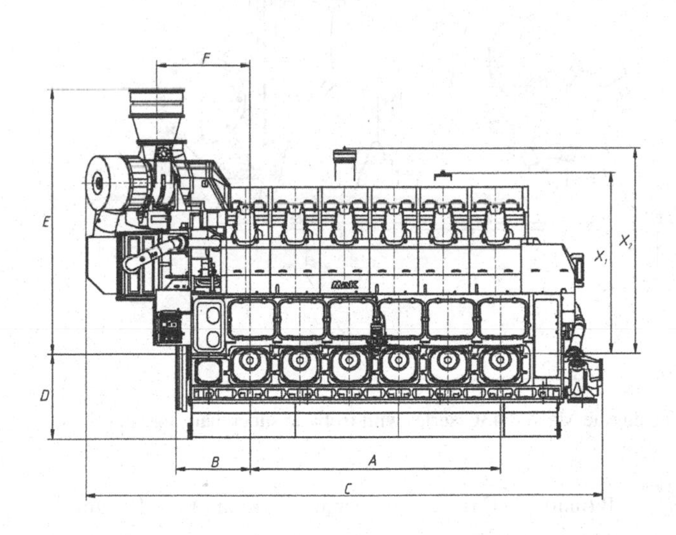

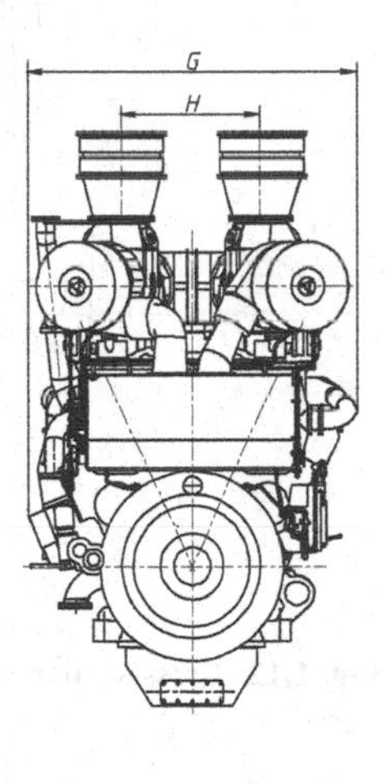

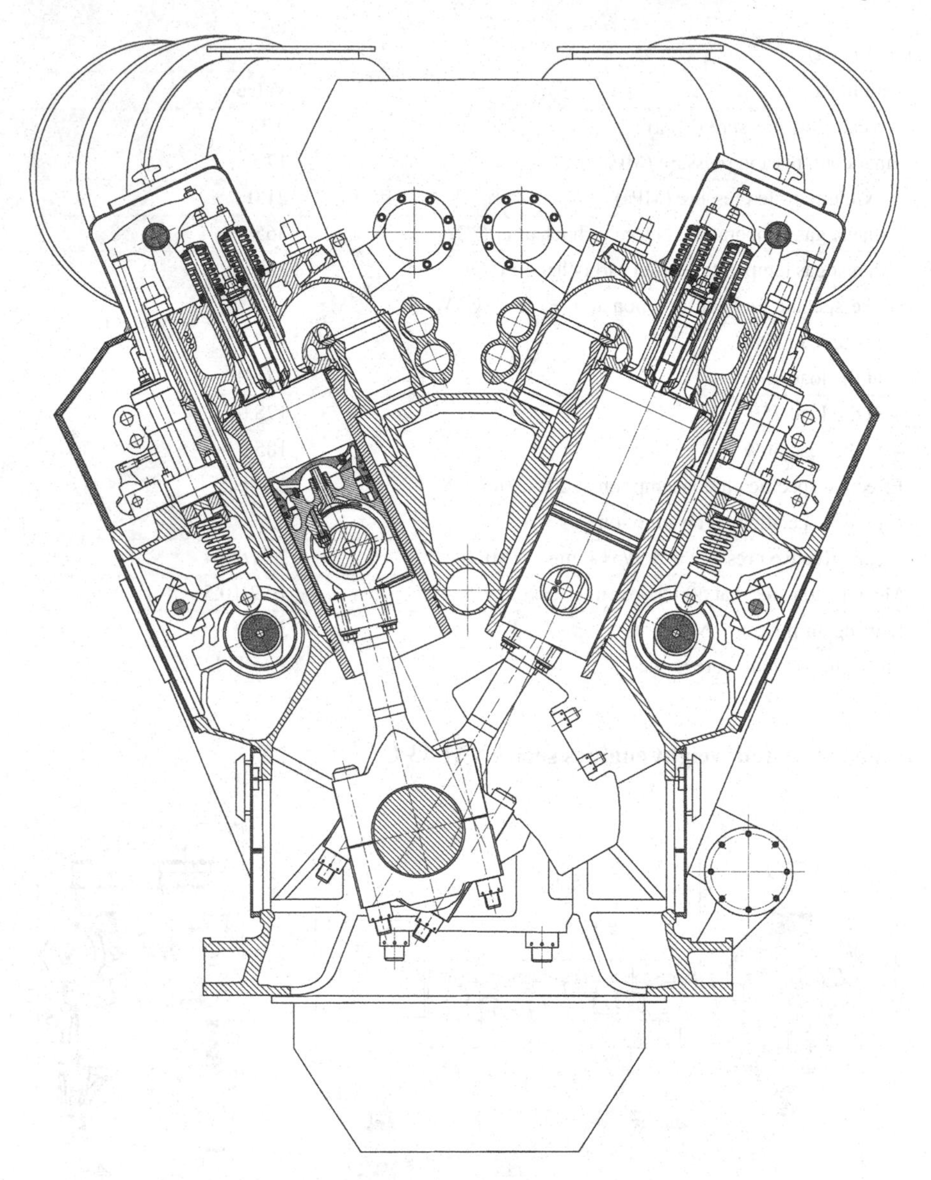

Fig. 1.12 Cross-section of the engine MaK V 43C series with isobaric supercharging [11]

Engine version	A (mm)	B (mm)	C (mm)	D (mm)	E (mm)	F (mm)
12M43C	4800	1440	9927	1625	5092	1788
16M43C	6720	1440	12,023	1625	5092	1773

(continued)

(continued)

Engine version	G (mm)	H (mm)	X_1 (mm)	X_2 (mm)	Weight (kg)
12M43C	3923	1685	3200	3700	162,000
16M43C	4027	1670	3200	3700	215,000

1.5 Daihatsu Motor Co., Ltd.

Daihatsu Motor Co., Ltd. was founded in March 1, 1907 in the city of Osaka, Japan, by professors of Osaka University Yoshinki and Turumi. The company was named Hatsudoki Seizo Co., Ltd. and specialized in the production of gas internal combustion engines for industrial use, which were popular in the Japanese market and successfully sold during 15 years. After World War I, industrialization began at a rapid pace in Japan, as a result of which demand for compact engines arose, and the emerging maritime industry demanded new fuel-efficient and reliable engines. In response to this request, the company, in cooperation with the American firm RM Bit Corporation, based in Chicago, by 1922, launched the production of compact single-cylinder diesel engines for universal use with an output of 3 horsepower and four-cylinder diesel engines for use on ships with an output of 60 horsepower.

Gradually, manufacturers of engines for various purposes in Japan became more and consumers began to use the word Daihatsu, a combination of the first kanji, denoting the location of Osaka (大阪), and the words Dai (大) and Hatsu (発) "engine production", to identify the company's products. In December 19, 1951 this name received official status, and the company became known as Daihatsu Motor Co. Ltd. By this time, the company expanded its product range and began production of various vehicles, including developing the production of cars.

In 1966, the management of the company decided to separate from the main company a division, specializing in the production of internal combustion engines for various purposes and other equipment for means of transport. The new division was called Daihatsu Diesel MFG. Co. Ltd. The management of the new company faced with the problem of lack of production facilities, located in Osaka, and their expansion was impossible, due to the fact that over the years the city has become a metropolis with a high density of development. In this regard, it was decided to build a new production base in the city of Moriyama, Shiga Prefecture. The construction of two plants began in November 1969. In July 1977, the construction of the first factory, Moriyama No. 1, was completed, and all production was transferred from Osaka to this newly created enterprise. In 1979 the plant "Moriyama №2" was commissioned.

In September 1993, Daihatsu Diesel moved its headquarters to the new "New Umeda City" complex, built on the site of an old factory in Osaka. In January 1993, the company launches medium-speed four-stroke engines with in-line cylinders of the DK-20 and DK-28 series (Fig. 1.13) for ships, which became the most popular among consumers. Based on these engines, the rest of the diesel and gas diesel engines produced by the company were subsequently developed. In March 1997, the production and sale of the most powerful 12DK-36 diesel engine in the engine lineup began. In July 2004, the production and sale of a diesel engine with reduced fuel consumption of the next generation DC-32 began.

In May 26, 2008, a research and development center was opened at the Moriyama No. 2 plant, the main task of which is the development of new generation engines. As a result, in June 2010 a new engine of the 6DK-20e series was launched, which formed the basis of the new "e-Diesel" engine series with improvements in environmental performance. In 2011, this series was supplemented with engines DE-18, DE-23.Today, the company produces diesel and gas-diesel engines and diesel generators with a power range from 500 to 5500 kW. All Daihatsu Diesel marine engines and equipment are meet the requirements EN ISO 9001, JIS Z9901 and BSEN ISO 900, Lloyd's Register Quality Assurance. Ltd.

1.5.1 The DK-28 Series Engine

The DK-28 series engine has an in-line arrangement of cylinders and is available in 6 and 8-cylinder versions (Fig. 1.13). Suspended-type crankshaft with an increased diameter of the connecting rod and main bearings, which can significantly reduce the specific pressure in the bearings. Cylinder covers are individual, fastened with four studs, each cap is equipped with two intake and two exhaust valves. Pistons are made of cast iron. The grooves of the piston rings are chrome-plated, as are the working surfaces of the rings themselves. Connecting rod marine type with removable connecting rod bearing housing. The engine is equipped with a fuel system with individual high-pressure pumps for each cylinder, closed-type injectors that inject fuel into the Hesselman-type combustion chamber, made in the piston. The air supply to the working cylinders is provided by a gas turbine supercharging system with intermediate air cooling.

Main technical parameters of engines series DK-28

Parameter	Value
Number and cylinders arrangement	6, 8 in-line
Cylinder bore (mm)	280
Piston stroke (mm)	390
Cylinder capacity (dm^3)	24.0
Rotation speed max (min^{-1})	720
Rotation speed min (min^{-1})	300
Cylinder power at 720 min^{-1} (kW)	302
Air charging pressure (MPa)	0.24
Maximum cycle pressure (MPa)	15.7
Brake specific fuel oil consumption (g/kWh)	193.0

(continued)

(continued)

Parameter	Value
Brake specific air consumption (m^3/kWh)	6.8
Lubrication system oil consumption (g/kWh)	0.8
Exhaust gas temperature at inlet of turbocharger (°C)	580
Exhaust gas temperature at outlet of turbocharger (°C)	380
Turbocharger hardware model	MET26SR
Mean effective pressure at 720 min^{-1} (MPa)	2.09
Mean piston speed at 720 min^{-1} (m/s)	9.4

Dimensions and weight engines series DK-28 with generator

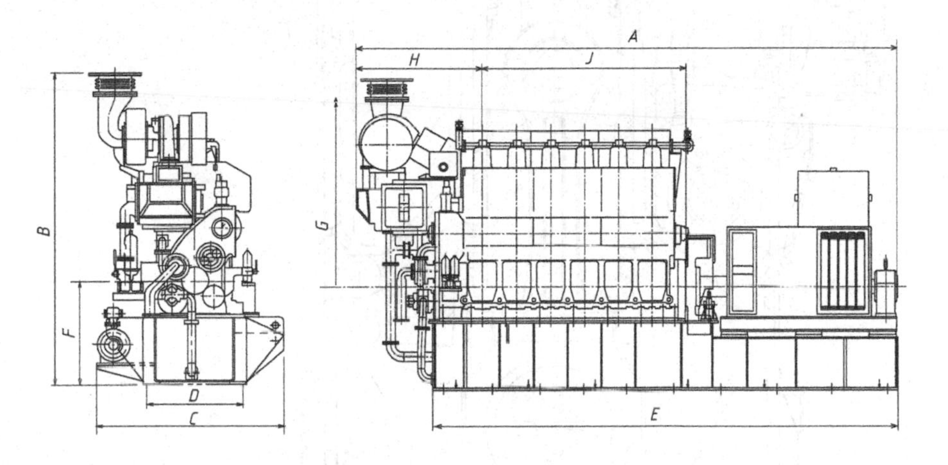

Engine series	A (mm)	B (mm)	C (mm)	D (mm)	E (mm)
6DK-28	6825	3710	2235	1230	6100
8DK-28	7865	3830	2235	1230	6780

Engine series	F (mm)	G (mm)	H (mm)	J (mm)	Weight (kg)
6DK-28	1300	2065	1095	2580	35,000
8DK-28	1300	2065	1095	3440	45,500

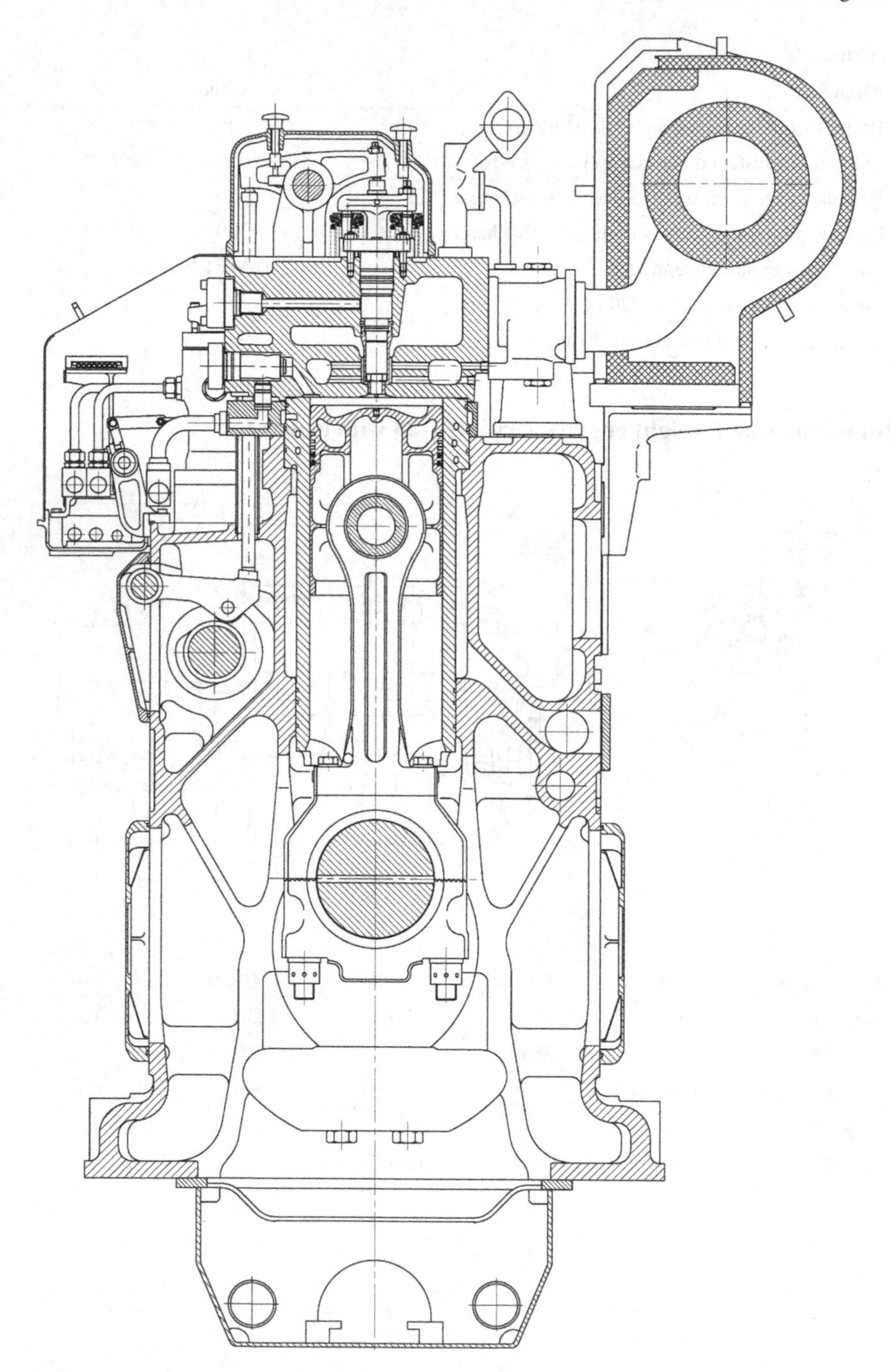

Fig. 1.13 Cross-section of the engine DK-28 series [12]

1.5.2 *The Engine of the DC-17Ae Series*

The engine of the DC-17Ae series has an in-line cylinder arrangement and is manufactured in 5 and 6-cylinder versions (Fig. 1.14). The air supply to the working cylinders is provided by a gas turbine supercharging system with intermediate air cooling. The engines of this series are mainly used to drive ship power plants and are characterized by a large S/D ratio of 1.59. It makes possible to optimize the shape of the combustion chamber, increasing its height, which provides high-quality mixing with minimal leakage of fuel spray onto the wall. The engine is made according to the classical scheme with a suspended-type crankshaft, installed in a solid block. The lower caps of the crankshaft main bearings, in addition to the bearing studs, have side fastenings that increase the rigidity of the whole structure. Each cylinder is closed by an individual cylinder cover equipped with two intake and two exhaust valves. Pistons are made of ductile iron. The grooves of the piston rings are hardened by high-frequency currents.

Main technical parameters of engines series DC-17Ae

Parameter	Value
Number and cylinders arrangement	5, 6 in-line
Cylinder bore (mm)	170
Piston stroke (mm)	270
Cylinder capacity (dm^3)	6.13
Rotation speed max (min^{-1})	900/1000
Cylinder power 5/6 cylinders (kW)	98.0/101.7
Compression ratio	15.0
Air charging pressure (MPa)	0.28
Maximum cycle pressure (MPa)	16.0
Brake specific fuel oil consumption (g/kWh)	198.0
Lubrication system oil consumption (g/kWh)	0.8
Exhaust gas temperature at inlet of turbocharger (°C)	435
Exhaust gas temperature at outlet of turbocharger (°C)	340
Turbocharger hardware model	AT 14
Mean effective pressure at 5/6 cylinders (MPa)	2.09/2.21
Mean piston speed at 900/1000 min^{-1} (m/s)	8.1/9.0

Dimensions and weight engines series DC-17Ae with generator

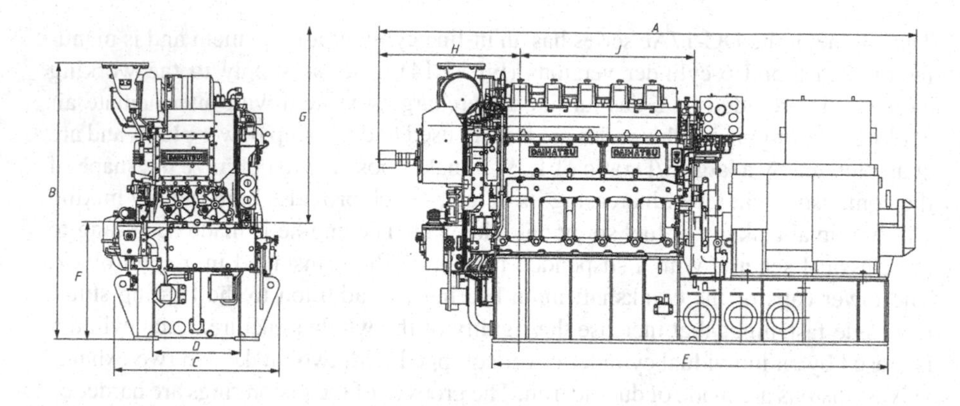

Engine series	A (mm)	B (mm)	C (mm)	D (mm)	E (mm)
6DC-17Ae	4070	2250	1350	950	3230
5DC-17Ae	4510	2250	1350	950	3565

Engine series	F (mm)	G (mm)	H (mm)	J (mm)	Weight[a] (kg)
6DC-17Ae	950	1435	845	1420	10,000
5DC-17Ae	950	1435	845	1690	11,000

[a]Generator set weight

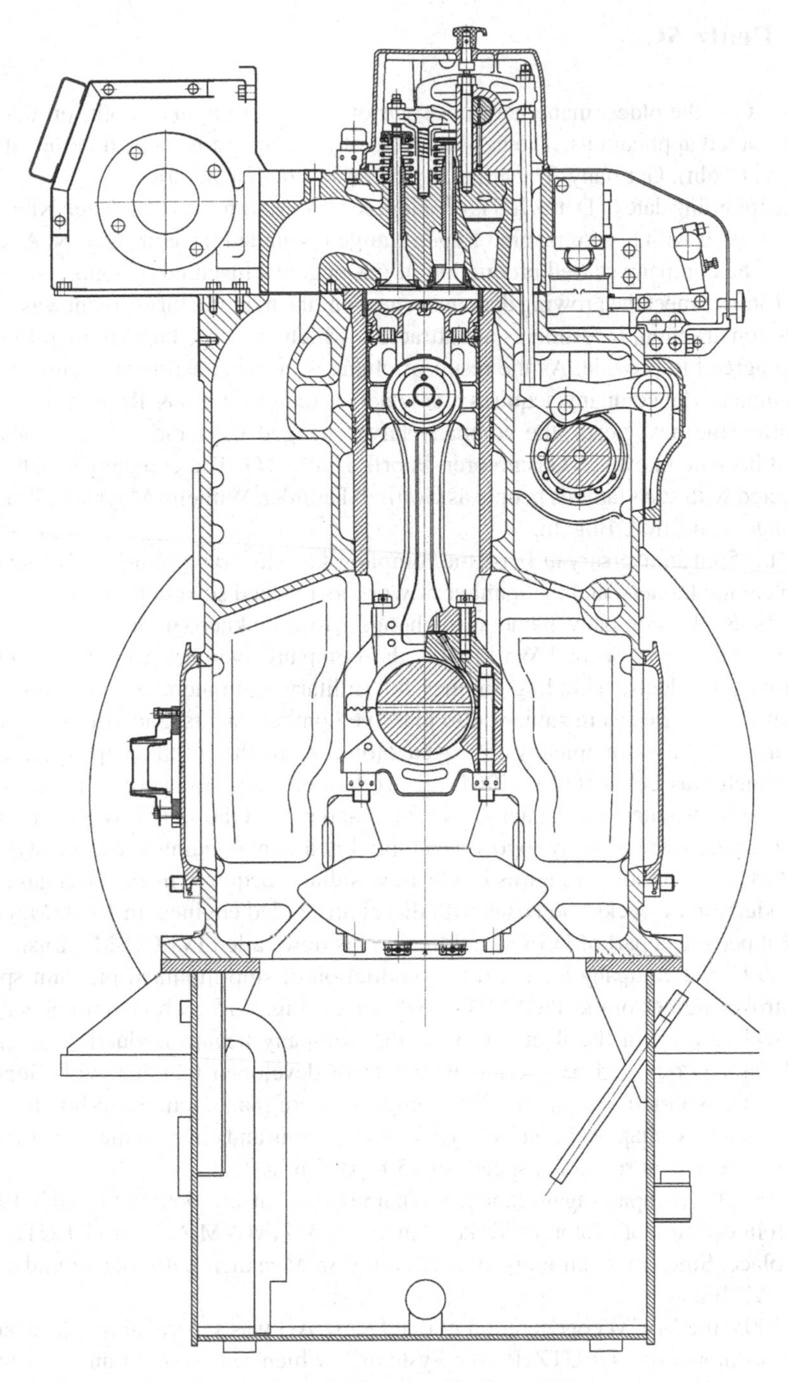

Fig. 1.14 Cross-section of the engine DC-17 series of Daihatsu [13]

1.6 Deutz AG

Deutz AG is the oldest manufacturer of various types of engines, both for air- and liquid-cooled applications. The headquarters of the concern is located in the city of Cologne (Köln), Germany. The company owns several subsidiaries.

The founding date of Deutz AG is considered to be March 31, 1864, when Nikolaus Otto, together with his companion Eugen Langen, founded the company, N. A. Otto & Cie. The company started production of gas engines, invented by Otto.

In 1869, to meet the growing demand, a new engine manufacturing plant was built. For its construction, investments are attracted, which the large businessman Rosen-Runge agreed to provide. As the result of changes in the constituent documents of the company, the company acquires a new name, Langen, Otto & Roosen.

During the next years, the company have changed its name several times. In 1872 it became known as Gasmotoren-Fabrik Deutz AG. The company's history is connected with such famous people as Gottlieb Daimler, Wilhelm Maybach, Prosper L'Orange, and Ettore Bugatti.

By the 50th anniversary in 1914, the company's product range consisted of several dozen engine models, and the number of workers reached several thousand.

In 1938, the company name was changed to Klockner-Humboldt-Deutz AG (KHD). During the Second World War, the company switched to the production of military products, including engines for military equipment. At that time, the company's factories were subjected to massive bombardments, and after the end of the war were almost completely destroyed. However, by the 100th company anniversary, which was celebrated in 1964, all production facilities were fully restored, production was launched, capacities were expanded, and the staff was increased. In the same year, the company logo was adopted and is now being used: the stylized letter "M" is a symbol of Magirus Deutz, a subsidiary (acquired in 1938), engaged in the production of trucks and buses with diesel air-cooled engines. In 1974, Magirus Deutz it became a part of IVECO holding and is now called IVECO-Magirus.

In 1993, the company launched the production of ship-mounted medium-speed four-stroke engines of the TBD MWM 645 series (Fig. 1.15), which until now have been widely used in the fleet. In 1994, the company began production of high-speed engines 628 and 632 series, which were developed together with General Electric Transportation Systems. These engines were manufactured in-line (6, 8, 9 cylinders) and V-shaped (12 and 16 cylinders) versions and covered the power range of 995–3600 kW at rotational speeds of 750–1000 min^{-1}.

In 1997, the company again changed its name to the current Deutz AG, and in 1998, the full integration of Motoren-Werke Mannheim AG (MWM AG) into DEUTZ AG takes place. Since then, all units, manufactured in Mannheim, are issued under the "DEUTZ" brand.

In 2005, the MWM cogeneration unit at Deutz AG was separated into a separate division, known as "DEUTZ Power Systems", which was sold to an investment company in 2007, and the original name MWM was returned to the engines. Since this time, Deutz AG concentrated on the production and sale of compact engines only under the Deutz brand. In 2008, MWM GmbH (Motoren-Werke Mannheim AG) left the concern DEUTZ AG, and in October 2010, MWM GmbH (Motorenwerke Mannheim) was acquired by Caterpillar as an independent brand within Caterpillar Power Division.

The engine of the TBD 645 series has an in-line arrangement of cylinders and is available in 6 and 8-cylinder versions (Fig. 1.15). There are suspended-type crankshaft with counterweights. Cylinder covers are individual, fastened with four studs, each cap is equipped with two intake and two exhaust valves. Exhaust valves are mounted in valve units, that can be removed from the cover without removing it from the engine. Pistons are composite, the head is made of alloyed heat-resistant steel, and the skirt is made of ductile iron. The connecting rod with a slanting lower bearing connector, which makes it easy to dismantle through the crankcase side hatches. The engine is equipped with a fuel system with individual high-pressure pumps for each cylinder that inject fuel through a closed-type injector directly into the combustion chamber. The air supply to the working cylinders is provided by a gas turbine supercharging system with intermediate air cooling.

Main technical parameters of engines series TBD 645

Parameter	Value
Number and cylinders arrangement	6, 8, 9 in-line
Cylinder bore (mm)	330
Piston stroke (mm)	450
Cylinder capacity (dm^3)	38.5
Rotation speed max (min^{-1})	600/650
Rotation speed min (min^{-1})	315
Cylinder power at 600/650 min^{-1} (kW)	425/460
Air charging pressure (MPa)	0.22
Compression ratio	13.5
Start-to-open injector pressure (MPa)	33.0
Brake specific fuel oil consumption (g/kWh)	178/181
Lubrication system oil consumption (g/kWh)	0.6
Turbocharger hardware model for 6/8/9 cylinders	VTR 304/VTR 354
Mean effective pressure at 650 min^{-1} (MPa)	2.21
Mean piston speed at 600/650 min^{-1} (m/s)	9.0/9.75

Dimensions and weight engines series TBD 645

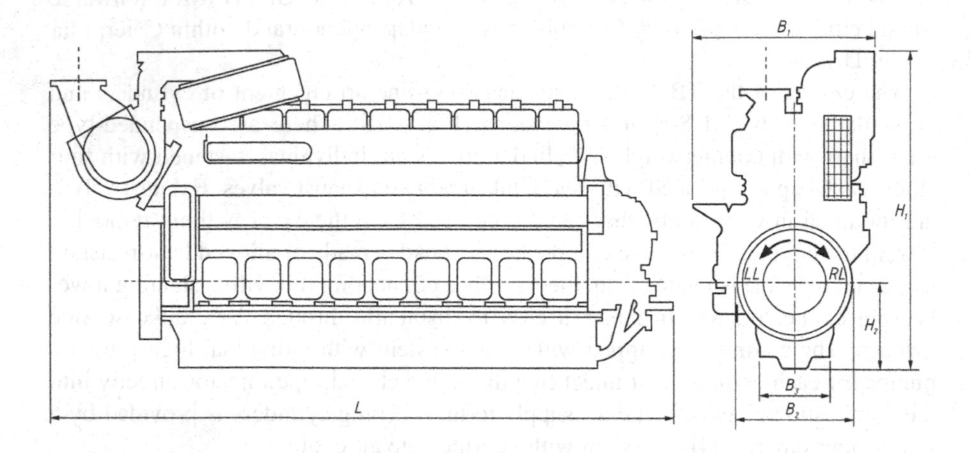

Engine version	L (mm)	B_1 (mm)	B_2 (mm)	B_3 (mm)	H_1 (mm)	H_2 (mm)	Weight (kg)
TBD 645 L6	5565	2110	1360	870	3690	990	26,500
TBD 645 L8	6530	2110	1360	870	3570	990	34,000
TBD 645 L9	7020	2150	1360	870	3570	990	37,600

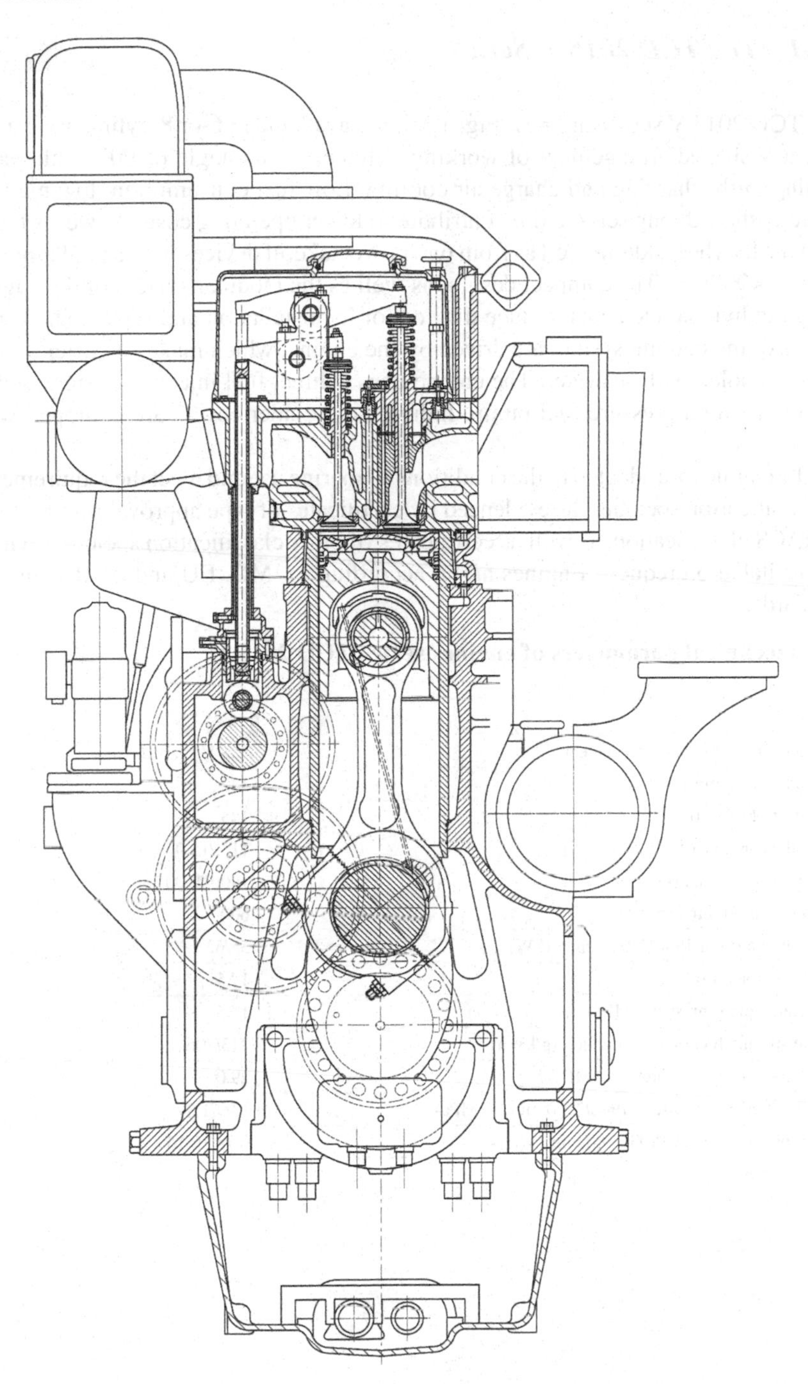

Fig. 1.15 Cross-section of the engine TBD 645 series [14]

1.6.1 The TCD 2015 V Series

The TCD 2015 V series engines (Fig. 1.16) are available in 6-or 8-cylinder versions with a V-shaped arrangement of working cylinders at an angle of 90°, with water cooling, turbocharging and charge air cooling. Low fuel consumption, low maintenance costs and long service life contribute to lower operating costs. Power is taken from the flywheel side, as well as from two power takeoff devices with a total moment of up to 400 N m. The compact design, as well as the modular system of the engine design individual elements, reduce the cost of its installation and repair. The water pumps of the cooling system are driven by the engine, which makes it easier to use, when it cooled with seawater. The engine uses a direct fuel injection system with a block-type high-pressure fuel pump. Individual cylinder covers are equipped with four valves.

The engines are adapted to the conditions of marine use and meet the requirements of classification societies, as evidenced by certificates of type approval according to the IACS classification, as well as certificates of other classification societies, which are available on request. Engines are in according to IMO, EU and ZKR emission standards.

Main technical parameters of engines series TCD 2015 V

Parameter	Value
Number and cylinders arrangement	6, 8 V-shaped
Cylinder bore (mm)	132
Piston stroke (mm)	145
Cylinder capacity (dm^3)	11.9/15.9
Rotation speed max (min^{-1})	1900/2100
Rotation speed min (min^{-1})	600
Cylinder power at 1900/2100 min^{-1} (kW)	55/62
Compression ratio	17.5
Maximum cycle pressure (MPa)	17.5
Brake specific fuel oil consumption (g/kWh)	213/206
Start-to-open injector pressure (MPa)	29.0
Mean effective pressure at 1900/2100 min^{-1} (MPa)	1.99/1.80
Mean piston speed at 1900/2100 min^{-1} (m/s)	9.18/10.15

Dimensions and weight engines series TCD 2015 V

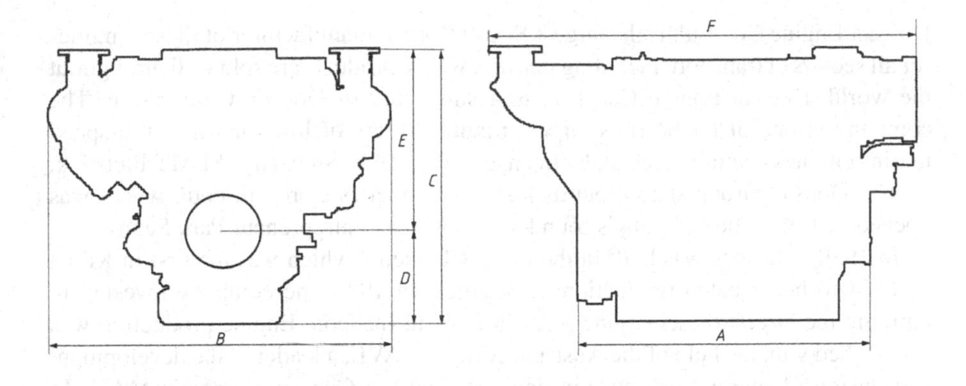

Engine version	A (mm)	B (mm)	C (mm)	D (mm)	E (mm)	F (mm)	Weight (kg)
TCD 2015 V06	1045	1315	1230	440	790	1520	1260
TCD 2015 V08	1210	1330	1230	440	790	1680	1480

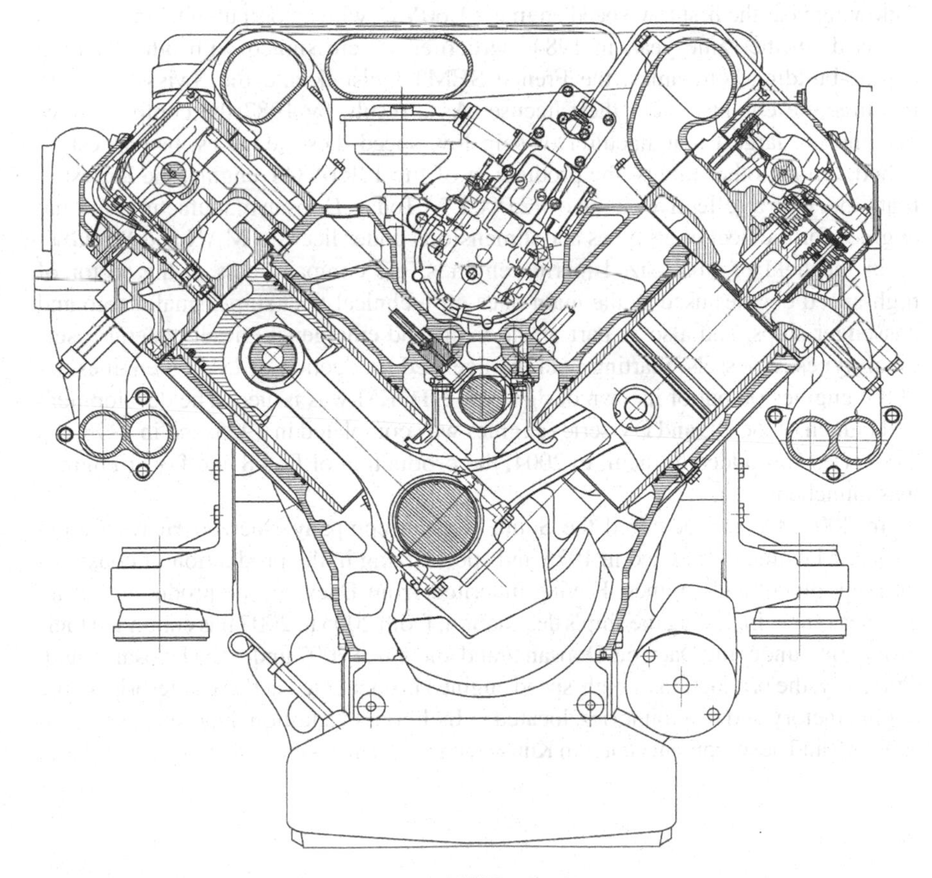

Fig. 1.16 Cross-section of the engine TCD 2015 V series [15]

1.7 Doosan Engine Co., Ltd.

Doosan Engine Co., Ltd. is the largest South Korean manufacturer of diesel engines for all sectors of transport, including marine, whose products are sold well throughout the world. Doosan Engine Co., Ltd. is a subsidiary of Doosan Corporation. The company is one of the world's largest manufacturers of low- and medium-speed marine engines manufactured under licenses from MAN, Sulzer and SEMT-Pielstick.

The Doosan Group started out as a small cosmetics store in Seoul, which was opened in 1896. The company's founder was Korean entrepreneur Park Seung.

In 1930, a factory was built in the city of Incheon, which was the first in Korea in 1958 to begin batch production of engines. In 1975, the company invested in building the largest diesel engine plant in Asia in Incheon. Engine production was established with the help of the Austrian company AVL, a leader in the development and design of internal combustion engines, and the German company MAN. In 1979, the production of small-sized diesel engines was started under license from the Japanese company ISUZU.

In 1983, the company adopted a large-scale program for the development of the domestic ship engine building. As part of this program, a licensing agreement was signed with MAN for the production of low- and medium-speed engines, and the following year the first low-speed engine, 6L60MC, with an output of 9185 kW, was released. In the same year, in 1984, agreements were signed with other leading engine-building companies, the French SEMT-Pielstick and the Swiss Sulzer to manufacture engines under their license. As a result, by 1987 the company have become the largest manufacturer of ship low-speed diesel engines, having established, among other things, the production of the 12RTA84T engine, the largest at that time, with an effective output of 42,000 kW. Today, Doosan manufactures marine engines for the fleet of all types and dimensions under licenses MAN and WinGD.

In the field of small-sized marine engines, the company acts in the sector of high-speed engines used as the main ones on technical fleet ships, small cargo and passenger ships, and also as part of the main and emergency diesel generator sets on all types of vessels. Starting with the licensed production of ISUZU engines, by 1985, engines output of its own design type STORM was issued. The development and launch of the DE and DV series engines was completed in 1995, and in 1998, the TIS series was added to them. In 2004, the production of DL08 and DV11 engines was launched.

In 2005, Doosan acquired the South Korean company Daewoo Heavy Industries & Machinery, founded in 1976 and specializing in the production of construction equipment and engines. Having inherited from Daewoo, the production of an engines range, including the ship's destination, from 2005 to 2007 has continued their production under the Daewoo-Doosan brand and since 2007 under the Doosan brand. Currently, the production of high-speed engines is conducted at three enterprises, one engine factory and two foundries, located in Incheon. In addition, Doosan has its own research and development center in Kunsan, engaged in design and engineering work.

Doosan represents in the market its own line of marine high-speed engines with an improved and modified fuel system, a new turbo-supercharger design, an improved exhaust manifold and an air cooler. The engines are presented in 15 different basic configurations that cover the power range from 51 to 883 kW. Engines are manufactured in-line, L and D series, and V-shaped (V series). All Doosan marine engines comply with current IMO standards for nitrogen oxide emissions into the atmosphere and are distinguished by high fuel efficiency.

Based on existing engines, Doosan supplies diesel generator sets in stationary and ship versions, as well as diesel gear units for ships of small and medium displacement. In addition, a range of Doosan engine controls may be included. All equipment, supplied by the company, certified by ISO 9001 and ISO 14001 certificates.

Based on the specific operating conditions, the engines are available in three versions, designed for operation:

- in hard conditions; the operating time is not limited, the average engine load is 85% of the rated power, the operating time at maximum power is no more than 50%;
- under conditions of moderate loads; the operating time is up to 2000 h per year (10 h per day), the average engine load is 70% of the rated power, the operating time at maximum power does not exceed 30%.
- under conditions of low loads; the operating time is up to 1000 h per year (5 h per day), the average engine load is 50% of the rated power, the operating time at maximum power does not exceed 20%.

1.7.1 The Engine of the V222TI Series

The engine of the V222TI series is a twelve-cylinder diesel engine, made according to the classical scheme with a rigid block, a V-shaped arrangement of cylinders and a suspended crankshaft (Fig. 1.17).

In the block the bores are made for installation of wet cylinder liners. Cylinder covers are paired, each cap is equipped with valves, two for each cylinder, one inlet and one exhaust ones. The engine is equipped with a fuel system with a block-type high-pressure pump with an all-mode regulator, closed-type injectors that make direct fuel injection into the combustion chamber, made in the piston. The air supply to the working cylinders is provided by a gas turbine supercharging system with intermediate air cooling.

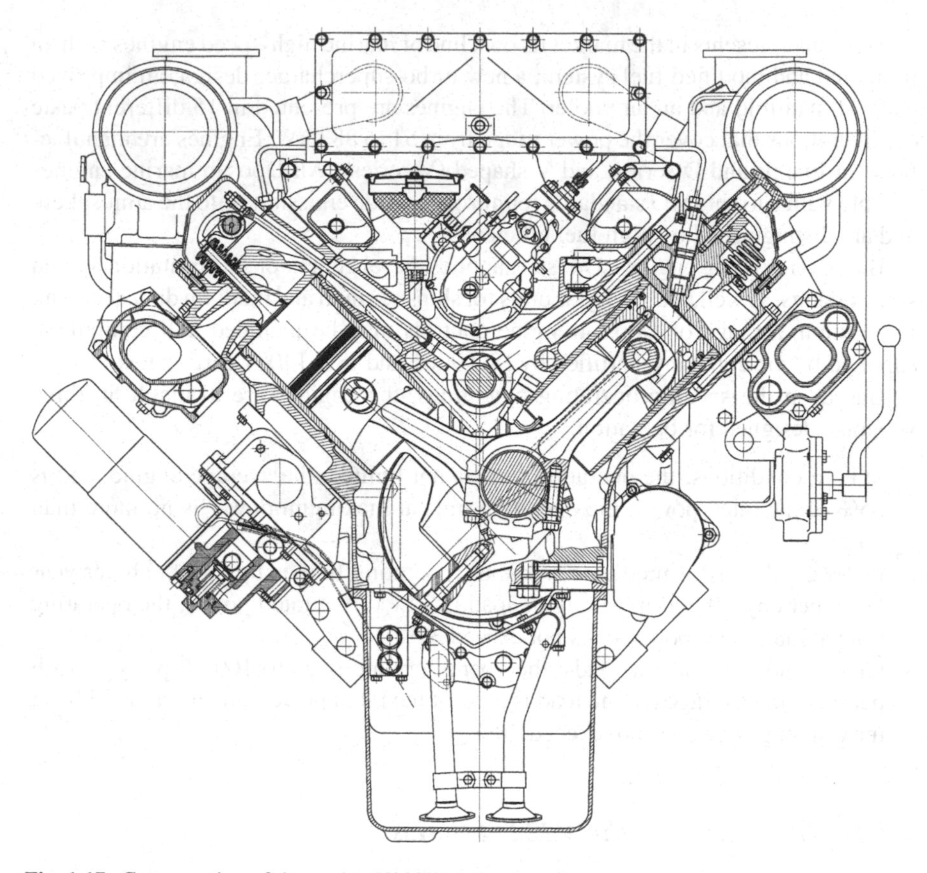

Fig. 1.17 Cross-section of the engine V222TI series [16]

Main technical parameters of engines series V222TI

Parameter	Value
Number and cylinders arrangement	12 V-shaped
Cylinder bore (mm)	128
Piston stroke (mm)	142
Cylinder capacity (dm^3)	21.93
Rotation speed max (min^{-1})	2300
Rotation speed min (min^{-1})	725
Cylinder power at 2300 min^{-1} (kW)	61.33
Compression ratio	15.0
Compression pressure (at 200 min^{-1}) (MPa)	2.8
Pressure pump timing angle, deg to TDC	22
Start-to-open injector pressure (MPa)	28.6

(continued)

(continued)

Parameter	Value
Maximum cycle pressure (MPa)	15.0
Brake specific fuel oil consumption (g/kWh)	258
Brake specific air consumption (кg/kWh)	8.15
Exhaust gas temperature (°C)	400
Mean effective pressure at 2300 min^{-1} (MPa)	1.785
Mean piston speed at 2300 min^{-1} (m/s)	10.89
Dry engine weight without reduction gear box (kg)	1750
Dry engine weight with reduction gear box type MGN 86E (kg)	2460

Engine V222TI dimensions without reduction gear box

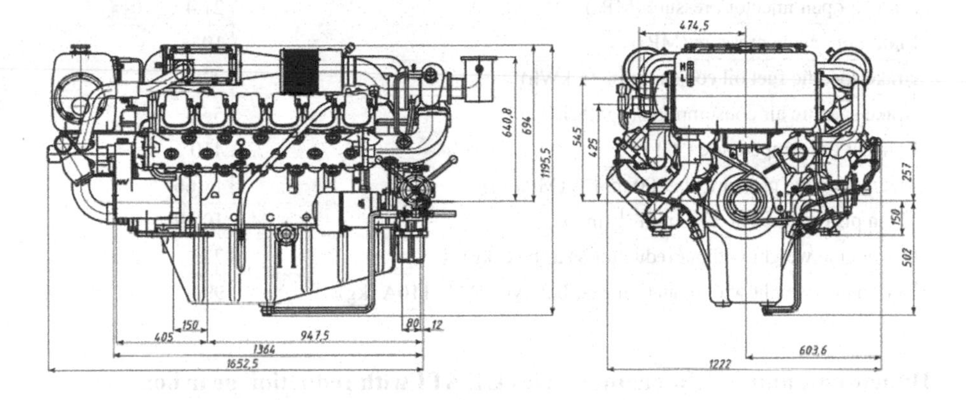

1.7.2 *The Engine of the L136TI Series*

The engine of the L136TI series is made according to the classical scheme with a rigid block, in-line arrangement of cylinders and a suspended crankshaft (Fig. 1.18). In the crankcase, cast from high-quality cast iron, there are bore holes for installing dry cylinder liners. Cylinder covers are paired, each cap is equipped with valves, two for each cylinder, one inlet and one exhaust. The engine is equipped with a fuel system with a block-type high-pressure pump with an all-mode regulator, closed-type injectors that provide direct fuel injection into the combustion chamber made in the piston. The air supply to the working cylinders is provided by a gas turbine supercharging system with intermediate air cooling. The engine is equipped with a starting system, using an electric starter.

Main technical parameters of engines series L136TI

Parameter	Value
Number and cylinders arrangement	6 in-line
Cylinder bore (mm)	111
Piston stroke (mm)	139
Cylinder capacity (dm^3)	8.07
Rotation speed max (min^{-1})	2200
Rotation speed min (min^{-1})	725
Cylinder power at 2200 min^{-1} (kW)	28.1
Compression ratio	16.7
Compression pressure (at 200 min^{-1}) (MPa)	2.8
Pressure pump timing angle, deg to TDC	14
Start-to-open injector pressure (MPa)	21.4
Maximum cycle pressure (MPa)	10.6
Brake specific fuel oil consumption (g/kWh)	220
Brake specific air consumption (кg/kWh)	5.95
Exhaust gas temperature (°C)	410
Mean effective pressure at 2200 min^{-1} (MPa)	1.166
Mean piston speed at 2200 min^{-1} (m/s)	10.19
Dry engine weight without reduction gear box (kg)	773
Dry engine weight with reduction gear box type DMT110A (kg)	994

Dimensions and weight engines series L136TI with reduction gear box

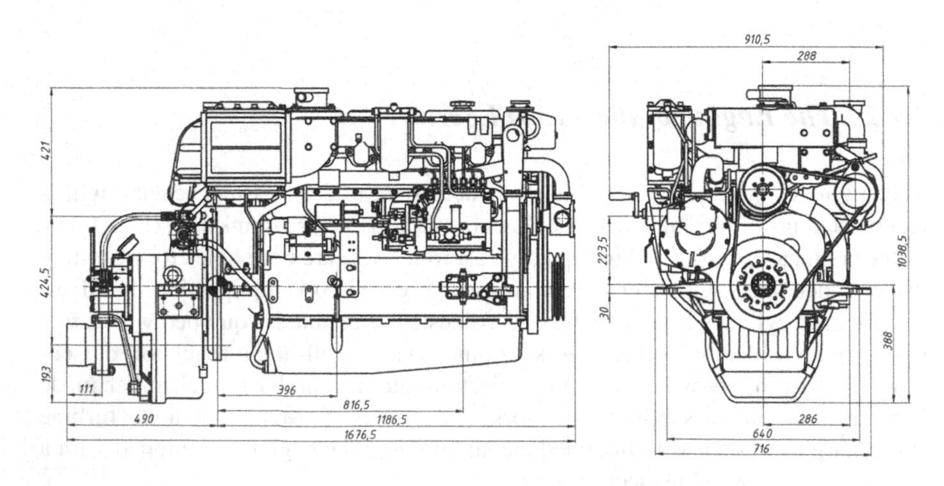

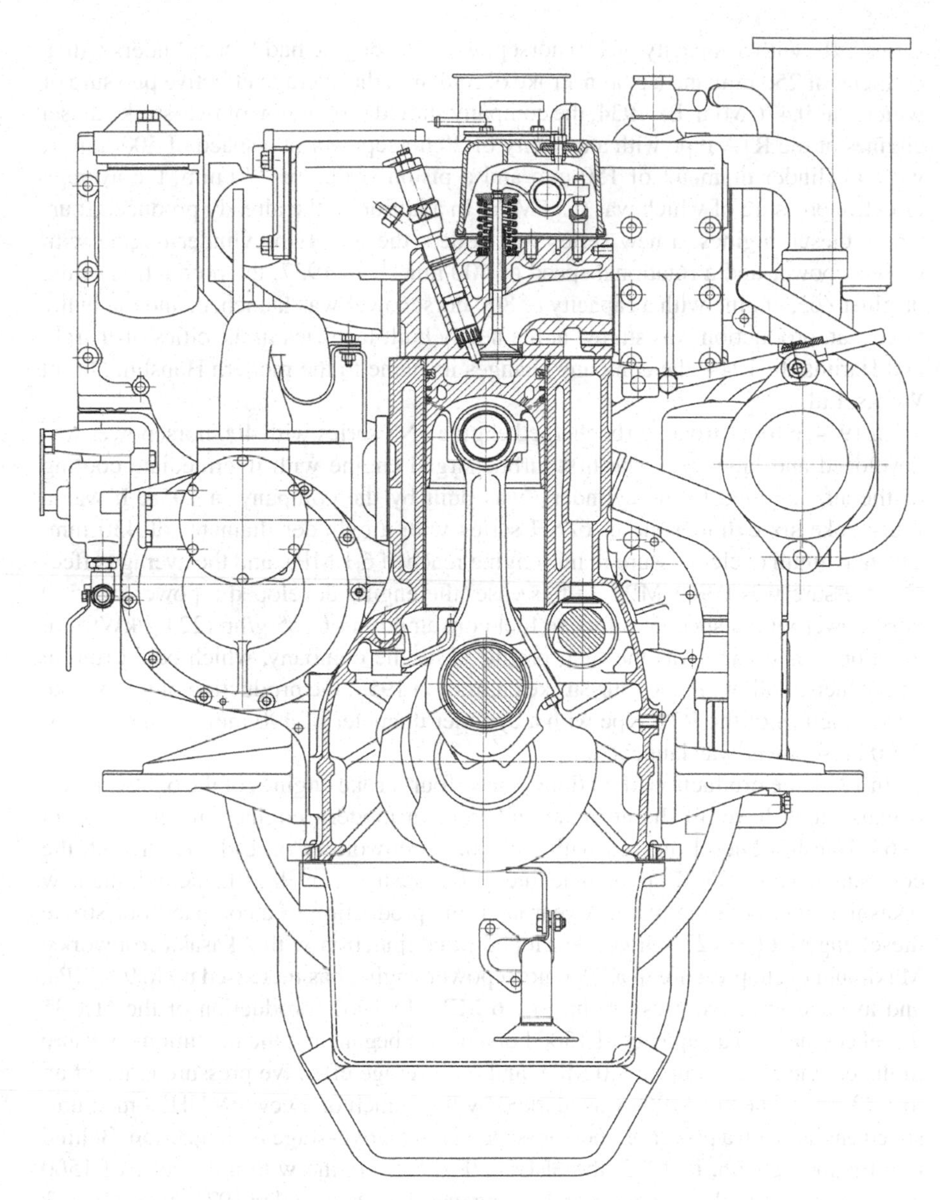

Fig. 1.18 Cross-section of the engine L136TI [17]

1.8 Hanshin Diesel

A company called Hanshin Ironworks Co., Ltd. was founded in 1918 to manufacture hot-bulb engines in the Japanese city Kobe, Hyogo Prefecture. By 1929, the company was reoriented to the production of four-stroke diesel engines of its own design series T4E, with a capacity of 150 horsepower. The engine had four cylinders with a diameter of 250 mm and a piston stroke of 380 mm, the average effective pressure of which was 0.476 MPa. By 1934, the company started production of two-stroke diesel engines of the R1A type with a capacity of 20 horsepower at a speed of 500 min^{-1}, with a cylinder diameter of 150 mm and a piston stroke of 270 mm, the average effective pressure of which was 0.38 MPa. In addition to the already produced four-stroke diesel engines, a new series was added, the S2F two-cylinder engine with 66 horsepower and a rotational speed of 400 min^{-1}. In 1937, the production of the original Z6K engine with a capacity of 800 horsepower was launched, and the in the next year, production was started at the newly built factories in the cities of Ibaraki and Harima. In 1944, the company changes its name to the modern Hanshin Diesel Works, Ltd.

In 1954, a four-stroke, turbocharged engine 6NS series with 400 horsepower was developed and launched. The first turbocharged engine with intermediate cooling of the air, supplied to the cylinders, was built by the company in 1958. It was a four-stroke six-cylinder diesel 6ZSH series with a cylinder diameter of 430 mm. The maximum cycle pressure in this engine reached 6.1 MPa, and the average effective pressure was 0.961 MPa. In this case, the engine developed a power of 1500 horsepower with a specific effective fuel consumption of 165 g/hp (224 g/kWh). In addition, this engine was the first, produced by the company, which could run on heavy fuels. Following the four-stroke engines in 1959, the production of two-stroke diesel engines of the R7E type with a cylinder diameter of 490 mm and a power of 2400 horsepower was launched.

In 1963, the production of medium-speed four-stroke engines of the 620SH series began, which have all the attributes inherent in modern engines of this class. In 1964, Hanshin Diesel merged with Kinoshita Ironworks Co., Ltd., as a result, the company began to develop the machine-tool industry, as well as launching the new Akashi engine-building plant. A year later, the production of a compact four-stroke diesel engine 6LUK27 started, developed in conjunction with Akasaka Ironworks. Maximum cycle pressure in a 1000 horsepower engine has increased up to 9.5 MPa, and average effective pressure up to 1.6 MPa. In 1967, production of the 6LU35 diesel engine with a capacity of 1500 horsepower began, and the maximum pressure in the engine reached up to 9.0 MPa, and the average effective pressure reached up to 1.33 MPa. The year 1969 was marked by the launch of a new 6MUH28 medium-speed engine with a power of 1600 horsepower, with two-stage fuel injection. Behind this Engine version, in 1972, the 6LUD32F diesel engine with a capacity of 1500 horsepower and high level of supercharging was launched, and in 1977, an 6LUD26T engine with exhaust gas recirculation. 1979 was a turning point in the history of the company, which produced a low-speed four-stroke long-stroke engine 6EL32 with a capacity of 2200 horsepower, with an average effective pressure of 2.29 MPa and a maximum cycle pressure of 13 MPa. In 1982, in the development of the concept of low-speed four-stroke diesel engines, a new 6LF58 engine with a cylinder diameter of 580 mm and a piston stroke of 1050 mm was developed. This 6000 horsepower engine has become the largest in its class. Further development of this concept was the launch of the series LH28 in 1986 and the series LH28L in 1989, and nine years later in 1998 the series LA 34 with improved economic and environmental indicators. In addition to engines of its own design, in cooperation with Kawasaki Heavy Industries Ltd. since 1986, Hanshin has been manufacturing two-stroke low-speed engines under a MAN B&W licence.

1.8.1 The Hanshin Engine of the LA Series 30

The LA engine series has replaced the well-established but outdated LH series. In the engines of the new series, the boost pressure and the maximum cycle pressure were increased, which, together with a decrease in the rotational speed, allowed the specific fuel consumption to be reduced by 3.6%. In connection with the growth of mechanical loads, a number of structural and technological solutions were applied. In particular, the piston head, which was previously made of gray iron, was manufactured from high-strength cast iron with a spherical structure. This allowed not only to increase the strength of the piston, but also to reduce its mass. Hydraulic transmission is used to drive distribution of gas valve timing in LA series engines, which can significantly reduce noise operation and simplify maintenance. Each cylinder has one inlet and one exhaust valve, which are located in separate housings. Each of valves can be removed from the cylinder head without removing it (Fig. 1.19).

Main technical parameters of engines Hanshin LA 30 series

Parameter	Value
Number and cylinders arrangement	6 in-line
Cylinder bore (mm)	300
Piston stroke (mm)	600
Cylinder capacity (dm^3)	42.41
Rotation speed (min^{-1})	290
Cylinder power at 290 min^{-1} (kW)	220.5
Maximum cycle pressure (MPa)	14.7
Mean effective pressure at 290 min^{-1} (MPa)	2.15
Cylinder outlet temperature (°C)	480
Turbocharger outlet temperature (°C)	335
Brake specific fuel oil consumption (g/kWh)	185
Lubrication system oil consumption (g/kWh)	0.3
Cylinder oil consumption (g/kWh)	0.7
Mean piston speed at 290 min^{-1} (m/s)	5.8

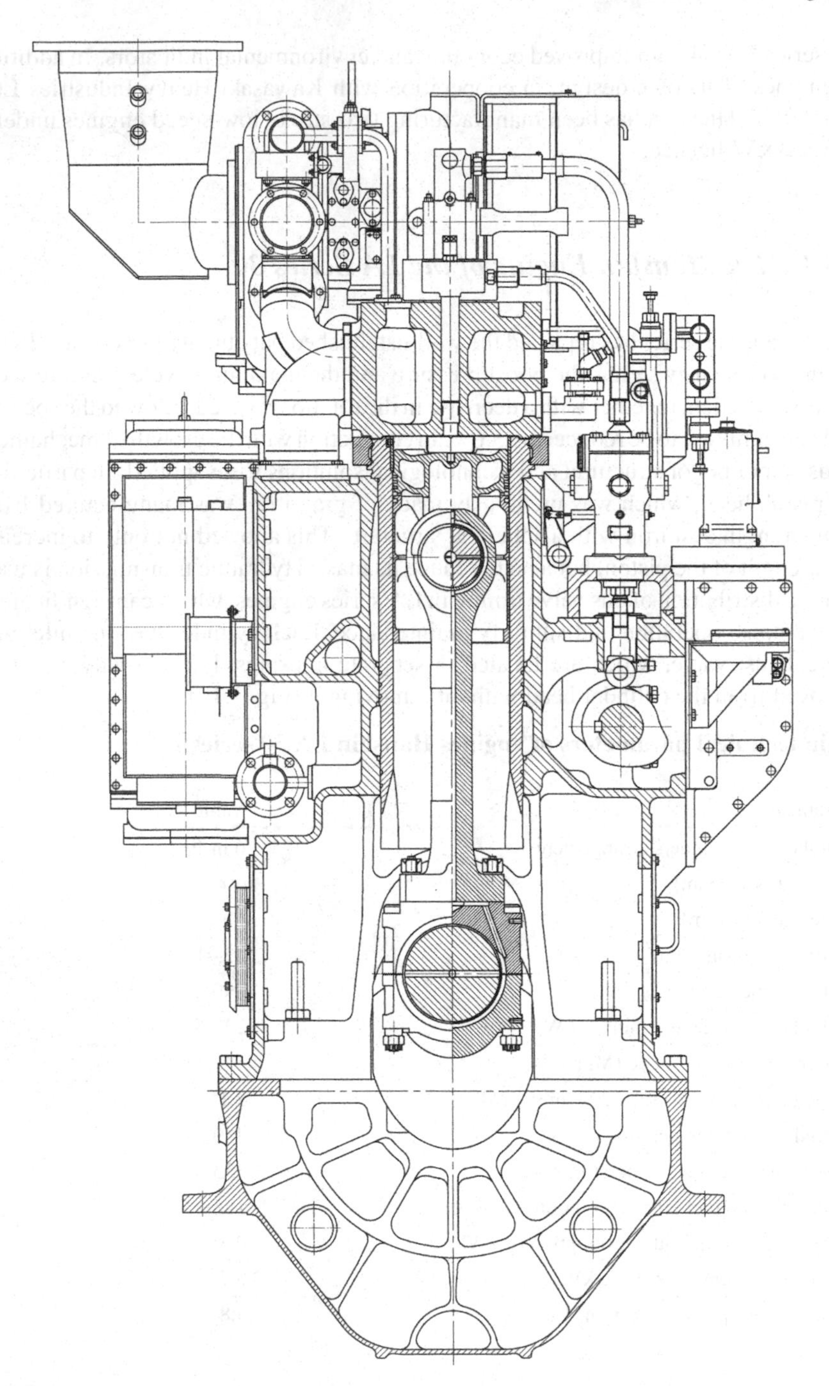

Fig. 1.19 Cross-section of the engine Hanshin LA 30 series [18]

Dimensions and weight engines series Hanshin LA 30

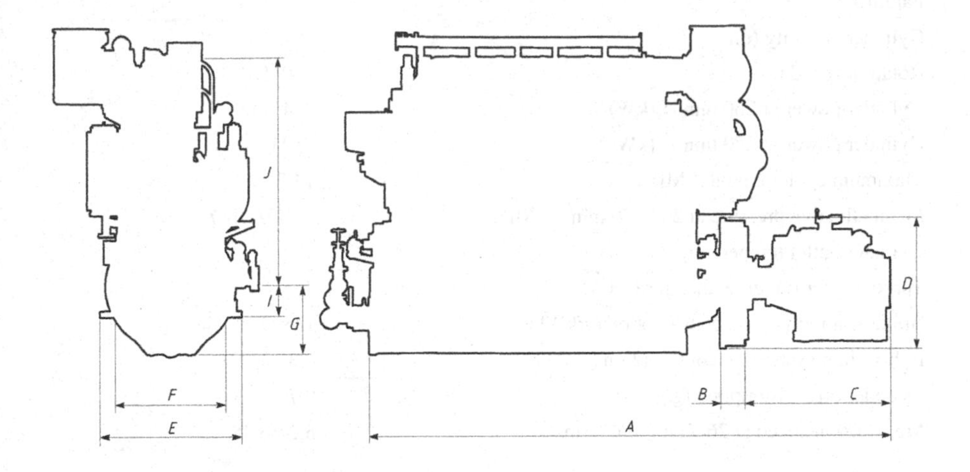

Engine version	A (mm)	B (mm)	C (mm)	D (mm)	E (mm)
LA30	5572	345	1540	1400	1400

Engine version	F (mm)	G (mm)	I (mm)	J (mm)	Weight (kg)
LA30	1070	685	310	2415	29,000

1.8.2 The Hanshin Series LA 34 Engine

The Hanshin series LA 34 engine is a low-speed, four-stroke in-line engine with gas turbine supercharging and charge-air cooling, left or right-hand rotation (Fig. 1.20). Based on the experience of the development and operation of the previous LH and LH-L series, the design of the LA series was improved by increasing the stroke of the piston and reducing the rotational speed, which allows these engines to be used both as part of diesel-geared units and with direct power transmission to the screw. LA Series engines have low exhaust emissions with NO_x and lower fuel consumption, which leads to CO_2 emissions decrease. To reduce the wear rate of the working cylinder bushing and reduce the oil consumption, an anti-polishing ring is installed in its upper part. LA Series engines meet all requirements for emissions of exhaust gases in accordance with IMO Tier-II, CCNR-2 and EU3A.

Main technical parameters of engines Hanshin series LA 34

Parameter	Value
Number and cylinders arrangement	6 in-line
Cylinder bore (mm)	340
Piston stroke (mm)	720

(continued)

(continued)

Parameter	Value
Cylinder capacity (dm^3)	392.2
Rotation speed (min^{-1})	260/270
Cylinder power at 260 min^{-1} (kW)	294.0
Cylinder power at 270 min^{-1} (kW)	319.0
Maximum cycle pressure (MPa)	14.2/14.7
Mean effective pressure at 260/270 min^{-1} (MPa)	2.077/2.167
Cylinder outlet temperature (°C)	465
Turbocharger outlet temperature (°C)	325
Brake specific fuel oil consumption (g/kWh)	185
Lubrication system oil consumption (g/kWh)	0.3
Cylinder oil consumption (g/kWh)	0.7
Mean piston speed at 260/270 min^{-1} (m/s)	6.24/6.48

Engines Hanshin series LA 34 dimensions and weight

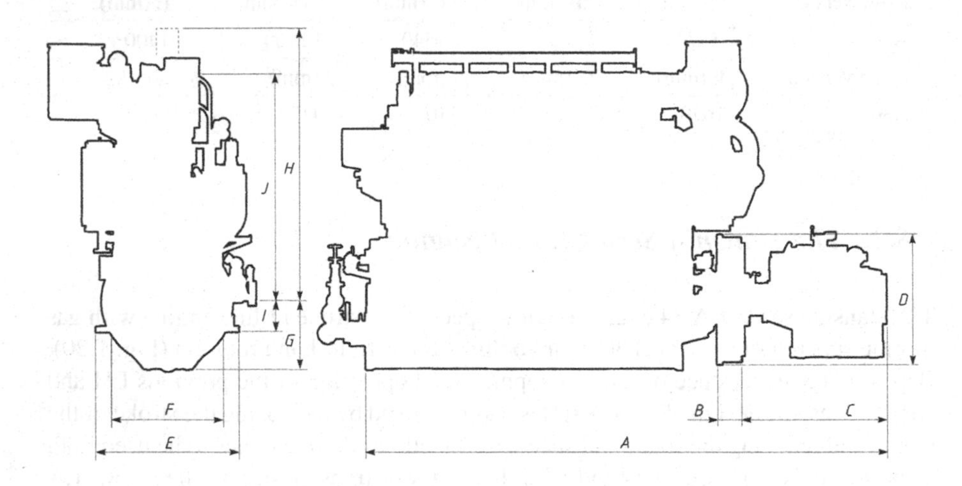

Engine version	A (mm)	B (mm)	C (mm)	D (mm)	E (mm)	F (mm)
LA34	6304	297	1760	1560	1700	1320

Engine version	G (mm)	H (mm)	I (mm)	J (mm)	Weight (kg)
LA34	830	3357	370	2753	40,000

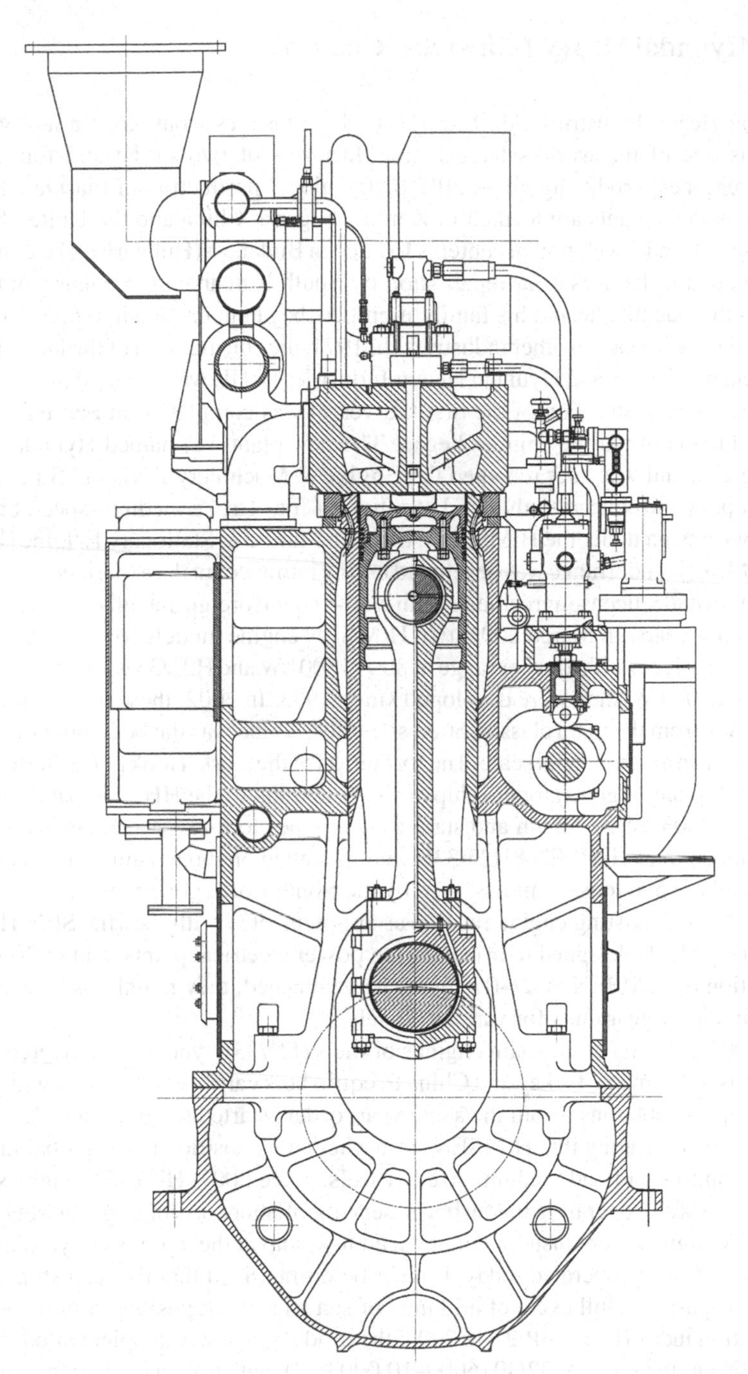

Fig. 1.20 Cross-section of the engine series Hanshin LA 34 [19]

1.9 Hyundai Heavy Industries Co., Ltd.

Hyundai Heavy Industries Co., Ltd. (Hyundai translates from Korean as "Modernity") is one of the world's largest manufacturers of two-stroke and four-stroke marine engines, producing about 30% of the global production of marine engines. The company's plants are located in Korea, Bulgaria, China and the United States. The research and development center is located in Budapest (Hungary). The company was founded in 1947 as a car repair shop by South Korean entrepreneur Chung Ju-yung. Subsequently, he and his family members began to engage in other activities, expanding the impact on other industries. In 1971, the construction of the largest shipyard began in Ulsan, and Hyundai Heavy Industries (HHI) was founded in 1973. The next year, the construction of the first sea vessel was completed. In February 1975, the production of marine engines began. The new plant was named Hyundai Engineering Co., and was later renamed the Engine & Machinery Division. Since 2001, the company introduced to the market a line of four-stroke medium-speed engines of its own design under the HiMSEN (Hi-touch Marine & Stationary ENgine) brand. The HiMSEN model range was developed taking into account the experience of engine production under license agreements with well-known foreign manufacturers for more than twenty years. In March 2001, two HiMSEN engine models were first released: H21/32 (Fig. 1.22) with a power range of 800–1800 kW and H25/33 with a power range of 1500–2700 kW, they were developed since 1993. In 2002, these models received certificates from the main classification societies, which was the beginning of engine production on a commercial scale. The formation of the market took place due to deliveries of ship diesel generators to shipyards, belonging to the HHI companies group, and since 2003 sales of main and stationary engines to foreign markets began. The first main engines HiMSEN 8H25/33P were installed on three fishing vessels with a deadweight of 500 tons, commissioned by the South Korea government.

In 2004, the existing engine range was supplemented with the HiMSEN H17/21 model (Fig. 1.21) designed to drive marine power electrical plants, and in 2006, the production of HiMSEN H32/40 engines was launched, they mainly used as part of the main diesel-gear units for various vessels.

In 2006, 50 units of the main engines of the 9H25/33P type were delivered to the shipyards of Vietnam, Turkey and China to equip bulk carriers and tankers with deadweight up to 6000 tons. From the same year, orders started to come from European shipowners, indicating that HiMSEN strengthened its position in the global market. In addition to small and medium-sized vessels, since 2007, HiMSEN engines have been successfully operating in tow boats, serving offshore drilling rigs. In May 2009, HiMSEN engines were installed on the lead tow ship at the Tebma shipyard and are being successfully operated today. It must be emphasized that the lead ship of this series has passed a full cycle of all kinds of sea trials with positive results in terms of vibration index (DNV-DP-2). In 2008, the model range was supplemented with an HiMSEN engine series V 32/40 (6000–10,000 kW), with a V-shaped arrangement of cylinders, designed for use on commercial and passenger ships with a deadweight of up to 20,000 tons. The version of the HiMSEN engine 35/33 with a V-shaped arrangement of cylinders appeared on the market in 2011, fully completing the formation of the model range.

At present, the HiMSEN production program covers the entire range: marine diesel generators, main engines, propulsion plants and stationary generators. The engines of this model range cover the power range from 575 to 10,000 kW, they are distinguished by high fuel efficiency, reliability and ease of maintenance. All diesel engines comply with IMO-Tier II emissions requirements for NO_x.

1.9.1 *HiMSEN Series H17/28 Engines*

HiMSEN series H17/28 engines are a specialized high-speed four-stroke engine with gas turbine supercharging and charge air cooling, designed for marine generator sets with an innovative design concept (Fig. 1.21). High specific power contributes to the creation of light and compact diesel generator sets on their basis. This engine uses the most advanced HiMSEN technology, called "Hi-touch", the use of which has allowed to create an engine with low fuel consumption and low emissions. The work process in the engine is organized according to the Miller cycle with the ability to optimize the valve timing depending on the mode of engine operation. The high S/D ratio (1.65) made it possible to optimize the shape of the combustion chamber, which contributes to more efficient mixing and burning of the fuel. This allows to reduce the maximum temperature of the cycle, and reduce the amount of harmful emissions in the exhaust gases, primarily NO_x.

Main technical parameters of engines HiMSEN series H17/28

Parameter	Value
Number and cylinders arrangement	5, 6, 7, 8 in-line
Cylinder bore (mm)	170
Piston stroke (mm)	280
Cylinder capacity (dm^3)	6.36
Rotation speed (min^{-1})	1000
Cylinder power (kW)	115.0
Air charging pressure (MPa)	2.5
Compression ratio	15.0
Maximum cycle pressure (MPa)	17.0
Mean effective pressure (MPa)	2.41
Brake specific fuel oil consumption (g/kWh)	191.0
Lubrication system oil consumption (g/kWh)	0.6
Mean piston speed (m/s)	9.33

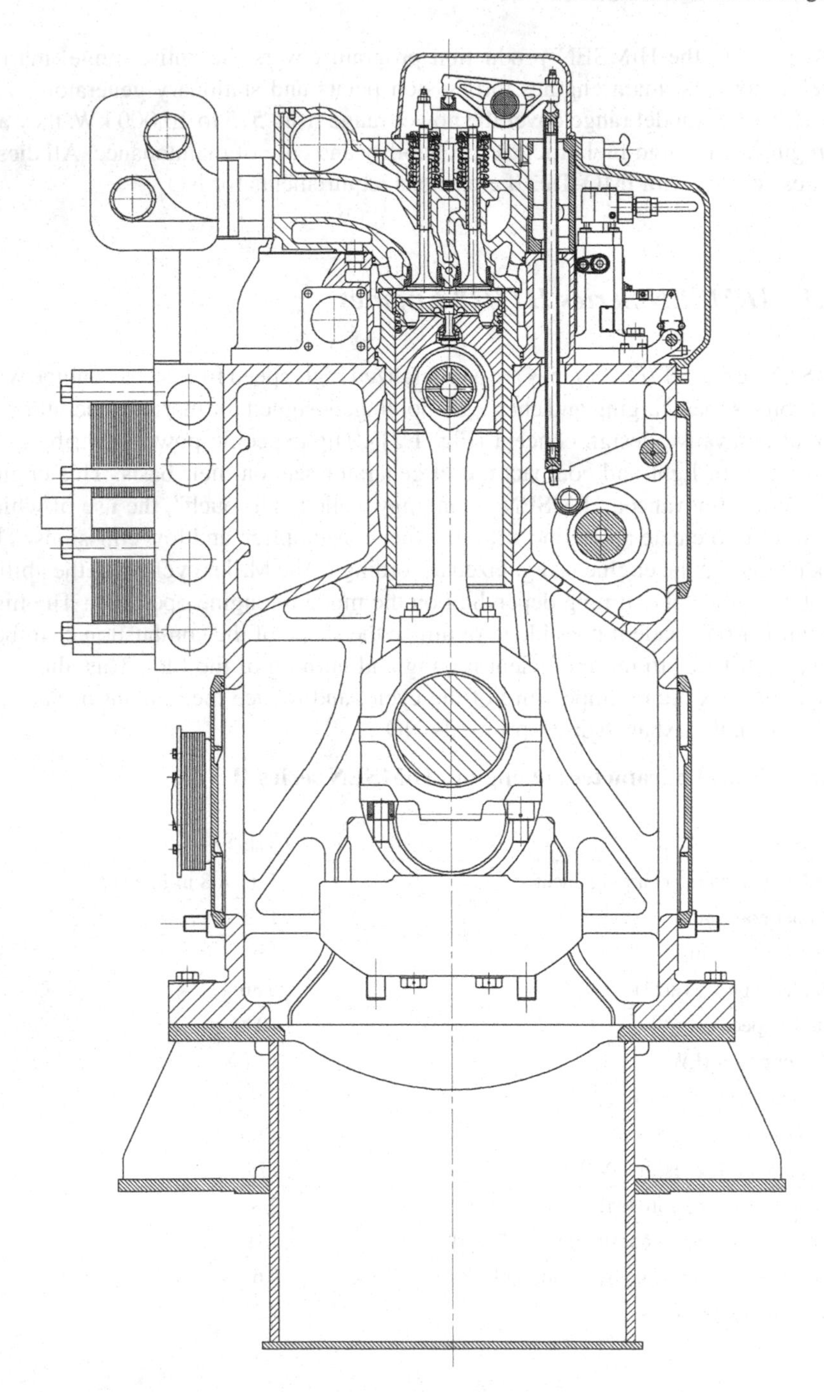

Fig. 1.21 Cross-section of the engine series HiMSEN H 17/28 [20]

Dimensions and weight engines HiMSEN series H17/28

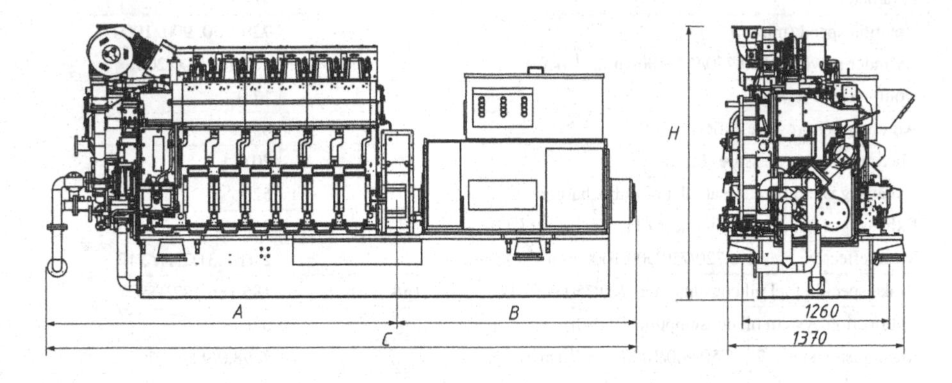

Engine version	A (mm)	B (mm)	C (mm)	H (mm)	Weight (kg)	Diesel generator weight (kg)
5 H17/28U	2791	2200	4991	2314	7700	13,600
6 H17/28U	3071	2200	5271	2314	8500	14,500
7 H17/28U	3351	2200	5551	2314	9400	15,600
8 H17/28U	3631	2320	5951	2314	10,400	16,700

1.9.2 HiMSEN Series H21/32 Engines

HiMSEN series H21/32 engines are high-speed four-stroke engines with turbine supercharging and charge air cooling, intended for use as part of diesel-gear and diesel generator sets (Fig. 1.22) of marine vessels. Like other HiMSEN engines, this engine version was developed in accordance with the concept of "Hi-touch", which included a number of key points to improve the efficiency and environmental performance of the engine. In particular, the high S/D ratio (1.52) made it possible to optimize the shape of the combustion chamber, high air-charging pressure (up to 0.38 MPa) in combination with a high compression ratio (17), allowed to organize the work process in the engine according to Miller's cycle with internal cooling air charge. The last approach made it possible to reduce the maximum cycle temperature and reduce the NO_x content in the exhaust gases.

Main technical parameters of engines HiMSEN series H21/32

Parameter	Value
Number and cylinders arrangement	6, 8, 9 in-line
Cylinder bore (mm)	210
Piston stroke (mm)	320
Cylinder capacity (dm^3)	11.1

(continued)

(continued)

Parameter	Value
Rotation speed (min^{-1})	720, 750, 900, 1000
Cylinder power at 720/750/900/1000 min^{-1} (kW)	160/160/200/200
Compression ratio	17.0
Air charging pressure (MPa)	0.32
Maximum cycle pressure (MPa)	20.0
Exhaust gas temperature at inlet of turbocharger (°C)	520
Exhaust gas temperature at outlet of turbocharger (°C)	350
Mean effective pressure 720/750/900/1000 min^{-1} (MPa)	2.41/2.31/2.41/2.17
Brake specific fuel oil consumption, 720/750/900/1000 min^{-1} (g/kWh)	186/186/187/189
Lubrication system oil consumption (g/kWh)	0.6
Mean piston speed 720/750/900/1000 min^{-1} (m/s)	7.7/8.0/9.6/10.7

Engines HiMSEN series H21/32 dimensions and weight

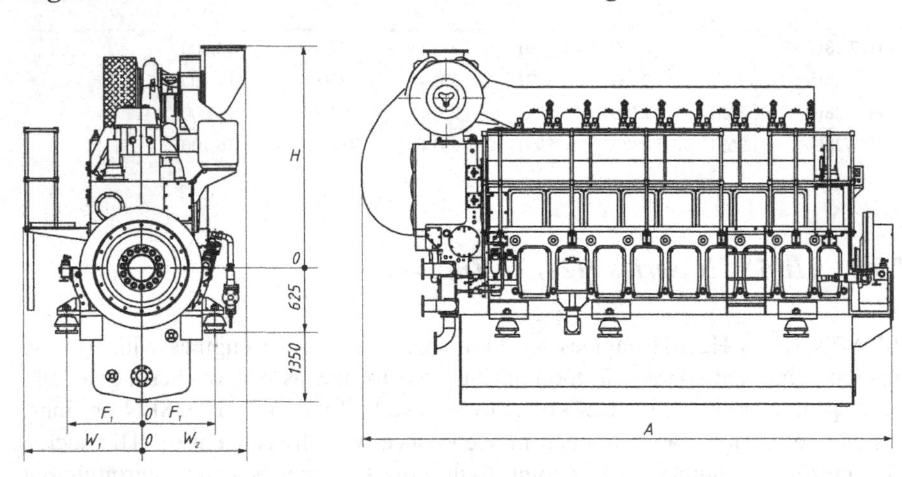

Engine version	A (mm)	F_1 (mm)	W_1 (mm)	W_2 (mm)	H (mm)	Weight (kg)
6 H21/32	3904	595	955	1126	2287	18,000
8 H21/32	4634	595	955	1214	2541	21,000
9 H21/32	4994	595	955	1214	2541	23,000

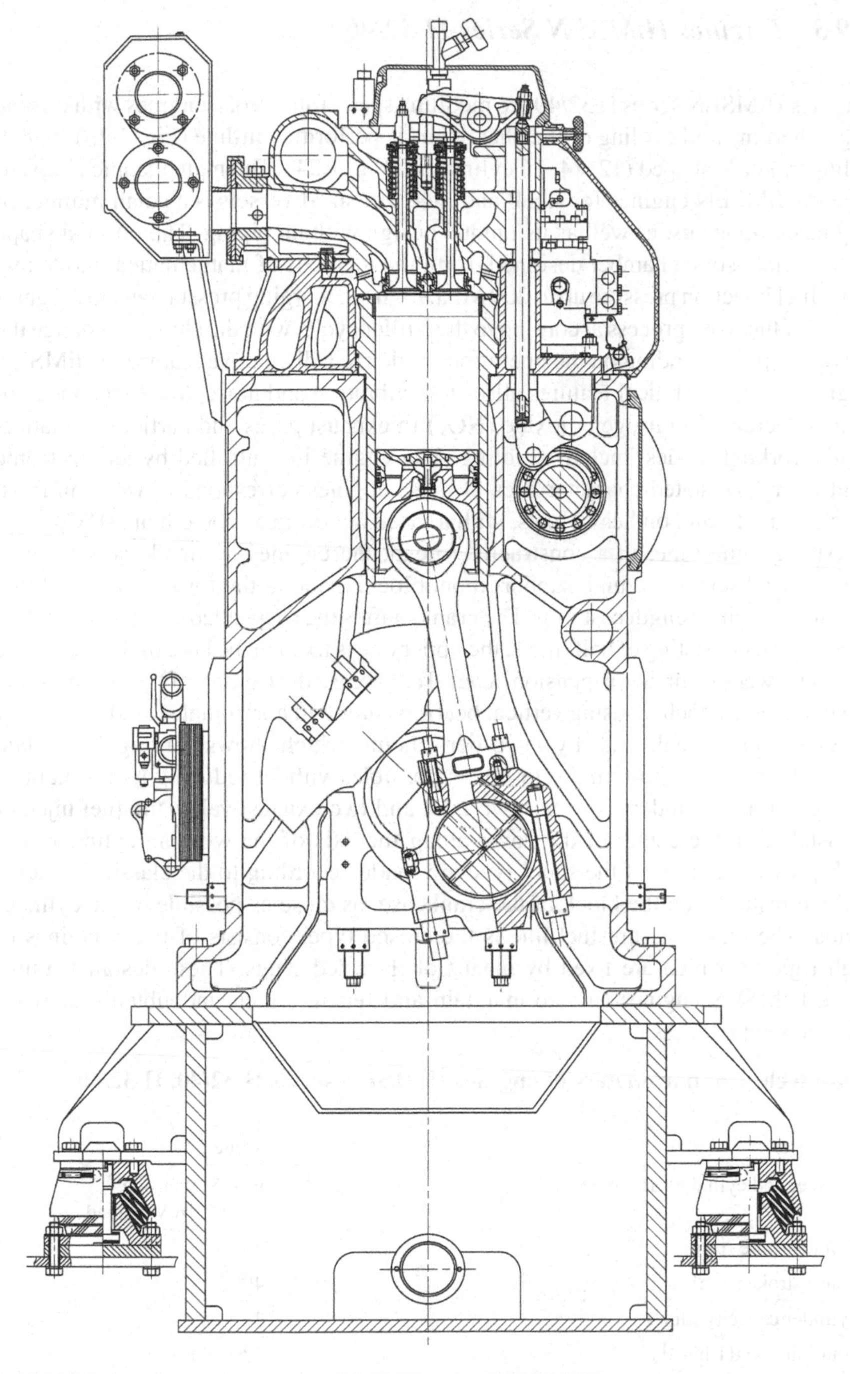

Fig. 1.22 Cross-section of the engine HiMSEN series H 21/32 [21]

1.9.3 Engines HiMSEN Series H 32/40

Engines HiMSEN series H 32/40 are medium-speed four-stroke engines with turbine supercharging and cooling of the charge air, is performed in-line (Fig. 1.23) (6, 8, 9 cylinders) or V-shaped (12, 14, 16 cylinders) (Fig. 1.24). The main technical advantage of HiMSEN engines for small and medium-sized vessels is a small number of engine components, as well as its unique design without piping. The optimal shape of the combustion chamber designed using the methods of mathematical modeling. High fuel injection pressure up to 200 MPa, high air charging pressure and the organization of the work process according to the Miller cycle, with the ability to change the valve timing, depending on the operation mode are a distinctive feature of HiMSEN engines. These technical features allow low vibration and noise, low fuel consumption, a decrease in nitrogen dioxide (NO_x) in exhaust gases and particulate matters in all working modes, including idling. The engine is controlled by an electronic controller. The stated characteristics of these engines correspond to work on light, distillate fuels, and on heavy fuels, with a viscosity of up to 700 cSt at 50 °C.

At the same time, in a constructive mean, the engines are made according to the classical scheme, with a support monoblock, made in the form of a monolithic casting of high-strength cast iron. The crankshaft is the suspended type; it is installed in bearings, consisting of half liners, the upper one is fixed in the bore of the crankcase and the lower one in the suspension cover. To increase the rigidity of the bearings, the lower cover is attached, using vertical bearing studs and horizontal screeds. Cylinder covers are individual, made by a two-tier scheme, which allows to bring the coolant channels to the fire bottom as closely as possible, without reducing its mechanical strength. Each cylinder cover has two inlet and two exhaust valve, the fuel injector is installed in the center of the cover along the axis of the working cylinder. The high-pressure pumps of the spool type are made according to the classical scheme and are installed on the block of the crankcase, as close as possible to the cylinder heads. The high-pressure fuel line is the closed type, consists of two pipelines of high rigidity, which are fixed by means of threaded plugs. These design features make HiMSEN engines easy to maintain and repair, which undoubtedly increase their competitiveness.

Main technical parameters of engines HiMSEN series H 32/40, H 32/40 V

Parameter	Value
Number and cylinders arrangement	6, 7, 8, 9 in-line 12, 14, 16 V-shaped
Cylinder bore (mm)	320
Piston stroke (mm)	400
Cylinder capacity (dm^3)	32.17
Rotation speed (min^{-1})	720/750
Cylinder power at 720/750 min^{-1} (kW)	500

(continued)

(continued)

Parameter	Value
Compression ratio	15.0
Air charging pressure (MPa)	0.35–0.45
Maximum cycle pressure (MPa)	21.0
Exhaust gas temperature at inlet of turbocharger (°C)	450–580
Exhaust gas temperature at outlet of turbocharger (°C)	280–400
Average effective pressure, 720/750 min^{-1} (MPa)	2.59/2.49
Specific fuel oil consumption, 720/750 min^{-1} (g/kWh)	179/181
Lubrication system oil consumption (g/kWh)	0.7
Average piston speed, 720/750 min^{-1} (m/s)	9.6/10.0

Dimensions and weight engines HiMSEN series H 32/40

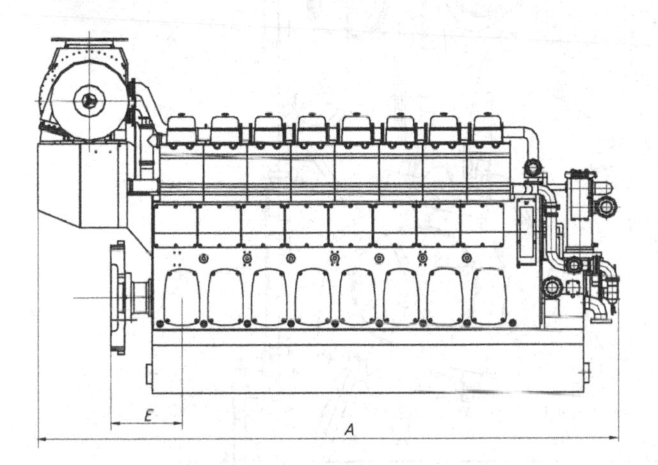

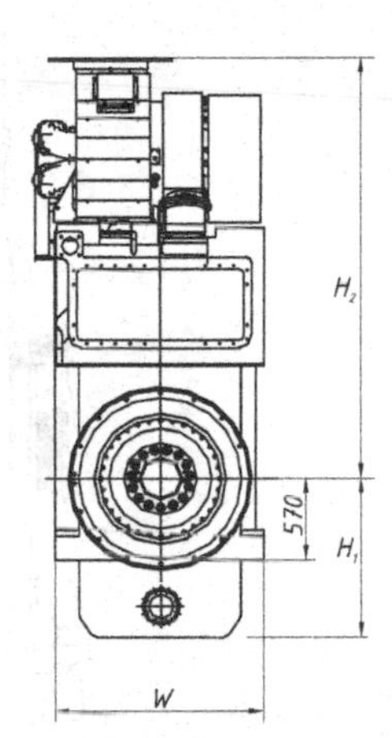

Engine version	A (mm)	E (mm)	H_1 (mm)	H_2 (mm)	W (mm)	Weight (kg)
6 H32/40	5515	800	1110	3295	1460	35,700
7 H32/40	6045	800	1110	3295	1460	39,600
8 H32/40	6545	800	1110	3495	1460	43,500
9 H32/40	7085	800	1110	3495	1460	46,600

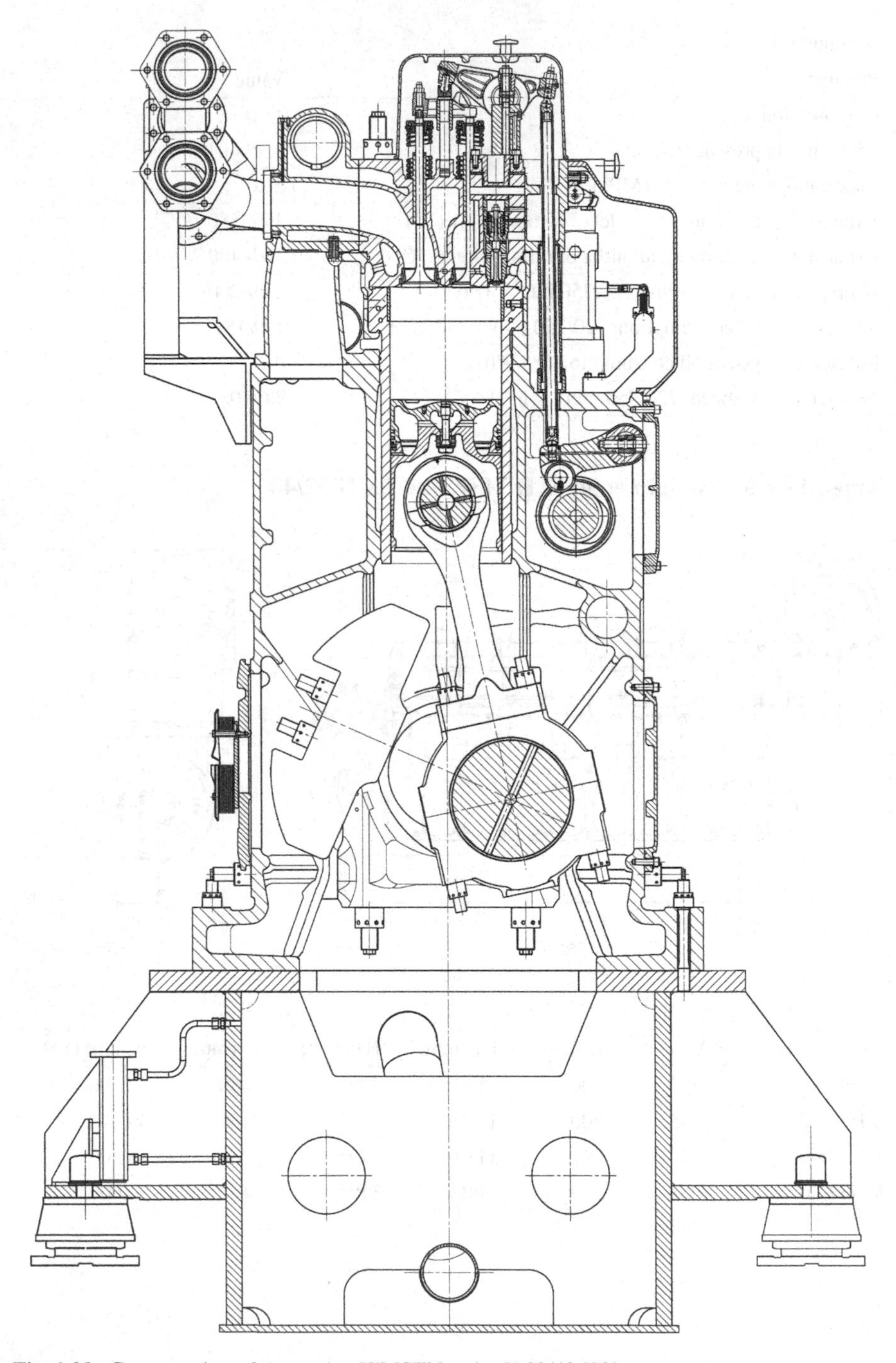

Fig. 1.23 Cross-section of the engine HiMSEN series H 32/40 [22]

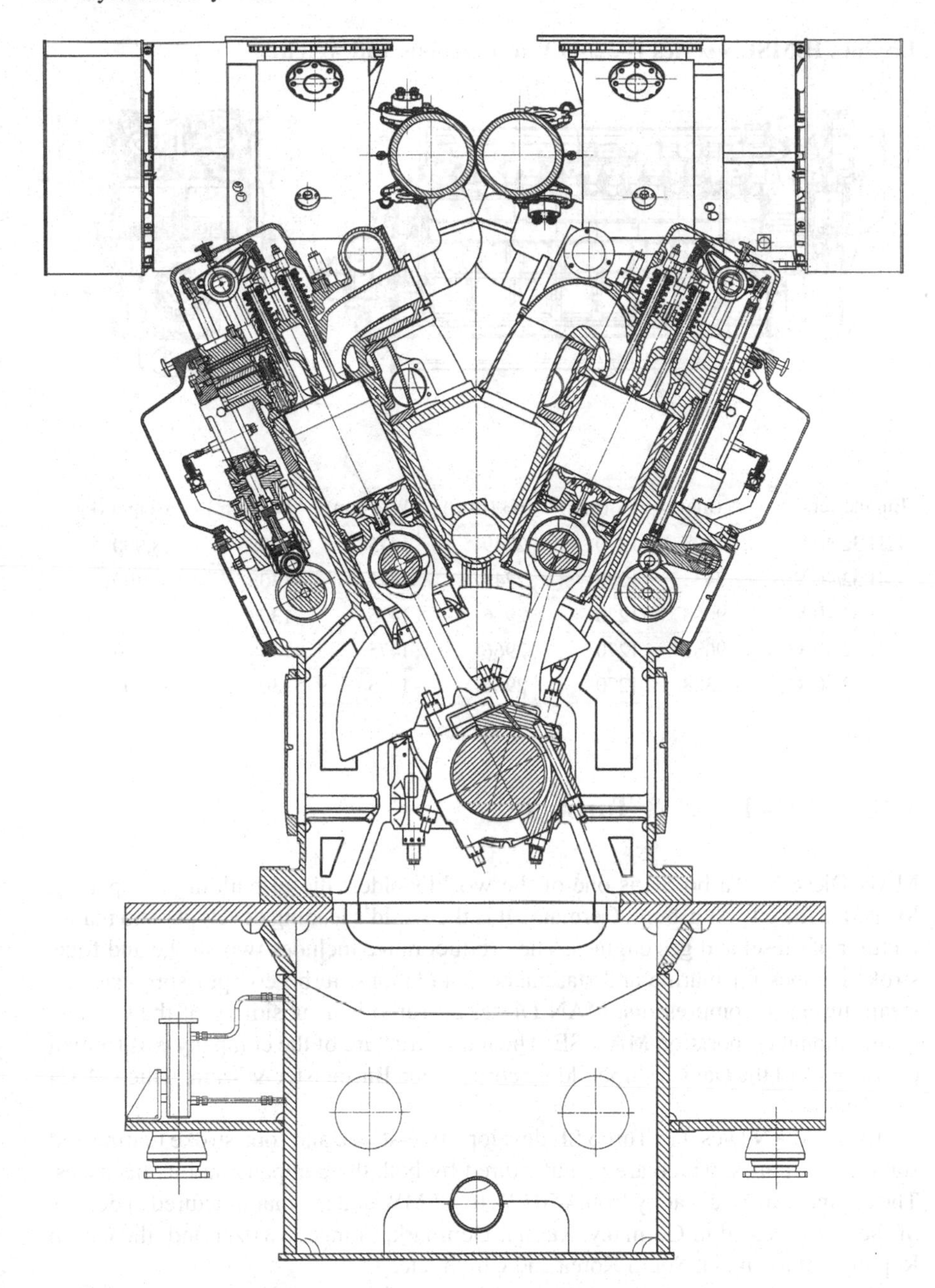

Fig. 1.24 Cross-section of the engine series HiMSEN H 32/40 [22]

Engines HiMSEN series H 32/40 V dimensions and weight

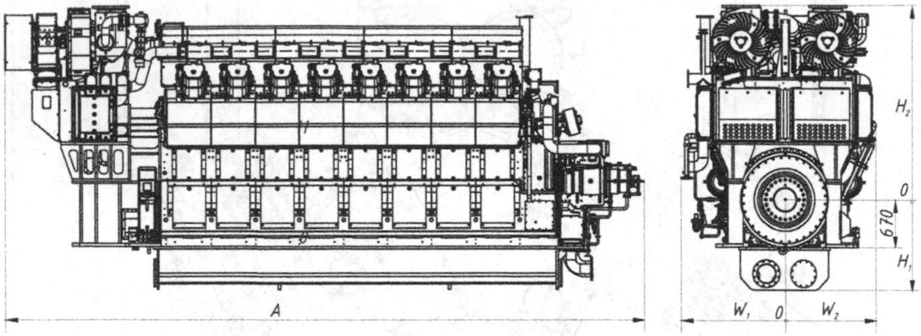

Engine version	A (mm)	H_1 (mm)	H_2 (mm)	W_1 (mm)	W_2 (mm)	Weight (kg)
12H 32/40 V	7808	1270	2749	1475	1308	58,000
14H 32/40 V	8433	1270	2749	1475	1308	65,300
16H 32/40 V	9058	1270	2966	1475	1308	71,100
18H 32/40 V	9683	1270	2966	1475	1308	78,300
20H 32/40 V	10,308	1270	2966	1475	1308	86,000

1.10 MAN Diesel & Turbo SE

MAN Diesel & Turbo SE is one of the world's oldest diesel-building companies, located in the city Augsburg, Germany. It is the world's leading developer and manufacturer of diesel and gas engines. The product range includes two-stroke and four-stroke engines for marine and stationary installations, turbo-compressors, gas and steam turbines, compressors. MAN Diesel & Turbo is a subsidiary of the German multinational corporation MAN SE. The main structure of the company was formed on the basis of the Danish shipbuilding corporation Burmeister & Wain, which MAN SE acquired.

Today, MAN Diesel & Turbo SE develops two-stroke and four-stroke marine and stationary engines, which are manufactured by both the company and its licensees. The engines have a capacity from 450 kW to 87 MW and are manufactured at dozens of factories located in Germany, Russia, Denmark, France, Switzerland, the Czech Republic, Italy, India, South Korea and China, etc.

At the beginning of the MAN concern origins there were production, located in the Ruhr region in southern Germany. The oldest predecessor enterprise was the St. Anthony Metallurgical Plant (Eisenhütte St. Antony), founded in 1758 in Oberhausen. In 1840, Ludwig Sander and Jean Gaspard Dollfus founded an engineering factory of Ludwig Zander (Sander'sche Maschinenfabrik) in Augsburg. That was the MAN foundation. Later, it was renamed Aghsburg Machine-Building Factory (Maschinenfabrik Augsburg). In 1898, as a result of a merger with a steel mill and the Klett & Comp engineering plant, founded in 1841, in the city of Nürnberg was established machine-building factory Maschinenfabrik Augsburg-Nürnberg AG, which in 1908 became abbreviated as MAN.

From 1893 to 1897, Rudolf Diesel created his first efficient engine together with the engineers of the company in the factory laboratory in Augsburg. From this point on, the company occupies a leading position in the production of diesel engines for transport and stationary purposes. So in 1904, the company supplies six diesel engines with a total capacity of 2400 horsepower for the Kiev power plant (Ukraine), which provided electricity to the tram lines.

By 1921, due to the defeat of Germany in World War I, the financial condition of MAN was very difficult. As a result its controlling stake was acquired by Gutehoffnungshütte (GHH). Until 1986, MAN was part of the GHH concern.

In 1924, the company produced the first diesel engine for transport purposes with direct fuel injection into the combustion chamber, and in the following 1925 four four-stroke six-cylinder diesel engines with a total capacity of 20,600 kW were installed on the "Augustus" line boat. The engines had a cylinder diameter of 700 mm and a piston stroke of 1200 mm. In 1927, the first four-stroke supercharged ship engines were installed on two passenger vessels "Preussen" and "Hansestadt Danzig", which had ten cylinders with a diameter of 540 mm and a piston stroke of 600 mm. The use of supercharging allowed to increase their power from 1250 to 1765 horsepower. These engines used Brown Boveri turbochargers with constant gas pressure before the turbine, and already in 1934 the development of their own turbocharger models began.

In the period from 1934 to 1945, thanks to military orders, the company developed intensively, producing, among other military products, engines for merchant and military fleets too.

At the end of the Second World War, all GHH enterprises were taken under the control of the Allies. Restoring production capacity quickly, the company launched production of two-stroke and four-stroke engines for merchant ships of its own design, which were distinguished by reliability and ease of maintenance, and therefore, they quickly gained popularity among consumers.

In 1965, the MAN Holeby Diesel division began production of the most popular medium-speed auxiliary marine engine of the L23/30 type, and in 1972 a more powerful version of this engine—the L28/32 series—entered the market. These sredneforsirovannye engines quickly gained popularity among customers, as they combined a good fuel economy, reliability and ease of maintenance and repair. These factors contributed to the fact that the engines have become the most popular in their sector. Their production with minor modifications continues to the present and has already exceeded 10,000 units.

The energy crisis of the early 70s of the 20th century showed, that the problem of energy resources will worsen and it will require an increase in fuel efficiency in all areas, including ship power engineering. In this regard, in the early 80s a program to create a new line of marine four-stroke engines was launched, which, in addition to the existing advantages, would have good fuel economy.

At this moment, in 1980, the company acquired the production assets of another leading marine engine manufacturer, the Danish company Burmeister & Wain (B&W), which, in turn, inherited diesel production from another Danish company Holeby Diesel, founded in 1663 as a blacksmith shop, making machines for farmers in Denmark. In 1901, having gained experience in the field of mechanical engineering, Holeby Diesel acquired a license from Rudolf Diesel for its engines. The release of the first model, was established in 1903. It had a power of four horsepower. Despite the achieved progress in production, the unfavorable financial condition forced the company in 1933 to become part of Burmeister & Wain. Over the next 50 years, Burmeister & Wain was the largest supplier of marine engines in the world. However, the growing global competition led to the fact, that the company was forced to sell its assets and become part of the MAN concern. From that moment all engines were produced under the MAN B&W Diesel logo. Combining the experience of its own developments and developments of other manufacturers, which were part of the concern by that time, MAN B&W Diesel began to create a model range of new-generation marine medium-speed diesel engines with diameters of 400, 480 and 580 mm, which were designed according to the similarity principle. As a result, the L40/54, L48/60 and L58/64 models were launched to the market within ten years. In 1992, this model range was supplemented by the series L32/40 engine, which was designed in the same scheme as the earlier models. In the early 90s, the power range was extended by creating V-shaped versions of the V48/60 and V40/50 engines, and in the early 2000s, a V-shaped version of the V32/40 engine appeared. To this date, these four models form the basis of the program for the production of marine medium-speed engines, covering the power range from 2880 to 21,600 kW.

In the early 90s, the company began to develop a new generation of high-speed and medium-speed engines with good effective and mass-dimensional performance. The first engine of the new generation was the compact high-speed diesel engine L16/24, launched on the market in 1995, designed to drive the generator sets. Later, this model range was supplemented with the L27/38 series in 1998 and the L21/31 in 2002, which are used on ships, both for driving diesel generators and as main ones, as part of diesel gear units. In 2007, MAN introduced a new generation, series L32/44, medium-speed engine to the market, which, in its basic version, is equipped with a "Common Rail" type fuel injection system. This engine has an electronic control system, based on microprocessor technology, which allows to adapt the workflow to the specific conditions of its work.

1.10.1 Engines MAN Series L23/30

These are middle-powered in-line engines with a number of cylinders from 5 to 8, with gas turbine supercharging and cooling of the charging air, cover the power range of 525–1280 kW. The engine is made according to the classical scheme, it has a rigid crankcase that includes the cylinder block, crankcase, charge air receiver, cooling jacket, camshaft housing, crankshaft bed. Suspended crankshaft is mounted by means of main bearing caps, mounted on two vertical studs. The engines use a charging system with a constant pressure of exhaust gases in front of the turbine. To speed up the turbocharger response to a load changing on the engine, an original compressor spin-up system with compressed air from the ship-wide air system is installed. This system, called the lambda controller, allows to improve the combustion of fuel in transients and low load modes. As a result, the engine runs without visible smoke in the entire range of operating modes (Fig. 1.25).

Main technical parameters of engines MAN series L23/30

Parameter	Value
Number and cylinders arrangement	5, 6, 7, 8 in-line
Cylinder bore (mm)	225
Piston stroke (mm)	300
Cylinder capacity (dm^3)	11.9
Rotation speed (min^{-1})	720/750/900
Cylinder power at 720/750/900 min^{-1} (kW)	130/135/160
Air charging pressure (MPa)	0.20–0.25
Compression ratio	13.5
Maximum cycle pressure (MPa)	13.0
Mean effective pressure at 720/750/900 min^{-1} (MPa)	1.82/1.81/1.79
Brake specific fuel oil consumption (g/kWh)	194.0
Brake specific air consumption (кg/kWh)	8.0
Mean piston speed at 720/750/900 min^{-1} (m/s)	7.2/7.5/9.0

Dimensions and weight engines MAN series L23/30

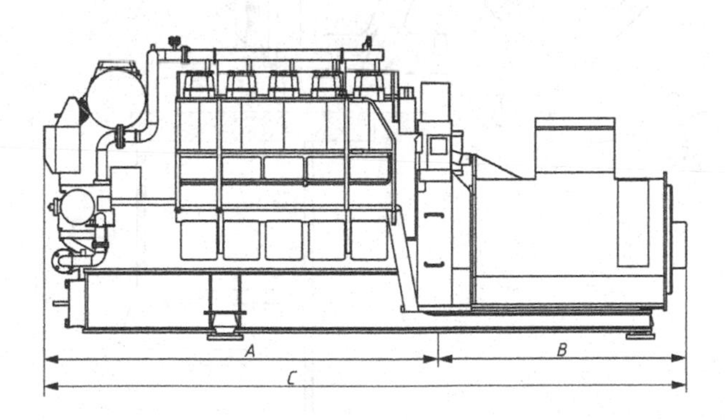

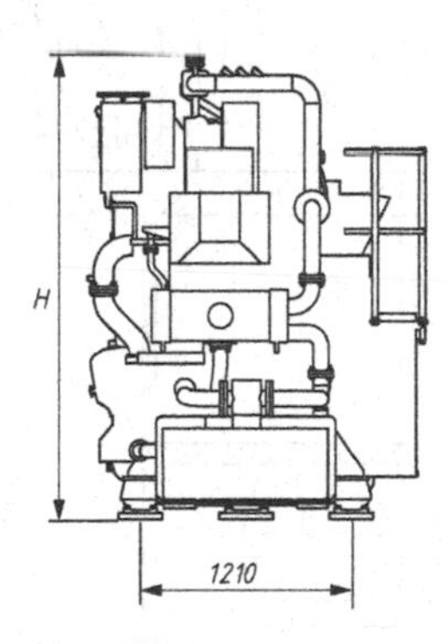

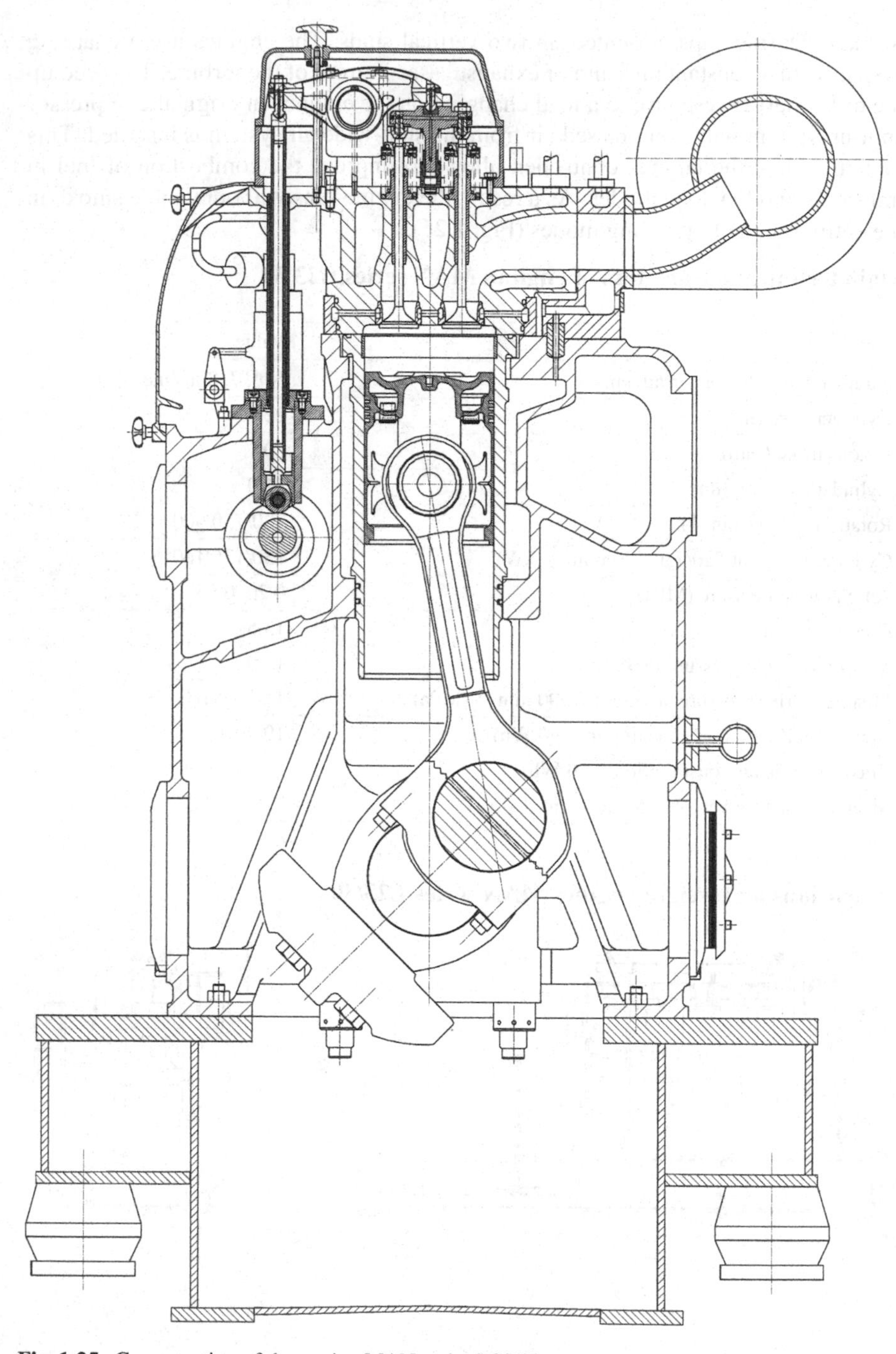

Fig. 1.25 Cross-section of the engine MAN series L23/30 as a generator set [23]

Engine version	A (mm)	B (mm)	C (mm)	H (mm)	Diesel generator weight (kg)
5 L23/30H	3415	2130	5545	2625	16,800
6 L23/30H	3785	2130	5915	2625	18,300
7 L23/30H	4155	2130	6285	2625	20,100
8 L23/30H	4525	2130	6655	2625	21,400

1.10.2 MAN Engines Series L16/24

MAN engines series L16/24 is a high-speed four-stroke engines with gas turbine supercharging and charge air cooling, intended for use as part of a diesel-geared and diesel-generator sets (Fig. 1.26) of ships of 450 to 990 kW. The engine is designed according to a compact scheme with maximum integration of all pipelines into the crankcase and other basic parts. Details of the cylinder-piston group are combined into a single unit, which is supplied as a separate, factory assembled kit. It allows to greatly simplify the maintenance and repair of the engine under operating conditions. The presence of piston relief for the valve timing eliminates their contact with the pistons even if the opening phases of the valves are disturbed. The engine is designed to work on heavy fuels. For better fuel combustion at lower loads, a system for heating the charge air is provided.

Main technical parameters of MAN engines series L16/24

Parameter	Value
Number and cylinders arrangement	5, 6, 7, 8, 9 in-line
Cylinder bore (mm)	160
Piston stroke (mm)	240
Cylinder capacity (dm^3)	4.82
Rotation speed (min^{-1})	1000, 1200
Cylinder power at 1000/1200 min^{-1} (kW)	95/110
Compression ratio	15.2
Air charging pressure (MPa)	0.27–0.31
Maximum cycle pressure (MPa)	17.0
Exhaust gas temperature at inlet of turbocharger (°C)	500–550
Exhaust gas temperature at outlet of turbocharger (°C)	325–375
Mean effective pressure 1000/1200 min^{-1} (MPa)	2.36/2.28
Brake specific fuel oil consumption, 1000/1200 min^{-1} (g/kWh)	195.0
Lubrication system oil consumption (g/kWh)	0.8
Mean piston speed 1000/1200 min^{-1} (m/s)	8/9.6

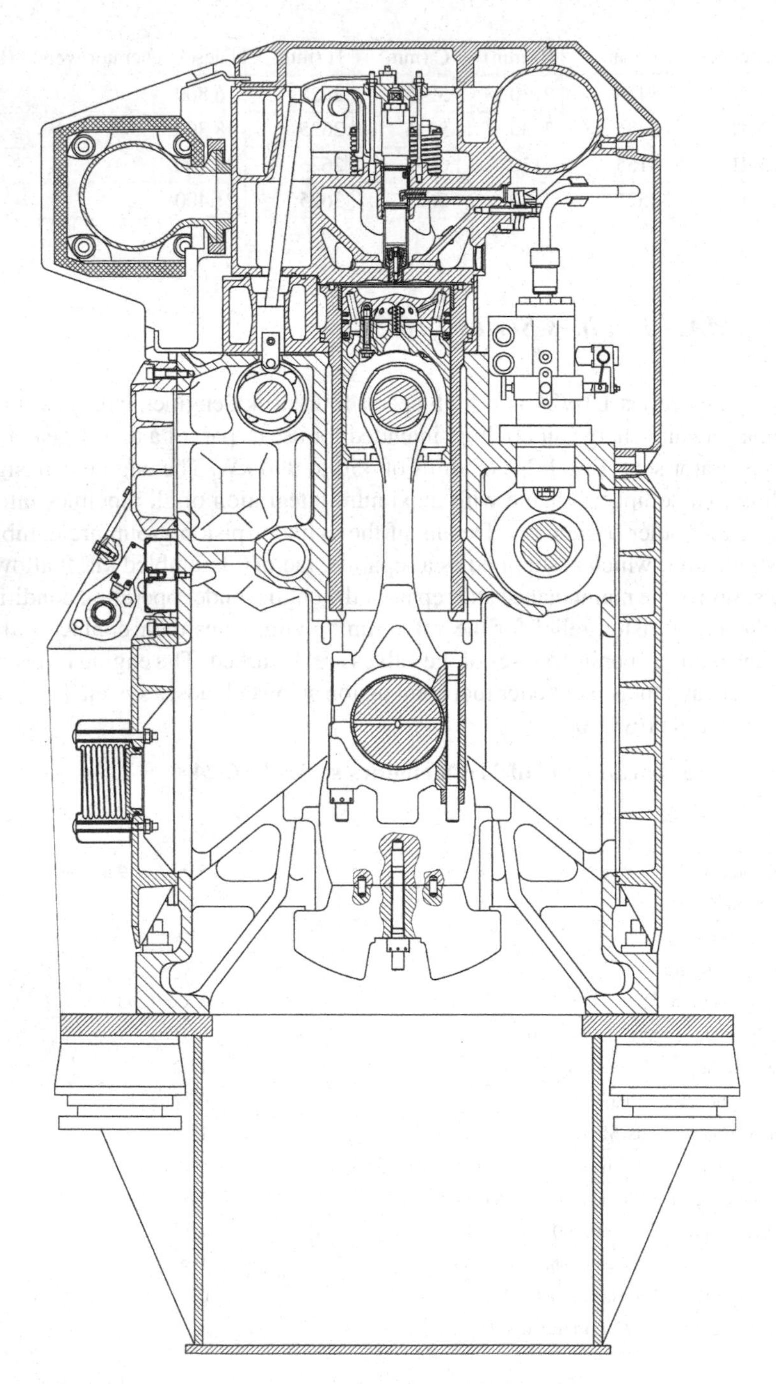

Fig. 1.26 Cross-section of the engine MAN series L16/24 as a generator set [24]

Dimensions and weight engines MAN series L16/24

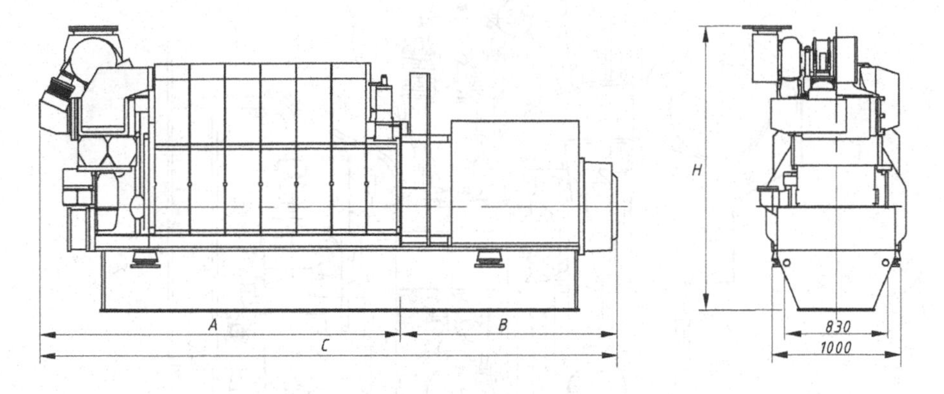

Engine version	A (mm)	B (mm)	C (mm)	H (mm)	Diesel generator weight (kg)
5 L16/24	2751	1400	4151	2457	9500
6 L16/24	3026	1490	4516	2457	10,500
7 L16/24	3501	1585	5086	2495	11,400
8 L16/24	3776	1680	5456	2495	12,400
9 L16/24	4051	1680	5731	2495	13,100

1.10.3 Engines MAN Series L21/31

Engines MAN series L21/31 is a compact, high-speed four-stroke engine with gas turbine supercharging and charge air cooling, designed for use as part of a diesel-gear and diesel generator sets (Fig. 1.27) of marine vessels. It was developed within the framework of the same concept as the engines of the L16/24 and L27/38 series, to cover the power range from 1000 to 1980 kW. Like the rest of the series, it is distinguished by its high compactness. The modular design provides ease of maintenance and repair, the presence of protective shields over the entire surface of the engine significantly reduces the noise and vibration, transmitted to the external space. The optimized shape of the combustion chamber, combined with high fuel atomization pressures, have made it possible to achieve good fuel economy and lower emissions of harmful substances with exhaust gases.

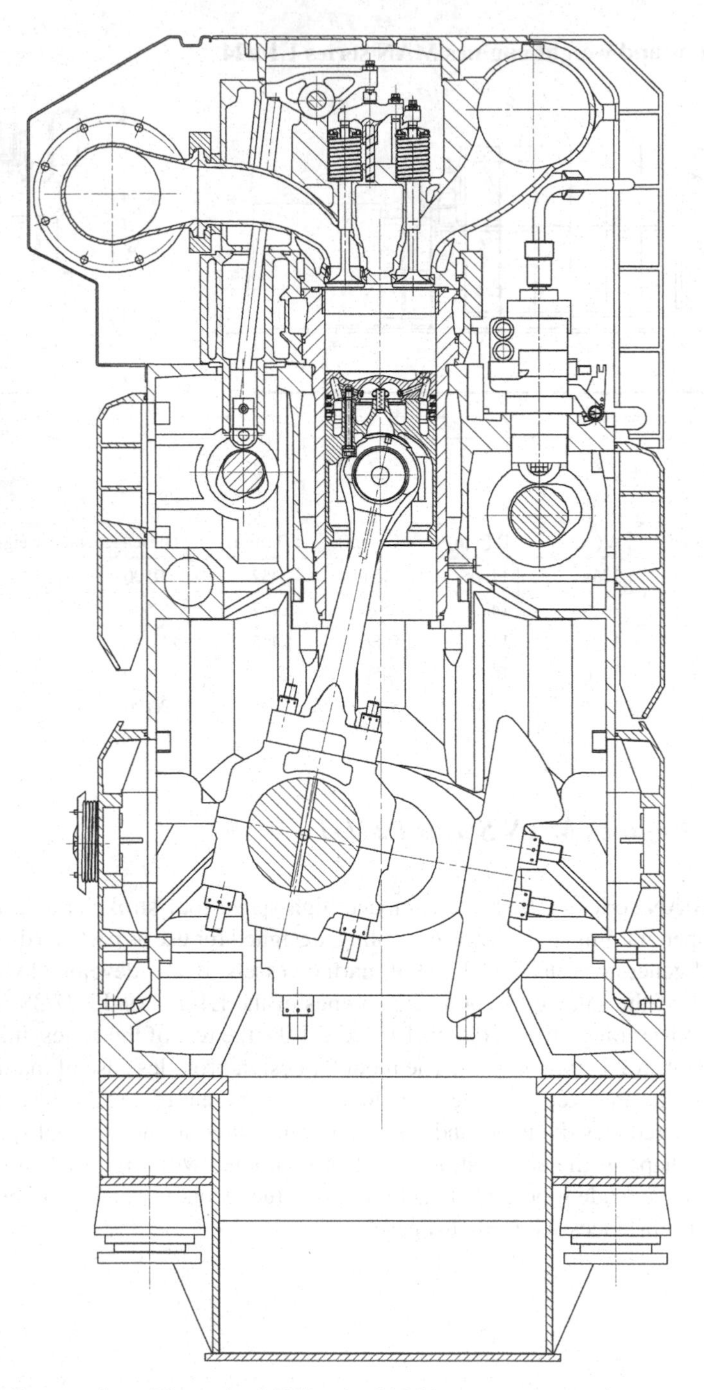

Fig. 1.27 Cross-section of the engine MAN series L21/31 as a generator set [25]

Main technical parameters of engines MAN series L21/31

Parameter	Value
Number and cylinders arrangement	6, 7, 8, 9 in-line
Cylinder bore (mm)	210
Piston stroke (mm)	320
Cylinder capacity, dm^3	10.08
Rotation speed (min^{-1})	900, 1000
Cylinder power at 900/1000 min^{-1} (kW)	200/220
Compression ratio	15.5
Air charging pressure (MPa)	0.30–0.32
Maximum cycle pressure (MPa)	20.0–21.0
Exhaust gas temperature at inlet of turbocharger (°C)	480–530
Exhaust gas temperature at outlet of turbocharger (°C)	300–350
Mean effective pressure 900/1000 min^{-1} (MPa)	24.8/24.0
Brake specific fuel oil consumption 900/1000 min^{-1} (g/kWh)	181
Lubrication system oil consumption (g/kWh)	0.5–0.8
Mean piston speed 900/1000 min^{-1} (m/s)	9.3/10.3

Dimensions and weight engines MAN series L21/31

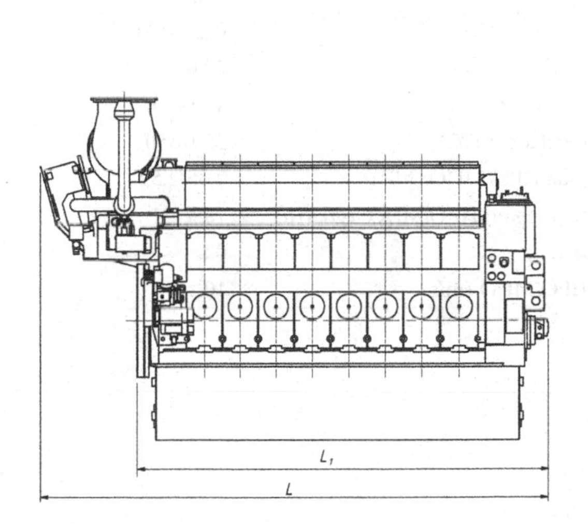

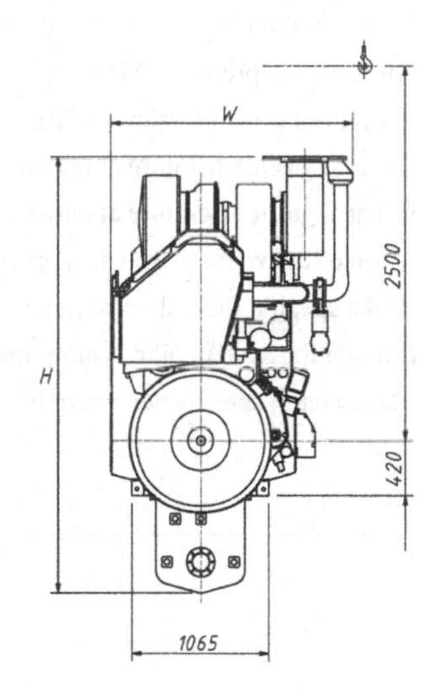

Engine version	L_1 (mm)	L (mm)	H (mm)	W (mm)	Weight (kg)
6 L21/31	3424	4544	3113	1695	16,000
7 L21/31	3779	4899	3267	1695	17,500
8 L21/31	4134	5254	3267	1820	19,000
9 L21/31	4489	5609	3267	1820	20,500

1.10.4 MAN Series L27/38 Engine

MAN series L27/38 engine is the largest high-speed four-stroke diesel engine from the model range of compact engines, designed for use as part of diesel-gear and diesel generator sets (Fig. 1.28) of ships in the power range 2040–3060 kW. The engine is equipped with a gas turbine supercharged with constant gas pressure before the turbine, as well as a system for controlling the temperature of the charge air before of the working cylinders. The use of the “marine” type connecting rod makes it possible to maintain a separate cylinder-piston group and a separate connecting rod bottom bearing, which greatly simplifies the engine maintenance and repair.

Main technical parameters of engines MAN series L27/38

Parameter	Value
Number and cylinders arrangement	6, 7, 8, 9 in-line
Cylinder bore (mm)	270
Piston stroke (mm)	380
Cylinder capacity (dm^3)	21.8
Rotation speed (min^{-1})	800
Cylinder power at operating on HFO/MDO (MGO) (kW)	340/365
Compression ratio	16.5
Air charging pressure (MPa)	0.28–0.31
Maximum cycle pressure (MPa)	20.0
Exhaust gas temperature at inlet of turbocharger (°C)	480–530
Exhaust gas temperature at outlet of turbocharger (°C)	250–350
Mean effective pressure when operating on HFO/MDO (MPa)	2.35/2.52
Brake specific fuel oil consumption (operating on HFO/MDO) (g/kWh)	188/191
Lubrication system oil consumption (g/kWh)	0.8
Mean piston speed when operating on HFO/MDO (m/s)	10.1

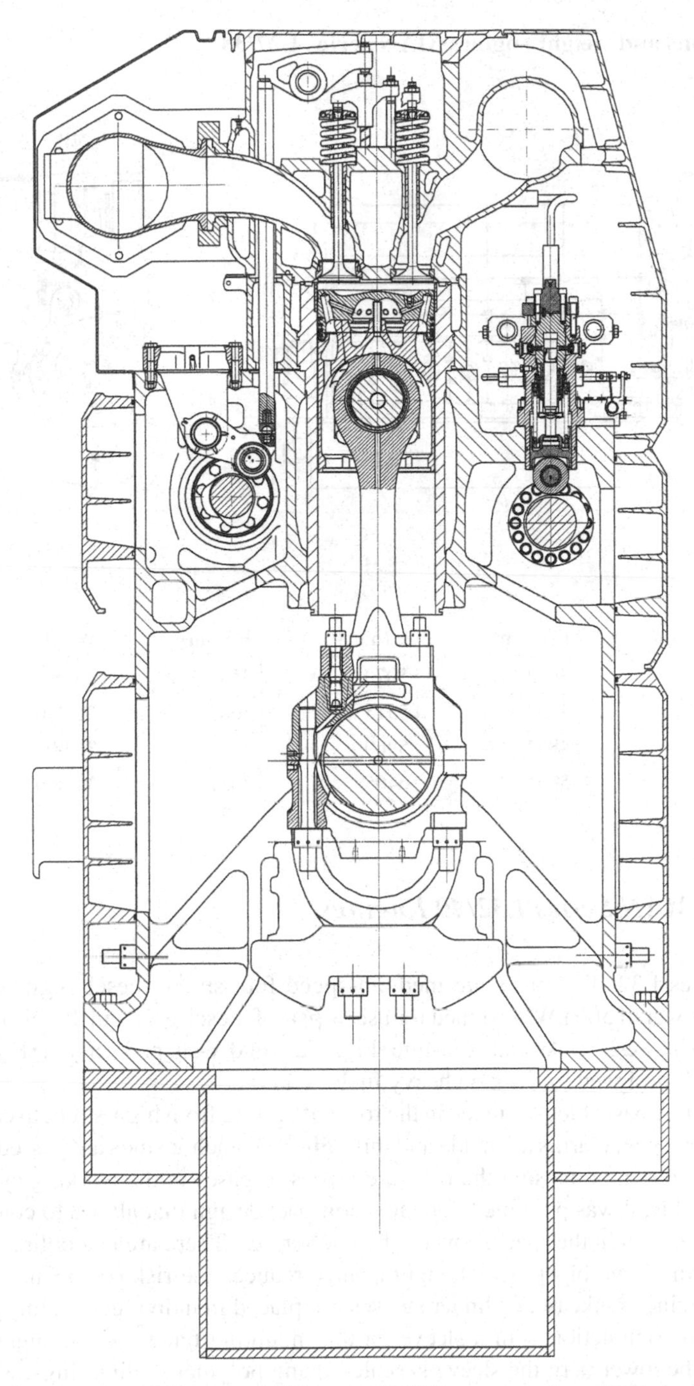

Fig. 1.28 Cross-section of the engine MAN series L27/38 as a generator set [26]

Dimensions and weight engines MAN series L27/38

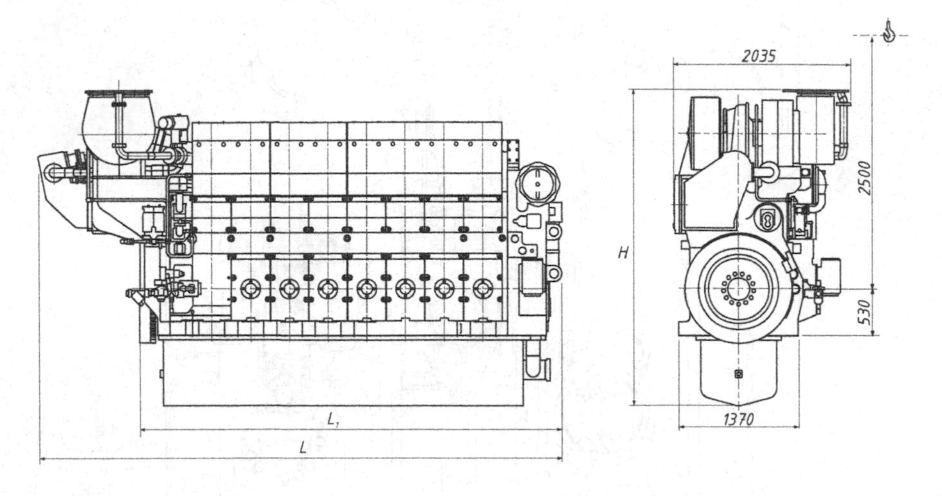

Engine version	L_1 (mm)	L (mm)	H (mm)	Weight (kg)
6 L27/38	3962	5070	3555	29,000
7 L27/38	4407	5515	3687	32,000
8 L27/38	4852	5960	3687	36,000
9 L27/38	5263	6405	3687	39,000

1.10.5 MAN Series L32/40 Engines

MAN series L32/40 engines are medium-speed four-stroke diesel engines with a cylinder power of 500 kW, designed for use as part of diesel-gear and diesel generator sets of marine vessels. Available in-line (Fig. 1.29) and V-shaped version (Fig. 1.30). Engines are designed to work on heavy fuels (HFO).

The engine basic block is made in the form of casting from high-strength cast iron. To stiffen it, special drills are made and through them anchor studs are passed, which create prestressing opposing the pressure forces of gases in the working cylinders. Thanks to this, it was possible to create a compact design that allows to connect all the power effects in the space, covered by anchor ties. There are no cooling cavities in the engine basic block, which significantly reduces the risk of coolant entering into the engine crankcase. Cylinder sleeves are placed in individual cooling jackets, installed in conjunction with a sleeve on the mounting flange of the engine basic block. In the lower part, the sleeve is sealed using polymer sealing rings, which do not stop its axial movement in the lower level during thermal expansion. An anti-polish ring is installed in the upper level of the sleeve. The cylinder cover is made by a two-tier scheme, which allows to organize effective cooling of the firing bottom

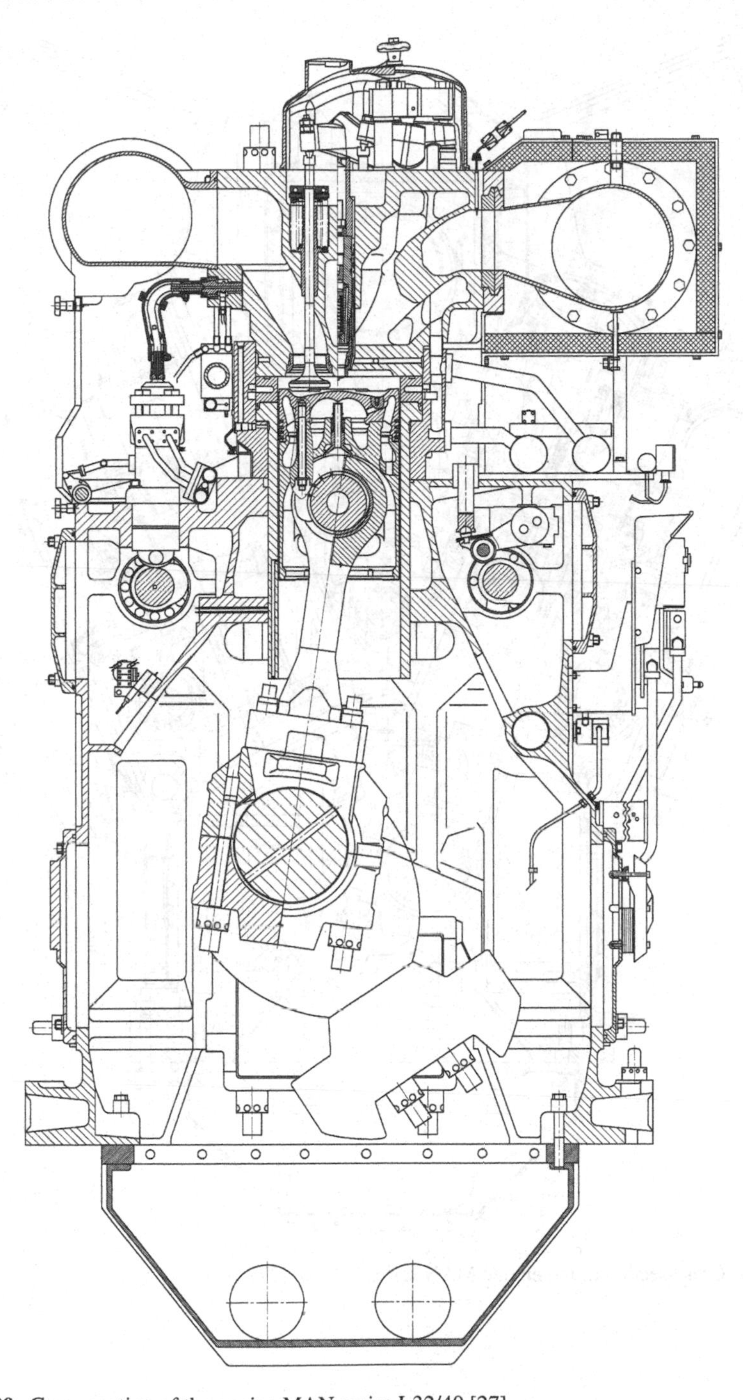

Fig. 1.29 Cross-section of the engine MAN series L32/40 [27]

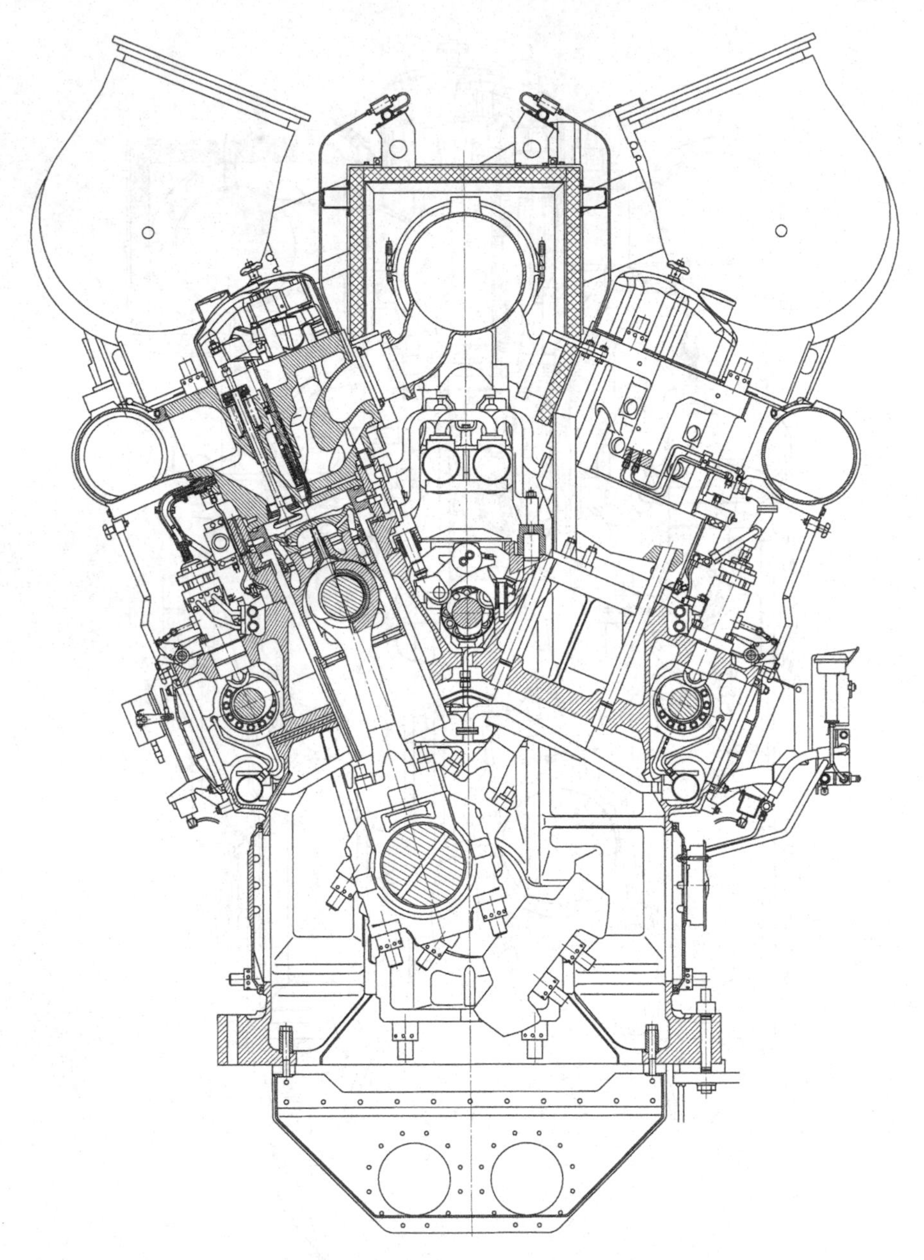

Fig. 1.30 Cross-section of the engine MAN series V32/40 [27]

without reducing its mechanical strength. The piston is of composite type, the head is forged from high-strength heat-resistant steel, the piston skirt is cast from spheroidal graphite cast iron. A “shaker cavity” for cooling oil is located beneath the piston firing head. Chrome-ceramic coating on the working surface of the first piston ring leads to minimal wear, which provides an extremely long period between maintenance.

There are two camshafts in engine, one of which controls the operation of the valves, and the other controls fuel injection. Both shafts are equipped with mechanisms for changing the phases of fuel supply and gas distribution, which provide an opportunity to optimize the working process, both for obtaining maximum efficiency and reducing harmful emissions. The engine is supplied with a highly efficient gas turbine supercharged with constant gas pressure before the turbine. The valve timing is made with reinforced sealing edges, and the exhaust valve seats are water-cooled. The impellers on the exhaust valve spindles allow them to rotate in an exhaust gas flow, which results in more uniform wear.

Main technical parameters of engines MAN series L32/40

Parameter	Value
Number and cylinders arrangement	6, 7, 8, 9 in-line 12, 14, 16, 18 V-shaped
Cylinder bore (mm)	320
Piston stroke (mm)	400
Cylinder capacity (dm^3)	32.17
Rotation speed (min^{-1})	720/750
Cylinder power at 720/750 min^{-1} (kW)	500
Compression ratio	15.2
Air charging pressure (MPa)	0.313
Maximum cycle pressure (MPa)	18.0
Exhaust gas temperature at inlet of turbocharger (°C)	380–500
Exhaust gas temperature at outlet of turbocharger (°C)	318–400
Mean effective pressure at 720/750 min^{-1} (MPa)	2.59/2.49
Brake specific fuel oil consumption (g/kWh)	186.0
Lubrication system oil consumption (g/kWh)	0.5
Mean piston speed at 720/750 min^{-1} (m/s)	9.6/10

Dimensions and weight engines MAN series L32/40

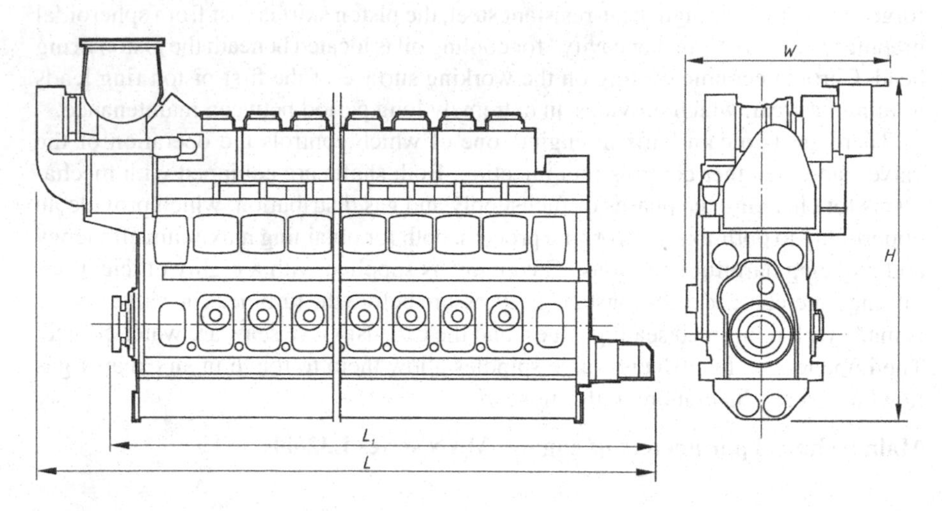

Engine version	L_1 (mm)	L (mm)	H (mm)	W (mm)	Weight (kg)
6 L32/40	5140	5940	4010	2630	38,000
7 L32/40	5670	6470	4010	2630	42,000
8 L32/40	6195	7000	4490	2715	47,000
9 L32/40	6725	7530	4490	2715	51,000

Dimensions and weight engines MAN series V32/40

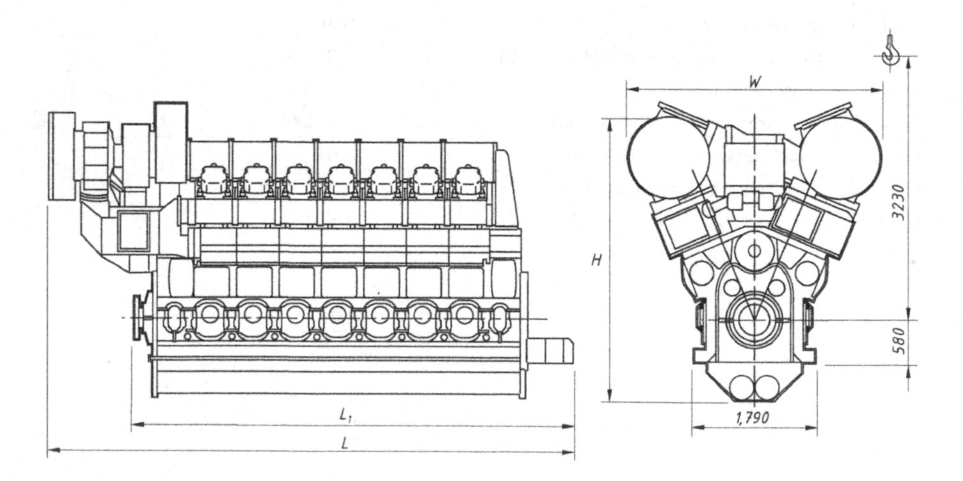

Engine version	L_1 (mm)	L (mm)	H (mm)	W (mm)	Weight (kg)
12 V32/40	5890	6915	4100	3140	61,000
14 V32/40	6520	7545	4100	3140	68,000
16 V32/40	7150	8365	4420	3730	77,000
18 V32/40	7780	8995	4420	3730	85,000

1.10.6 MAN Engine Series L40/54

Engines of the L40/54 series appeared on the market in 1987, having absorbed MAN B&W Diesel's experience for more than 20 years in producing engines with a cylinder diameter of 400 mm. The engine has a rigid engine basic block with cross-cutting linkages, extending from the upper bearing surface through the crankshaft frame bearing covers. This design allowed to create a prestress and optimally distribute the forces, opposing the pressure of gases in the cylinders, evenly distributing them throughout the structure of the core. This made it possible to minimize the deformation of the working sleeves, ensuring reliable operation of the piston and crankshaft bearings. The use of the fire insert in the upper part of the cylinder bushing has made it possible to minimize its cavitation wear. Intensive cooling of only the fire insert and the upper part of the cylinder liner made it possible to evenly distribute temperatures along the working surface of the sleeve, thereby reducing the likelihood of low-temperature corrosion (Fig. 1.31).

Main technical parameters of engines MAN series L40/54

Parameter	Value
Number and cylinders arrangement	6, 7, 8, 9 in-line
Cylinder bore (mm)	400
Piston stroke (mm)	540
Cylinder capacity (dm^3)	67.82
Rotation speed (min^{-1})	550/514/500
Cylinder power at 550/514/500 min^{-1} (kW)	720/720/700
Compression ratio	14.2
Air charging pressure (MPa)	0.277
Maximum cycle pressure (MPa)	15.8
Exhaust gas temperature at inlet of turbocharger (°C)	550
Exhaust gas temperature at outlet of turbocharger (°C)	365
Mean effective pressure operating at 550/514/500 min^{-1} (MPa)	2.31/2.48/2.48
Average specific fuel oil consumption at a load 100/80% min^{-1} (g/kWh)	183/181
Lubrication system oil consumption (g/kWh)	1.0
Mean piston speed operating on 550/514/500 min^{-1} (m/s)	9.9/9.2/9.0

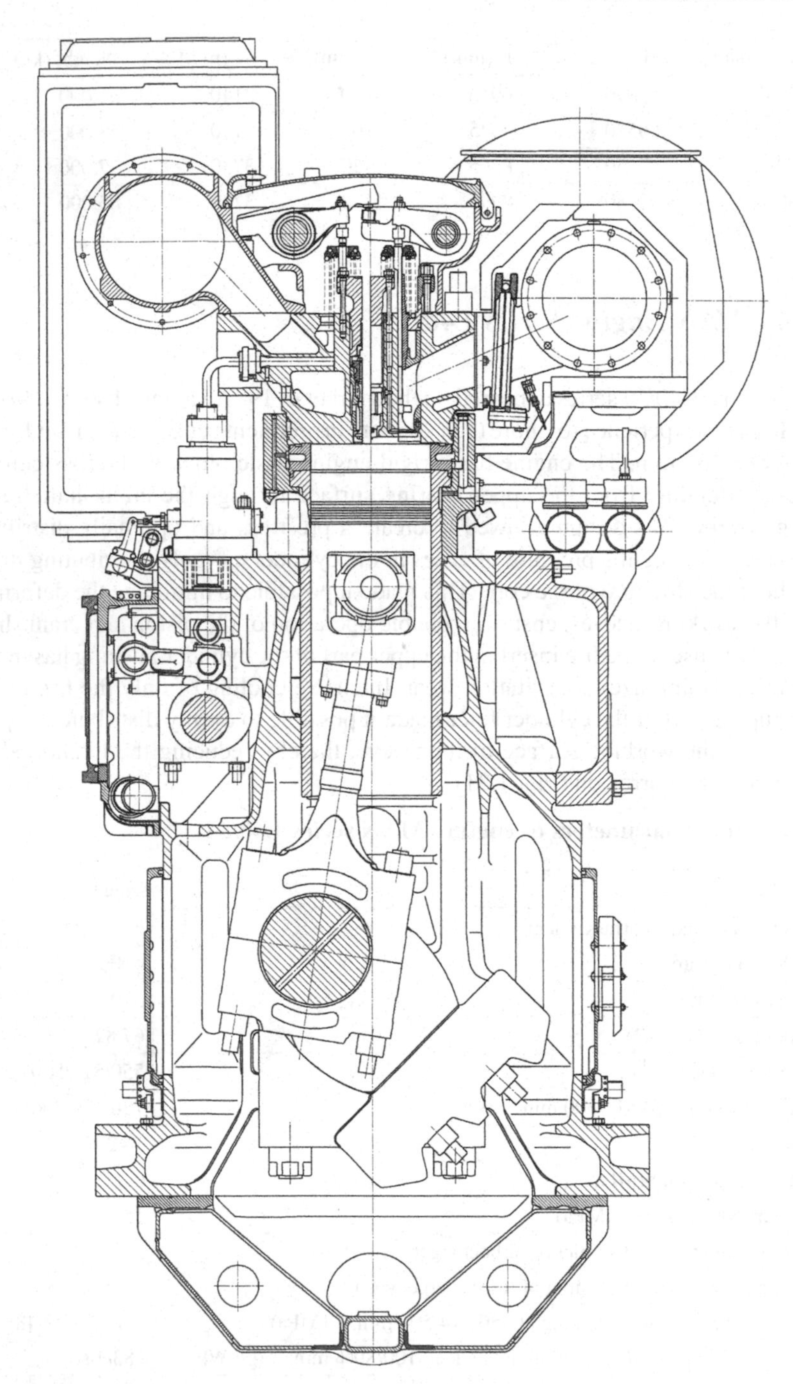

Fig. 1.31 Cross-section of the engine MAN series L40/54 [28]

Dimensions and weight parameters of engines MAN series L40/54

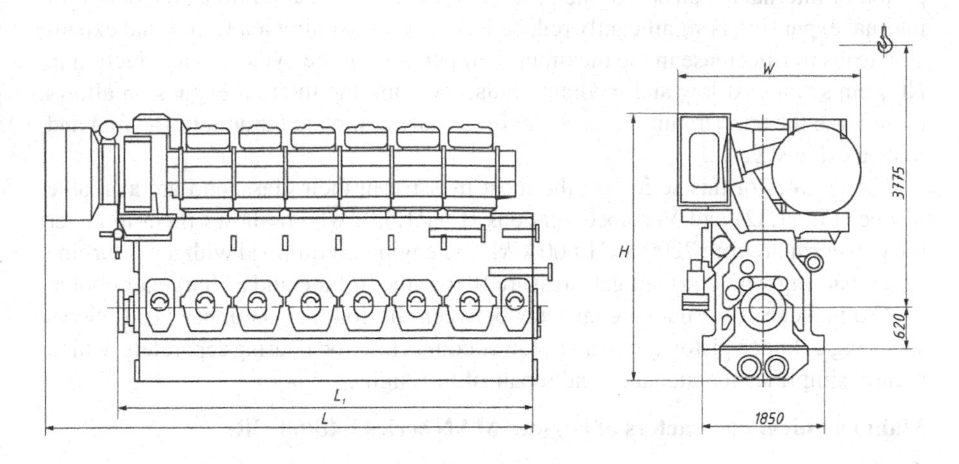

Engine version	L_1 (mm)	L (mm)	H (mm)	W (mm)	Weight (kg)
6 L40/54	5910	7520	2600	4345	70,000
7 L40/54	6610	8600	2750	4380	80,000
8 L40/54	7310	9155	2750	4380	89,000
9 L40/54	8010	10,000	2750	4380	97,000

1.10.7 MAN Series L48/60 CR Engines

MAN series L48/60 CR engines are a further development of 48/60 medium speed engines. Unlike previous models, engines of this series are equipped with an electronic Common Rail fuel injection system. All elements of the fuel system are designed with the concept of full integration into the existing engine design. As a result of this modernization, it was possible to combine proven design solutions with new technologies in the organization of fuel supply and work processes. The electronic engine control system known as $\text{SaCoS}_{\text{one}}$ (Safety and Control System) allows automatically changing the valve timing during engine operation, as well as change parameters and the type of the fuel supply to the injector. The combustion chamber has also been modernized, based on the results of computer simulation and optimization of the combustion processes in it. The piston design was changed, which allowed to increase the compression ratio compared to previous models. As a result, for each mode of operation, an optimal combination of parameters are selected to ensure high performance and/or high environmental performance. Therefore, at operating modes close to the maximum power, the valve timing is install in such a way, that the workflow is carried out according to the Miller cycle with a large period of internal expansion of the charge, and at low load conditions, the period of internal expansion is significantly reduced. As a result, at high loads, internal expansion leads to a decrease in the maximum temperature of the cycle and a reduction in NO_x emissions. At low and medium loads, reducing the internal expansion allows to increase the temperature of the cycle for a more complete burnout of the fuel and reduce exhaust gas.

The engines of this series are the most massive in their class, and are available in-line (Fig. 1.32) and V-shaped versions (Fig. 1.33), which allows them to cover the power range from 7200 to 21,600 kW. The engine is equipped with a gas turbine supercharging with constant gas pressure before the turbine and a charge air cooler.

The presence of a flange connector between the rod and its upper head allows servicing cylinder-piston group and a lower connecting rod bearing separately, which greatly simplifies maintenance and repair of the engine.

Main technical parameters of engines MAN series L48/60 CR

Parameter	Value
Number and cylinders arrangement	6, 7, 8, 9 in-line 12, 14, 16, 18/V-shaped
Cylinder bore (mm)	480
Piston stroke (mm)	600
Cylinder capacity (dm^3)	108.6
Rotation speed (min^{-1})	500/514
Cylinder power at 500/514 min^{-1} (kW)	1200
Compression ratio	17.0
Air charging pressure (MPa)	0.35
Maximum cycle pressure (MPa)	19.0
Exhaust gas temperature at inlet of turbocharger (°C)	480
Exhaust gas temperature at outlet of turbocharger (°C)	350
Mean effective pressure operating on 500/514 min^{-1} (MPa)	2.65/2.58
Brake specific fuel oil consumption on 500/514 min^{-1} (g/kWh)	180.0
Lubrication system oil consumption (g/kWh)	0.8
Mean piston speed operating on 500/514 min^{-1} (m/s)	10.0/10.3

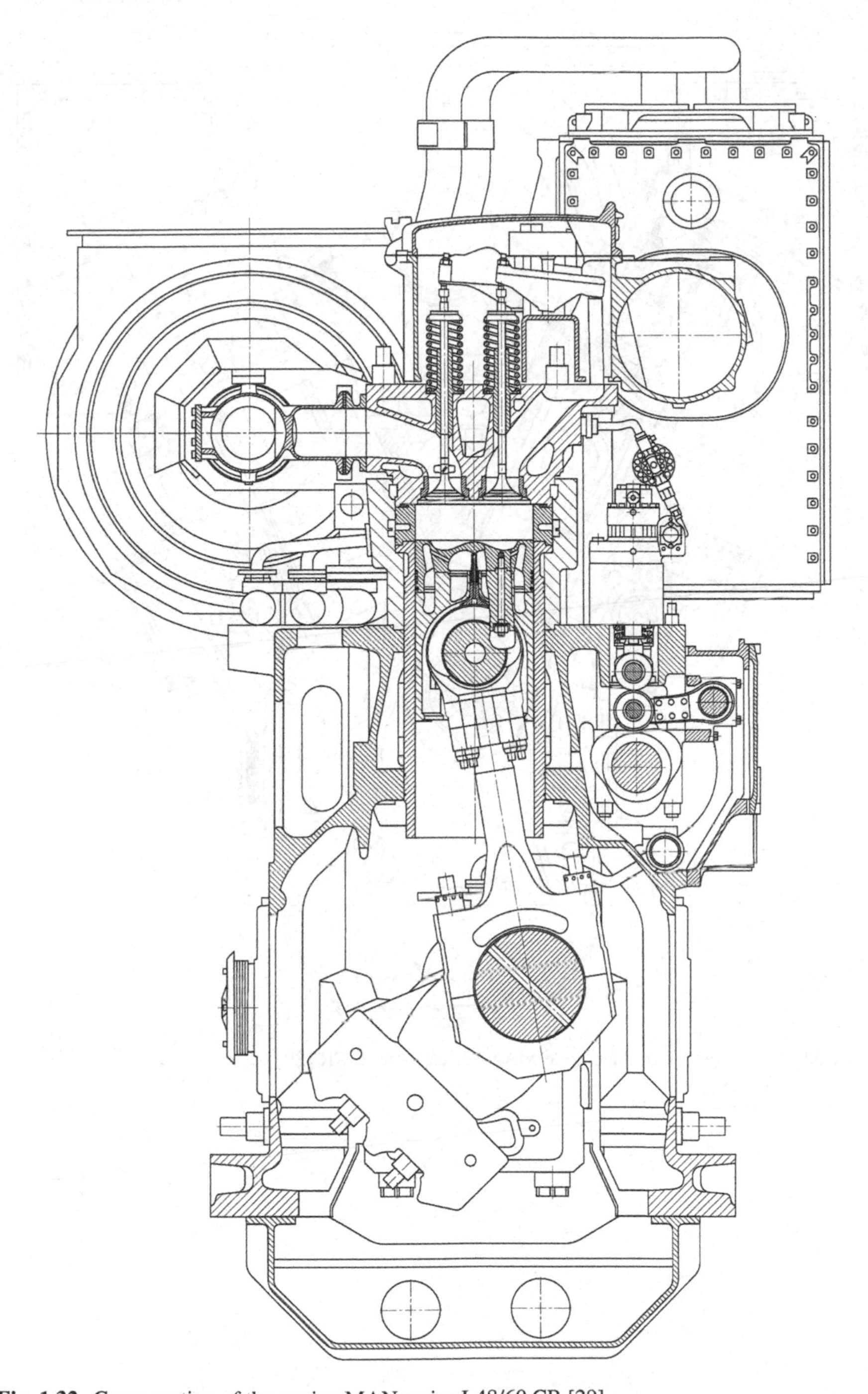

Fig. 1.32 Cross-section of the engine MAN series L48/60 CR [29]

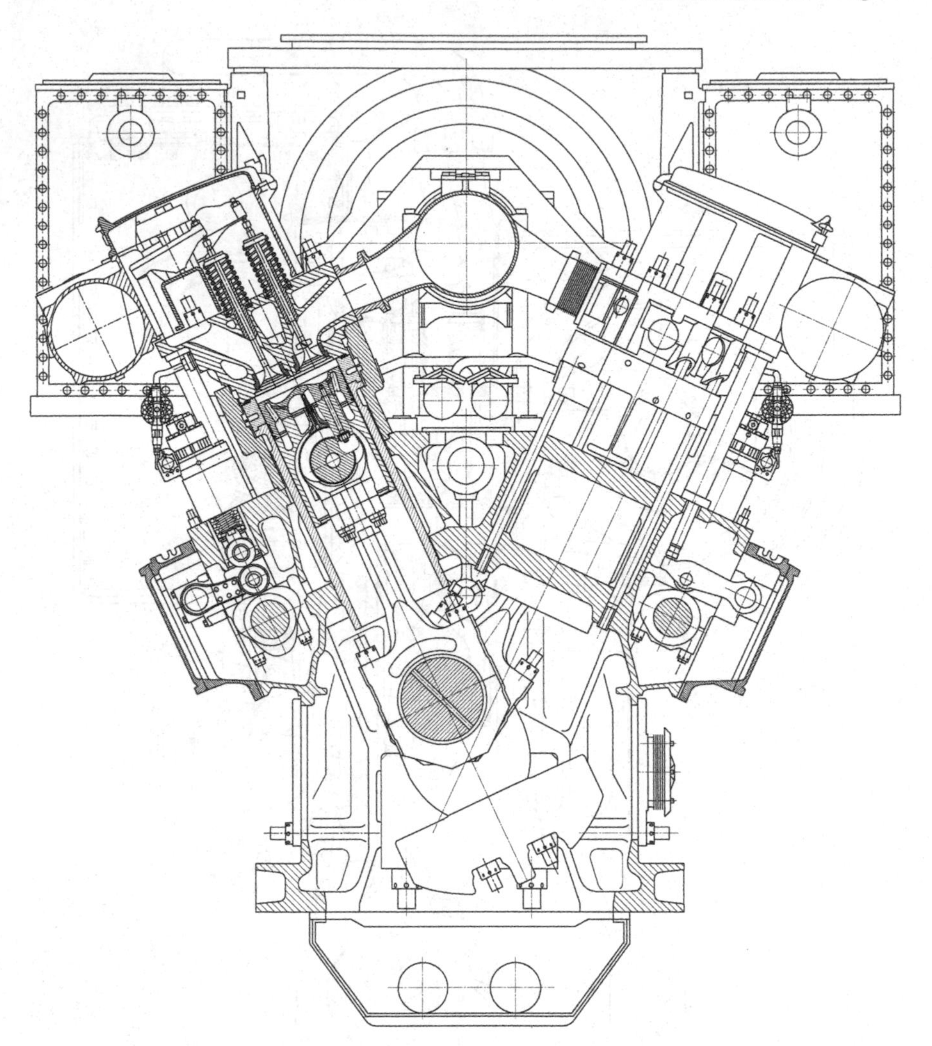

Fig. 1.33 Cross-section of the engine MAN series V48/60 CR [29]

Dimensions and weight engines MAN series L48/60 CR

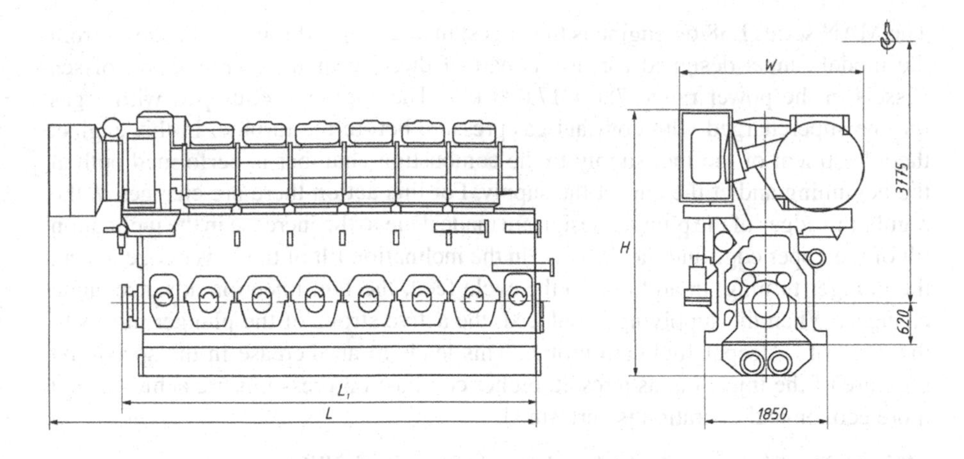

Engine version	L_1 (mm)	L (mm)	H (mm)	W (mm)	Weight (kg)
6 L40/54	5910	7520	2600	4345	70,000
7 L40/54	6610	8600	2750	4380	80,000
8 L40/54	7310	9155	2750	4380	89,000
9 L40/54	8010	10,000	2750	4380	97,000

Dimensions and weight engines MAN series V48/60 CR

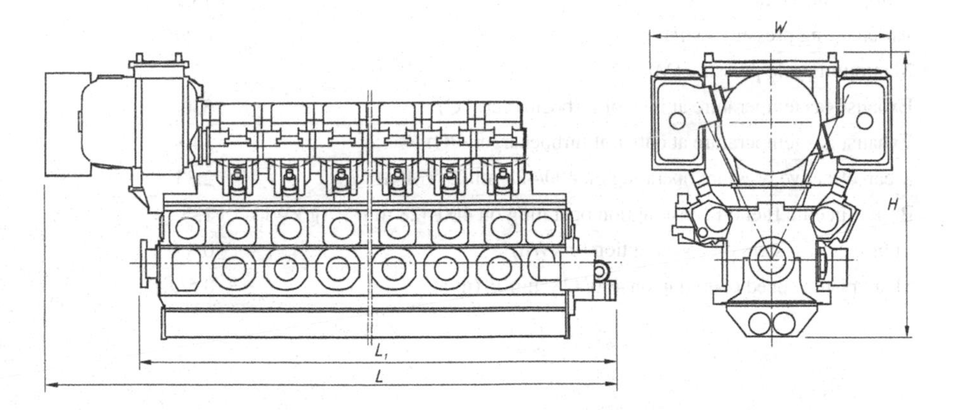

Engine version	L_1 (mm)	L (mm)	H (mm)	W (mm)	Weight (kg)
12 L48/60CR	8915	10,760	5355	4700	189,000
14 L48/60CR	9915	11,760	5355	4700	213,000
16 L48/60CR	10,915	13,100	5355	4700	240,000
18 L48/60CR	11,915	14,100	5355	4700	265,000

1.10.8 The MAN Series L58/64 Engine

The MAN series L58/64 engine is the largest medium-speed four-stroke diesel from the model range, designed for use as part of diesel-gear units (Fig. 1.34) of sea vessels in the power range 7800–17,000 kW. The engine is equipped with a gas turbine supercharged with constant gas pressure before the turbine. In this engine, the adjustment of the fuel supply to the combustion chamber is performed both at the beginning and at the end of the supply. For this action there are changes in the regulating edges of the plunger design are made. Due to the increase in the inclination tilt of the upper edge and the decrease in the inclination tilt of the lower edge during the plunger turning, in addition to the cycle feed, the fuel injection advance angle changes. When the supplying is reduced, the active stroke of the plunger moves to the zone of a steeper fuel cam profile. This leads to an increase in the speed and pressure of the injection, as a result, higher combustion pressures are achieved and more economical operation is performed.

Main technical parameters of engines MAN series L58/64

Parameter	Value
Number and cylinders arrangement	6, 7, 8, 9 in-line
Cylinder bore (mm)	580
Piston stroke (mm)	640
Cylinder capacity (dm^3)	169.0
Rotation speed (min^{-1})	400/428
Cylinder power at 400/428 min^{-1} (kW)	1310/1400
Compression ratio	13.2
Air charging pressure (MPa)	0.297
Maximum cycle pressure (MPa)	15.8
Exhaust gas temperature at inlet of turbocharger (°C)	530
Exhaust gas temperature at outlet of turbocharger (°C)	340
Mean effective pressure operating on 400/428 min^{-1} (MPa)	2.33
Brake specific fuel oil consumption operating on 400/428 min^{-1} (g/kWh)	185
Lubrication system oil consumption (g/kWh)	0.8
Mean piston speed operating on 400/428 min^{-1} (m/s)	8.5/9.1

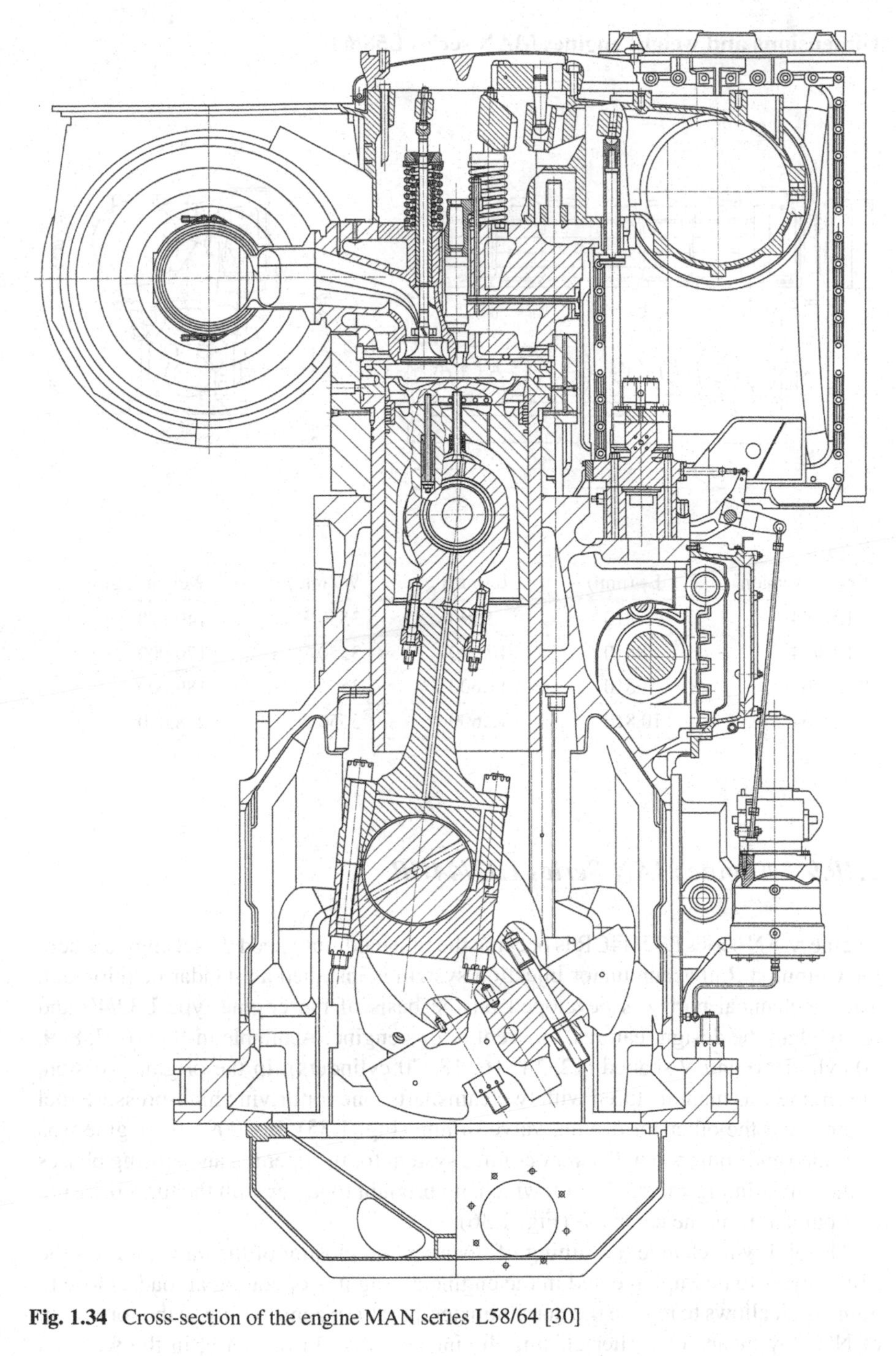

Fig. 1.34 Cross-section of the engine MAN series L58/64 [30]

Dimensions and weight engines MAN series L58/64

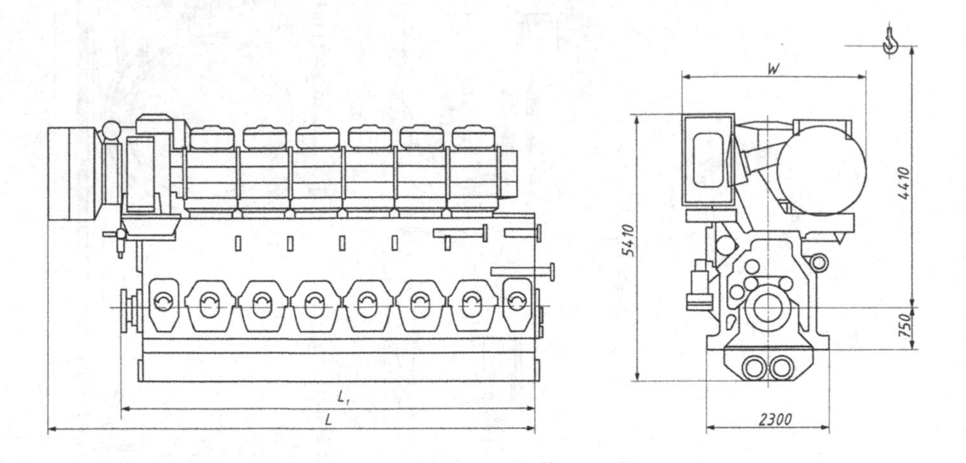

Engine version	L_1 (mm)	L (mm)	W (mm)	Weight (kg)
6 L58/64	7810	9190	3550	149,000
7 L58/64	8810	10,600	3550	170,000
8 L58/64	9810	11,600	3550	189,000
9 L58/64	10,810	12,600	3550	208,000

1.10.9 Engine MAN Series L32/44CR

Engine MAN series L32/44CR is became the first medium-speed diesel engine where the Common Rail accumulator injection system is installed as standard equipment. The mechanical part was developed on the basis of the engine type L32/40 and retained all the design features, inherent in this engine. Available in-line (6, 7, 8, 9, 10 cylinders) and V-shaped (12, 14, 16, 18, 20 cylinders). In the original version, the engine was manufactured with two camshafts, one for driving high-pressure fuel pumps, and the other for driving valve timing (Fig. 1.35). In 2009, the engine was upgraded and equipped with a new control system for the opening and closing phases of the valve timing valves (VVT), which are brought together with the high-pressure fuel pumps from one camshaft (Fig. 1.36).

The ability to change the timing of opening and closing of the valve, allows the Miller cycle to be implemented in the engine during it is operating at loads, close to nominal. It allows to reduce the maximum temperature, contributing to the formation of NO_x, by means of earlier closing the intake valve. In this case, in the working cylinder, the air charge is expanded and cooled before the start of compression, which

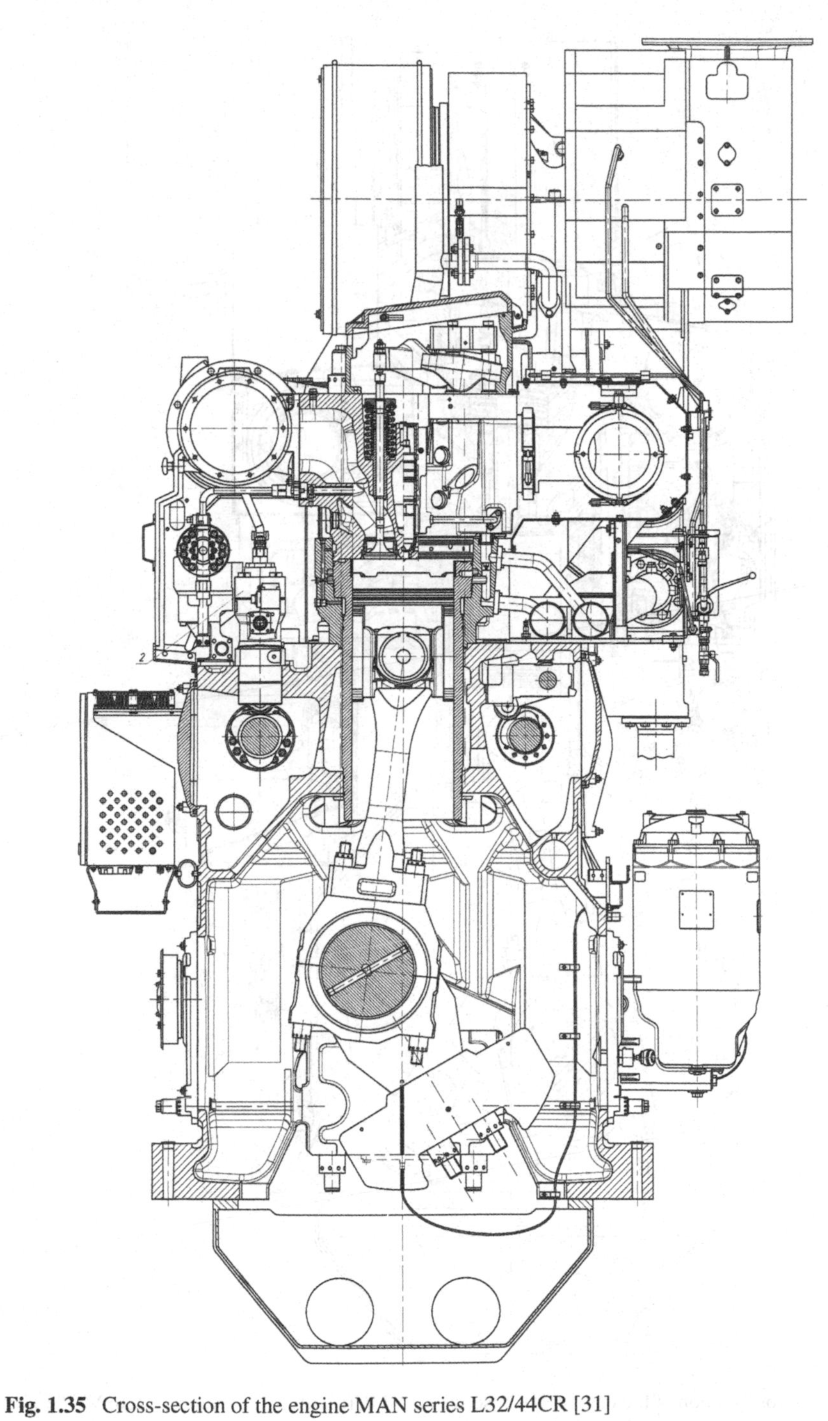

Fig. 1.35 Cross-section of the engine MAN series L32/44CR [31]

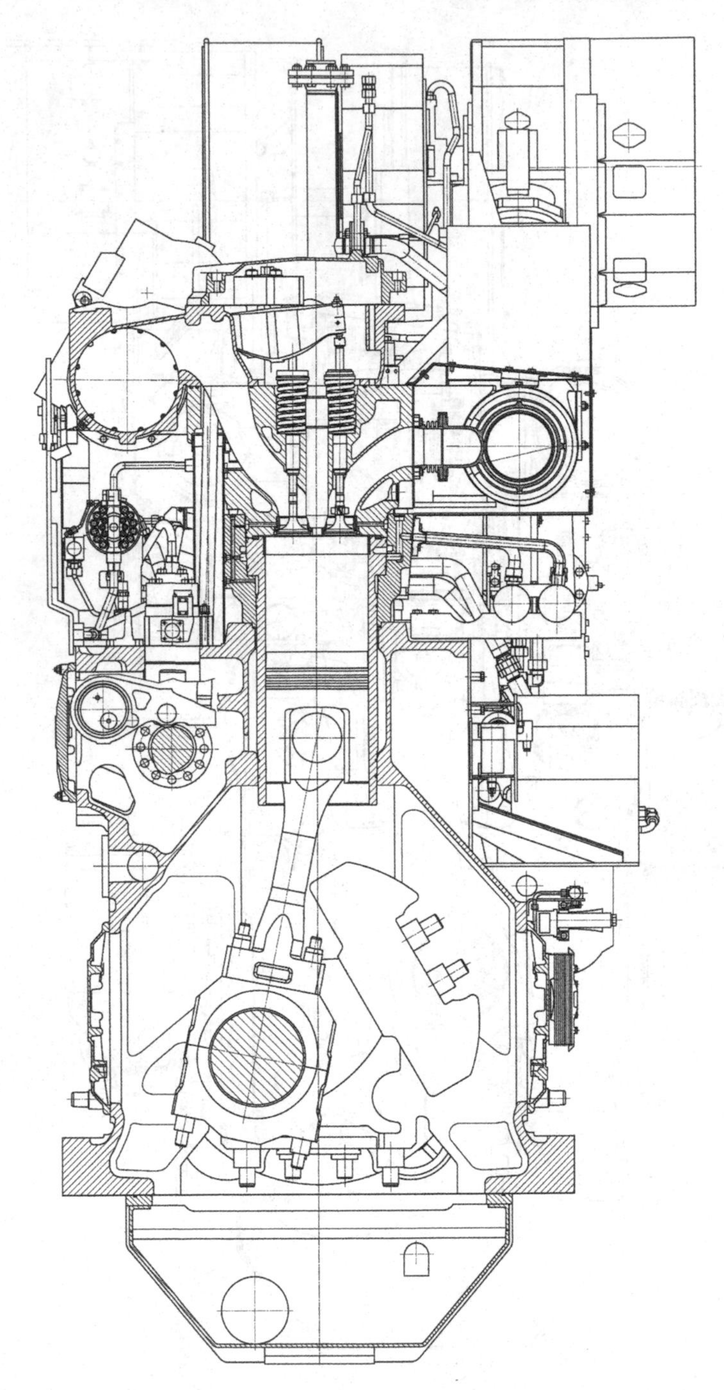

Fig. 1.36 Cross-section of the engine MAN series L32/44CR with valve timing system VVT [32]

leads to a decrease in the combustion temperature. When switching to partial loads, the opening time of the intake valve increases and the maximum temperature of the cycle increases, providing more complete fuel burning.

The Common Rail accumulator injection system is designed to operate on heavy fuel with a viscosity of up to 700 cSt at 50 °C. It is built according to a modular principle and consists of high-pressure fuel accumulators, each of them serves one or two cylinders, fuel supply control valves, located on the covers of accumulators, fuel pumps, injectors and auxiliary valves.

The Common Rail system provides flexible control of parameters and the mode of fuel injection for each cylinder over the entire load range. This flexibility allows to optimize fuel consumption and emissions of harmful substances in all modes of the 32/44CR engine. Due to the increased performance of high-pressure fuel pumps, the system does not lose its working ability even if a failure of one of them. At the same time the engine can continue to work without power reducing.

As a control fluid in the system, high pressure fuel is used, which, on the one hand, simplifies the design of the fuel system, and on the other hand, increases the energy consumption for fuel pumping. All elements of the fuel system under high pressure (about 160 MPa) are placed in protective covers (batteries, high-pressure pipes) designed in a way that in case of rupture of the main element to withstand the residual pressure.

Like other MAN electronic engines, this series has an electronic $SaCoS_{one}$ control system that allows to change automatically the valve timing during engine operation, as well as change parameters and the mode of the fuel injectors. All elements of the system are installed on the engine and is a single unit with a system of actuators.

The main advantages of the system $SaCoS_{one}$ include:

- integrated self-diagnosis functions;
- maximum operational availability;
- easy operation and diagnostics;
- fast replacement of modules and a short period of work commissioning.

High-performance boosting system is made with constant gas pressure before the turbine. New engine modifications are equipped with turbochargers with a changeable geometry of the nozzle arrays of a VTG turbine (Variable Turbocharger Geometry).

Main technical parameters of engines MAN series L32/44CR

Parameter	Value
Number and cylinders arrangement	6, 7, 8, 9, 10 in-line
Cylinder bore (mm)	320
Piston stroke (mm)	440
Cylinder capacity (dm^3)	35.37
Rotation speed (min^{-1})	720/750
Cylinder power at 720/750 min^{-1} (kW)	560

(continued)

(continued)

Parameter	Value
Compression ratio	16.3
Specific air volume flow rate (m^3/kWh)	5.92
Specific air mass flow rate (кg/kWh)	6.50
Air charging pressure (MPa)	0.299
Maximum cycle pressure (MPa)	23.0
Exhaust gas temperature at inlet of turbocharger (°C)	490
Exhaust gas temperature at outlet of turbocharger (°C)	320
Specific exhaust gas volume flow rate (m^3/kWh)	11.80
Specific exhaust gas mass flow rate (кg/kWh)	6.70
Mean effective pressure operating on 720/750 min^{-1} (MPa)	2.83/2.71
Brake specific fuel oil consumption operating on 720/750 min^{-1} (g/kWh)	177.0
Lubrication system oil consumption (g/kWh)	0.5
Mean piston speed operating on 720/750 min^{-1} (m/s)	10.56/11.0

Dimensions and weight engines MAN series L32/44CR

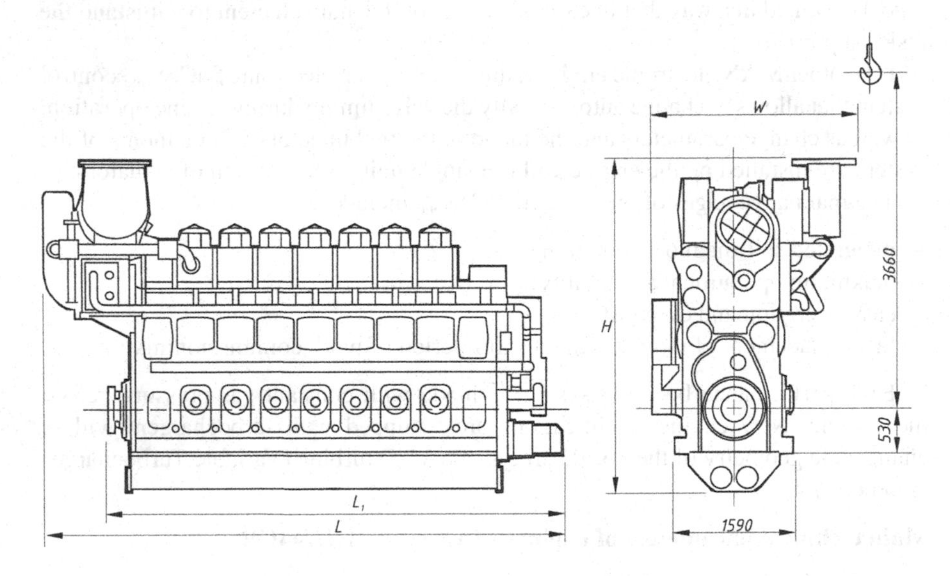

Engine version	L_1 (mm)	L (mm)	H (mm)	W (mm)	Weight (kg)
6 L32/44CR	5265	6312	4163	2174	39,500
7 L32/44CR	5877	6924	4369	2359	44,500
8 L32/44CR	6407	7454	4369	2359	49,500
9 L32/44CR	6937	7984	4369	2359	53,500
10 L32/44CR	7556	8603	4369	2359	58,000

1.10.10 The MAN Series D2876 Engines

The MAN series D2876 engines are a six-cylinder, in-line, high-speed diesel engine, designed for use as part of diesel-gear units of small vessels, diesel generator sets and emergency diesel generators (Fig. 1.37). The engine is equipped with a gas turbine with a charge air cooler. Bosch fuel injection system with integral direct-acting electromagnetic drive with direct fuel injection into the combustion chamber.

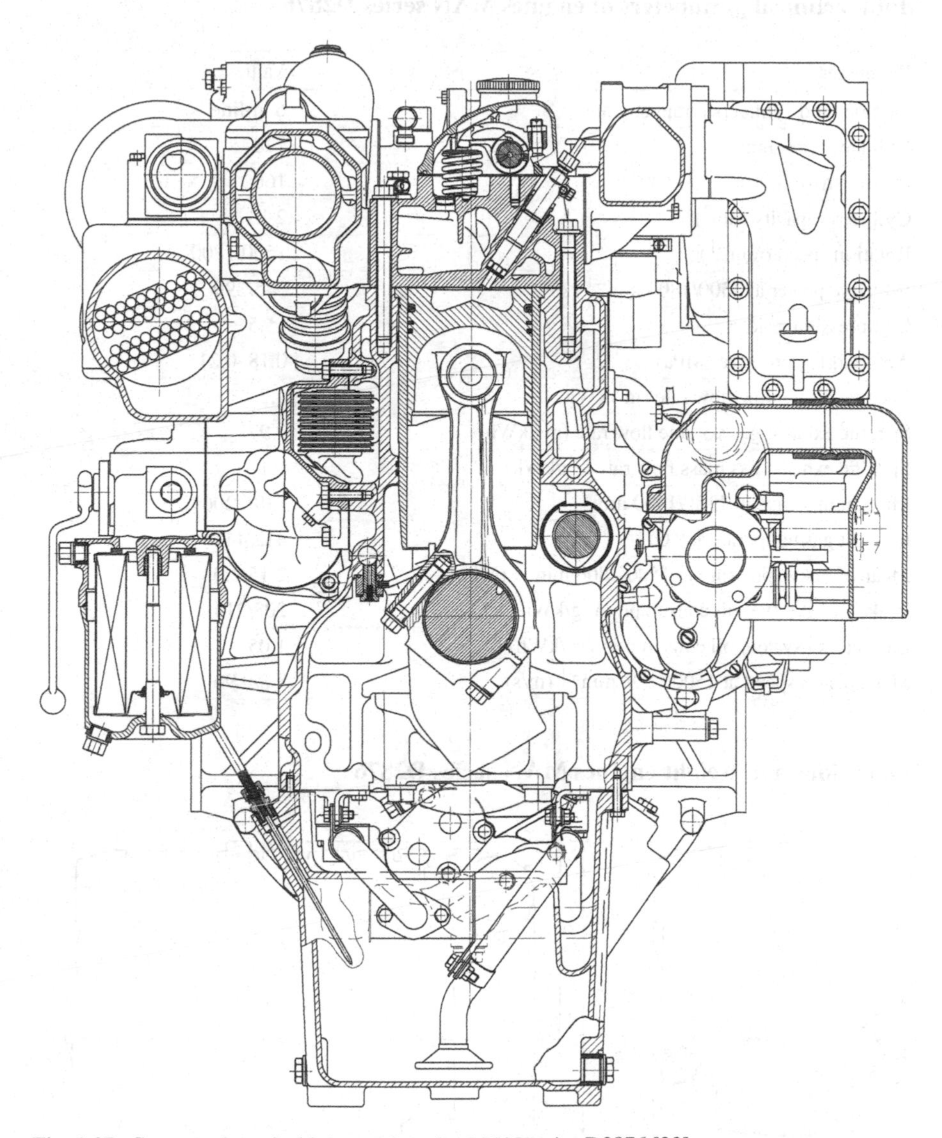

Fig. 1.37 Cross section of a high-speed engine MAN series D2876 [33]

The engine basic block is made of high-quality cast iron with wet-type cylinder liners, installed in it. Cooling water is pumped through the cooling jacket by a pump, installed directly on the engine. Forced lubrication of moving parts is performed by means of gear pump. The stock of lubricating oil is located in the pan of the engine block, which can be either standard size or with increased capacity. The engine is started using a 5.4 kW electric starter. The emissions of harmful substances with exhaust gases, the engine meets the requirements of IMO Tier II.

Main technical parameters of engines MAN series D2876

Parameter	Value
Number and cylinders arrangement	6 in-line
Cylinder bore (mm)	128
Piston stroke (mm)	166
Cylinder capacity (dm^3)	2.13
Rotation speed (min^{-1})	1500/1800
Effective power at 1500/1800 min^{-1} (kW)	345/390
Compression ratio	15.5
Air charging pressure (MPa)	0.18–0.21
Specific air volume flow rate (m^3/kWh)	4.2
Specific exhaust gas volume flow rate (m^3/kWh)	9.97
Specific exhaust gas mass flow rate (кg/kWh)	5.1
Maximum torque at 1150/1400 min^{-1}	2196/2069
Exhaust gas temperature (°C)	412/444
Mean effective pressure at 1500/1800 min^{-1} (MPa)	2.15/2.03
Brake specific fuel oil consumption (g/kWh)	208/214
Lubrication system oil consumption (g/kWh)	1.05
Mean piston speed at 1500/1800 min^{-1} (m/s)	8.3/9.96

Dimensions and weight engines MAN series D2876

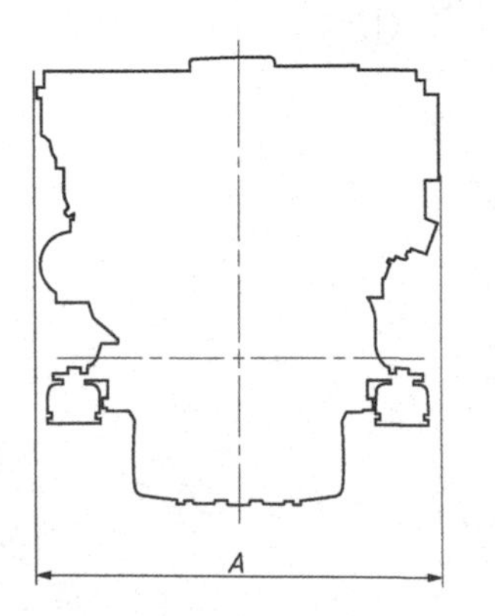

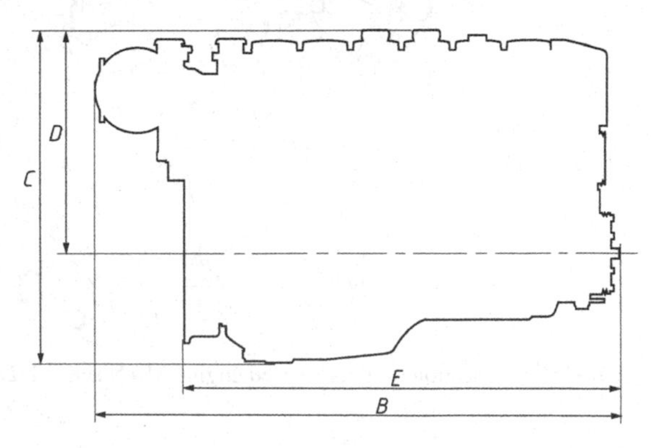

Engine version	A (mm)	B (mm)	C (mm)	D (mm)	E (mm)	Weight (kg)
D2876	830	1565	992	650	1320	1160

1.11 Mitsui & Co., Ltd.

Mitsui & Co., Ltd. was established by the Japanese entrepreneur Takashi Masuda on July 1, 1876, shortly after Japan opened its home market to foreign trade after several centuries of isolation. At the time of its establishment, the company had 16 employees, but it was developing rapidly and by the early 1880s, branches were opened in Shanghai, Paris, New York and London. At that time, the company took a solid position as an exporter of rice and coal and as an importer of modern industrial equipment.

Today, Mitsui & Co., Ltd., is one of the world's most diversified corporations working in the production of goods, trade, investment and the provision of services. Mitsui operates in many business sectors and has 14 structural divisions, including the shipyard Mitsui Engineering & Shipbuilding, founded in 1917 in the city of Tamano. In the same year, the first vessel "Kaisei Maru" was launched, and in 1924 the construction of the first vessel, the Akagisan Maru, equipped diesel engine as a main, was completed in Japan.

Since 1926, a licensing agreement about technical cooperation and co-production of marine diesel engines with the Danish company Burmeister & Wain came into effect. As a result of such cooperation, the production of licensed engines was launched next year to meet the domestic market of Japan and the countries of the Pacific region.

In 1942, the company name was changed to Mitsui Shipbuilding & Engineering Co., Ltd, and in 1952 the head office was moved from Tamano to Tokyo (Tokyo). In place of the old office in Tamano, a shipbuilding research center is being created, which also deals with the development of ship power plants.

By the early 1970s, the company was becoming the largest producer of licensed low-speed engines in the Far Eastern region, having reached the total capacity of 10 million horsepower by the engines produced by 1976.

In addition to low-speed diesel engines, the company brings to market the medium-speed engines of its own design.

The first engine was the diesel engine series 60 M with a cylinder diameter of 600 mm and a cylinder capacity of 1120 kW. The launch of this engine coincided with the energy crisis of the early 1970s, which adversely affected its prospects for widespread introduction in the fleet, therefore, soon, its production program was shut down.

The 42MA series diesel engine with a cylinder diameter of 420 mm, a piston stroke of 450 mm and a cylinder capacity of 558 kW at a rotational speed of 530 min^{-1} became the second engine, developed by the company itself. This engine was produced in-line and V-shape and was widely spread due to its high reliability and ease of maintenance.

In 1992, as part of the expansion of the production program, the construction of the largest diesel engine plant in Tamano was completed.

From 1992 to 1996, Mitsui Engineering & Shipbuilding Co. Ltd. in collaboration with another Japanese company Advanced Diesel Engine Development Co. (ADD) have developed, started the production and sale of engines type ADD30V. In 1996, the company received the first order for the supply of eight compact diesel engines of the type ADD, designed to operate as part of the main diesel generator set on a large oceanographic vessel, designed for the Maritime Safety Agency.

This engine has an original design, which, with minimum dimensions, covers the power range of 3300–10,300 kW. High reliability rates were confirmed during operational tests that lasted 5 000 h, during which 2200 h the engine worked with 110% load.

1.11.1 The MITSUI Series ADD30V

The MITSUI series ADD30V engine is a medium-speed four-stroke engine with gas turbine supercharging and charge air cooling (Fig. 1.38). The valve timing system is made according to the original scheme, which allows controlling the intake and exhaust processes, using a single large-diameter central valve. Gas flows are controlled by a distribution valve, that alternately connects the cavity above the valve with the inlet or outlet receivers. Fuel injection is performed through two injectors, located on the periphery of the cylinder. Combustion characteristics are optimized by increasing the fuel injection pressure to about 200 MPa, which improves the mixture formation and contributes to reducing NO_x emissions, without reducing fuel economy. The cylinder liner and piston rings are covered with wear-resistant ceramics, which significantly increases their service life. The piston is of composite type, the head is made of high-strength forged alloy steel, and the firing plate is made of heat-resistant steel; they are interconnected by diffusion bonding. The floating piston skirt is made of globular graphite cast iron.

Main technical parameters of engines MITSUI series ADD30V

Parameter	Value
Number and cylinders arrangement	6, 8, 10, 12, 14, 16, 18/V-shaped
Cylinder bore (mm)	300
Piston stroke (mm)	480
Cylinder capacity (dm^3)	33.93
Rotation speed (min^{-1})	720/750
Cylinder power at 720 min^{-1} (kW)	550.0

(continued)

(continued)

Parameter	Value
Cylinder power at 750 min^{-1} (kW)	575.0
Maximum fuel injection pressure (MPa)	200.0
Air charging pressure (MPa)	0.30
Maximum cycle pressure (MPa)	19.70
Mean effective pressure (MPa)	2.70
Brake specific fuel oil consumption (g/kWh)	184.0
Exhaust gas temperature at inlet of turbocharger (°C)	550.0
Exhaust gas temperature at outlet of turbocharger (°C)	380.0
Mean piston speed at 720/750 min^{-1} (m/s)	11.5/12.0

Dimensions and weight engines MITSUI series 12ADD30V

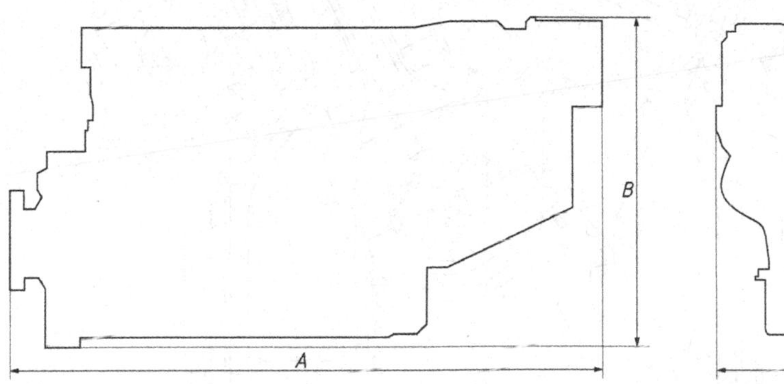

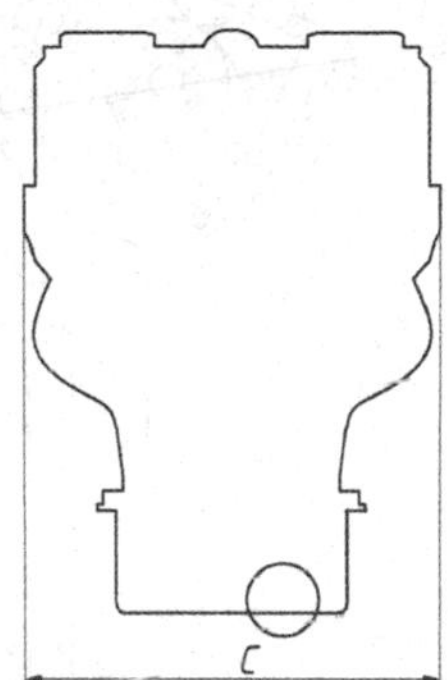

Engine version	A (mm)	B (mm)	C (mm)	Weight (kg)
12ADD30V	6383	3373	2385	-----

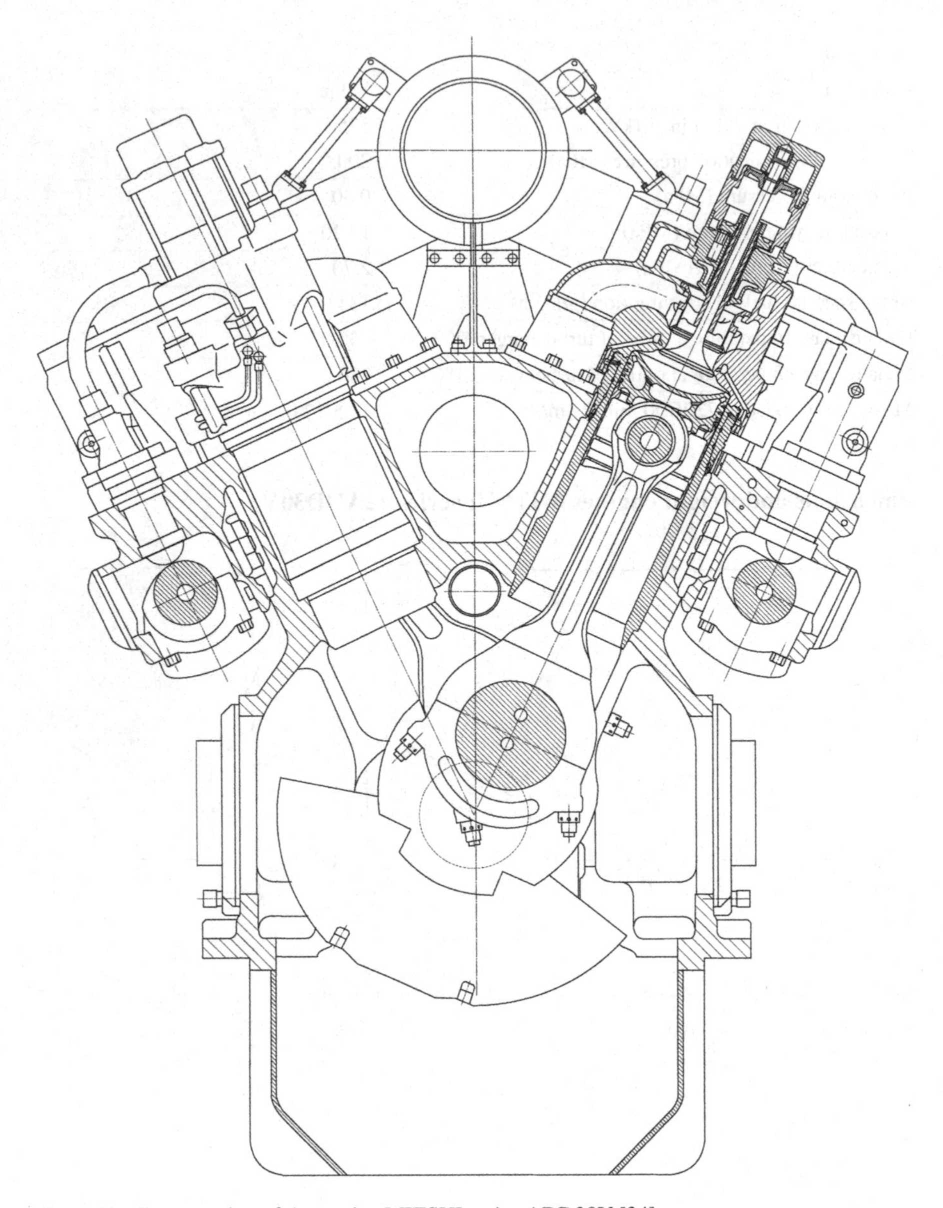

Fig. 1.38 Cross-section of the engine MITSUI series ADD30V [34]

1.12 MTU Friedrichshafen GmbH

In 1900, the first airship was built by the German earl Ferdinand Zeppelin. There are two engines of 12 hp each were installed. They were built specifically for this occasion by Daimler concern under the leadership of Wilhelm Maybach. In 1908, after the accident of the LZ 4 airship due to engine failure, Wilhelm Maybach, together with Zeppelin, on March 23, 1909, founded Luftfahrzeug-Motorenbau GmbH in Bissingen to produce airship engines, developed by Wilhelm Maybach's son Karl. The first engines, built by Karl Maybach, were designed and manufactured at the factory Grozz in Bissingen. This production was the beginning of the company later named MTU.

The first six-cylinder engine of the AZ series with 145 hp was tested in 1910. Its main feature was good maintainability, many of its elements could be replaced with an idle engine even during flight. In 1911–1912 the company moved to Friedrichshafen to the place, where the MTU plant is located today. In May 1912, the company changed its name to Motorenbau GmbH. In 1916, the six-cylinder gasoline aviation engine of the IVa series was developed. The first serial aircraft engine was 260 hp.

On May 18, 1918, the company changed its name from Motorenbau to Maybach Motorenbau. In 1919, the company began to develop a universal diesel engine of type G1 with a capacity of 150 horsepower at a speed of 1,300 min^{-1}, which was supposed to be used for motor boats and locomotives. In 1924, Maybach Motorenbau, together with Eisenbahn-Verkehrsmittel-AG, developed a diesel locomotive and presented it for the first time at a railway exhibition. On October 12, 1924, the German airship LZ 127 crossed the Atlantic in 81 h. There were five twelve-cylinder V-shaped engines installed. They capacity was 421 hp at 1400 min^{-1} type VL 1 Maybach. That test was successfully passed. In 1934, the company introduced a twelve-cylinder V-shaped diesel engine to the market, equipped with a series GO 6 turbocharger and a direct fuel injection system into the combustion chamber. The engine developed 600 hp at 1300 min^{-1}. Since then, gas turbine supercharging technology has become an integral part of all MTU engines. In 1943, Friedrichshafen became a target for numerous Allied forces bombing, and in this connection the design division was moved to Wangen-im-Allgau. After the end of the Second World War, the company fell under the control of the Allied forces, and some of the objects were partially dismantled. After World War II, the production of engines was restored, and by 1949 the production of MD 650 diesel engines (MD from Maybach Diesel) with a capacity of 1100 horsepower was restored. These engines had a number of design features, among them: the use of a tunnel-type engine basic block with roller bearings of the crankshaft, three valves per cylinder (two intake and one exhaust), etc.

On October 28, 1966, the Maybach-Motorenbau and Mercedes-Benz Motorenbau merged. The new company was named Maybach Mercedes-Benz Motorenbau GmbH. On July 11, 1969, Maybach Mercedes-Benz GmbH and MAN Turbo GmbH founded MTU (Motoren und Turbinen-Union). In September 1994, MTU entered into cooperation with Detroit Diesel Corporation (DDC), and in 1995, MTU acquired all shares of L'Orange GmbH, specializing in the production of diesel fuel equipment.

By the joint efforts of the companies, in 1996, the 2000 and 4000 series engines were introduced to the market. The latest series in the basic version comes with a Common Rail fuel injection system. In September 2000, the company introduced a new 8000 series, which in 20-cylinder design develops a power of 9000 kW.

After a series of transformations, from the MTU and L'Orange the Rolls-Royce Power Systems AG brand separates, which is a subsidiary of Rolls-Royce Power Systems Holding GmbH, a joint venture of the Rolls-Royce Group plc. and Daimler AG.

1.12.1 Engines MTU V 4000

Engines MTU V 4000 are high-speed, highly accelerated four-stroke engines with two-stage gas turbine supercharging (Fig. 1.39). Air cooling system has charge inter-cooling after the first stage turbocharger. Directly before the working cylinder, a second-stage thermostatic valve is installed, which allows adjusting the temperature of the charge air, depending on the engine's operating mode. All engines of this series are equipped with a Common Rail microprocessor-controlled combustion system. This allows all engines in this series to meet the most stringent global emission requirements for emissions. At the request of the customer, the engines can be programmed for maximum efficiency or minimum toxicity of exhaust gases.

Structurally, the engine is made according to the classical scheme with a solid engine basic block, outboard crankshaft and individual covers of each working cylinder, where two intake and two exhaust valves are installed.

Main technical parameters of engines MTU series V4000 M60

Parameter	Value
Number and cylinders arrangement	8, 12, 16/V-shaped
Cylinder bore (mm)	165
Piston stroke (mm)	190
Cylinder capacity (dm^3)	4.06
Rotation speed (min^{-1})	1800
Cylinder power at 1800 min^{-1} (kW)	110.0
Maximum fuel injection pressure (MPa)	200.0
Compression ratio	16.6
Air charging pressure (MPa)	0.36
Maximum cycle pressure (MPa)	18.50
Mean effective pressure (MPa)	2.39
Exhaust gas temperature at inlet of turbocharger (°C)	455
Brake specific fuel oil consumption (g/kWh)	210.0
Mean piston speed at 1800 min^{-1} (m/s)	10.4
Lubrication system oil consumption (g/kWh)	0.2

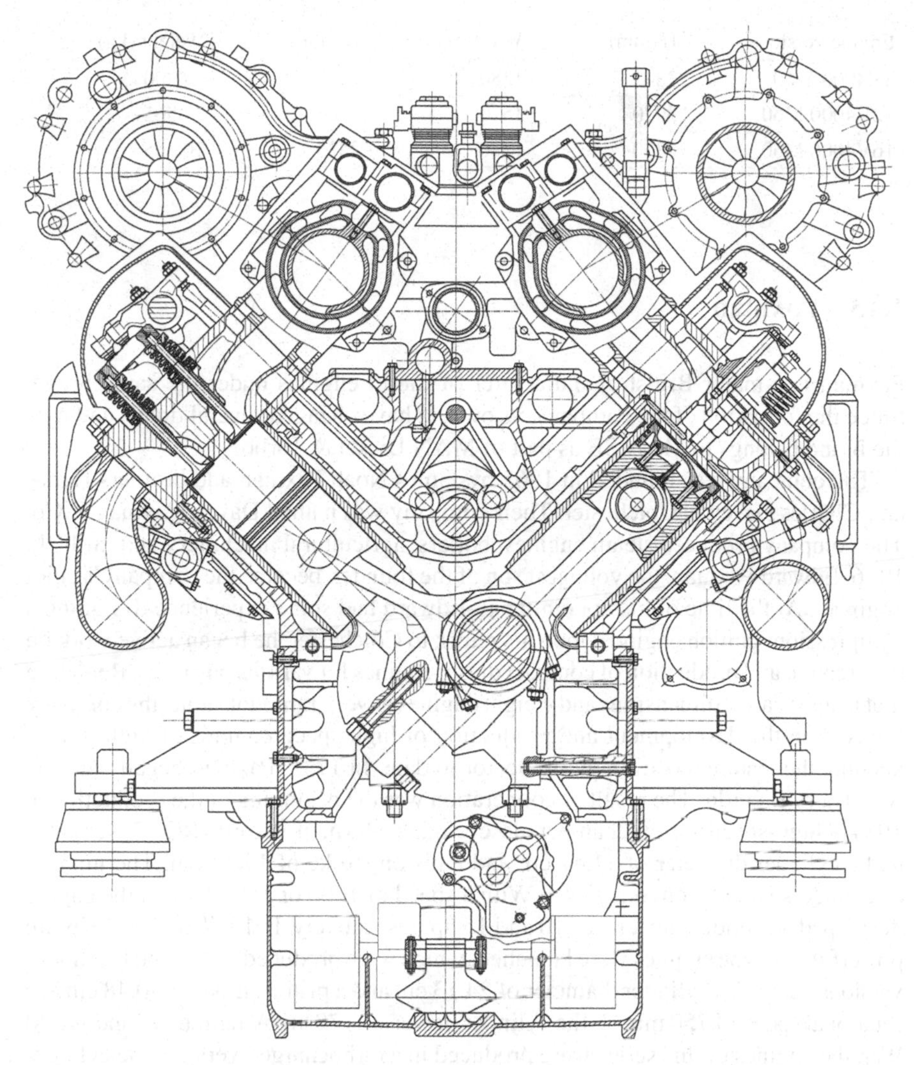

Fig. 1.39 Cross-section of the engine MTU series V4000 M60 [35]

Weight and dimensions of an Engine MTU Series V4000 M60 with a Diesel-Reducing Unit

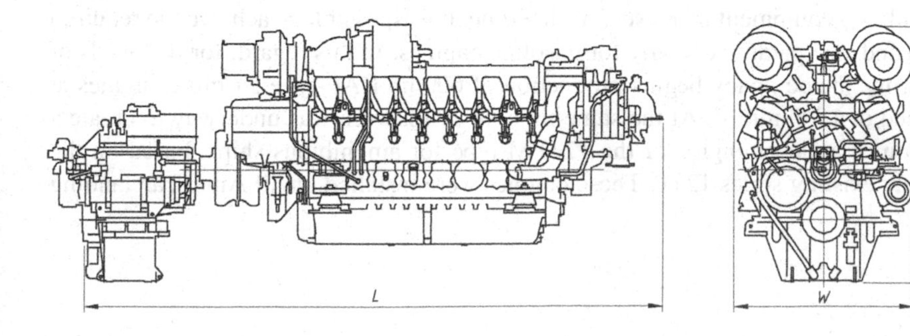

Engine version	L (mm)	W (mm)	H (mm)	Weight (kg)
8V4000 M60	3305	1380	2250	6400
12V4000 M60	3970	1520	2125	8565
16V4000 M60	4330	1520	2225	10,365

1.13 Paxman

Paxman is a major British manufacturer of diesel engines under the same brand. Since the formation of the company, its owners have changed several times, and now the brand belongs to MAN SE as part of MAN Diesel & Turbo.

The company was founded in 1865 by James Noah Paxman and brothers Henry and Charles Davey in Colchester. Their company was named Davey, Paxman & Co. The company produced steam engines, boilers, agricultural machinery and mills. In 1926, Edward Paxman, the youngest son of the founder, became the company's chief engineer. At the time of his appointment, Edward had some experience designing a compression ignition engine. Upon his return to Colchester, he began active work on the creation and production of compact diesel engines for various purposes. Realizing that in most cases, dimensions and weight engines have paramount value, the company focused on the development and production of high-speed engines of high power, keeping the leading position in this sector to date. In 1932, Paxman began working with Harry Ricardo. The result of cooperation was the beginning of the production in 1934 of high-speed vortex-chamber diesel engines Paxman-Ricardo RQ. The engines had a cylinder diameter of 11.47 cm and a piston stroke of 14.48 cm. The number of cylinders varied from one to six. With a speed of 1000 or 1500 min^{-1}, the engine developed a cylinder capacity of 10 and 15 hp. respectively. In 1937, a series of more powerful RX-type engines were launched, which were produced in 4, 6 and 8 cylinder versions. Having a cylinder diameter of 24.13 cm and a piston stroke of 30.48 cm at a rotational speed of 750 min^{-1}, the cylinder power was 70 hp. After the Second World War, the engines of this series were produced in a turbocharged version, the cylinder power in this case increased to 93 hp. These engines were installed on British military vessels for various purposes, as well as on submarines of the V and U series.

In 1940, Ruston & Hornsby Ltd. acquired a controlling stake in Paxman. As a result of this merger, a new industrial group Ruston-Paxman was organized. With the beginning of the Second World War, the need for powerful and compact engines for military equipment increases. At that time, it waspossible to achieve the required mass and size parameters only for gasoline engines. In this regard, for the needs of the army, the company begins production of 600 hp RAF-type gasoline engines at a speed of 1500 min^{-1}. At the same time, in 1942, work was underway to create a version of a diesel engine of the 12TPM type for amphibious ships, based on the already existing series 12TP. These engines were installed on all American landing craft, built in Britain at that the time. Later, these vessels played a key role in the landing of Allied forces in Normandy.

After the war, the company returned to the release of high-power diesel engines, having introduced a new series of vortex-shaped twelve-cylinder V-shaped RPH engines to the market in 1947. In the supercharged version with intermediate air cooling, this engine developed a power of 680 hp. In 1952, on the basis of vortex-chamber engines of the RPH type, a new YH series was created, with direct fuel injection into the combustion chamber with a power up to 900 hp at 1500 min^{-1}.

In 1954, Paxman's engine building division was transformed into a subsidiary of Ardleigh Engineering Ltd. In 1962, Paxman acquired shares of the Curtiss-Wright Corporation, combining both enterprises under the name Regulateurs Europa.

The increase in competition on the market of high-speed engines in the 1970–80s forced the company to create new models of compact diesel engines. In 1987, a team of designers began developing the VP185 series engine, where the computer-aided design systems was first used. The first 12-cylinder engine of the new series was released on August 31, 1991.

1.13.1 The Paxman VP185 Engine

The Paxman VP185 engine is a highly accelerated four-stroke high-speed engine with a V-shaped arrangement of cylinders at an angle of 90°, which is available in 12 and 18 cylinder design. The engine uses the original patented two-stage gas turbine supercharging system, using several turbo-compressors, placed in one gas-tight case with water cooling. The system provides high boost pressures up to the lowest rotational speeds, providing the necessary air/fuel ratio for efficient combustion of fuel and high torque on the engine shaft. Three turbo compressors of the first stage operate in a pulsed mode (for a 12-cylinder engine), supplying air to the working cylinder through the final cooler. After each first-stage turbocharger, the exhaust gases are supplied at constant pressure to two second-stage turbochargers, which supply air through the intercooler to the first-stage compressor. Automobile-type turbochargers are used as pressurization units, which can significantly reduce the initial cost of the engine and reduce maintenance and repair costs (Fig. 1.40).

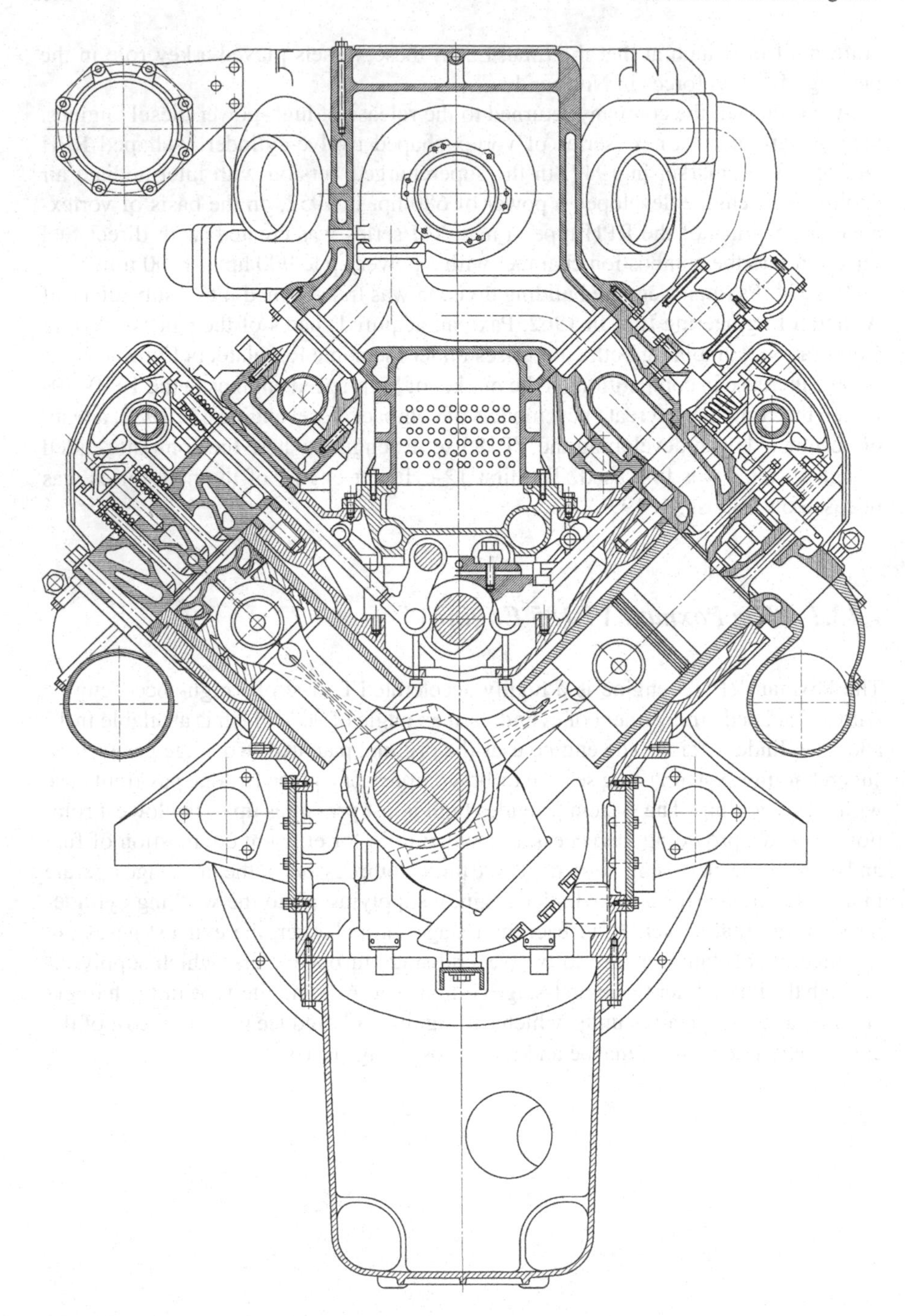

Fig. 1.40 Cross-section of the engine Paxman series VP185 [36]

Main technical parameters of engines Paxman series VP185

Parameter	Value
Number and cylinders arrangement	12, 18/V-shaped
Cylinder bore (mm)	185
Piston stroke (mm)	196
Cylinder capacity (dm^3)	5.27
Rotation speed (min^{-1})	1800
Cylinder power at 1800 min^{-1} (kW)	166.6
Maximum fuel injection pressure (MPa)	140.0
Maximum cycle pressure (MPa)	20.7
Compression ratio	13.0
Mean effective pressure (MPa)	2.53
Brake specific fuel oil consumption (g/kWh)	208.0
Mean piston speed at 1800 min^{-1} (m/s)	11.76

Dimensions and weight of engines Paxman series VP185

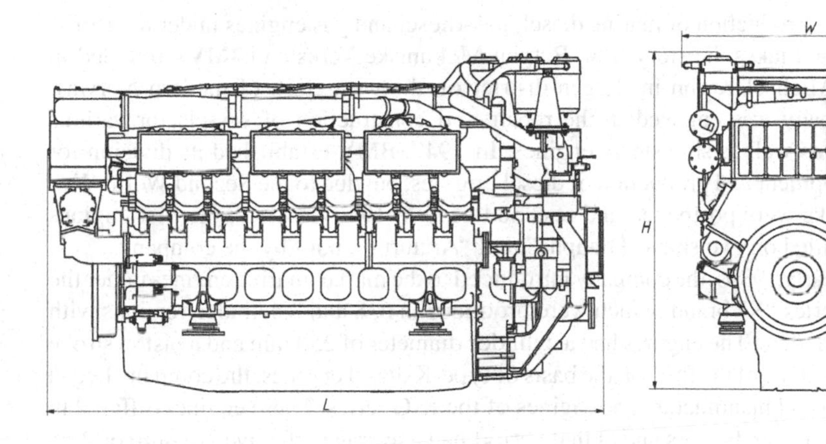

Engine version	L (mm)	H (mm)	W (mm)	Weight (kg)
Paxman 12VP185	3200	2312	1692	7460
Paxman 18VP185	4039	2447	1692	10,580

1.14 Rolls-Royce Group Plc.

Rolls-Royce Group plc. is a British company, specializing in the manufacture of equipment for aviation, ships and power equipment. Formed by the nationalization in 1987 of the company Rolls-Royce Limited.

Rolls-Royce company begins its history from F.H. Royce & Company for the production of mechanical and electrical equipment, created in Manchester in 1884 by self-taught mechanic and entrepreneur Frederick Henry Royce. In 1901, Henry Royce made several cars, according to his own designs that interested Charles Stewart Rolls, an athlete, an aristocrat, and the heir to a multi-million dollar fortune, who in 1902 opened a French car trading company.

In 1906, the base of the production of cars under the sign Rolls-Royce. Since 1907, Rolls-Royce cars have successfully performed in races and gained a reputation that continues to this day. At the beginning of World War I, in response to the needs of the country, Henry Royce developed an aircraft engine, the Eagle, which provides half of the total power, used in the air war by the allies. In the late 1920s, the engineers of the company developed the "R" series engine for the Schneider Trophy seaplanes. During the Second World War, Rolls-Royce became the largest producer of aircraft piston engines, but after the war, it turned its production in favor of gas turbine engines.

Modern production of marine diesel, gas-diesel and gas engines under the Rolls-Royce brand takes the story from Bergen Mekaniske Verksted (BMV), founded in 1855 by Michael Krohn in Bergen district on the west coast of modern Norway. The company was engaged in the repair and construction of vessels for various purposes, as well as ship steam engines. In 1942, BMV established its division for the development and production of diesel engines, but due to the Second World War, these works were postponed and resumed only in 1946. In 1947, the first engines were installed on two ships "Draupne" and "Arcturus", built by the company.

In the early 1980s, the company introduced to the market marine engines under the Bergen Series "K" brand, which were produced in 6, 8 and 9 cylinder versions with in-line cylinders. The engines had a cylinder diameter of 250 mm and a piston stroke of 300 mm. One of the first, on the basis of type-K diesel engines, the company began to develop and manufacture gas engines of the K-G series. These engines differed in high efficiency indicators and at that period had a average effective pressure of 2.0–2.2 MPa (1.8 MPa for gas engines) covering the power range from 1190 to 3970 kW. In the same period, the company manufactures Pielstick PA6 STC and PA6B STC engines manufactured under the SEMT license from Pielstick (France), which are manufactured with a number of cylinders from 6 to 20 in-line and V-shaped.

In 1984, the division for the production of diesel engines was allocated to a separate company, BMV Maskin AS, which a year later acquired the industrial group Ulstein Group and re-registered under the new name as Bergen Diesel AS. In 1986, the company began the production of ship type "B" diesel engines (Fig. 1.41) covering the power range of 2545–5300 kW, with a rotational speed of up to 750 min^{-1}.

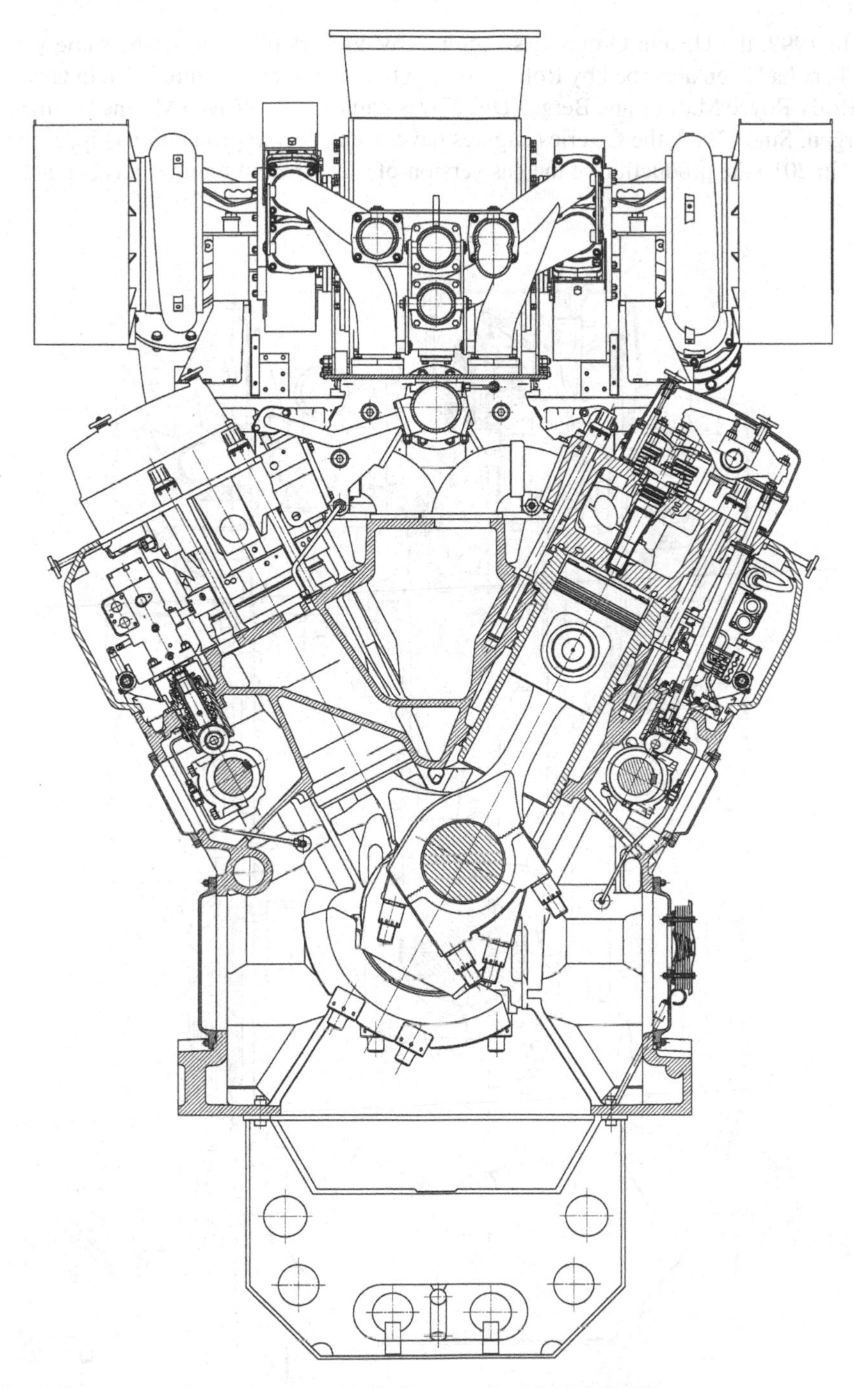

Fig. 1.41 Cross-section of the engine Rolls-Royce series B32:40VP [37]

In 1999, the Ulstein Group was acquired by Vickers plc., and in the same year Vickers had been absorbed by Rolls-Royce, which renamed the units Ulstein Group to Rolls-Royce Marine, and Bergen Diesel was named Rolls-Royce Marine Engines-Bergen. Since 2002, the C-series engines have been put into production (Fig. 1.42), and in 2010 the production of the gas version of this series of engines has begun.

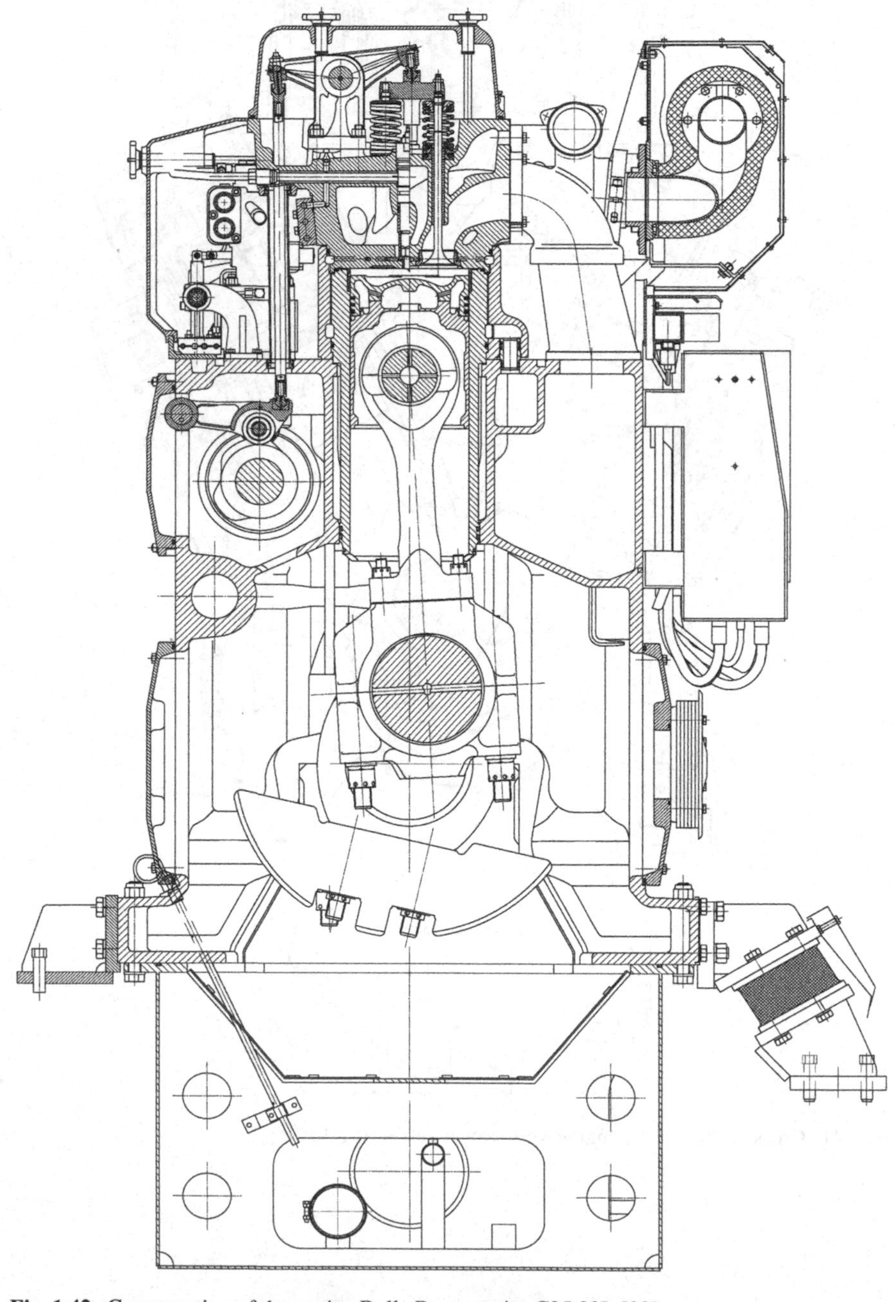

Fig. 1.42 Cross-section of the engine Rolls-Royce series C25:33L [38]

In March 2011, Rolls-Royce and Daimler took over Tognum company, which currently operates as a joint venture with Rolls-Royce, and promotes Bergen diesel and gas engines to the markets.

1.14.1 The Rolls-Royce B32:40R Engine

The Rolls-Royce B32:40R engine is a four-stroke medium-speed engine with in-line or V-shaped (Fig. 1.41) cylinder arrangements. The engine uses a pulse gas turbine supercharging system with air cooling after the compressor.

The engine basic block is a monolithic structure of nodular cast iron. Suspended crankshaft is fixed by means of bearing caps, fixed with vertical and horizontal studs. Large hatches on both sides of the engine basic block provide easy access to the details of the crank mechanism. The crankshaft is solid forged, made of chrome-molybdenum steel and equipped with removable counterweights. At the free end of the crankshaft mounted torsion damper of torsional vibrations. Cranks forged with a slash connector lower bearing that allows to dismantle the connecting rod with the piston through the sleeve of the cylinder. Composite-type pistons with steel forged heads and nodular cast iron skirts. Pistons cooling is of the "shake" type with oil from the engine lubrication system.

Main technical parameters of engines Rolls-Royce B32:40P

Parameter	Value
Number and cylinders arrangement	6, 8, 9 in-line, 12, 16/V-shaped
Cylinder bore (mm)	320
Piston stroke (mm)	400
Cylinder capacity (dm^3)	32.17
Rotation speed (min^{-1})	750
Cylinder power at 750 min^{-1} (kW)	500.0
Air charging pressure (MPa)	0.32
Compression ratio	15.5
Maximum cycle pressure (MPa)	16.20
Brake specific air consumption (кg/kWh)	6.77
Mean effective pressure (MPa)	2.49
Brake specific fuel oil consumption (g/kWh)	184.0
Exhaust gas temperature at inlet of turbocharger (°C)	375.0
Exhaust gas temperature at outlet of turbocharger (°C)	325.0
Mean piston speed at 750 min^{-1} (m/s)	10.0

Dimensions and weight engines Rolls-Royce series B32:40VL

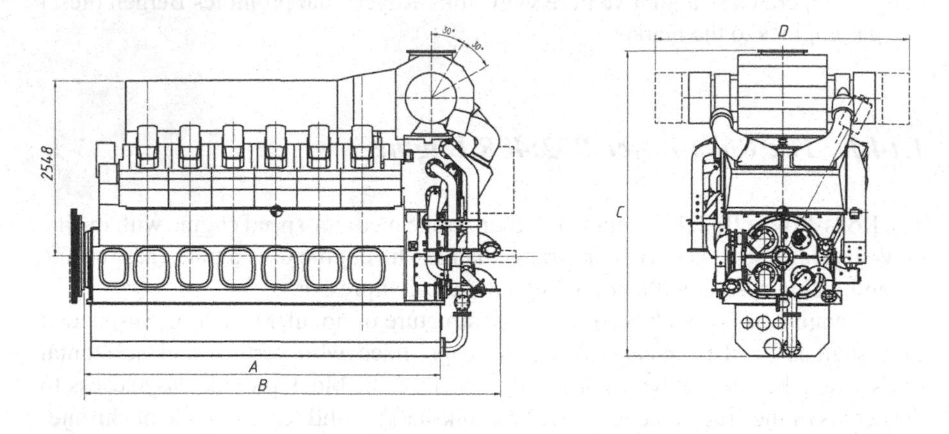

Engine version	A (mm)	B (mm)	C (mm)	D (mm)	Weight (kg)
B32:40 V12P	5176	6040	4526	2712	56,000
B32:40 V16P	6426	7489	4830	3192	73,000

1.14.2 Rolls-Royce C25:33L

Rolls-Royce C25:33L engine is a four-stroke medium-speed engine with in-line or V-shaped (Fig. 1.42). On the engine applied modular assembly system of the main components. Parts of the piston/cylinder group assembled into a single piston set, which can be replaced as an assembly without disassembling the crankshaft bearings. This possibility is achieved by using the "marine" type connecting rod with a removable rod and upper head. A distinctive feature of this series of engines is the presence of a variable valve timing mechanism VVT (Variable Valve Timing). To change the phases, the pusher arms are mounted on an eccentric shaft, turning it leads to a change in the position of the pusher roller relative to the cams of the camshaft. This feature allows to organize the engine workflow according to the Miller cycle at modes of loads close to the nominal, when switching to partial loads, the engine operates according to the "classical" cycle.

Main technical parameters of engines Rolls-Royce C25:33L

Parameter	Value
Number and cylinders arrangement	6, 8, 9 in-line, 12, 16/V-shaped
Cylinder bore (mm)	250
Piston stroke (mm)	330

(continued)

(continued)

Parameter	Value
Cylinder capacity (dm^3)	16.2
Rotation speed (min^{-1})	900
Cylinder power at 900 min^{-1} (kW)	320
Compression ratio	17.0
Air charging pressure (MPa)	0.43
Maximum cycle pressure (MPa)	109.80
Mean effective pressure (MPa)	2.634
Brake specific fuel oil consumption (g/kWh)	182.0
Exhaust gas temperature at inlet of turbocharger (°C)	545.0
Exhaust gas temperature at outlet of turbocharger (°C)	356.0
Mean piston speed at 900 min^{-1} (m/s)	9.90

Dimensions and weight engines Rolls-Royce series C25:33L

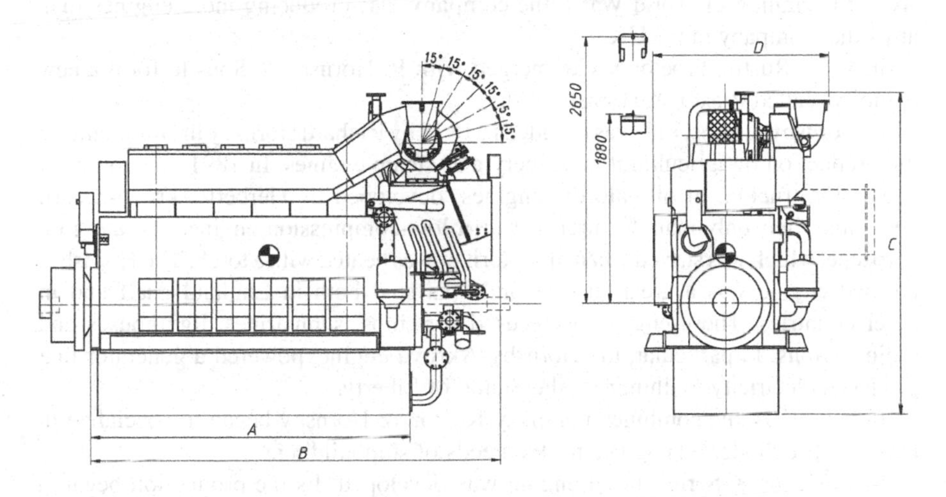

Engine version	A (mm)	B (mm)	C (mm)	D (mm)	Weight (kg)
C25:33 L6P	3170	4036	3179	1775	18,300
C25:33 L8P	3930	4796	3195	1873	23,200
C25:33 L9P	4310	5176	3230	1873	25,000

1.15 Ruston & Hornsby

Ruston & Hornsby is a major British manufacturer of diesel engines under the same brand. The company was founded in 1857 by Joseph Ruston. Having inherited a small fortune from his father, a small landowner, Ruston began to look for a business, where he could invest the available money. Thus, he met with the owners of a small workshop for the production and repair of agricultural machinery and steam engines, Burton and Proctor from Lincoln. Having invested money in their enterprise Burton & Proctor, Ruston proposed to reassign the company to the production of warehouse equipment. This proposal did not suit Burton and he, having sold his part to Ruston, left the company. Later it became known as Ruston, Proctor & Co.

The company developed rapidly under the professional guidance of Ruston, expanding the range of produced products and, by 1889, had become one of the largest engineering companies in the country.

Joseph Ruston died in 1897, then the company was headed by his eldest son Joseph Seward Ruston. Under his leadership the oil engine was designed, and the fuel injector, developed by the company in 1912, is used almost unchanged to date. By the beginning of World War I, the company was producing more engines than any other company in the UK.

In 1918, Rustin, Proctor & Co. merged with R. Hornsby & Sons to form a new company called Ruston & Hornsby Ltd.

R. Hornsby & Sons, Ltd. was founded in 1828 by Richard Hornsby in Grantham for the production of agricultural machinery and steam engines. In 1891, the company began construction of oil calorific engines, developed by Herbert Akroyd-Stuart, becoming their only manufacturer. It was a low-compression engine, operating on kerosene, which was started when the calorizer was heated with a torch. The Hornsby-Akroyd engine was a great success and was sold both in England itself and in other countries. These engines were used for ships, submarines, lighthouses and radio stations. In particular, the Hornsby-Akroyd engine, powered a generator that produced electricity to illuminate the Statue of Liberty.

In the 1920s, the combined company Ruston & Hornsby began to specialize in large multi-cylinder oil engines for the needs of shipbuilding.

In 1934, the K-Series diesel engine was developed. Its the production began a year later at the Rugby plant. This engine had a cylinder diameter of 254 mm (10 in.). Shortly after the end of the Second World War in 1945, a V-shaped 16-cylinder version of this engine was developed. Further improvements were made over the next two years, and in 1947 the engine began to be produced with the RK index. Initially, this engine had two valves in the cylinder head, but in 1951 the number of valves was increased to four with the start of production of the RK Mk 2 engine. Further development led to the introduction of the RK Mk3 engine in 1962.

In 1966, Ruston & Hornsby was taken over by the company English Electric Co, which, in turn, was absorbed by the GEC group in a few years.

The first engine Ruston RK270 (Fig. 1.43) was commissioned in 1982. Good weight and size indicators allowed this engine to win strong positions on high-speed

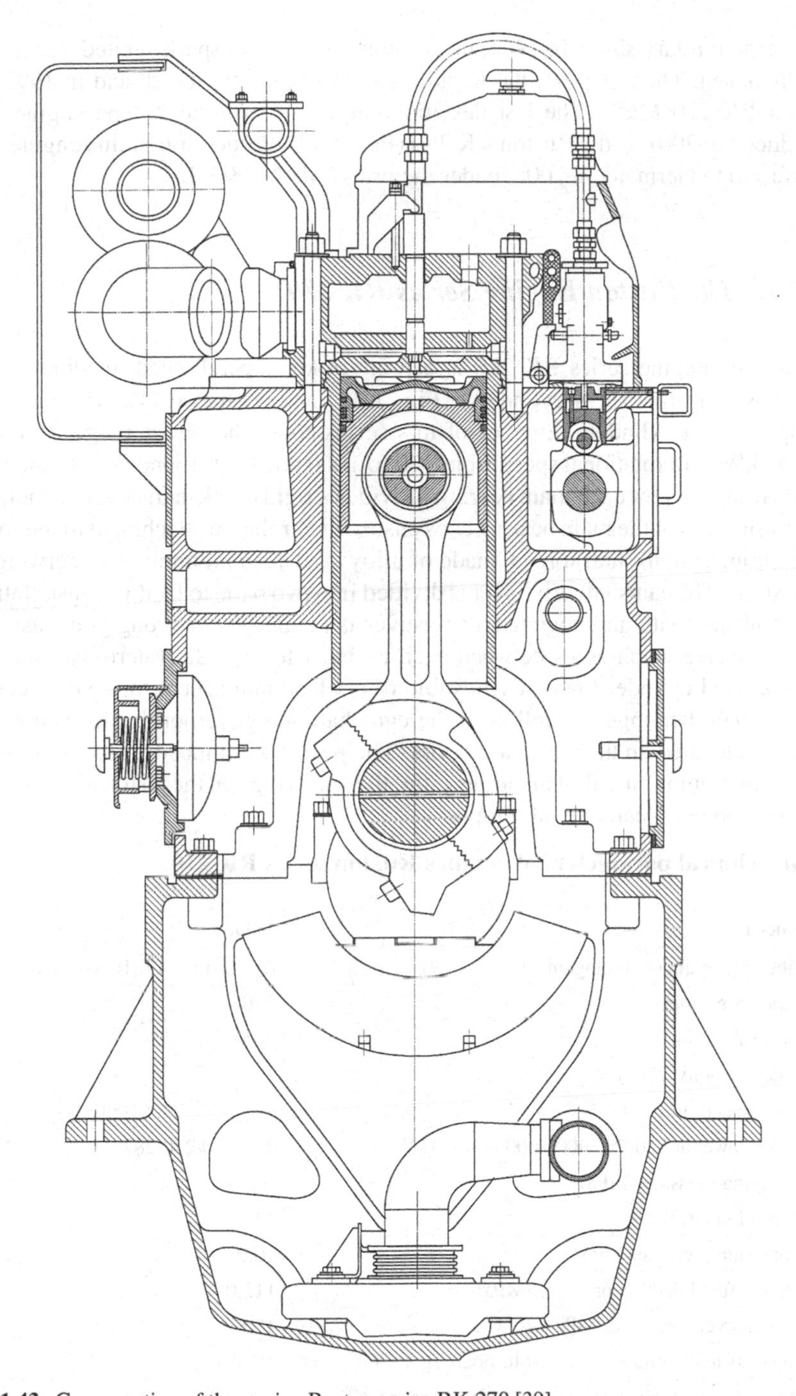

Fig. 1.43 Cross-section of the engine Ruston series RK 270 [39]

ferries and military ships. In 1985, the Ruston RK 270 GS spark-ignited gas engine was launched. Then in 1991, the Ruston RK 215 was introduced, and in 1992 the Ruston RK 270 Mk 2. The last development of the original K-type engine was introduced in 2001 as the Ruston RK 280 (Fig. 1.44). Production of this engine was transferred to Germany in 2006, under the brand MAN 28/33D.

1.15.1 The Ruston Engine Series RK 270

The Ruston engine series RK 270 is a four-stroke, supercharged, medium-speed engine with in-line (6, 8 cylinders) (Fig. 1.43) and V-shaped (12, 16 cylinders) arrangement of cylinders. Engines of this series cover the power range from 880 to 7650 kW with rotational speed from 720 to 1000 min^{-1}. A foundation frame with transverse beams for each main bearing provides a rigid crankshaft support. Inclined mating surfaces of the main bearing caps ensure their reliable attachment to the foundation frame. Crankshaft forged, made of alloy steel with mounting counterweights using studs. The camshaft drive gear is divided into two parts to facilitate installation. The crankcase with an integrated air receiver is molded from strong gray cast iron with transverse diaphragms between each cylinder to provide watertight compartments around cylinder liners. Removable doors facilitate access to the connecting rod and main bearings, as well as to the camshaft and governor mechanism. Blast valves are located on the crankcase doors. The piston is composite, has a steel head and an aluminum skirt. To seal the gas joint, there are rings on the piston are installed; three compression and one oil scraper.

Main technical parameters of engines Ruston series RK 270

Parameter	Value
Number and cylinders arrangement	6, 8 in-line, 12, 16 V-shaped
Cylinder bore (mm)	270
Piston stroke (mm)	305
Cylinder capacity (dm^3)	17.3
Rotation speed (min^{-1})	720; 750; 900; 1000
Cylinder power at 720/750/900/1000 min^{-1} (kW)	235/240/270/287
Air charging pressure (MPa)	0.22
Compression ratio	12.8
Compression pressure (MPa)	10.0
Maximum fuel injection pressure (MPa)	112.0
Maximum cycle pressure (MPa)	13.8
Exhaust gas temperature at inlet of turbocharger (°C)	580.0
Exhaust gas temperature at outlet of turbocharger (°C)	390.0
Mean effective pressure (MPa)	2.24/2.2/2.06/1.97

(continued)

(continued)

Parameter	Value
Brake specific fuel oil consumption (g/kWh)	202/202/207/208
Mean piston speed at 720/750/900/1000 min^{-1} (m/s)	7.32/7.62/9.15/10.7

Dimensions and weight engines Ruston series RK 270

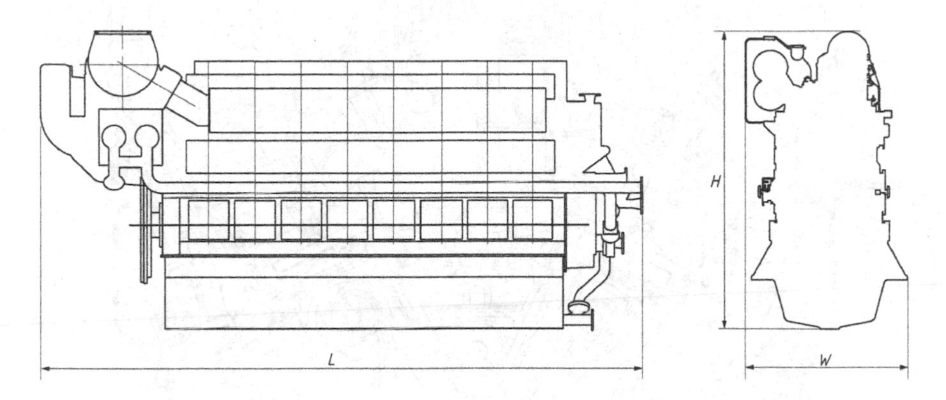

Engine version	L (mm)	H (mm)	W (mm)	Weight (kg)
Ruston 6RK 270	4020	2490	1325	28,200
Ruston 8RK 270	4750	2490	1325	37,900

1.15.2 The Ruston Series RK 280 Engine (MAN V28/33D STC)

The Ruston series RK 280 engine (MAN V28/33D STC) is a high-efficiency four-stroke high-speed engine with a V-shaped arrangement of cylinders at an angle of 52°, which is available in 12, 18 and 20 cylinder versions (Fig. 1.44). The engine uses the STC (Sequential Turbo Charging) gas turbine supercharging system, which allows for partial loads to disconnect one turbocharger, thus ensuring the necessary fuel-air ratio for efficient combustion of fuel and obtaining high torque on the engine shaft throughout the operating mode range. Two high-performance axial turbochargers TCA 33, specially designed for the V28/33D STC, are used as boosting units. Electronic fuel injection system with electronic control EFI (electronic fuel injection). The fuel supply is controlled by a solenoid valve, mounted on the housing of the high pressure pump. Short pressure pipes to the nozzles with a double wall ensure feed stability and safety. The electronic control unit of fuel pumps is mounted directly on the engine.

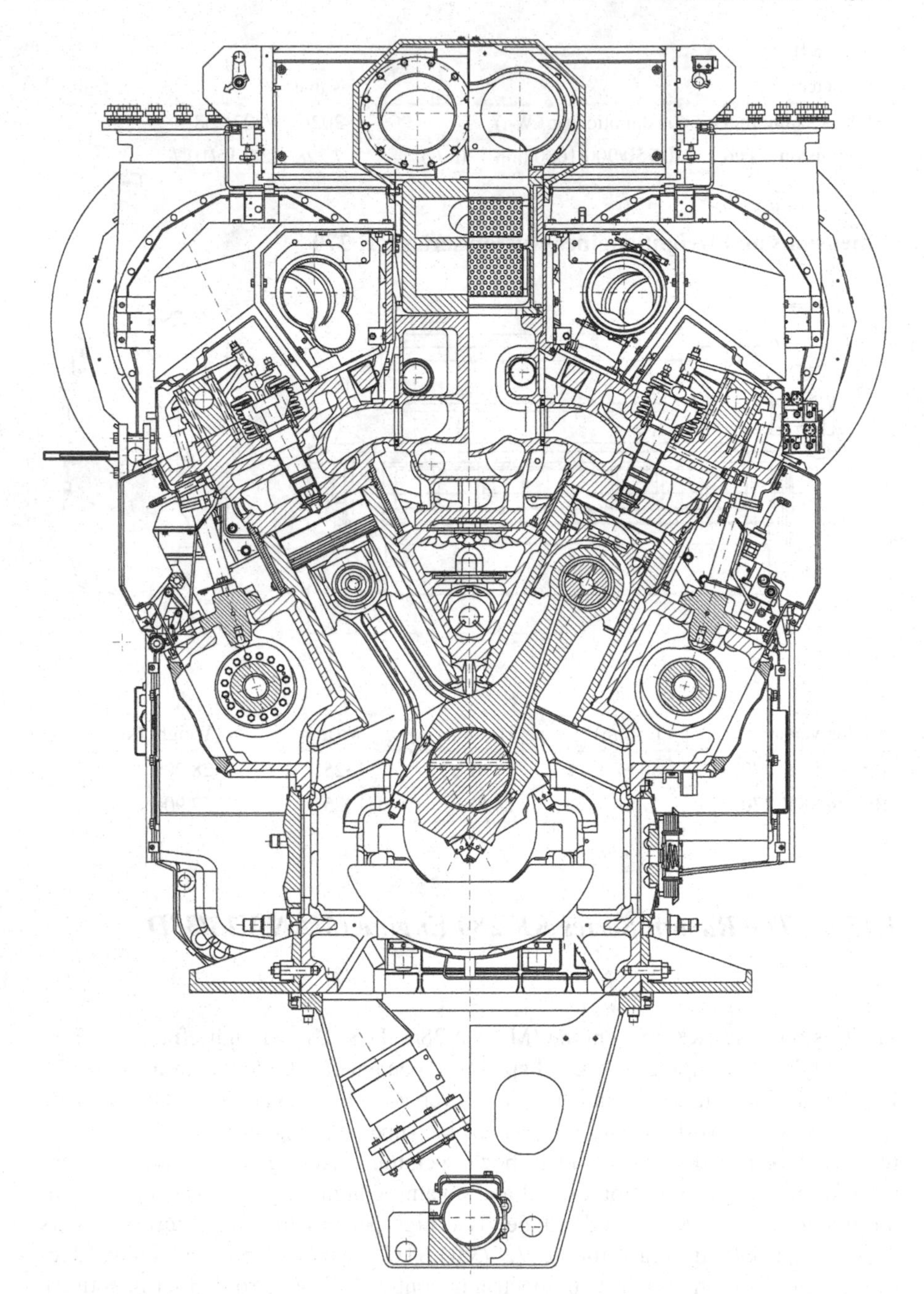

Fig. 1.44 Cross-section of the engine Ruston series RK 280 (MAN V28/33D STC) [40]

Main technical parameters of engines series RK 280 (MAN V28/33D)

Parameter	Value
Number and cylinders arrangement	12, 16, 20 V-shaped
Cylinder bore (mm)	280
Piston stroke (mm)	330
Cylinder capacity (dm^3)	20.32
Rotation speed (min^{-1})	1000
Cylinder power at 1000 min^{-1} (kW)	450
Compression ratio	14.2
Air charging pressure (MPa)	0.32
Maximum cycle pressure (MPa)	21.0
Compression ratio	14.2
Exhaust gas temperature at inlet of turbocharger (°C)	565.0
Exhaust gas temperature at outlet of turbocharger (°C)	460.0
Mean effective pressure (MPa)	2.65
Brake specific fuel oil consumption (g/kWh)	183.0
Mean piston speed at 1000 min^{-1} (m/s)	11.0

Dimensions and weight engines Ruston series RK 280 (MAN V28/33D STC)

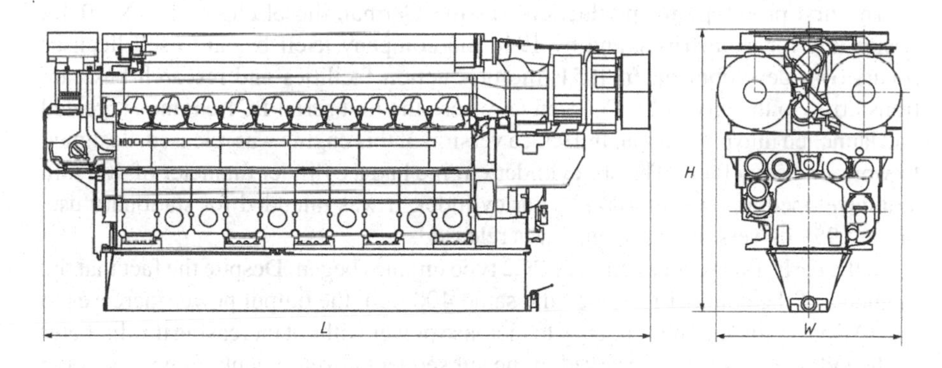

Engine version	L (mm)	H (mm)	W (mm)	Weight (kg)
Ruston 12RK 280	5713	3734	2473	31,900
Ruston 16RK 280	6633	3734	2473	39,900
Ruston 20RK 280	7543	3734	2473	48,000

1.16 Pielstick

Pielstick is a French diesel engine company located in the Villepinte district of Paris. Currently, the company belongs to MAN Diesel & Turbo, a division of the German company MAN.

The company was founded in 1946 on the initiative of the French Ministry of Industrial Production with the support of five national companies with the goal of developing diesel engines for ships, railway locomotives and power plants. Initially, the company was named Société d'Études de Machines Thermiques (heat engine research company, from the French language) abbreviated S.E.M.T. Since the company was founded, it was headed by Gustav Pielstick, a German engineer and a specialist in marine engines.

Gustav Pilstic at the time of the company creation had considerable experience in designing various engines. As early as October 1911, he began his career as a design engineer at MAN in Augsburg, where he worked on the development of engines for merchant ships and submarines. In the late 1920s, there was a need for high-performance diesel engines with high weight and dimensions. In 1931, Pilstic led the development of such engines, and in 1934 he was appointed director of the direction for the production of MAN diesel engines. Under his leadership, diesel engines were designed for ships, locomotives and stationary energy. After the Second World War, Pilstic accepts the offer of the French government and actually becomes the creator of the company S.E.M.T., which is successfully developing and, paying tribute to its founder, receives a new name S.E.M.T. Pielstick.

The first prototype for production was the German diesel engine MAN 40/46, designated for submarines, and by 1948 the company itself began to sell licenses for its own development. In 1951, the production facilities and research base are transferred to the city of La Courneuve, where the first serial PC1 engine is created and launched into production. In 1953, a version of this engine was released, designed to work on heavy fuels. The six-cylinder engine had a cylinder diameter of 400 mm and developed a power of 180 kW per cylinder. It was intended for stationary use, but in 1955 its version in for ships was released.

In the early 1960s, production of PC2 type engines began. Despite the fact that the diameter of the cylinder remained the same 400 mm, the output power increases to 310 kW per cylinder. This tendency to increase power without increasing the diameter of the cylinder has been preserved in the subsequent development of engines of the PC2.3 series in 1971 and PC2.5 in 1973. In 1981, production of engines of the PC2.6 type with a cylinder power of 550 kW was started, and in 1995 the production of the most mass series of PC2.6B engines (Fig. 1.45) with a power of 750 kW per cylinder was launched. Engines are made both in in-line, and V-shaped type.

PC3 series engines with a cylinder diameter of 480 mm will be produced in 1969 with an output of 700 kW per cylinder.

PC4 series engines with a diameter of 570 mm appear in 1972. Initially, they develop power of 990 kW/cyl., but already in 1981 in version PC4.2 the power was increased to 1215 kW/cyl. In 1985, a version of the engine PC4.2B with an

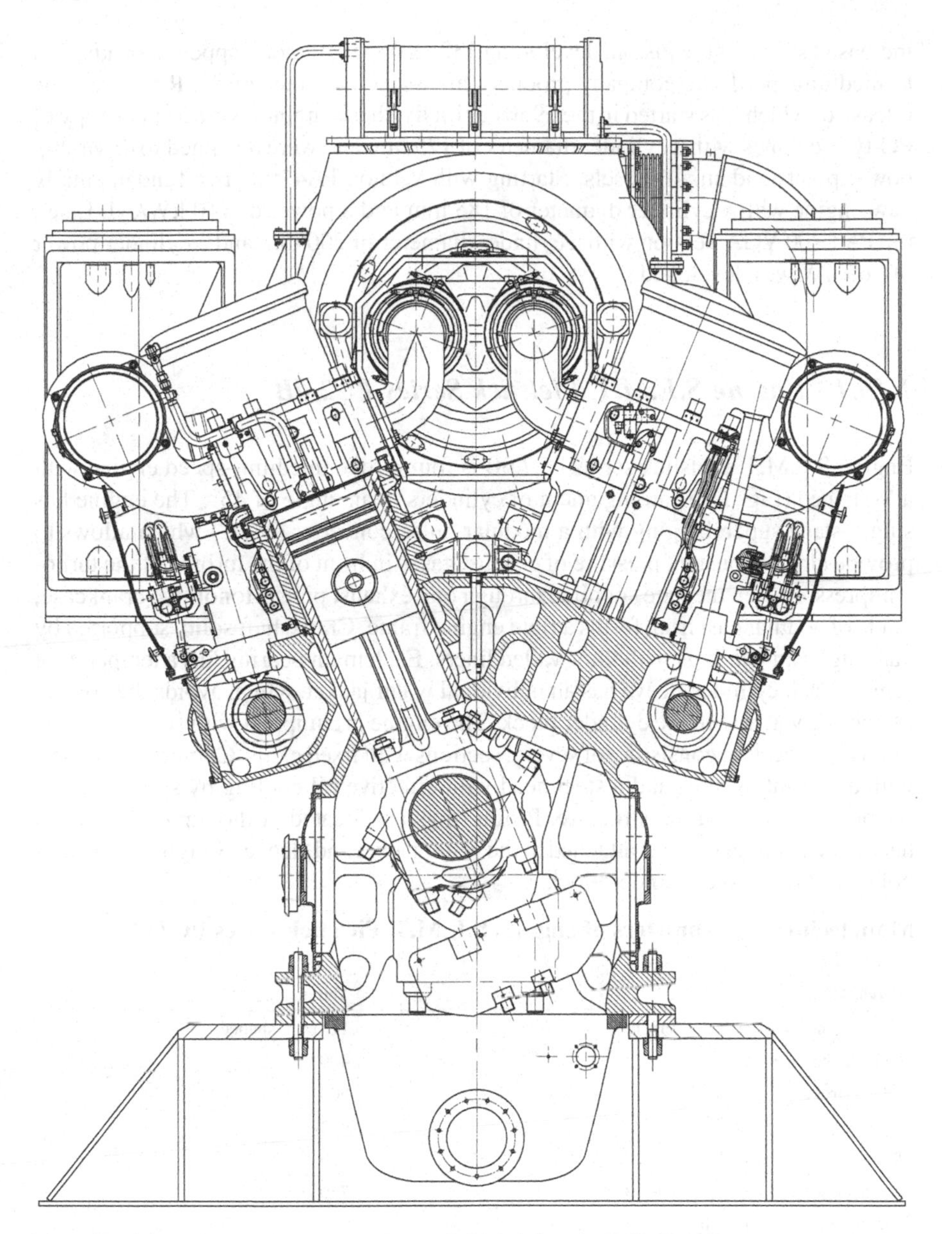

Fig. 1.45 Cross-section of the engine S.E.M.T. Pielstick series PC2.6B [41]

increased stroke of the piston, developing 1325 kW per cylinder, appears. In addition to medium-speed, the company produces high-speed engines of the RA series, the release of which was started in the 1950s. Initially, these engines were a small copy of PC-type engines with a cylinder diameter of 175 mm and were designed to drive ship power plants and small vessels. Starting with version PA4, this is a fundamentally new engine with a cylinder diameter of 185 mm and a power of 110 kW/cyl. Later, the PA4 200 VGA version with a cylinder diameter of 200 mm and a cylinder power of 165 kW was developed.

1.16.1 Engine S.E.M.T. Pielstick Series PC2.6B

Engine S.E.M.T. Pielstick series PC2.6B is four-stroke, medium-speed engine with a V-shaped (Fig. 1.45) arrangement of cylinders at an angle of 45°. The engine has single-stage supercharging with a modular pulse converter (MPC) which allows to provide almost constant pressure of exhaust gases in front of the turbine of the turbo-compressor when they are supplied through one exhaust pipe. Monolithic crankcase, made of nodular cast iron, forms a rigid engine frame. Crankshaft solid, supported by hanging main bearings with thin-walled liners. Each main bearing has a temperature sensor. Each cylinder sleeve has an individual water jacket, which avoids the contact of cooling water with the cylinder block. The engine is equipped with two camshafts of modular design, consisting of several sections with fixed cams. Composite pistons, with a light-alloy skirt and a steel head with effective oil cooling by spraying. The piston pin has a floating structure. Piston rings are located at the top of the piston head. The cylinder cover is attached to the water jacket and sleeve by eight connecting bolts attached to the cylinder block.

Main technical parameters of engines S.E.M.T. Pielstick series PC2.6B

Parameter	Value
Number and cylinders arrangement	12, 16 V-shaped
Cylinder bore (mm)	400
Piston stroke (mm)	500
Cylinder capacity (dm^3)	62.83
Rotation speed (min^{-1})	600
Cylinder power at 600 min^{-1} (kW)	750
Air charging pressure (MPa)	0.3
Compression ratio	13.8
Maximum cycle pressure (MPa)	15.0
Mean effective pressure (MPa)	2.39
Brake specific fuel oil consumption (g/kWh)	184
Mean piston speed at 600 min^{-1} (m/s)	10.0
Lubrication system oil consumption (g/kWh)	0.3–0.6

Dimensions and weight engines S.E.M.T. Pielstick series PC2.6B

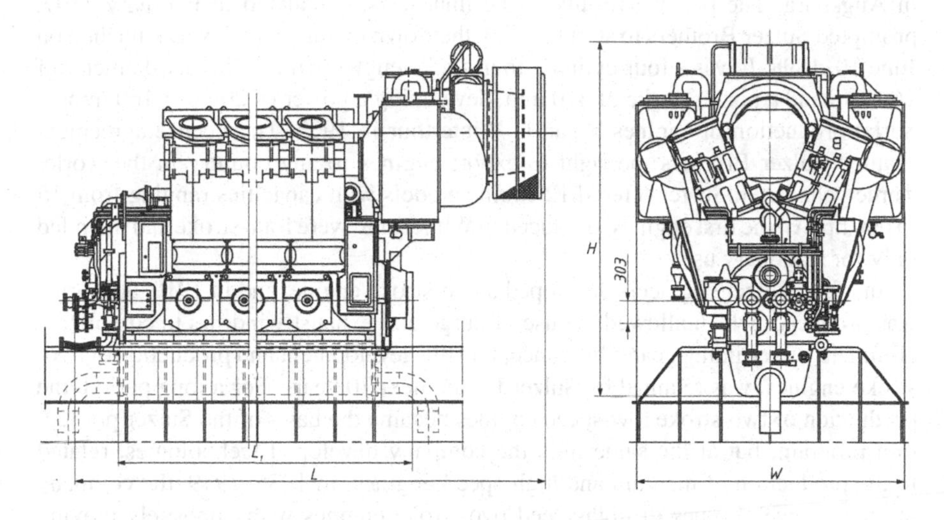

Engine version	L (mm)	L_1 (mm)	H (mm)	W (mm)	Weight (kg)
S.E.M.T. Pielstick 12PC2.6B	8520	5460	4770	3580	100,000
S.E.M.T. Pielstick 16PC2.6B	10,000	6940	4770	3580	120,000

1.17 Sulzer Brothers Ltd.

Sulzer Brothers Ltd. was founded in 1834 by brothers Johann Sulzer-Hirsel and Salomon Sulzer-Sulzer and their father Johann Jacob Sulzer-Neuffer. The company began as a iron foundry on the outskirts of the Swiss town of Winterthur. In 1839, a machine shop was established, which began to produce pumps, presses, machines for the textile industry and steam piston machines.

By the 1880s, the company had become the world leader in the manufacture of steam engines for the needs of shipbuilding. In 1879, a 21-year-old Rudolf Diesel arrived at the company in Winterthur after studying at the Munich Technical High School. On the eve of his illness, he missed the final exams and in order to better prepare for their passing the following year, one of the professors arranged for him an internship at the Sulzer Brothers plant. After receiving a patent for the engine, which he invented in 1892, R. Diesel requested to the management of the company with a proposal to purchase his patent. The head of the company, Johann Jakob Sulzer-Imhof, the son of one of the company founders, visited the Diesel laboratory in Augsburg several times. As a result of these visits, on May 16, 1893, an agreement was signed between Sulzer Brothers and R. Diesel for the exclusive use of its patents in Switzerland with a further opportunity to start production of engines. The agreement gave Sulzer Brothers the right to receive all materials, related to engine research in Augsburg. The positive results of the final tests, conducted in February 1897, prompted Sulzer Brothers to start building their own engine, which was launched on June 10, 1898. It was a four-cylinder four-stroke engine with a cylinder diameter of 260 mm and a piston stroke of 410 mm, developing a power of 20 hp at 160 min^{-1}. Serial production of engines began in Winterthur in 1903. The license agreement granted Sulzer Brothers the right to export engines to any country in the world. Three years later, Sulzer offered 12 engine models with capacities ranging from 15 to 600 hp. All the first engines, produced in Winterthur, were four-stroke and intended only for stationary use.

In 1905, Sulzer engineers developed a two-stroke engine, and in 1910, an engine was produced, which allowed the use of purge and exhaust windows to completely eliminate valve-timing gear. This concept, in the development and production of two-stroke engines, was retained by Sulzer for the next 70 years. For a long period, the production of two-stroke low-speed engines became the basis of the Sulzer production program, but at the same time the company developed technologies, related to the production of medium and high-speed engines. In 1935–1949, the company produced several types of high-speed two-stroke engines with oppositely moving pistons for merchant and military ships, submarines and stationary energy.

In the early 1960s, the medium-speed two-stroke engines series ZH40 with a cylinder diameter of 400 mm were developed for use in multi-machine power plants of icebreakers, ferries and cruise ships. In 1972, on the basis of this engine, the four-stroke version of the Z40 was created (Fig. 1.46), which was completely redone and replaced in 1982 with the series ZA40 engine. In particular, the piston stroke has been increased from 480 to 560 mm.

The ZA40S engine (Fig. 1.47) was launched in 1986. More powerful engine ZA50S, was released in 1996. Having a cylinder diameter of 500 mm and a piston stroke of 660 mm, it develops a power of 980 kW/cyl. In 1988, the S20 series was launched; it became widely used on ships as auxiliary engines.

Currently, all of Sulzer's production assets have been transferred to Wartsila NSD, which has stopped producing the above diesel models, but a large number of these engines are still in operation.

1.17.1 The Sulzer Series Z40 Engine

The Sulzer series Z40 engine is a four-stroke, medium-speed engine with an in-line (6, 8 cylinder) (Fig. 1.46) and V-shaped (10, 12, 14, 16 and 18 cylinder) arrangement of cylinders. Engines of this series cover the power range from 3200 to 9900 kW at rotational speeds of 530 and 560 min^{-1}. The crankcase of in-line engines had a welded or monolithic structure, and the V-shaped engine was made in the form of a solid casting. The main design feature of the Z-type engine is the use of rotating pistons, the design of which Sulzer patented in 1937. The ratchet mechanism transforms the rocking of the connecting rod into a smooth rotation of the piston. With the piston,

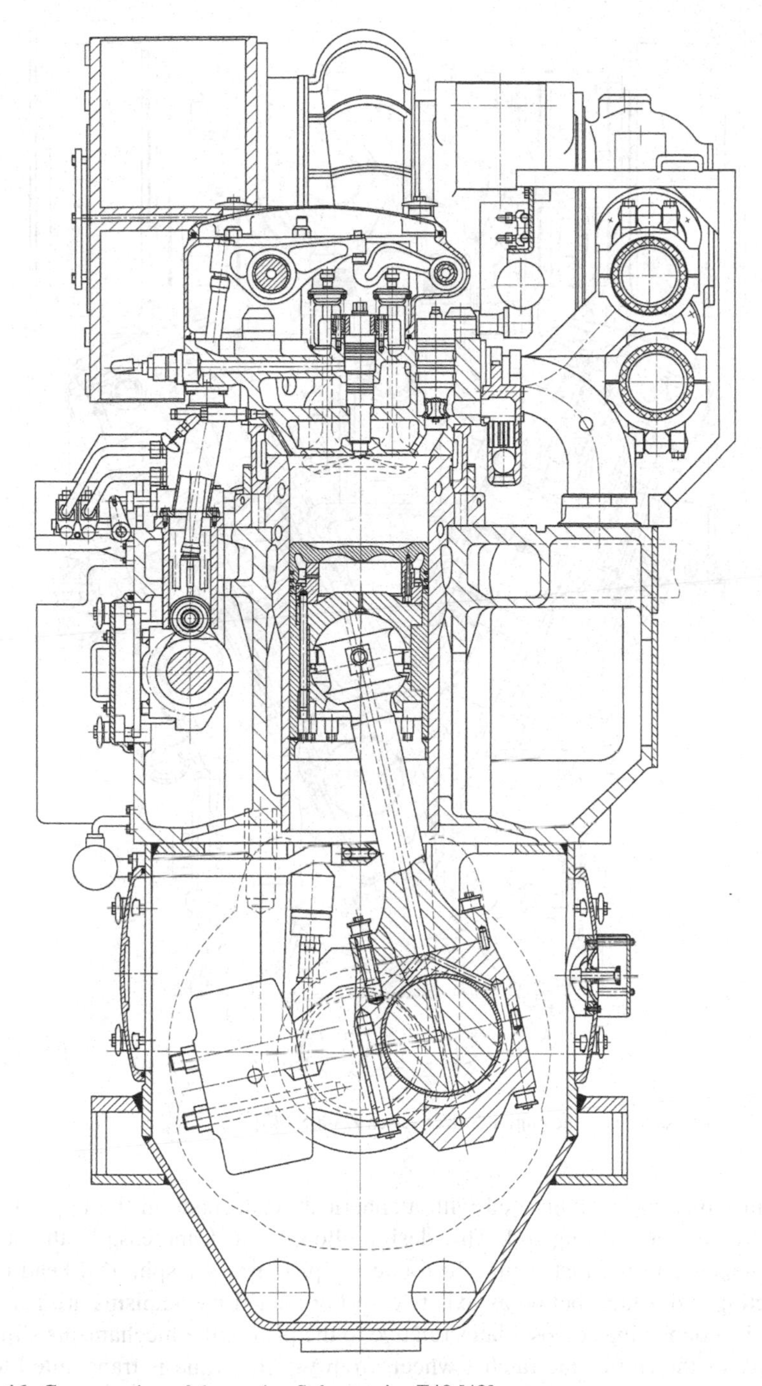

Fig. 1.46 Cross-section of the engine Sulzer series Z40 [42]

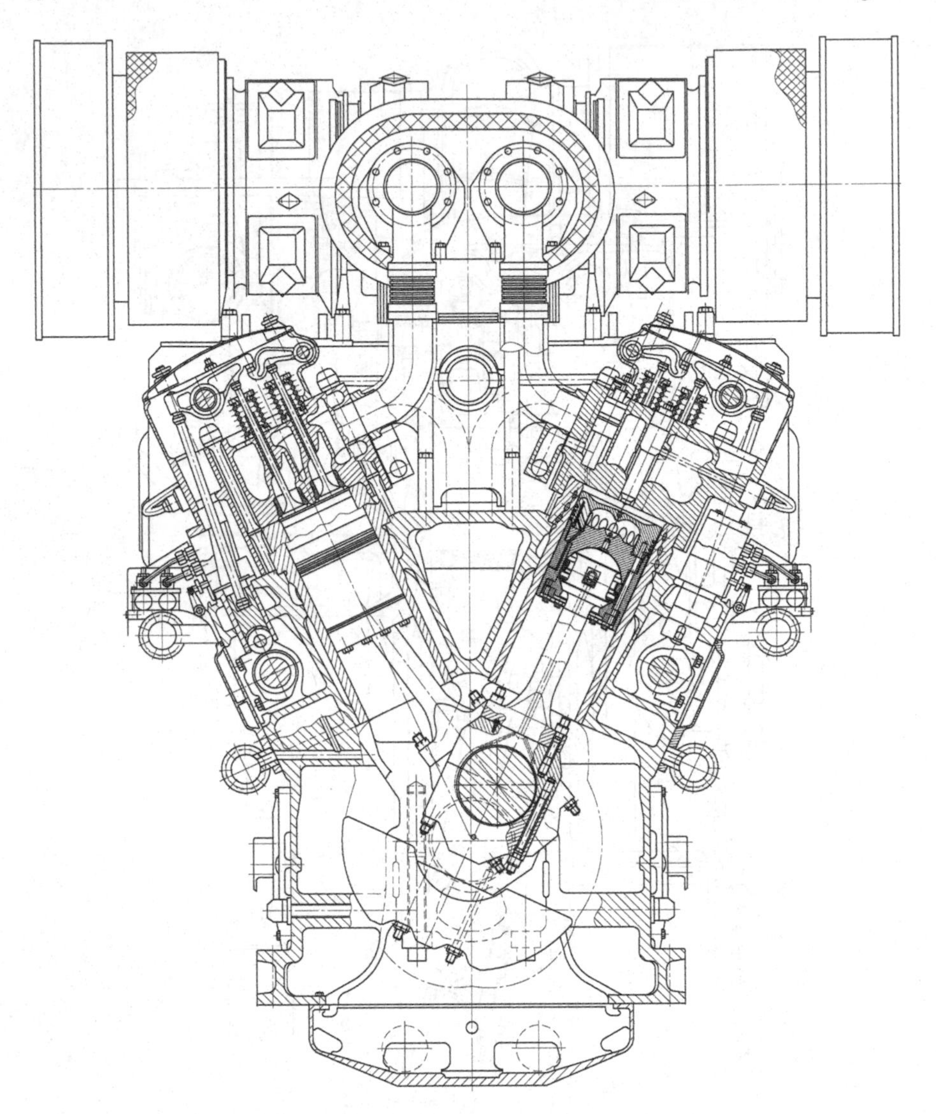

Fig. 1.47 Cross-section of the engine Sulzer series ZA40S [43]

the connecting rod is connected with a spherical head, made in the upper part of the rod of the connecting rod. This design allows a 40% increase in the surface area compared to the piston pin. To rotate the piston in the spherical head of the connecting rod a little below its axis two spring-loaded mechanisms are installed. When the connecting rod oscillates relative to the piston, the mechanisms slip over the teeth of the rim of the ratchet wheel, from which torque is transmitted to the piston by an annular spring. The elastic connection maintains the forces, required to rotate the piston at a constant level. During 67 oscillations of the connecting rod, the piston makes a complete revolution about its axis.

Main technical parameters of engines Sulzer series Z40

Parameter	Value
Number and cylinders arrangement	6, 8 in-line, 10, 12, 14, 16 and 18 V-shaped
Cylinder bore (mm)	400
Piston stroke (mm)	480
Cylinder capacity (dm^3)	60.3
Rotation speed (min^{-1})	530, 560
Cylinder power at 530/560 min^{-1} (kW)	533/550
Air charging pressure (MPa)	0.18
Compression pressure (MPa)	8.6
Start-to-open injector pressure (MPa)	32.5
Maximum cycle pressure (MPa)	13.5
Exhaust gas temperature at inlet of turbocharger (°C)	550–615
Exhaust gas temperature at outlet of turbocharger (°C)	385–425
Mean effective pressure (MPa)	1.83
Brake specific fuel oil consumption (g/kWh)	209.0
Mean piston speed at 530/560 min^{-1} (m/s)	8.48/8.96

Dimensions and weight engines Sulzer series Z40

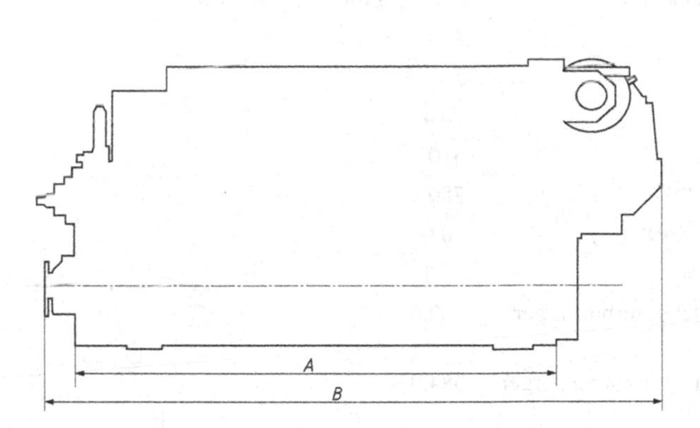

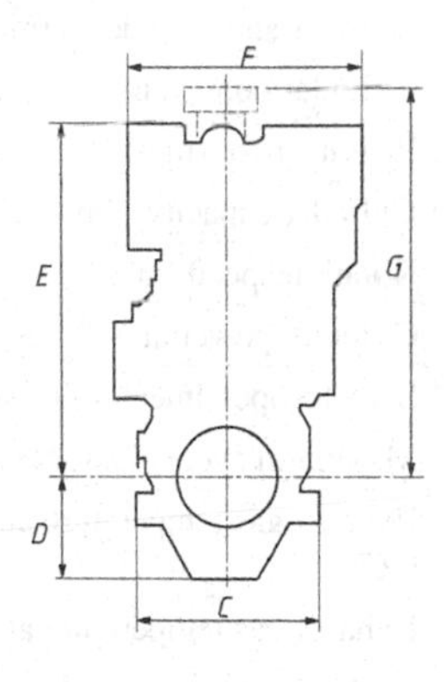

Engine version	A (mm)	B (mm)	C (mm)	D (mm)	E (mm)	F (mm)	G (mm)	Weight (kg)
6ZL40	5010	6570	1540	820	2910	2015	3200	50,000
8ZL40	6110	7930	1540	820	2910	2070	3200	65,000

1.17.2 The Sulzer Series ZA40S Engine

The Sulzer series ZA40S engine is a four-stroke, medium-speed engine with in-line (6, 8, 9 cylinders) and V-shaped (12, 14, 16 and 18 cylinders) arrangement of cylinders (Fig. 1.47). The basic part of the engine is a massive engine basic block with outboard main bearings, which are attached to the rack and frame with anchor bolts and studs. Cylinder sleeves are fixed in the upper part of the block. The bushings and cylinder covers are cooled with water that circulates through the internal channels, washing the exhaust valve seats. The exhaust valves are made of heat resistant nickel alloy (Nimonic) and are equipped with a valve turning mechanism (Rotocap). The engine is equipped with rotating pistons, which are continuously rotated under the action of the ratchet mechanism, with each rocking of the connecting rod. To accomplish this rotation, the upper head of the connecting rod is equipped with a spherical hinge. Rotation provides leveling of the temperature field in the piston crown, and also improves the lubrication of the piston rings. The integral piston is cooled with oil through channels in the connecting rod. Three compression rings are installed in the steel head of the piston and one oil scraper in the lower part of the cast-iron skirt. Thin-walled liners are used as crankshaft bearings. Plunger type fuel pumps have fuel angle adjustment. A gas turbine supercharging system with a pulse converter is a combination of a pulse system and a constant pressure system.

Main technical parameters of engines series ZA40S

Parameter	Value
Number and cylinders arrangement	6, 8, 9 in-line; 12, 14, 16 and 18 V-shaped
Cylinder bore (mm)	400
Piston stroke (mm)	560
Cylinder capacity (dm^3)	70.4
Rotation speed (min^{-1})	510
Cylinder power at 510 min^{-1} (kW)	750
Start-to-open injector pressure (MPa)	40.0
Maximum cycle pressure (MPa)	15.5
Exhaust gas temperature at inlet of turbocharger (°C)	560.0
Exhaust gas temperature at outlet of turbocharger (°C)	384.0
Mean effective pressure (MPa)	2.41
Brake specific fuel oil consumption (g/kWh)	181.0
Mean piston speed at 510 min^{-1} (m/s)	9.52

Dimensions and weight engines Sulzer series ZA40S

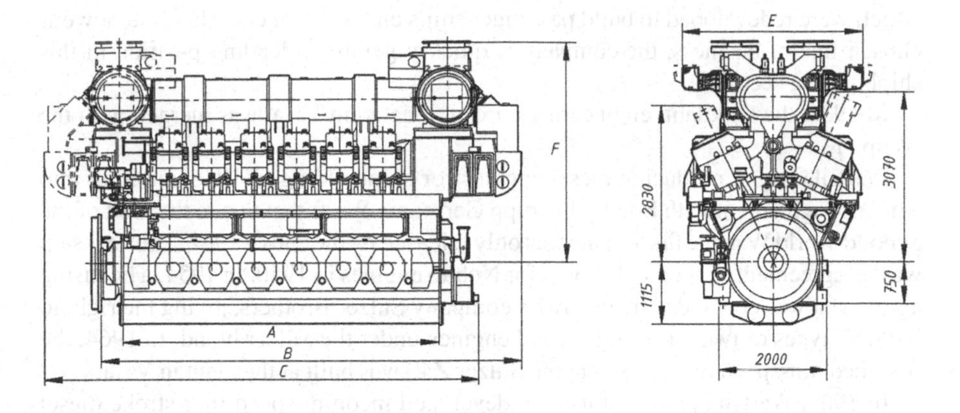

Engine version	A (mm)	B (mm)	C (mm)	E (mm)	F (mm)	Weight (kg)
Sulzer 12 ZA40S	5740	7650	7960	3464	4008	102,000
Sulzer 14 ZA40S	6520	8605	8915	4190	4152	119,000
Sulzer 16 ZA40S	7300	9385	9695	4190	4152	132,000
Sulzer 18 ZA40S	8080	10,165	10,475	4190	4152	145,000

1.18 Wärtsilä NSD

Wärtsilä NSD is a Finnish corporation that produces and maintains power and other equipment for the ship and stationary power industry. The main products of the Wärtsilä are small and medium-speed internal combustion engines, used on ships of all types and especially on cruise ships and ferries.

At the end of 2017, the share of Wärtsilä in the global market in the sector of ship main medium-speed engines was 47%, and auxiliary was 10%. The share of Wärtsilä in the market of gas, gas diesel and diesel power plants was 19%. The company's headquarters is located in the city Helsinki.

Wärtsilä was founded in 1834 as a sawmill in the county of Karelia in the eastern part of Finland at the time when Finland was an autonomous republic in the Russian Empire. The company's founder was a local entrepreneur, Nils Ludwig Arppe, who is considered one of the first industrialists in Finland.

In 1851, metallurgical production was added to the sawmill and the company received the name Wärtsilä Ab. The name Wärtsilä is come from the name of the locality, where enterprises were located. A lots of rivers were there with rapids, where water wheels were installed, which served as a source of energy for industrial activities.

Until the 1930s, the company's activities were not related to the maritime industry, but since this period, Wärtsilä has acquired two shipyards in Helsinki and Turku, which were redeveloped to build passenger ships and icebreakers. Thanks to a well-chosen line of business, the company is quickly gaining a leading position in this shipbuilding sector.

In 1936, the Onkilahti engineering bureau, located in Vasa, was merged with the company.

Wärtsilä began producing diesel engines for shipbuilding in 1938, entering into a license agreement with Friedrich Krupp Germania Werft, but due to the start of the Second World War, the first engine was only released by the end of 1942. The next step was an agreement to acquire licenses for Nohab engines in 1950. In 1954, a licensing agreement was concluded by the Swiss company Sulzer Brothers, giving the right to build all types of two-stroke low-speed engines under the Sulzer brand. In 1964, the first medium-speed two-stroke engine Sulzer Z40 was built at the plant in Vasa.

In 1960, Wärtsilä produced its own developed medium-speed four-stroke diesel engine Vasa 14. In 1975, the Vasa 22 engine was introduced, and in 1978, Wärtsilä introduced the Vasa 32 engine, which quickly gained a good reputation from consumers. In 1979, Wärtsilä acquired most of the Nohab Diesel company located in Trollhattan, Sweden, which, together with the Vaasa Factory, was integrated into the Wärtsilä Diesel company. The next major events were the acquisition in 1986 of the Norwegian company Wichmann and the creation of a joint venture with the company Echevarria, located in Bermeo, Spain.

In 1988, the company built an engine assembly plant in Khopoli, India, bringing production closer to the main consumers from southeast Asia. The final transformation of Wärtsilä Diesel into a multinational company was due to the acquisition of shares of SACM companies in France and Stork-Werkspoor Diesel in the Netherlands in 1989. In 1997, Wärtsilä Diesel merged with New Sulzer Diesel and Diesel Ricerche. The merged company was named Wärtsilä NSD Corporation.

The first half of the 1990s was marked by the development of a new model line of medium-speed marine engines with cylinder diameters from 200 to 640 mm. These engines are capable of operating on both light and heavy fuels, with all engines fully complying with the IMO Tier II exhaust emission limits set out in Appendix VI to MARPOL 73/78 Convention.

1.18.1 The Wärtsilä Series L20 Engine

The Wärtsilä series L20 engine is a four-stroke, supercharged, medium-speed engine with in-line (4, 6, 8, 9 cylinders) cylinder arrangement (Fig. 1.48). The engines of this series cover the power range from 740 to 1980 kW and are used as main ones for small vessels, as well as auxiliary on all types of vessels.

Compact design, made in a modular type, with a supporting engine basic block, cast from high-strength cast iron. Suspended crankshaft with removable counterweights mounted with the help of the lower caps, in which the mounted thin-walled bearings. Connecting rods with an inciling lower bearing connector made it possible to increase

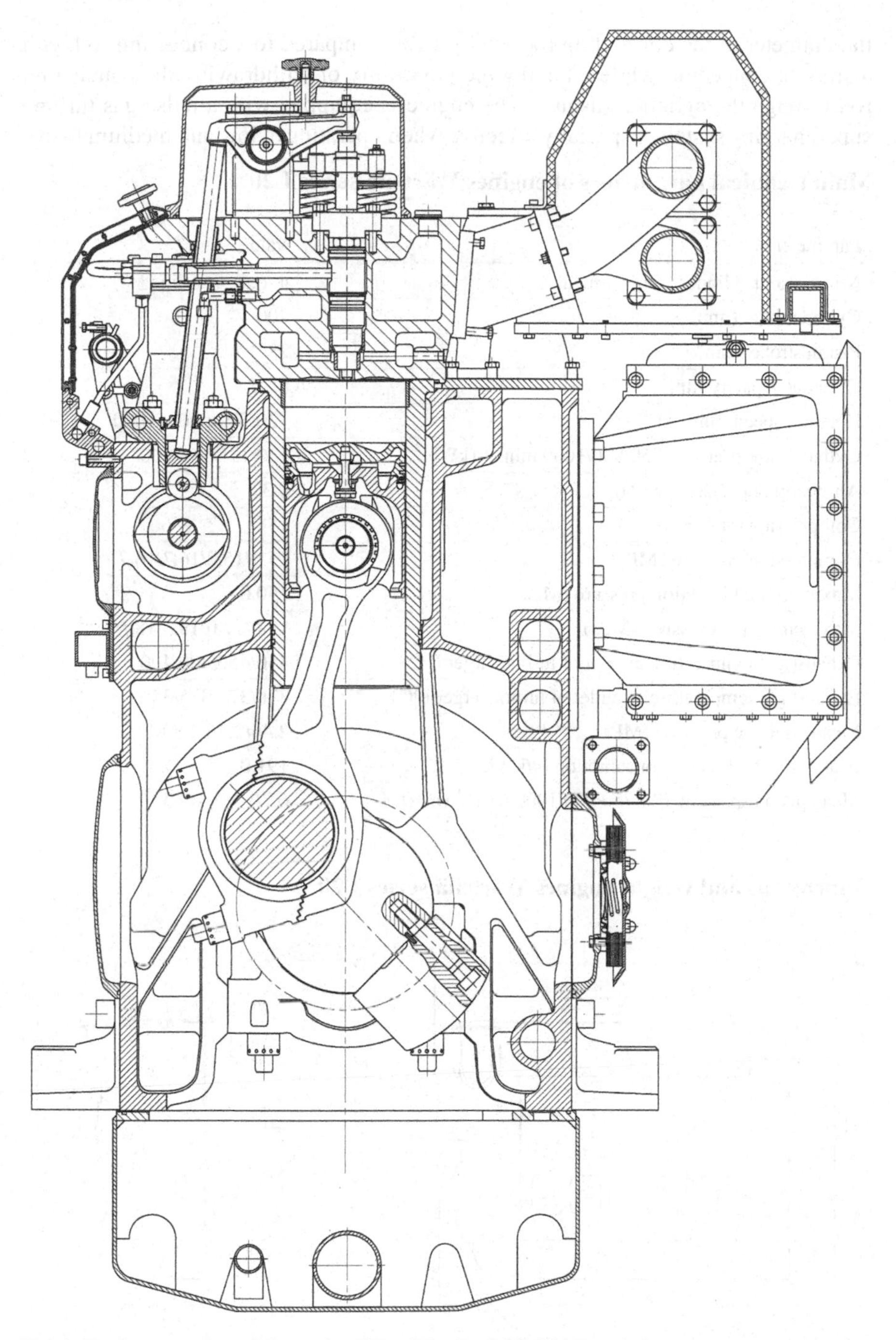

Fig. 1.48 Cross-section of the engine Wärtsilä series L20 C [44]

the diameter of the connecting rod pin by 15%, compared to a connecting rod with a straight connector, while retaining the possibility of withdrawing the connecting rod through the cylinder bushing. The engine is equipped with a pulse gas turbine supercharging system, especially effective when operating at low and medium loads.

Main technical parameters of engines Wärtsilä series L20

Parameter	Value
Number and cylinders arrangement	4, 6, 8, 9 in-line
Cylinder bore (mm)	200
Piston stroke (mm)	280
Cylinder capacity (dm^3)	8.80
Rotation speed (min^{-1})	720; 750; 900; 1000
Cylinder power at 720/750/900/1000 min^{-1} (kW)	130/135/185/200
Air charging pressure (MPa)	0.30
Compression ratio	15
Compression pressure (MPa)	15.0/15.0/16.7/16.7
Maximum fuel injection pressure (MPa)	150.0
Maximum cycle pressure (MPa)	17.0/17.0/18.5/18.5
Exhaust gas temperature at inlet of turbocharger (°C)	450/450/400/400
Exhaust gas temperature at outlet of turbocharger (°C)	370/370/335/335
Mean effective pressure (MPa)	2.46/2.46/2.80/1.73
Brake specific fuel oil consumption (g/kWh)	191.0
Mean piston speed at 720/750/900/1000 min^{-1} (m/s)	6.7/7.0/8.4/9.3

Dimensions and weight engines Wärtsilä series L20

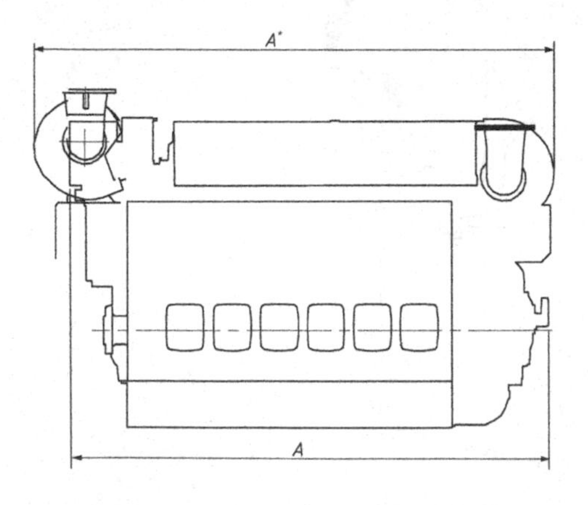

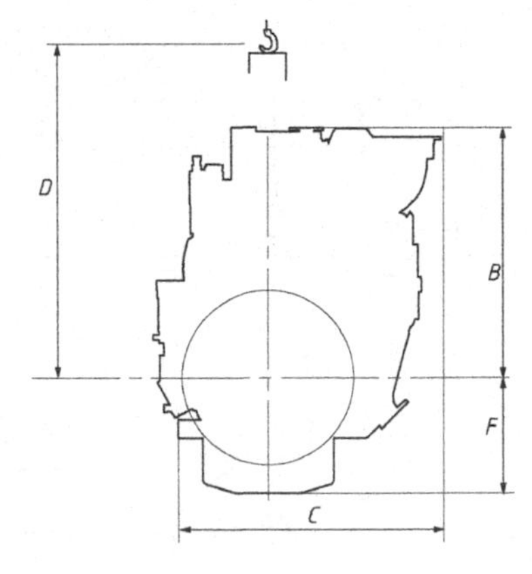

Engine version	A* (mm)	A (mm)	B* (mm)	B (mm)	C* (mm)
4L20	–	2510	–	1348	–
6L20	3254	3108	1528	1348	1580
8L20	3973	3783	1614	1465	1756
9L20	4261	4076	1614	1449	1756

Engine version	C (mm)	D (mm)	F (mm)	Weight (kg)
4L20	1483	1800	725	7200
6L20	1579	1800	624	9300
8L20	1713	1800	624	11,000
9L20	1713	1800	624	11,600

1.18.2 The Wärtsilä Series L/V32C Engine

The Wärtsilä series L/V32C engine is a four-stroke, supercharged, medium-speed diesel engine with in-line (6, 7, 8, 9 cylinders), vertical cylinders (Fig. 1.49) and V-shaped (12, 16, 18 cylinders) (Fig. 1.50) arrangement of the cylinder at an angle of 55°. The engine basic block is cast from nodular cast iron, using the latest casting technology, which allowed to integrate all channels into the casting for supplying and discharging process fluids. Composite pistons, with a steel head and cast-iron skirt, equipped with an original lubrication system for friction surfaces. Three compression rings are mounted on the piston, which results in a good sealing of the gas joint with minimal friction costs in the cylinder-piston group. Rod is of composite type with flange fastening. Lower head serves as a housing for the connecting rod bearing. The upper head of the connecting rod is made as one piece with the connecting rod. This design allows servicing elements of the cylinder-piston group without disassembling the connecting rod bearing. The thick-walled cylinder liner with the installation collar has a high rigidity, which allows it to withstand high pressures with virtually no deformation. Channels for circulation of coolant are made in the installation collar, which allows to reduce thermal stress and to avoid low-temperature corrosion of working surfaces. Cooling water is distributed around the bushing with distribution rings at the bottom end of the collar. In the upper part of the sleeve, on the side of the working surface an anti-polishing ring is installed. The function of this ring is to remove carbon deposits from the side of the piston head to prevent contact between the liner wall and the deposits in any position of the piston. The cylinder cover is attached to the crankcase with four studs, which allows to optimize the shape of the channels for the air supply to the two intake valves and exhaust gas from the two exhaust valves. The cylinder cover is made according to the multilevel scheme, which allows to provide the necessary rigidity with the minimum thermal stress of the firing bottom. The engine is equipped with a single-tube gas turbine supercharging system of the type Spex (Single Pipe EXhaust), which is similar in its characteristics to impulse systems. The fuel system of volumetric action with direct

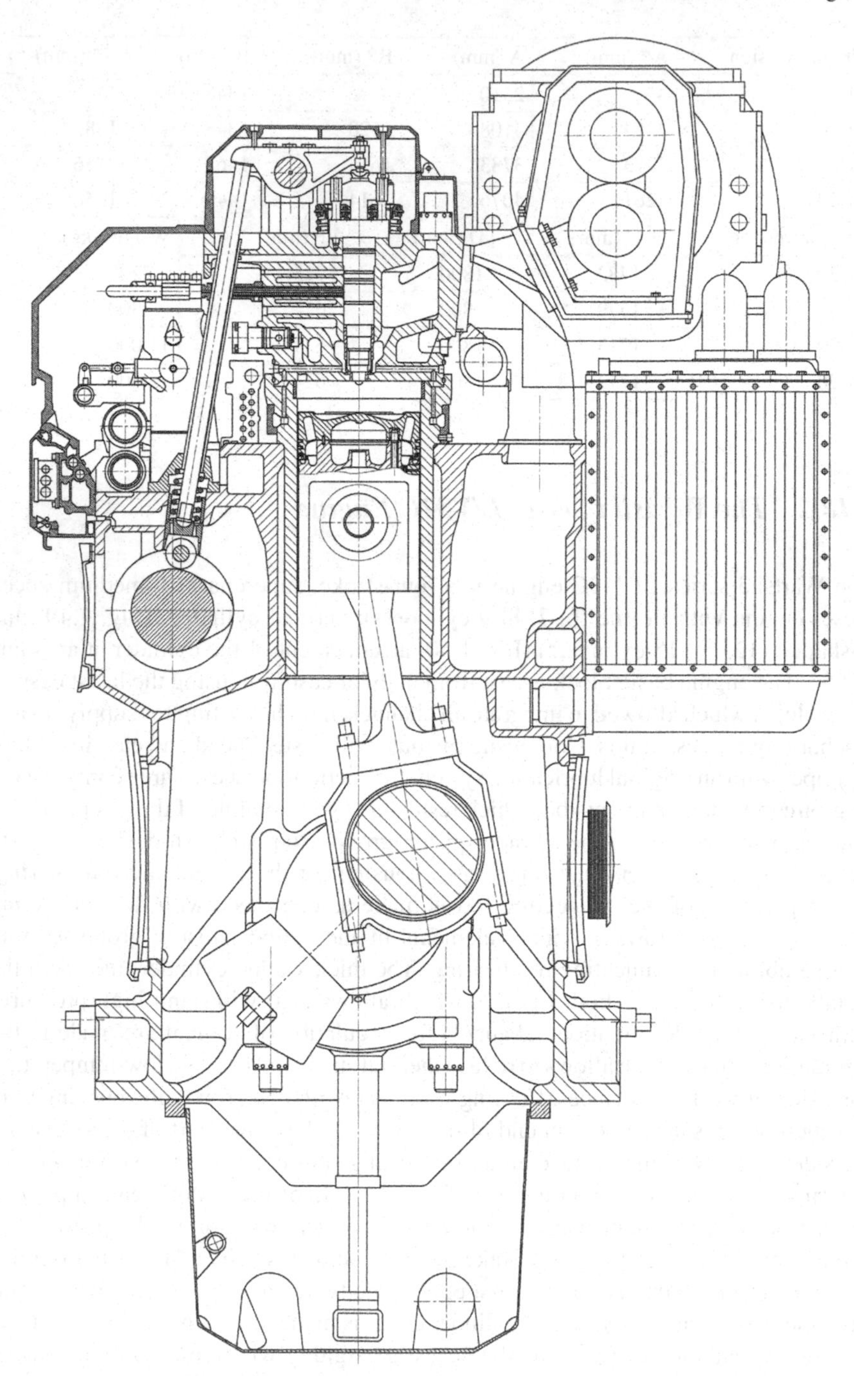

Fig. 1.49 Cross-section of the engine Wärtsilä series L32 C [45]

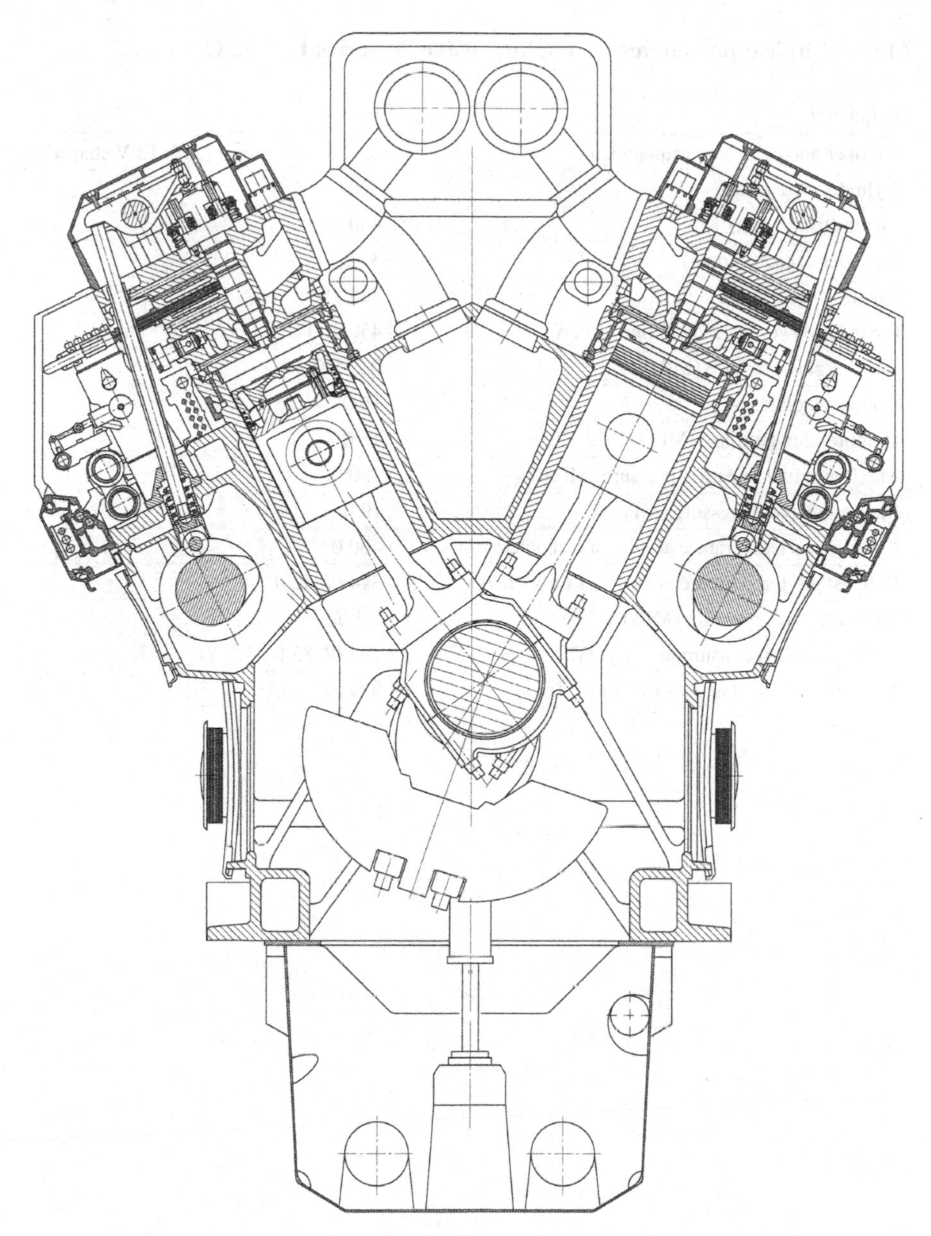

Fig. 1.50 Cross-section of the engine Wärtsilä series V32 C [45]

fuel injection into the combustion chamber of the engine from the high-pressure pump installed individually on each cylinder. Alternatively, a Common Rail type fuel injection system can be installed on the engine.

Main technical parameters of engines Wärtsilä series L/V 32 C

Parameter	Value
Number and cylinders arrangement	6, 7, 8, 9 in-line./12, 16, 18 V-shaped
Cylinder bore (mm)	320
Piston stroke (mm)	400
Cylinder capacity (dm^3)	32.2
Rotation speed (min^{-1})	720/750
Cylinder power at 720/750 min^{-1} (kW)	450/460
Air charging pressure (MPa)	0.265/0.27
Compression ratio	16.0
Compression pressure (MPa)	16.0
Maximum fuel injection pressure (MPa)	180.0
Maximum cycle pressure (MPa)	19.0
Exhaust gas temperature at inlet of turbocharger (°C)	460.0
Exhaust gas temperature at outlet of turbocharger (°C)	344.0/342.0
Mean effective pressure (MPa)	2.33/2.29
Specific fuel oil consumption (g/kWh)	182.0/183.0
Mean piston speed at 720/750 min^{-1} (m/s)	9.6/10

Dimensions and weight of engines Wärtsilä series L/V 32C

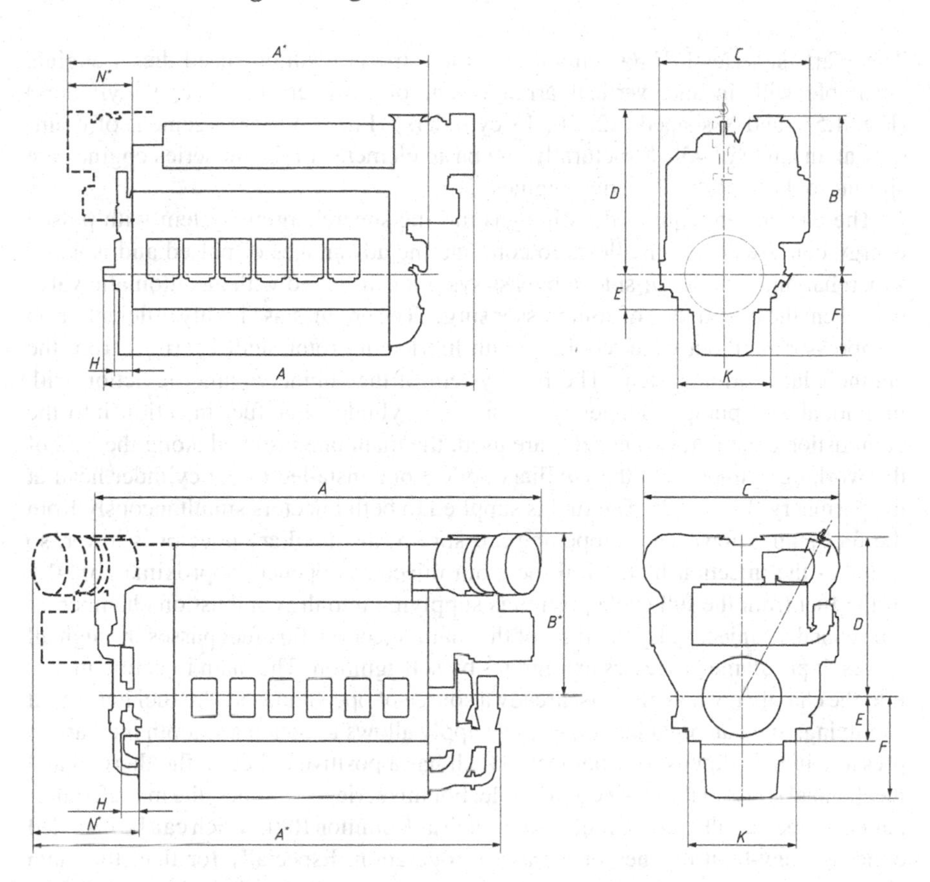

Engine version	A* (mm)	A (mm)	B* (mm)	B (mm)	C (mm)	D (mm)
6L32	5108	5267	2268	2268	2207	2345
8L32	6478	6480	2438	2418	2207	2345
9L32	6968	7086	2438	2418	2207	2345
12V32	6795	6435	2350	2390	2870	2120
16V32	–	7890	–	2523	3293	2120
18V32	–	8450	–	2523	3293	2120
Engine version	**E (mm)**	**F (mm)**	**H (mm)**	**K (mm)**	**N* (mm)**	**Weight (kg)**
6L32	500	1153	250	1350	877	35,500
8L32	500	1153	250	1350	1294	45,000
9L32	500	1153	250	1350	1294	48,500
12V32	650	1475	300	1590	1568	60,500
16V32	650	1475	300	1590	–	76,000
18V32	650	1475	300	1590	–	82,500

1.18.3 The Wärtsilä Series L/V46C Engine

The Wärtsilä series L/V46C engine is a four-stroke, medium-speed diesel engine. Available with in-line, vertical arrangement of cylinders (6, 7, 8, 9 cylinders) (Fig. 1.51) and V-shaped (12, 16, 18 cylinders) (Fig. 1.52) arrangement of cylinders at an angle of 45°. Structurally, the basic elements of the 46 series engines are similar to those of the 32 series engines.

The engines are equipped with a gas turbine supercharging system with pulsed energy converters, which allows to combine the advantages of pulsed and isobaric supercharging. On the air side, a bypass system is installed with an automatic valve to prevent the occurrence of compressor surge. The engines use highly efficient turbo compressors without water cooling, with lubricating rotor shaft bearings from the engine's lubrication system. The fuel system of the engine volumetric action with individual fuel pumps plunger type for each cylinder. For fuel injection into the combustion chamber, two nozzles are used, the main one installed along the axis of the working cylinder and the auxiliary (pilot) one installed in the cylinder head at the periphery (Fig. 1.52). The fuel is supplied to both injectors simultaneously from the fuel pump, however, the opening pressure of the auxiliary injector is lower, so it makes the injection first. Until the main injector is opened, approximately 10% of the fuel from the full cycle portion is supplying into the combustion chamber by the secondary injector. By the time of the main injection, this fuel passes through all stages of pre-flame processes and ignites by self-ignition. The main injection makes into the chamber, where there is already a source of open flame, so the fuel burns as it supplying. Such an organization of fuel supply allows to obtain a smooth increase in pressure in the combustion chamber, which has a positive effect on the thermal and mechanical stress of the engine as a whole. For this series of engines, the manufacturer has developed an alternative fuel system such as Common Rail, which can be installed either on new-built engines or already in operation. Especially for that, the main elements of the accumulatory system are maximally unified with the elements of the standard system. Regardless of the type, the pipes and the main components of the fuel systems are located in a protective insulated box, which ensures maximum safety in case of a break and protects personnel from high temperatures. Fuel leaks from pipes, injectors, valves and pumps are collected in a closed piping system, through which they are removed from the engine, allowing it to stay dry and clean.

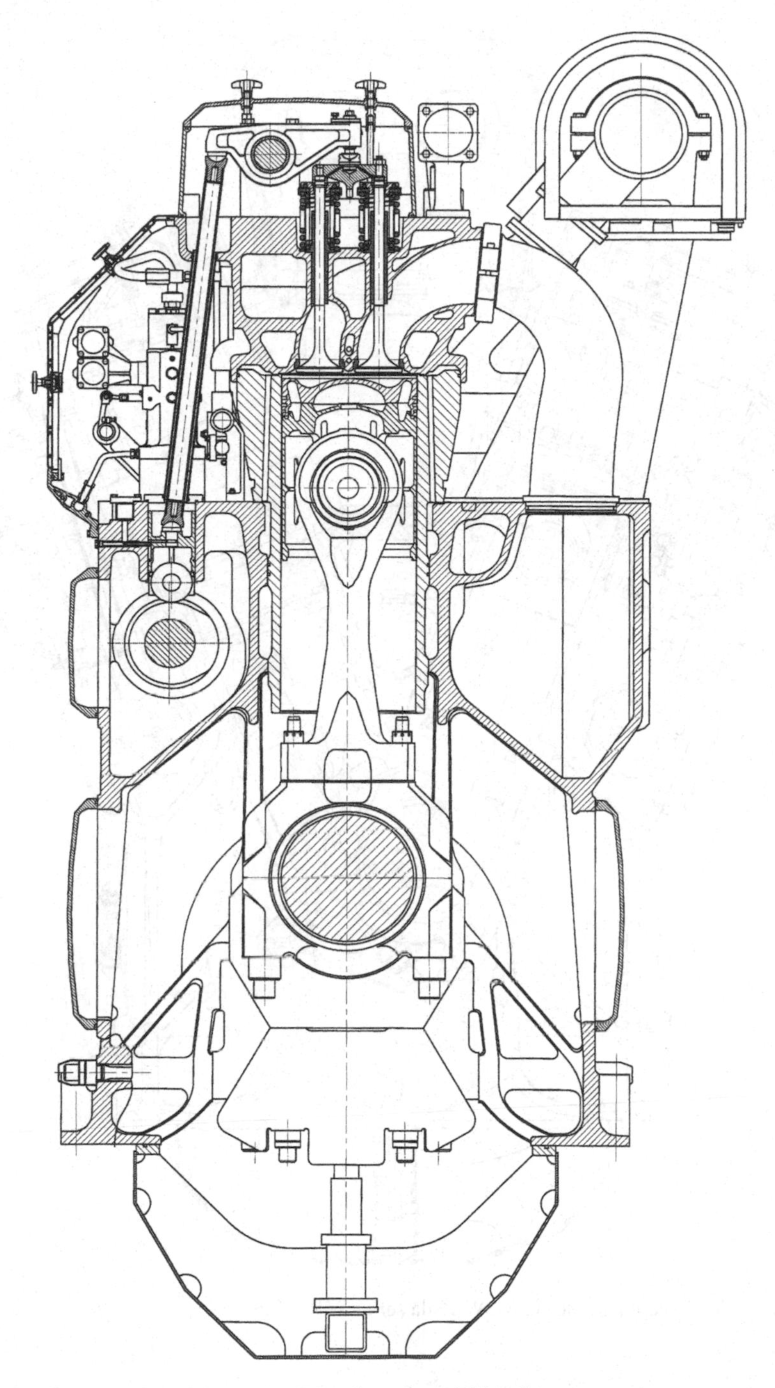

Fig. 1.51 Cross-section of the engine Wärtsilä series L46C [46]

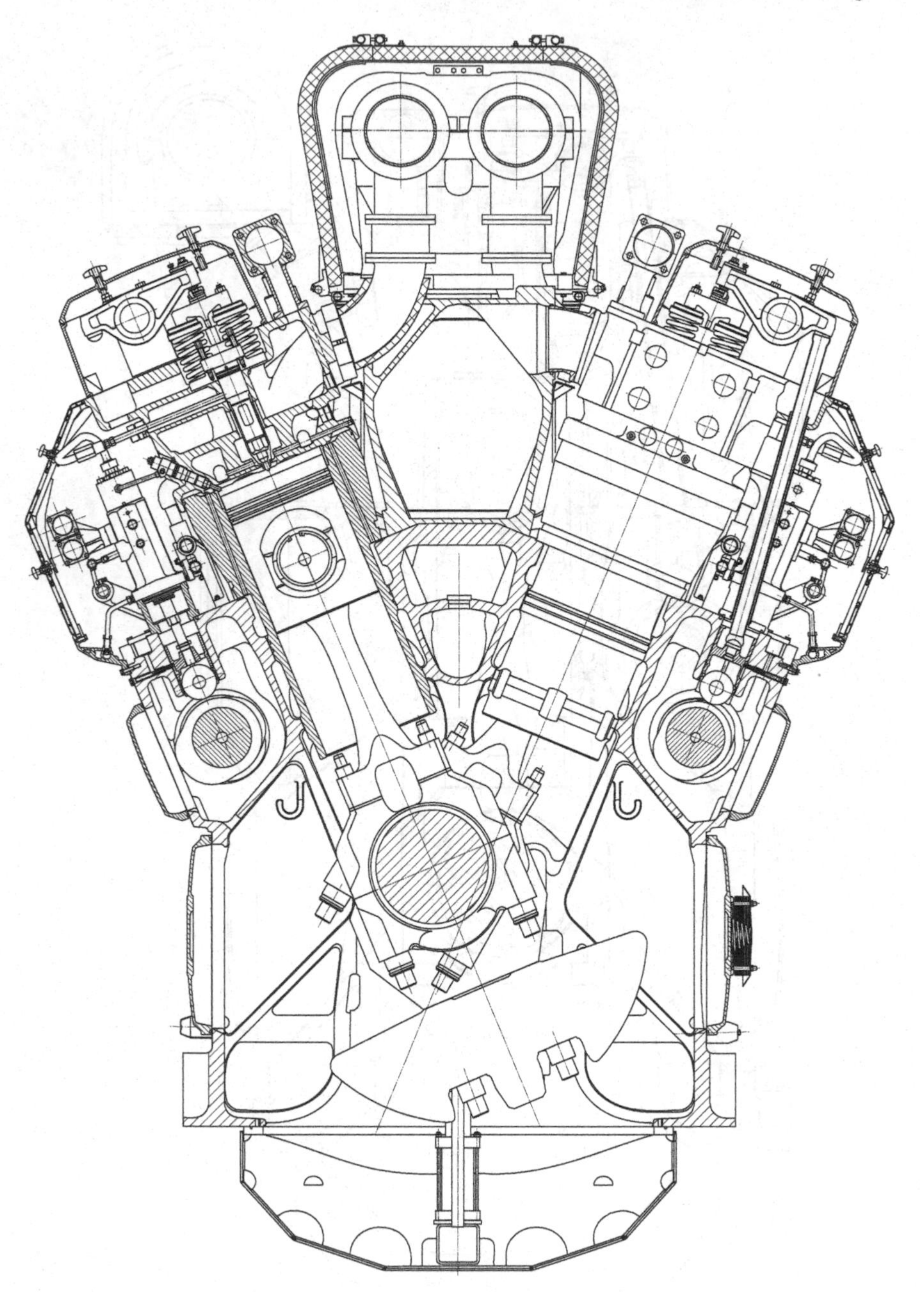

Fig. 1.52 Cross-section of the engine Wärtsilä series W46C [46]

Main technical parameters of engines Wärtsilä series L/V46C

Parameter	Value
Number and cylinders arrangement	6, 7, 8, 9 in-line./12, 16, 18 V-shaped
Cylinder bore (mm)	460
Piston stroke (mm)	580
Cylinder capacity (dm^3)	96.4
Rotation speed (min^{-1})	500, 514
Cylinder power at 500/514 min^{-1} (kW)	1050/1150
Compression ratio	16
Air charging pressure (MPa)	0.35
Maximum cycle pressure (MPa)	21.0
Specific air consumption (кg/kWh)	6.11/6.23
Specific exhaust gas flow (кg/kWh)	6.29/6.40
Exhaust gas temperature at inlet of turbocharger (°C)	455.0
Exhaust gas temperature at outlet of turbocharger (°C)	380/390
Mean effective pressure (MPa)	2.61/2.88
Brake specific fuel oil consumption (g/kWh)	174/182
Mean piston speed at 500/514 min^{-1} (m/s)	9.7/9.8

Dimensions and weight of engines Wärtsilä series L/V 46C

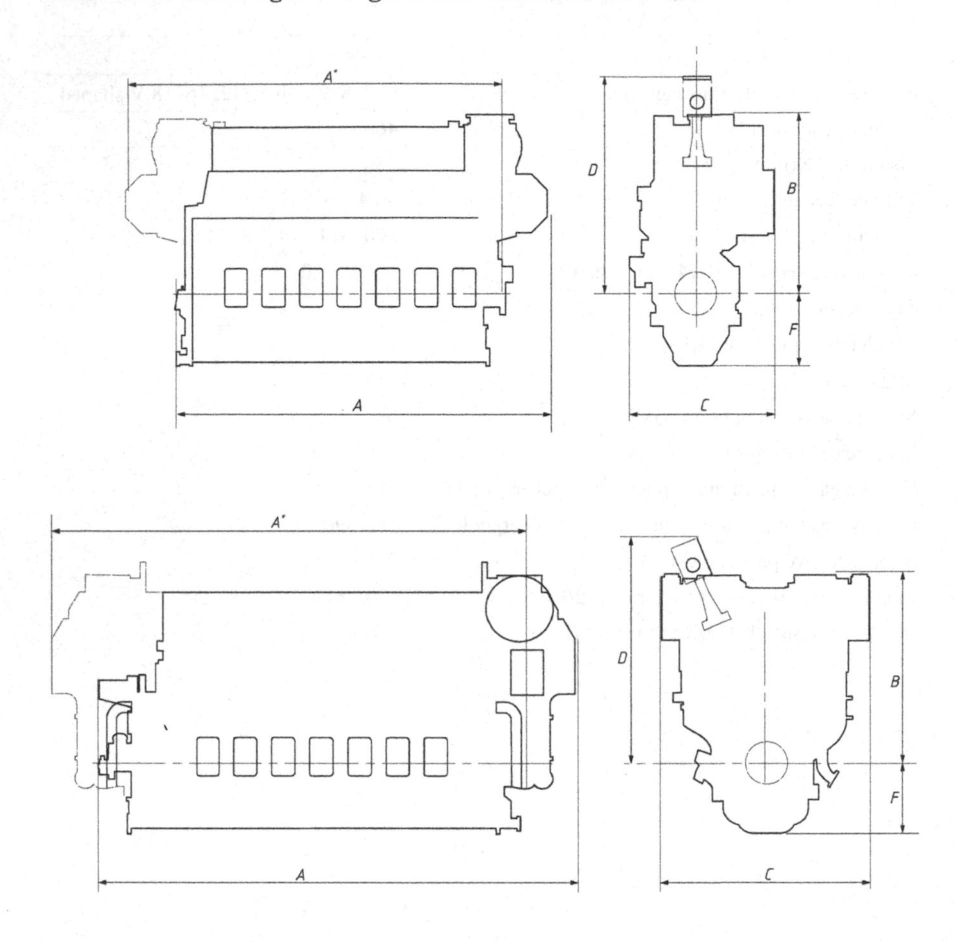

Engine version	A* (mm)	A (mm)	B (mm)	C (mm)	D (mm)	F (mm)	Weight (kg)
6L46	7580	8290	3340	2880	3820	1460	95,000
8L46	9490	10,005	3260/ 3600[a]	3180	3820	1460	120,000
9L46	10,310	10,830	3600	3270	3820	1460	137,000
12V46	10,260	10,210	3660	3810/ 4530[b]	3600	1500	169,000
16V46	12,345/ 12,460[a]	12,480/ 12,590[a]	3660/ 3990[a]	4530/ 5350[a]	3600	1500	214,000

*The turbocharger is mounted on the flywheel side
[a]Depending on the design of the exhaust pipe
[b]Depending on the type of turbocharger and the design of the exhaust pipe

1.18.4 Wärtsilä 64C Engines

Wärtsilä 64C engines are the largest four-stroke mid-speed engines in the world. Available in-line (6, 7, 8, 9 cylinders (Fig. 1.53)) and V-shaped arrangement of cylinders (12, 16, 18 cylinders). The engines of this series cover the power range

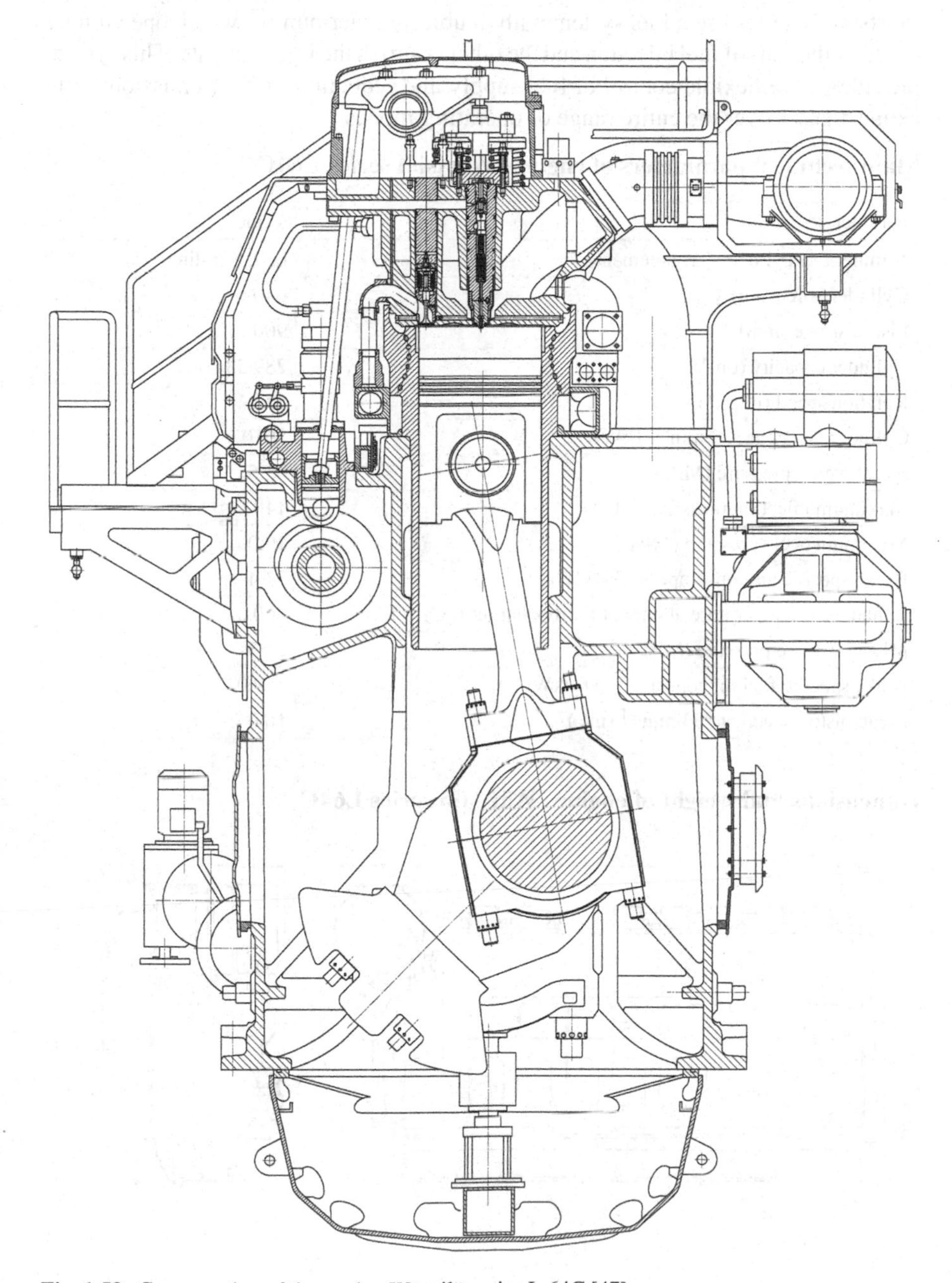

Fig. 1.53 Cross-section of the engine Wärtsilä series L 64C [47]

from 12,000 to 23,280 kW and are used as the main ones for ships, which have increased requirements for the weight and dimensions of the power plant. For the first time, the engines of this series appeared on the market in September 1996, and were first transferred to the customer in the autumn of 1997. This engine was the first four-stroke engine whose effective efficiency exceeded 50%. Engine production is carried out at a factory in the city of Triest in northern Italy. The engines of the Wärtsilä 64 series use a fuel system with double-plunger pumps, where one plunger controls the start of fuel injection and the other controls the injection rate. This system provides more flexible control of fuel supply and reduction of NO_x emissions with exhaust gases over the entire range of operating modes.

Main technical parameters of engines Wärtsilä series L64C

Parameter	Value
Number and cylinders arrangement	6, 7, 8 in-line
Cylinder bore (mm)	640
Piston stroke (mm)	900
Cylinder capacity (dm^3)	289.53
Rotation speed (min^{-1})	333
Cylinder power at 333 min^{-1} (kW)	2010
Air charging pressure (MPa)	0.355
Maximum injection pressure (MPa)	140
Maximum cycle pressure (MPa)	19.0
Brake specific air consumption (kg/kWh)	5.34
Exhaust gas temperature at outlet of turbocharger (°C)	360.0
Mean effective pressure (MPa)	2.55
Brake specific fuel oil consumption (g/kWh)	169.0
Mean piston speed at 333 min^{-1} (m/s)	10.0

Dimensions and weight of engines Wärtsilä series L64C

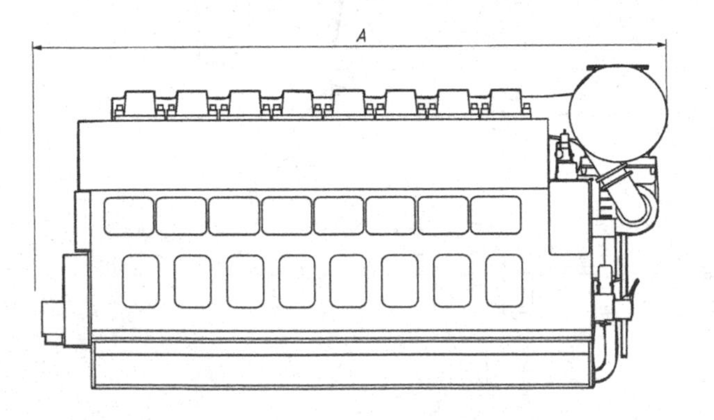

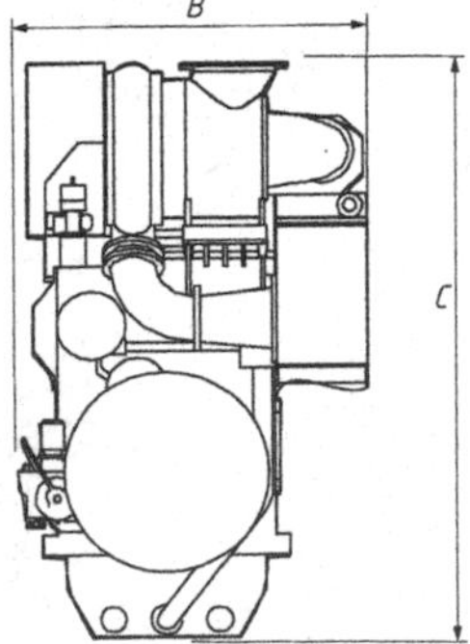

Engine version	A (mm)	B (mm)	C (mm)	Weight (kg)
6L64C	10,250	4065	6031	232,000
7L64C	11,300	4165	6269	264,000
8L64C	12,350	4165	6269	292,000
9L64C	13,670	4165	6637	325,000

1.19 Yanmar Co., Ltd.

Yanmar Co., Ltd. is a large Japanese manufacturer of diesel engines, which are widely used around the world in many industries, including ships, building equipment, agricultural machinery and stationary energy. The company was founded in March 22, 1912 in the Osaka by Japanese entrepreneur Magokichi Yamaoka (Magokichi Yamaoka) who registered it under the name Yamaoka Hatsudoki Kosakusho. Yanmar as a brand name appeared only 9 years later in 1921. This year, the company launched oil engines for the agricultural sector. In the summer of 1921, an abnormally large number of dragonflies was observed, which, according to Japanese omens, was a good sign. Therefore, it was decided to name the new engine as Yanma, after the name of one of the species of dragonflies. In English transliteration, the name was transformed into Yanmar.

In 1933, the company's engineers developed the most compact diesel engine at that time, which was mass-produced and actively used in the manufacture of various products. From the second half of the 1930s to the first half of the 1940s, two plants for the production of diesel engines under the Yanmar trademark were opened in Japan at once, one in the city of Amagasaki, and the second in the city of Nagahama. In 1947, a new activity was opened the production of small diesel engines for fishing vessels. In 1952, the company name was changed from Yamaoka Hatsudoki Kosakusho to Yanmar Diesel Engine. Since the late 1950s, the company began to open its own representative offices and service centers around the world.

In December 1952, a factory in Kanzaki began production of a four-cylinder engine with preignition chamber of type 4MS (L) with a cylinder diameter of 200 mm and a power of 120 hp, which became the first medium-speed engine, produced by the company. Then the construction of a new machine-building plant in Kanzaki followed. In November 1953, the company put up for sale a 6 MSL-T "Yanmar Super Diesel" type engine with a capacity of 270–300 hp, on which, it first used gas turbine boost with intermediate cooling of charge air, as well as a four-valve timing mechanism. The growth in demand for light, powerful and reliable engines in the fishing fleet was due to the fact, that in February 1963 the company launched the 300 hp 6M-T engine to the market. The first Yanmar engine with direct injection of fuel into the combustion chamber was the 400KP 6KE-HT engine, introduced in October 1975. In this engine, a dual-circuit cooling system was also used for the first time. After the oil crisis of the 1970s, the demand for medium-speed engines fell sharply, to which the company responded by introducing compact high-speed diesel engines series 6HA with direct fuel injection to the combustion chamber with a capacity of 200 hp to the market in May 1977. In subsequent years, the range of compact engines was supplemented with the S185 series, with a capacity of 550–600 hp and T260 with a power of 1400–1500 hp.

In June 1978, Yanmar began production of medium-speed engines 6ZL-UT series (1300 hp) capable of operating on heavy fuels, and in March 1985 supplemented them with the T260L-EX series with the 1500 hp. In 1987, the Yanmar Diesel Engine developed the world's first diesel engine for outboard engines.

In 2000, a new research institute was opened in the city of Maibara, and in 2002, by the ninetieth anniversary of the company, its name was changed to the current one "Yanmar Co.", Ltd. In 2009, the engine version 6EY18, launched in 2005, was the first in Japan which received a certificate from IMO for compliance with the Tier II standard. In the same year, the 6N18AEV engine, equipped with an electronic control system was launched. In 2010, the production of 6EY22 engines was launched, and in 2012 the series 6EY17. Engine version 6EYG26L was launched in 2014, and the following year, the Yanmar range of mid-speed engines was supplemented with the engine version 6EY33.

1.19.1 The Yanmar Series 6N21 Engine

The Yanmar series 6N21 engine is a four-stroke, six-cylinder, medium-speed, in-line engine (Fig. 1.53). The engine basic block is a box-shaped solid cast iron construction, where the channels for process fluids, the air receiver, the crankshaft and camshaft beds are integrated. The bottom of the engine basic block closes with an oil pan, that serves to collect and store lubricating oil. Cylinder sleeves cast iron. An anti-polishing ring is installed in the top of each sleeve. The crankshaft is made by stamping from special carbon steel with subsequent induction hardening of the working surfaces. To ensure dynamic balance, 12 counterweights are installed on the shaft. The piston is a thin-walled one-piece construction of ductile iron. The upper and second annular grooves are laser hardened. To cool the bottom of the piston, oil is supplied to the "shaker cavity" by a nozzle, located on the underside of the side of the cylinder. Three piston and one oil scraper ring, mounted on the piston, have a chrome plated work surface. Plunger-type fuel pumps, individual for each cylinder. The engine is equipped with a pulse system of pressurization with intermediate air cooling, which is supplying to the working cylinders (Fig. 1.54).

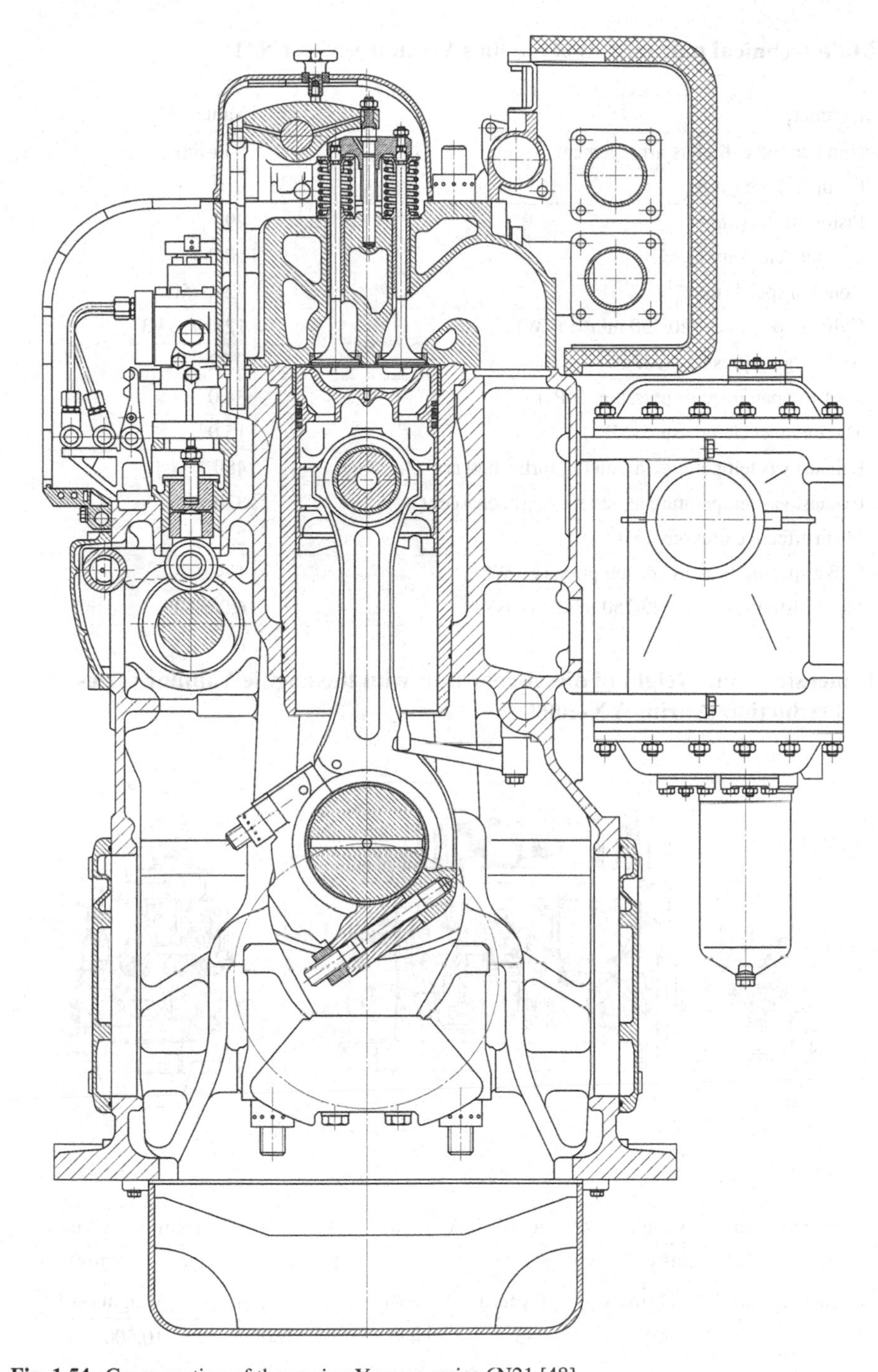

Fig. 1.54 Cross-section of the engine Yanmar series 6N21 [48]

Main technical parameters of engines Yanmar series 6N21

Parameter	Value
Number and cylinders arrangement	6 in-line
Cylinder bore (mm)	210
Piston stroke (mm)	290
Cylinder capacity (dm^3)	10.04
Rotation speed (min^{-1})	720, 750
Cylinder power at 720/750 min^{-1} (kW)	122.6/133.3
Air charging pressure (MPa)	0.225
Start-to-open injector pressure (MPa)	34.0
Maximum cycle pressure (MPa)	15.0
Exhaust gas temperature at inlet of turbocharger (°C)	480
Exhaust gas temperature at outlet of turbocharger (°C)	330
Mean effective pressure (MPa)	2.213
Brake specific fuel oil consumption (g/kWh)	194.0
Mean piston speed at 720/750 min^{-1} (m/s)	6.96/7.25

Dimensions and weight of diesel-gear unit with the engine Yanmar series 6N21 and reduction gearing YX-1000

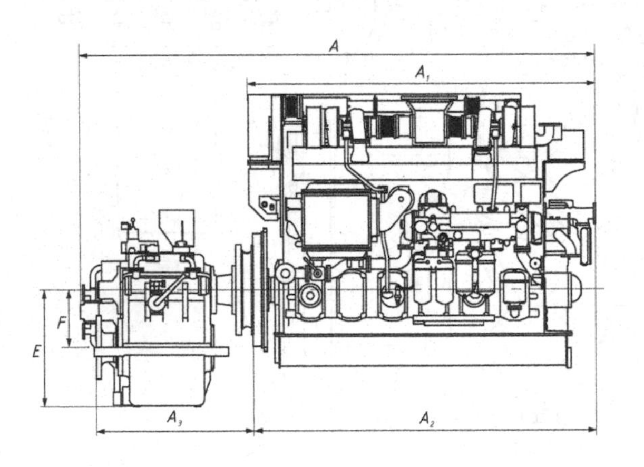

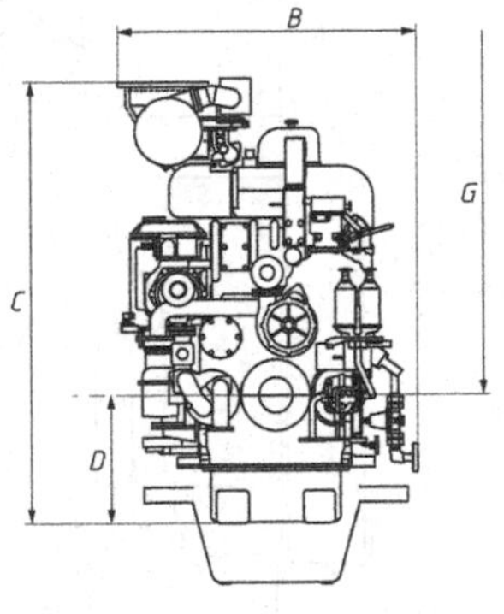

Engine version	A (mm)	A_1 (mm)	A_2 (mm)	A_3 (mm)	B (mm)	C (mm)
6N21	4053	2776	2733	1199	1420	2081

Engine version	D (mm)	E (mm)	F (mm)	G (mm)	Weight (kg)
6N21	601	885	435	1802	10,500

1.19.2 The Yanmar Series EY18L Engine

The Yanmar series EY18L engine is a four-stroke, medium-speed six-cylinder engine with in-line cylinder arrangement (Fig. 1.55). It is used as part of a diesel generator sets with a power of 360–800 kW with rotational speeds of 720, 750, 900 and 1000 min^{-1}. The cast engine basic block is made of nodular cast iron. Crankshaft bed with outboard main bearings, which are fastened to the crankcase with studs. Cylinder sleeves are installed in the crankcase through spacers. Cylinder covers for each individual cylinder have a multi-tiered design. The inlet channels in the cylinder heads are profiled to impart a rotational movement to the charge, entering the working cylinder. A fuel nozzle is installed along the cylinder axis in the lid with a two-level arrangement of injectors, openings of different diameters and a minimized volume of the subnozzle chamber. The exhaust valve seats are water cooled. Crankshaft is made as solid forged, with counterweights. The piston is monolithic, made of high-strength cast iron. The connecting rod with an inclining lower bearing connector is made of forged carbon steel.

Main technical parameters of engines Yanmar 6EY18ALW

Parameter	Value
Number and cylinders arrangement	6 in-line
Cylinder bore (mm)	180
Piston strokc (mm)	280
Cylinder capacity (dm^3)	7.125
Rotation speed (min^{-1})	1000
Cylinder power at 1000 min^{-1} (kW)	133.3
Air charging pressure (MPa)	0.30
Air consumption (at 25 °C) (m^3/h)	4570.0
Exhaust gas consumption (at 0 °C) (m^3/h)	4060.0
Exhaust gas temperature at inlet of turbocharger (°C)	475.0
Exhaust gas temperature at outlet of turbocharger (°C)	370.0
Mean effective pressure (MPa)	2.246
Brake specific fuel oil consumption (g/kWh)	195.0
Mean piston speed at 1000 min^{-1} (m/s)	9.33

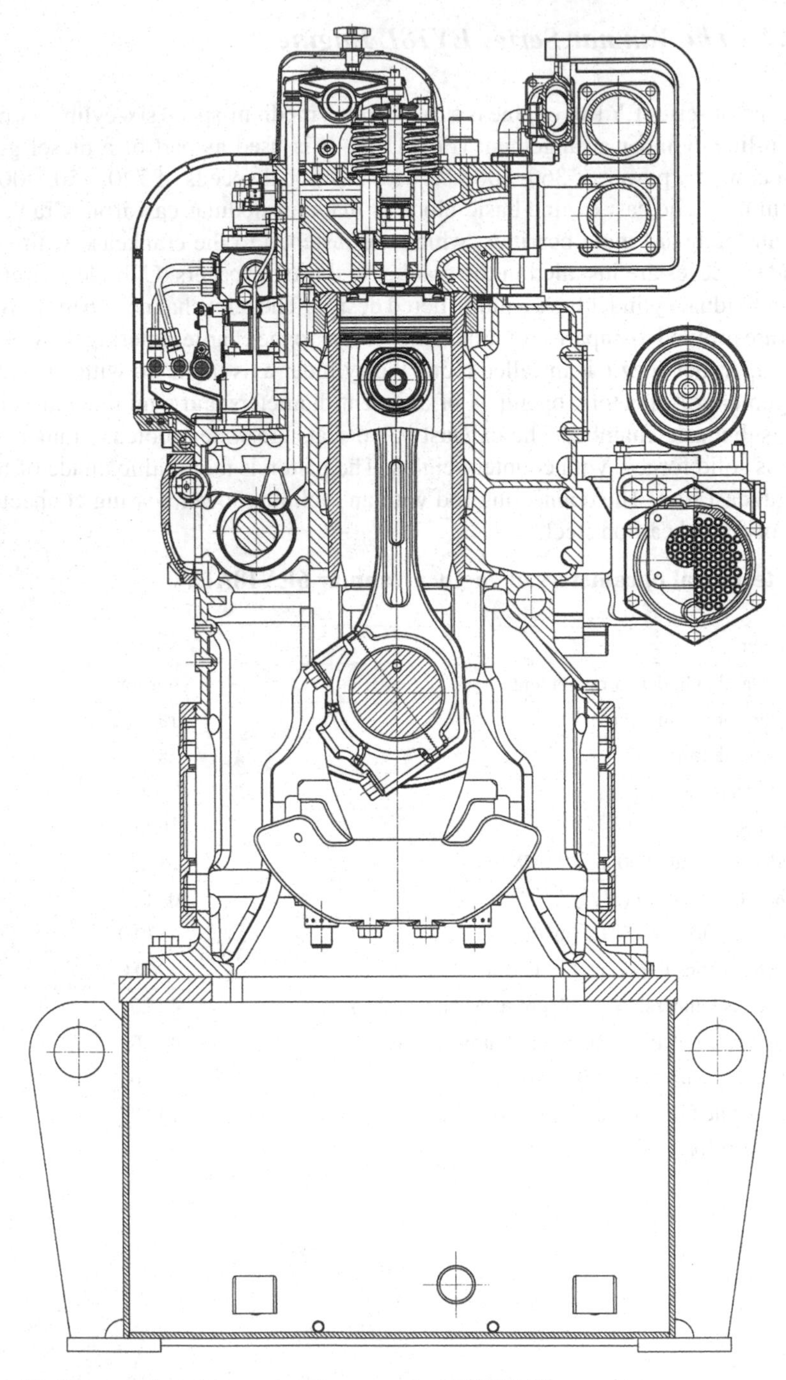

Fig. 1.55 Cross-section of the engine Yanmar series EY18L [49]

Dimensions and weight of diesel generator with an engine Yanmar series EY18ALW

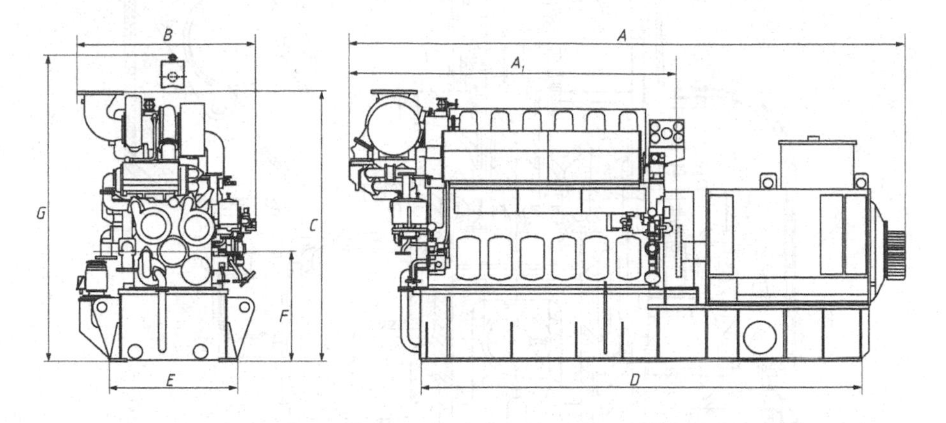

Engine version	A (mm)	A_1 (mm)	B (mm)	C (mm)	D (mm)
6EY18LW	4441	2751	1489	2255	3620

Engine version	E (mm)	F (mm)	G (mm)	Weight (kg)
6EY18LW	1070	915	2564	6600

The weight of the diesel generator is 11,200 kg

1.19.3 The Yanmar Series EY26L Engine

The Yanmar series EY26L engine is a four-stroke, medium-speed engine with an in-line (6, 8 cylinder) cylinder arrangement (Fig. 1.56). The cast engine basic block is made of nodular cast iron. The bed of the crankshaft with outboard main bearings, which are attached to the rack and the frame with anchor bolts and studs. Cylinder sleeves are installed directly into the crankcase. The bushings and cylinder covers are cooled with water from a high-temperature circuit. In the crankcase, the sleeve is sealed with three polymer o-rings. Cylinder covers are individual for each cylinder. Each cap is attached to the crankcase with four studs. There are two inlet and two exhaust valves in the cylinder cover, which are actuated from the crosspiece, which is affected by the double-arm drive lever. The exhaust valve seats are water cooled. The crankshaft is solid forged, with counterweights. The piston is monolithic, made of ductile cast iron, cooled with lubricating oil, supplied through a hole in the connecting rod. The rod consists of three parts, made of forged carbon steel.

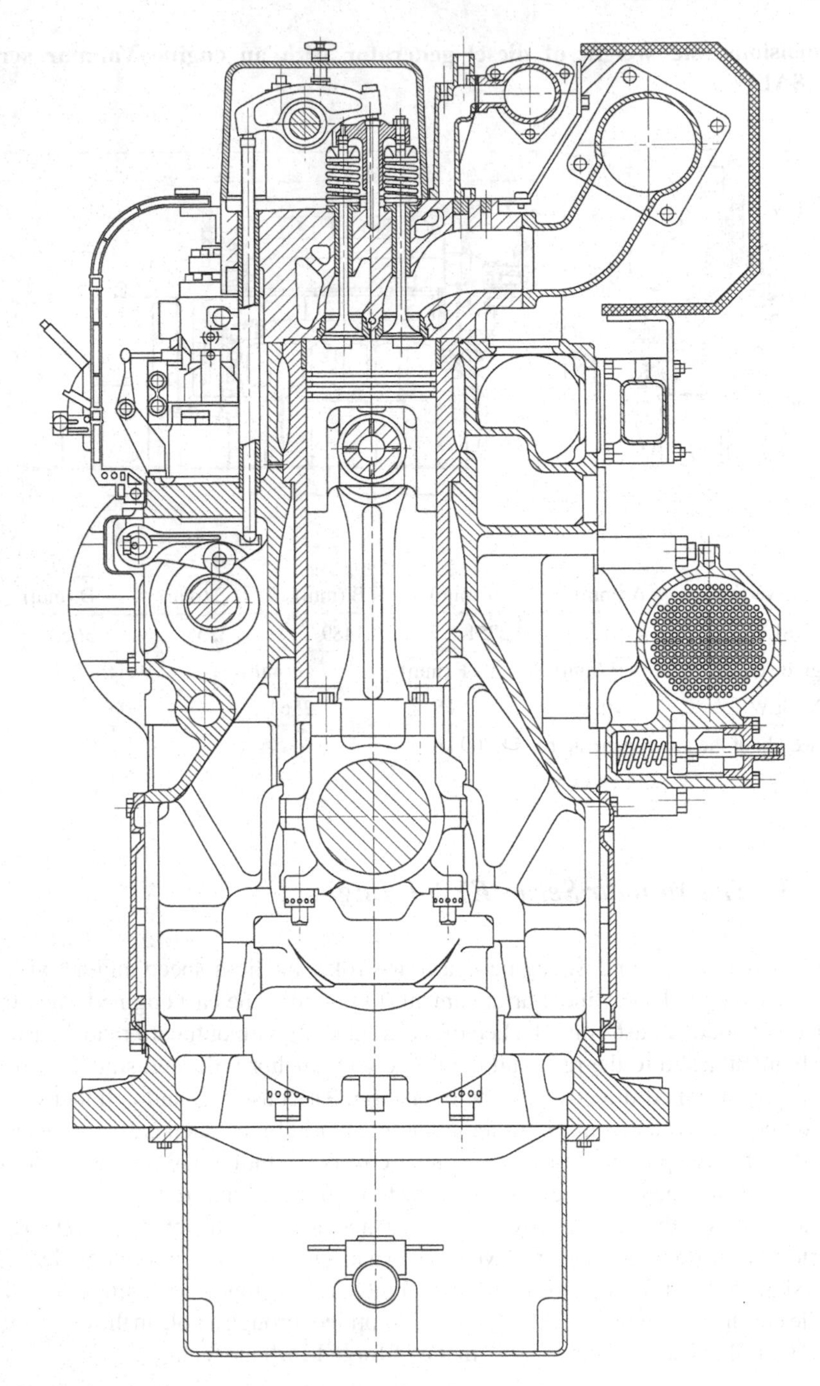

Fig. 1.56 Cross-section of the engine Yanmar series EY26L [50]

Main technical parameters of engines Yanmar series EY26L

Parameter	Value
Number and cylinders arrangement	6, 8 in-line
Cylinder bore (mm)	260
Piston stroke (mm)	385
Cylinder capacity (dm^3)	20.44
Rotation speed (min^{-1})	720, 750
Cylinder power at 720/750 min^{-1} (kW)	288/306
Compression ratio	15.0
Maximum cycle pressure (MPa)	15.0
Exhaust gas temperature at inlet of turbocharger (°C)	560.0
Exhaust gas temperature at outlet of turbocharger (°C)	384.0
Mean effective pressure (MPa)	2.11/2.41
Brake specific fuel oil consumption (g/kWh)	182.0
Mean piston speed at 720/750 min^{-1} (m/s)	9.24/9.62

Dimensions and weight of diesel-gear unit with the engine Yanmar series EY26L and reduction gearing YXH-2000 M/YX-3500 M

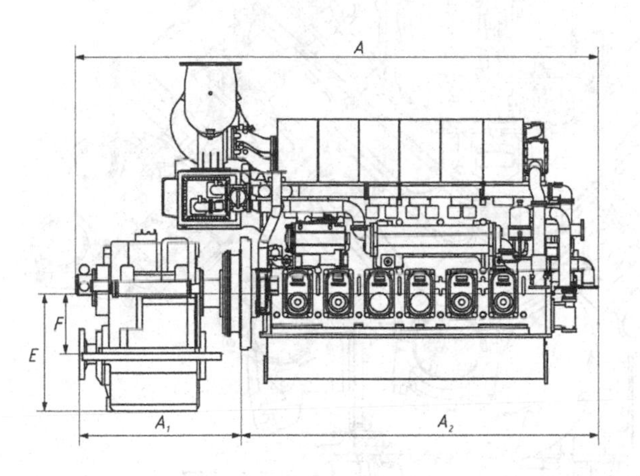

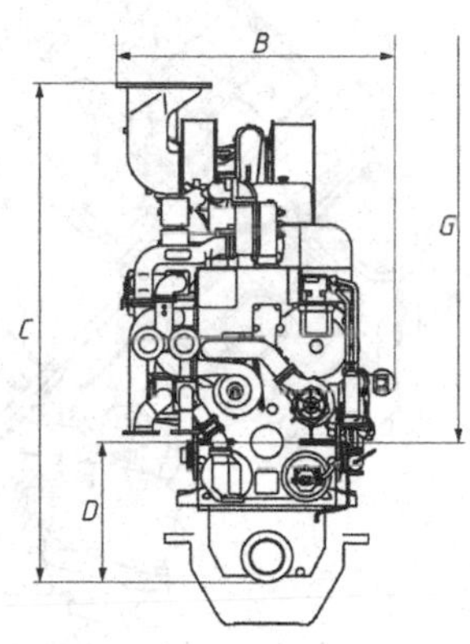

Engine version	A (mm)	A_1 (mm)	A_2 (mm)	B (mm)	C (mm)
6EY26L	5702	3563	1882	1804	3112
8EY26L	6912	5022	1890	2085	3257

Engine version	D (mm)	E (mm)	F (mm)	G (mm)	Weight (kg)
6EY26L	842	1145	590	1900	22,550
8EY26L	842	1427	777	1900	33,428

1.19.4 The Yanmar Series AYM Engine

The Yanmar series AYM engine is a four-stroke, high-speed, highly accelerated engine with in-line (6 cylinders) and V-shaped cylinders at an angle of 90° (12 cylinders) (Fig. 1.57). Engines of this series belong to the so-called group of eco-friendly diesel engines (Eco Diesel), developed under the concept of reducing harmful emissions, and, above all, NO_x, without increasing fuel consumption. To reduce harmful emissions, the engine used the original internal exhaust gas recirculation system (EGR), based on the fact, that the exhaust valve during the intake is in the ajar state, with the result, that some of the exhaust gases from the exhaust manifold is returned to the working cylinder. The use of injectors with ultra-thin nozzle openings of different diameters, combined with a carefully selected shape of the combustion

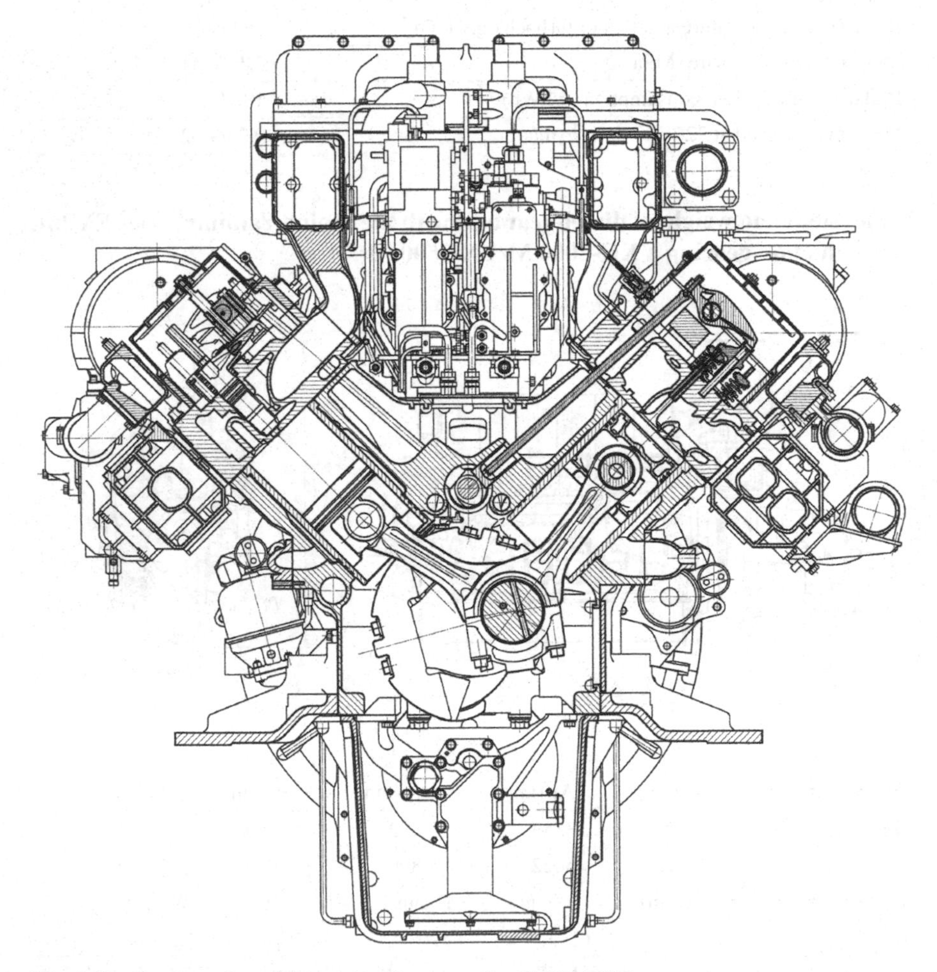

Fig. 1.57 Cross-section of the engine Yanmar series 12AYM [51]

chamber and the firing bottom of the cylinder head cover, made it possible to achieve high-quality mixture formation and combustion of fuel in all engine operating modes. In 12-cylinder engines, two high-pressure block pumps unified with in-line engines are used for fuel injection, which increases their reliability. To increase the engine's life, an artificial ceramic "Silicard" coating, developed on the basis of sintering silicon carbide powder (SiC), is applied to the cylinder liners. Piston rings made of stainless steel are nitrated. Carefully selected clearance between the piston skirt and the cylinder liner significantly reduced engine oil consumption. The use of separate cylinder covers for each cylinder, combined with the presence of large hatches on the side surfaces of the engine basic block, significantly simplify the maintenance and repair of the engine during operation.

Main technical parameters of engines Yanmar series 12AYM

Parameter	Value
Number and cylinders arrangement	12/V-shaped
Cylinder bore (mm)	155
Piston stroke (mm)	180
Cylinder capacity (dm^3)	3.4
Rotation speed (min^{-1})	1840
Cylinder power at 1840 min^{-1} (kW)	96
Compression ratio	13.2
Brake specific fuel oil consumption (g/kWh)	212.0
Mean piston speed at 1840 min^{-1} (m/s)	11.04
Weight (kg)	4950

Dimensions and weight of engines Yanmar series 12AYM

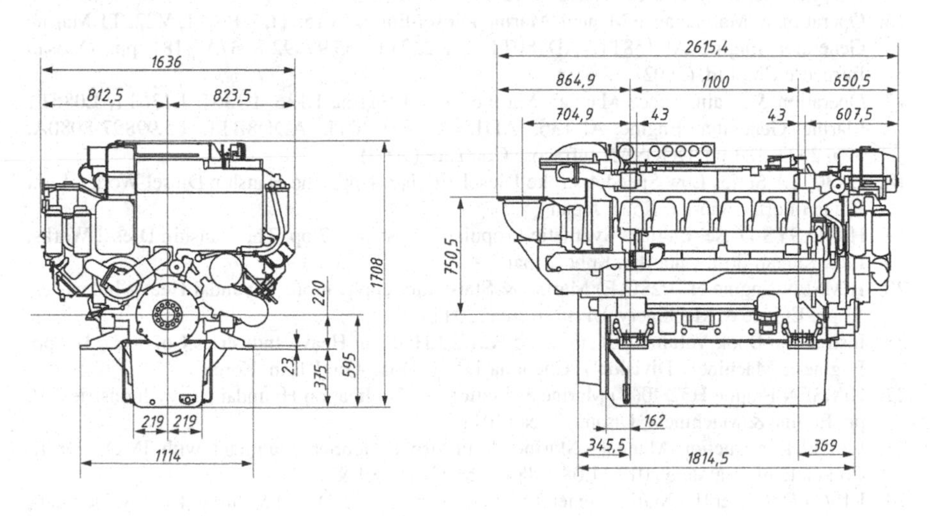

References

1. Anglo Belgian Corporation: We Power Your Future, 8 pp. Moteurs de traction, Wiedauwkaai 43 9000 Gent (Belgique) (2013)
2. Anglo Belgian Corporation: We Power Your Future Marine Engines, 12 pp. Propulsion & Power Solutions, Wiedauwkaai 43 9000 Gent (Belgique) (2013)
3. Product Catalog Akasaka Diesels Limited: Environmental Management System Eco Stage2-CMS EST-546-C CAT. No.1707A2,000b, 32 pp. (2016)
4. Curso Técnico de Motores Caterpillar serie 3500. Maquinarias Pesadas.org – Maravillas de la ingenierfa. finning.com 146 pp.
5. Конке, Г.А., Лашко, В.А.: Мировое судовое дизелестроение. Концепции конструирования, анализ международного опыта: Учебное пособие – М.: Машиностроение, 512 с (2005)
6. Project guide M 552C/601C, 104 pp. MaK Motoren GmbH & Co. KG A Caterpillar Company, Kiel (1998)
7. M25C Long-Stroke Diesel Engines for Maximum Efficiency and High Reliability 6, 8, 9, 16 pp. Caterpillar Marine A Division of Caterpillar Motoren GmbH & Co. KG Neumühlen Hamburg, Germany (2014)
8. M32C Low Emission Engine. Caterpillar Marine Power Systems, 12 pp. Neumühlen Hamburg, Germany (2009)
9. VM32 C Project Guide. Propulsion, 98 pp. Caterpillar Motoren GmbH & Co. KG, Kiel, Germany (2009)
10. M43 C Project Guide. Propulsion, 112 pp. Caterpillar Motoren GmbH & Co. Kiel, Germany (2008)
11. VM43 C Project Guide. Propulsion, 95 pp. Caterpillar Motoren GmbH & Co. Kiel, Germany (2008)
12. 6-8DK-28 Instruction manual (Operation), 109 pp. Daihatsu Diesel MFG Co., Ltd., Kita-ku, Osaka, Japan (2008)
13. DC-17A Instruction manual (Operation), 461 pp. Daihatsu Diesel MFG Co., Ltd., Kita-ku, Osaka, Japan (2010)
14. Operating Instructions TBD 645, 244 pp. Deutz-Müiheimer-Straße 147-149 D-51057 Köln, Motoren-Werke Mannheim AG (DEUTZ MWM), Germany (1997)
15. Руководство по эксплуатации TCD 2015, 68 pp. The engine company DEUTZ AG, Cologne, Germany. Номер заказа: 0312 3476 ru. (2007)
16. Operation & Maintenance Manual. Marine Diesel Engine V158TI, V180T1, V222T1 Marine Generator Engine AD158T1, AD180T1, AD222T1. 65.99892-8067A, 181 pp. Doosan Infracore Co., Ltd. (2002)
17. Operation & Maintenance Manual. Marine Diesel Engine L136, L136T, L136T1, L086T1 Marine Generator Engine AD136, AD136T, AD136TI, ADD8BT1. 65.99897-8080A. Sep. 2003, 183 pp. DOOSAN Infracore Co., Ltd. (2003)
18. LA30. LA Series Low Speed 4-stroke Diesel Engine, 1 pp. The Hanshin Diesel Works, Ltd., Kaigan-dori, Chuo-ku, Kobe, Japan
19. HANASYS Diesel Engines Synthetic Propulsive System, 47 pp. The Hanshin Diesel Works, Ltd., Kaigan-dori, Chuo-ku, Kobe, Japan
20. HiMSEN Engine H17/28U(E) Marine & Stationary applications. Hyundai Heavy Industries, 12 pp. Engine & Machinery, Ulsan, Korea (2011)
21. Instruction Book Volume I Engine type H21/32 Hyundai Heavy Industries Co., Ltd., 323 pp. Engine & Machinery Division 1, Cheonha-Dong, Dong-Gu, Ulsan, Korea
22. HiMSEN Engine H32/40(V) Marine & Stationary Application Hyundai Heavy Industries, 16 pp. Engine & Machinery, Ulsan, Korea (2011)
23. L23/30H Instruction Manual—Marine. Four-stroke GenSet compliant with IMO Tier II. Complete manual date 2012.11.08, 878 pp. MAN Diesel & Turbo
24. L16/24 IMO Tier II—Marine Generating Sets, 16 pp. MAN Diesel & Turbo, Holeby, Denmark

25. L21/31 IMO Tier II—Marine Generating Sets, 15 pp. MAN Diesel & Turbo, Holeby, Denmark
26. Project Guides L27/38, 188 pp. MAN B&W Diesel A/S Holeby Generating Sets, Denmark
27. L+V32/40 Project Guide—Marine Four-stroke diesel engines compliant with IMO Tier II, 450 pp. MAN Diesel & Turbo, Augsburg, Germany
28. Project Guide for Marine Plants Diesel Engine 40/54 Status: 04.2007, 372 pp. MAN Diesel SE, Augsburg, Germany
29. MAN 48/60CR Project Guide—Marine Four-stroke diesel engines compliant with IMO Tier III, 456 pp. MAN Diesel & Turbo, Augsburg, Germany (2016)
30. Project Guide for Marine Plants. Engine L 58/64. Status: 02.2006, 364 pp. MAN Diesel SE, Augsourg, Germany
31. Project Guide for Marine Plants Engine 32/44 CR Preliminary Version. "Engines in compliance with IMO I or Emission level DNV Clean Design" Status: 11.2008, 364 pp. Augsburg, Germany
32. MAN 32/44CR MAN Energy Solutions Future in the making Project Guide—Marine Four-stroke diesel engine compliant with IMO Tier II, 456 pp. MAN Energy Solutions SE, Augsburg, Germany (2019)
33. MAN marine Diesel engines Repair Manual D 2866 LE 401/402/403/405 D 2876 LE 301 D 2876 LE 403, 138 pp. MAN Nutzfahrzeuge Aktiengesellschaft Nuremberg Works (1999)
34. Mizuhara, S., Kunimitsu, M., Beppu, O., Takahashi, M., Sakane, A., Tanaka, M.: High-powered ADD3OV medium speed diesel engine. Bull. M.E.S.J. **27**(2), 9 pp
35. MTU Series 4000 Legendary. Since 1996, 5 pp. MTU Friedrichshafen GmbH I MTU Asia Pte Ltd I MTU America Inc. Part of the Rolls-Royce Group
36. Woodyard D. Marine diesel engines and gas turbines, Ninth Edition, 896 pp. Oxford OX2 8DP 200 Wheeler Road, Burlington, MA 01803 (2009)
37. Service manual Intrusions book. Engine type: Bergen B32:40L8P Engine, 370 pp. February 2007 Rolls-Royce Marine AS, Bergen, Norway
38. Project Guide. Bergen engine type C25:33, 128 pp. Edition: May 2009 (Rev. 05. February 2016). Bergen Engines AS, Bergen, Norway
39. Ruston RK 270 Diesel Engines, 14 pp. Ruston Diesels Limited Vulcan Works Newton-le-Willows Merseyside, England
40. V28/33D, V28/33D STC Project Guide—Marine Four-stroke diesel engine compliant with IMO Tier II and EPA Tier, 2312 pp. MAN Diesel & Turbo, Augsburg, Germany
41. Project Planning Manual for Marine Main Engines PC 2-6 B, 166 p. S.E.M.T. Pielstick Paris-Nord II, Roissy CDG Cedex, France. Edition: 02.03
42. Описание и инструкция по эксплуатации дизеля «ZGODA – SULZER» типа ZL 40. Zakłady Urządzeń Technicznych «ZGODA», 384 pp. Swiętochłowice, Польша
43. Maintenance manual ZAV40S, 313 pp. Wärtsilä Corporation (2004)
44. Wärtsilä 20—Project Guide, 154 pp. Wärtsilä Finland oy Ship Power Application Technology Vaasa (2005)
45. Wärtsilä 32—Project Guide, 192 pp. Wärtsilä Ship Power Technology, Product Support Vaasa (2008)
46. Wärtsilä 46—Project Guides, 210 pp. Wärtsilä Ship Power 4-stroke, Business Support Vaasa (2007)
47. Marine Project Guide W64, 124 pp. Application Technology, Wärtsilä NSD Finland Oy, Marine, Vaasa (1997)
48. Yanmar Diesel Engine. Marine Auxiliary Engine 6N21(A)L-V Series. Operation manual, 123 pp. Yanmar Co., Ltd. (2004)
49. Project Guide. Marine Auxiliary Diesel Engine Model 6EY18 Series, 240 pp. Yanmar Technical Service Co., Ltd., Japan (2011)
50. EY26 Operation Manual, 543 pp. Yanmar Co., Ltd. (2006)
51. Marine Diesel Engine 12AYM-WET, 2 pp. Yanmar Co., Ltd., Marine Operations Division, Tsukaguchi Honmachi Amagasaki, Hyogo, Japan

Chapter 2
Gas and Gas Diesel Four-Stroke Marine Engines

According to Lloyd's Marine Intelligence experts, the world merchant fleet as for 2018 has more than 50,000 vessels with a total carrying capacity of about 1.4 billion tons, which carry about 75% of the goods, transported all over the world. The part of harmful emissions from ship power plants into the atmosphere is relatively not high and does not exceed 5–7%. However, the requirements for the environmental performance of marine diesel engines are becoming more stringent from year to year. This is explained by the large aggregate capacity, which in some cases already exceeds 100 MW. Thus, in areas of intensive navigation, air pollution can reach critical values.

The process of legislative regulation of harmful emissions into the atmosphere by merchant ships began as early as 1973 with the adoption of the International Maritime Organization IMO, operating under the UN auspices, the Convention on the Prevention of Pollution from Ships, which was ratified in 1978 and received the name MARPOL 73/78. In 1997, the Convention was supplemented by the "Protocol 1997", which included Annex No. VI "Procedure for the Prevention of Atmospheric Pollution by Ships", which entered into force in 2005. The Convention limits the content in the fuel of sulfur combustion products (SO_x), nitrogen oxides (NO_x), particulate matter (PM), as well as greenhouse gases.

In addition to the general international standards that is everywhere, international maritime legislation also establishes special control zones (ECAs). In these zones, where today includes 200-mile zones of the United States and Canada, the waters of the North and Baltic Seas, the English Channel, the Caribbean Sea and other regions, even more stringent requirements are imposed on emissions. Every year the restrictions of MARPOL become tougher, so if in 2010 the sulfur content in the fuel should not exceed 1.0% in the ECA zones and 4.5% in other water areas, by 2020 the permissible amount of sulfur compounds will be limited to 0.1 and 0.5%, respectively.

I. Bilousov et al., *Modern Marine Internal Combustion Engines*, Springer Series on Naval Architecture, Marine Engineering, Shipbuilding and Shipping 8,
https://doi.org/10.1007/978-3-030-49749-1_2

To achieve the above standards, especially in ECA zones, for all types of marine engines, the use of combustible gas mixtures as the main fuel as the most promising areas now is considering. The most promising gas fuels are natural gas, consisting mainly of methane (CH_4) and oil gases, which are mainly mixtures of propane (C_3H_8) and butane (C_4H_{10}). On board a vessel, natural gas can be stored in a liquefied state at temperatures below 160 °C (LNG) in special cryogenic tanks, and propane-butane mixtures (LPG) can be stored in a liquid state at ambient temperatures under a pressure of 1.6 MPa. Technologically, gas fuels can be specially stored on a vessel's board, both the main fuel for the operation of its power plant, and to be a by-product of the process of transporting various types of fuels, such as gas carriers or oil tankers.

The use of gas fuel can significantly reduce the amount of harmful emissions in comparison with fuels of petroleum origin and to completely eliminate sulfur emissions, drastically (by 90%) reduce NO_x emissions and significantly (by 30%) reduce particulate matters emissions and carbon dioxide (CO_2).

The specific working conditions of the vessels left their imprints on the development of marine engines, operating on gas. This is primarily due to the need to maintain the ability of the engine to operate on liquid fuels, which occurs whenever a ship moves in ballast. In addition, depending on the type of cargo, navigation conditions and time, the composition of the gases, used in the power plant, may vary significantly, and therefore, the fuel system must adequately respond to such changes and ensure the engines operate at nominal conditions. On this basis, the bulk of ship engines today are created with dual-fuel (DF), that is, capable of operating on gas, liquid fuel, or both fuels in different proportions at once.

To the organization of the working process, using gas fuels in engines, there are three fundamentally different approaches:

- converting diesel engines into engines with external mixture formation and electric spark ignition, operating according to the Otto cycle;
- the use of external mixing with the ignition of the gas-air mixture from a small portion of the liquid fuel injected into the working cylinder;
- the use of internal mixing and ignition of the gas-air mixture from a small portion of the liquid fuel injected into the working cylinder.

The first two approaches are widely used in shipboard four-stroke engines for various purposes. The latter approach is more spread for low-speed two-stroke engines, used as the main ones.

2.1 Otto Cycle Gas Engines

Converting diesel engines into engines with external mixture formation and spark or ignition operating in the Otto cycle is the easiest way to convert the engine to gas fuel. Its benefits include:

- simplification of the engine design (it is possible to completely abandon the injection system with pilot oil, replacing it with more simple ones: the system of mixing air with gas and the system of spark or ignition flame);
- operation of the power supply system at low pressures, which reduces the requirements for ensuring safety in the operation of such systems.

The main disadvantages of using these systems are the loss of the fuel change, as well as the reduction in liter capacity, associated with a decrease in the weight filling of the cylinders. In addition, with the compression rates typical to diesel engines, detonation is observed in a wide range of loads, which prevents the use of such engines in installations with direct power transfer to the screw.

Stable operation of engines with high power and efficiency indices is observed only in a relatively narrow range on rather lean mixtures. In this regard, the use of these engines is very limited. Basically, such engines are used in stationary power engineering, including as a part of power installations of offshore facilities, gas and oil-producing offshore platforms, where it is possible to ensure their reliable supply with gas fuel throughout the entire period of operation. Less commonly, they are used as part of multi-machine generator sets, ships with electrical power transmission to the screw.

To obtain a stable ignition and efficient combustion of lean gas mixtures in engines with large sizes of working cylinders, the energy of an electric spark is often insufficient, therefore, a scheme with a so-called antechamber flare ignition is widely used in engines of this type. The antechamber is a cavity that connects to the main combustion chamber by means of several channels made in the tip of the antechamber protruding from the cylinder head. A spark plug and a gas supply valve, which is driven by a special actuator, are installed in the antechamber. The gas-air mixture is initially ignited in the antechamber, then the mixture in the form of plasma jets enters the combustion chamber, igniting its contents. Next, we consider some of the engines, made according to this scheme.

2.1.1 The V35/44G Engine

The V35/44G engine from MAN Diesel & Turbo (Fig. 2.1) is a new class of high-powered gas reciprocating engines with single-stage turbocharging, operating on the Miller cycle. The V35/44G engine was created on the basis of a V32/44CR diesel engine, and is available in a V-shaped 12 and 20-cylinder design. The output power of the engine reaches 10,600 kW, and the effective efficiency reaches 48.4%. The engine has implemented many innovative technological solutions, in addition, it meets the standard for the level of emissions of harmful substances TA-Luft. Engines have a highly efficient electric spark ignition system. A spark plug is installed in an antechamber where gas enters through a separate metering valve. Mixing with the lean mixture, which comes from the main combustion chamber during the compression stroke, the gas that enters into an antechamber forms a gas-air mixture similar in composition to the stoichiometric one, which is ignited by an electric spark. The streams of burning gas from the antechamber effectively ignite the lean mixture in the main chamber, contributing to its complete burnout. TCR type turbochargers are equipped with a VTA (Variable Turbine Area) unit with adjustable turbine nozzle geometry, which allows optimizing the performance of the turbocharger depending on the engine's operating mode, to maintain a given air-fuel ratio. For cooling the charge air, one air cooler is used for each row of cylinders. To extend the range of operating temperatures of the air at the engine inlet, bypassing of the compressor was used.

Main technical parameters of engines series V35/44G

Parameter	Value
Number and cylinders arrangement	12, 20/V-shaped
V-angle (°)	55
Cylinder bore (mm)	350
Piston stroke (mm)	440
Cylinder capacity (dm^3)	42.3
Rotation speed (min^{-1})	750
Cylinder power at 750 min^{-1} (kW)	530.0
Mean effective pressure (MPa)	2.34
Brake specific energy consumption (kJ/kWh)	7722
Mean piston speed at 750 min^{-1} (m/s)	11.0
Lube oil consumption (kg/h)	3.7

Dimensions and weight engines series V35/44G

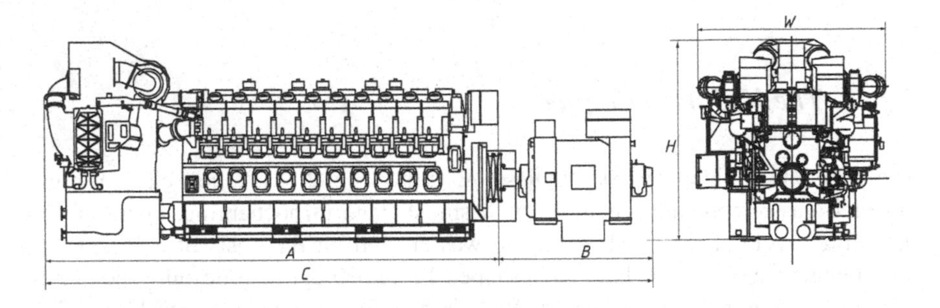

Engine version	A (mm)	B (mm)	C (mm)	W (mm)	H (mm)	Weight (kg)
12V35/44G	9028	4330	13,358	4925	5200	144,000
20V35/44G	11,549	4137	15,686	4925	5200	200,000

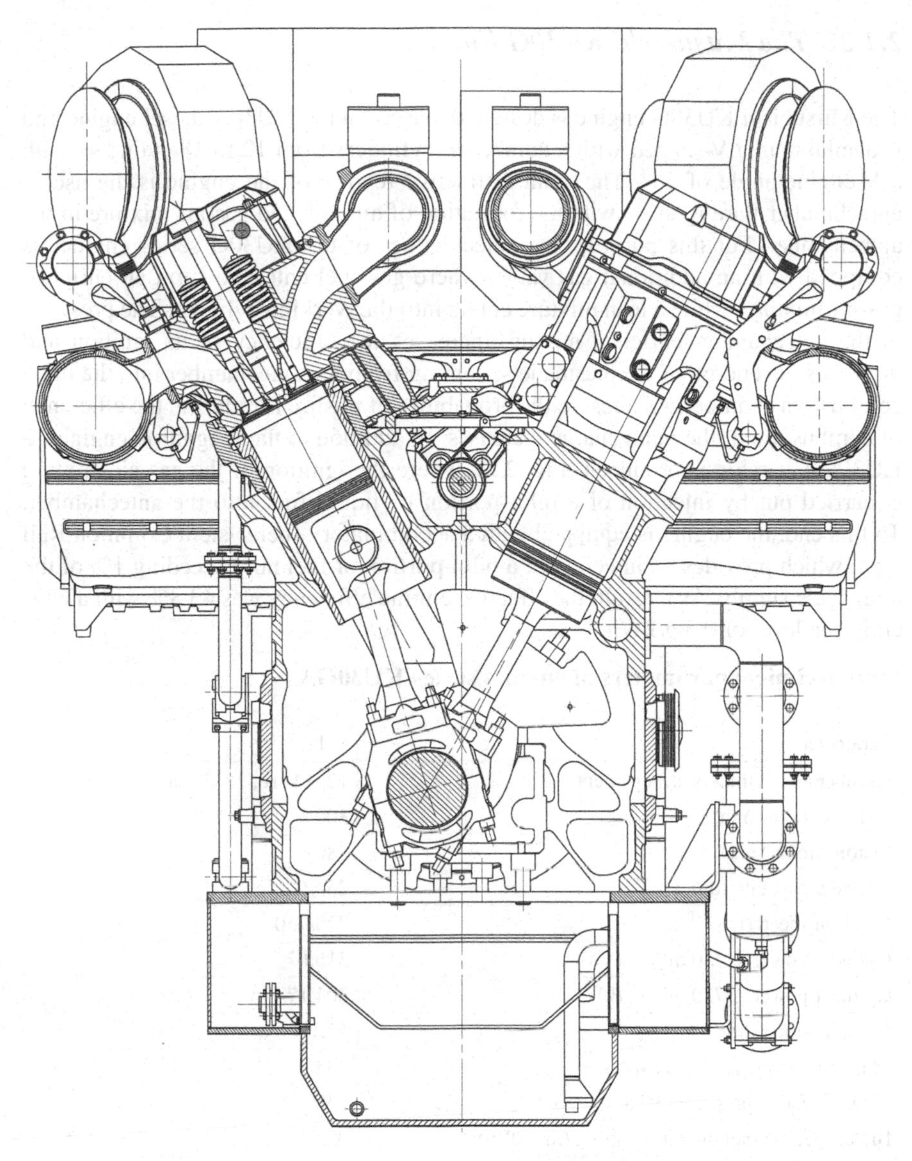

Fig. 2.1 Medium-speed gas engine with antechamber-flare electric flame ignition V35/44G of MAN Diesel & Turbo [1]

2.1.2 The Mitsubishi KU30G Engine

The Mitsubishi KU30G engine is designed, based on the KU30A diesel engine and is available in a V-shaped with a number of cylinders from 12 to 18 cylinders with a V-engine angle of 50°. The main distinctive feature of the engine is the use of antechamber-flare ignition with the formation of an enriched gas-air mixture in the antechamber. For this purpose, the antechamber of the KU30G series engines is equipped with an additional gas valve, where gas fuel enters to form an enriched gas-air mixture, while a lean mixture enters into the working cylinder. The presence in the antechamber enriched mixture increases the efficiency of its ignition and increases the energy of the plasma jets, coming from the antechamber into the main combustion chamber. As a result, the probability of misfire is reduced, the efficiency of combustion of the main charge increases. In addition to flame ignition engines, a KU30GA version is produced (Fig. 2.2), where the ignition of the gas-air mixture is carried out by injection of a pilot portion of liquid fuel into the antechamber. To this end, the engine is equipped with a accumulatory fuel system Common Rail type, which provides an injection of a pilot portion of fuel not exceeding 1% of the total cycle supply. As a result, the effective engine efficiency was 43.8% with a NO_x emission level of 0.5 g/kWh.

Main technical parameters of engines series KU30GA

Parameter	Value
Number and cylinders arrangement	12, 14, 16, 18 V-shaped
Cylinder bore (mm)	300
Piston stroke (mm)	380
Cylinder capacity (dm^3)	26.86
Rotation speed (min^{-1})	720/750
Cylinder power at 750 min^{-1} (kW)	316.67
Cylinder power at 720 min^{-1} (kW)	304.17
Compression ratio	12.1
Maximum cycle pressure (MPa)	18.6
Mean effective pressure (MPa)	1.96
Brake specific fuel oil consumption (m^3/kWh)	0.202[a]
Mean piston speed at 720/750 min^{-1} (m/s)	9.1/9.5

[a]The lowest calorific value of gas is 40.63 MJ/m^3 at atmospheric pressure and at a temperature of 20 °C

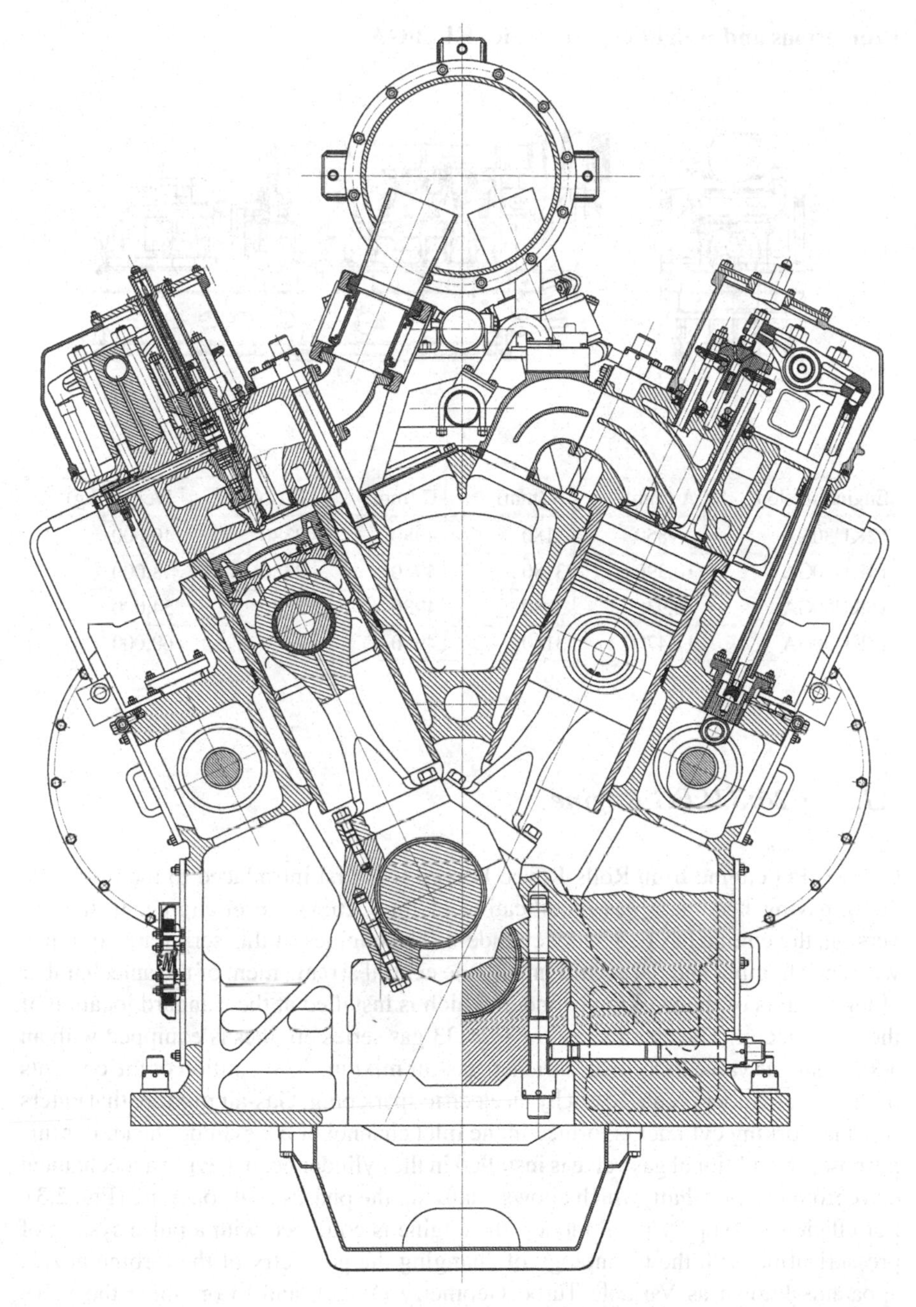

Fig. 2.2 Medium-speed gas engine with antechamber-flare electric flame ignition Mitsubishi KU30GA [2]

Dimensions and weight engines series KU30GA

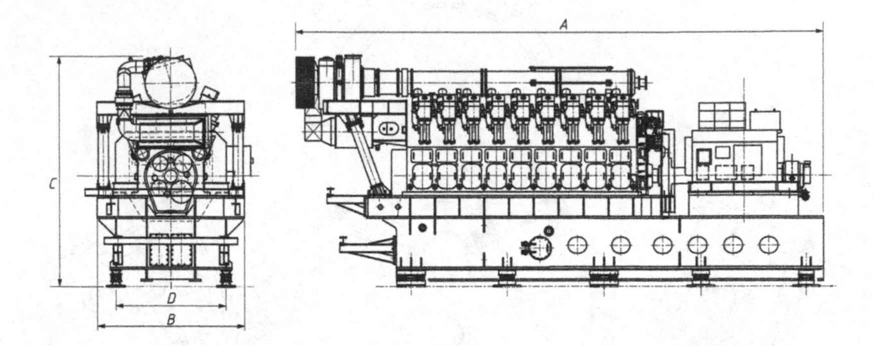

Engine version	A (mm)	B (mm)	C (mm)	D (mm)	Weight (kg)
12KU30GA	9850	3180	4980	2380	40,000
14KU30GA	10,390	3180	4980	2380	48,000
16KU30GA	10,930	3180	4980	2380	54,000
18KU30GA	11,470	3180	4980	2380	60,000

2.1.3 C26:33LPG Engine

C26:33LPG engine from Rolls-Royce Bergen was first introduced to the market in 2010, having become a gas modification of the C26:33 diesel engine. In the gas version, the engine has 6, 8 and 9 cylinders. The engines of this series are equipped with antechamber and flare ignition with the central arrangement of the antechamber along the axis of the working cylinder, which is installed at the standard location of the fuel injector. The chamber of the C26:33 gas series engines is equipped with an additional gas valve to form an enriched gas-air mixture. The ignition of the contents of the antechamber is carried out by an electric spark plug. Gas-air mixture that enters into the working cylinder is formed in the inlet channel of the cylinder head. For this purpose, an additional gas valve is installed in the cylinder head, having a mechanical drive from the camshaft, which allows changing the phases of its opening (Fig. 2.3). For efficient air supply in all modes, the engine is equipped with a pulse system of pressurization with the technology of changing the geometry of the turbine nozzle apparatus known as Variable Turbo Geometry (VTG), and to optimize the valve timing the engine is equipped with a system for changing the opening and closing angles of valves Variable Valve Timing.

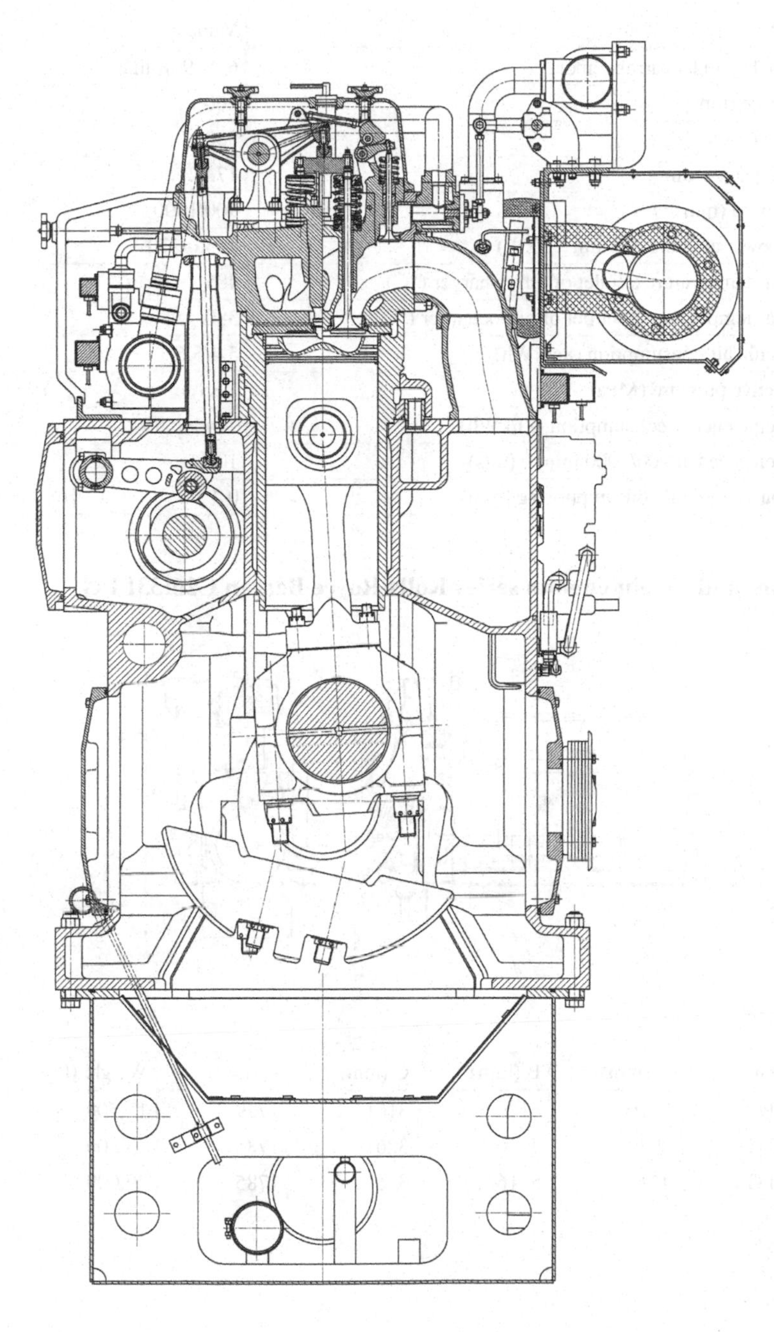

Fig. 2.3 Medium-speed gas engine with antechamber-flare electric flame ignition-Royce Bergen C26:33LPG [3]

Main technical parameters of engines Rolls-Royce Bergen C26:33LPG

Parameter	Value
Number and cylinders arrangement	6, 8, 9 in-line
Cylinder bore (mm)	260
Piston stroke (mm)	330
Cylinder capacity (dm^3)	17.52
Rotation speed (min^{-1})	900/1000
Cylinder power at 900/1000 min^{-1} (kW)	243.0/270
Exhaust gas temperature at inlet of turbocharger (°C)	485
Exhaust gas temperature at outlet of turbocharger (°C)	380
Brake specific air consumption (кg/kWh)	5.4/5.6
Mean effective pressure (MPa)	1.85
Brake specific energy consumption (kJ/kWh)	7445/7550
Mean piston speed at 900/1000 min^{-1} (m/s)	10/11
Lubrication system oil consumption (g/kWh)	0.4

Dimensions and weight engines series Rolls-Royce Bergen C26:33LPG

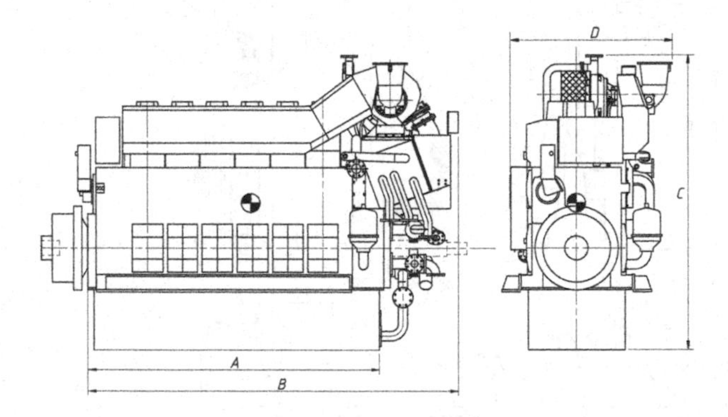

Engine version	A (mm)	B (mm)	C (mm)	D (mm)	Weight (kg)
C26:33L6PG	3170	4176	3161	1729	17,500
C26:33L8PG	3930	4936	3261	1785	25,800
C26:33L9PG	4310	5316	3161	1785	29,000

2.2 Gas Diesel Dual Fuel Engines

The specific operating conditions of ship engines left their imprints on the development of ship engines, capable of operating on gas fuels.

The main design feature of such engines is associated with the need to maintain the ability to work on liquid fuels, which occurs when moving outside the ECA areas or when the vessel moves in ballast. Based on this requirement, the bulk of marine engines used as both main and auxiliary engines are performed using a dual-fuel gas-diesel scheme (DF), that is, such engines are capable of operating on gas, liquid fuel or both fuels simultaneously in various proportions.

At the same time, diesel engines are taken as the basis for the creation of such engines, the design of which makes the appropriate changes.

To the organization of the working process in gas-diesel engines there are two main approaches:

- the use of external mixing with the ignition of the gas-air mixture from a small portion of the liquid fuel, injected into the working cylinder;
- the use of internal mixing and ignition of the gas-air mixture from a small portion of the liquid fuel injected into the working cylinder.

The first approach is widely used in shipboard four-stroke medium and high-speed engines, the second one in low-speed two-stroke engines.

To implement the gas-diesel cycle, dual-fuel engines are equipped with an additional system for preparing the gas-air mixture at the inlet to the working cylinders. Such a system practically does not differ from the previously considered flame ignition gas supply system. To ensure sustained ignition of the gas-air mixture, a small portion of the liquid fuel, called ignition or pilot feed, is injected into the combustion chamber.

For the injection of a pilot portion of fuel, a regular system of injection of liquid fuel or a special system, that only works when engine operated on gas fuel is used.

In the first case, the basic diesel engine is upgraded with components of the gas sys-tem without a significant rework of the standard power supply system. When operating on gas, fuel pumps are switched to the minimally stable feed mode, and the engine load is adjusted by changing the amount of incoming gas fuel or gas-air mixture.

The undoubted advantages of the first approach are: the need for a minimum rework of the engine itself and its fuel system, the ability to go to work on liquid fuel at any time.

The main disadvantage of this approach is the impossibility of a significant reduction in the consumption of liquid fuel for the organization of pilot injection. Typically, in engines of this type, the flow consumption for ignition is 15–20% of the flow consumption on liquid fuel. This is explained by the impossibility of obtaining with the help of regular fuel injection equipment of stable injection at low supply.

The second approach boils down to the additional equipment of the engine with an additional fuel system, specifically designed to supply the pilot portion of fuel.

An ex-ample of such an approach is the series of gas engines, produced by MAN, Wartsila, and others, created on the basis of the existing model range of diesel engines.

In some engines, to compensate for power losses when switching to gas fuel, manufacturers went to increase the diameter of the cylinder, for example, in the Wartsila 50DF engine, created on the basis of the 46th series diesel (cylinder diameter increased from 460 to 500 mm), MAN 35/44, created on the basis of MAN 32/44 (cylinder diameter in-creased from 320 to 350 mm).

2.2.1 The MAN 35/44DF Engine

The MAN 35/44DF engine was created on the basis of the well-proven MAN 32/44CR engine (Fig. 2.4), inheriting from it a full-featured Common Rail type fuel injection system, capable of delivering fuel to the injectors at 160 MPa pressure. As a result, when operating on liquid fuels, the engine ensures compliance with current emission standards of harmful substances with exhaust gases in accordance with the requirements of IMO Tier II.

For the preparation of a gas-air mixture, the engine is equipped with an external mixture formation system with electronic regulation of the gas fuel supply, using gas valves with an electromagnetic drive, that are individually installed on each cylinder. The valves provide gas supply at the time of filling the cylinder, thus avoiding the ingress of gas-air mixture into the intake receiver. As a result, the probability of explosions in the receiver, which can lead to the destruction of the engine, is excluded.

All gas pipelines on the engine are double for increased safety. The main gas pipeline is covered with a sealed casing. The enclosure between the casing and the line is vented with an inert gas mixture and sensors for monitoring the concentration of gas in the inert mixture are installed at the exit of the ventilation line. If the presence of gaseous fuel is detected in the gas mixture, an automatic protection system is triggered, which switches the engine to liquid fuel, cuts off the gas supply and turns on the purge of gas lines.

For the injection of a pilot portion of fuel, the engine is equipped with an additional accumulator system (Common Rail) with electronic control, and an additional small-sized injector is installed in the cylinder head. When working on the gas-diesel cycle, the MAN 35/44DF engine is capable of fulfilling the IMO Tier III standard. Compared to running on fuel oil in a gas-diesel cycle, CO_2 emissions are reduced by 25%, and other emissions are reduced to 99% or higher. During operation, engines of this series can switch from liquid to gas fuel and back, with loads ranging from 15 to 100% without loss of power.

The MAN 35/44DF series engines are equipped with a new generation of control system developed by MAN Diesel & Turbo, called the $SaCoS_{one}$ (Safety and Control System), which combines the functions of controlling engine systems in one system, including control of the injection fuel injection system as well as the gas supply system. Thanks to integration, the system allows not only to control the engine, but also to collect information for its diagnosis and self-diagnosis.

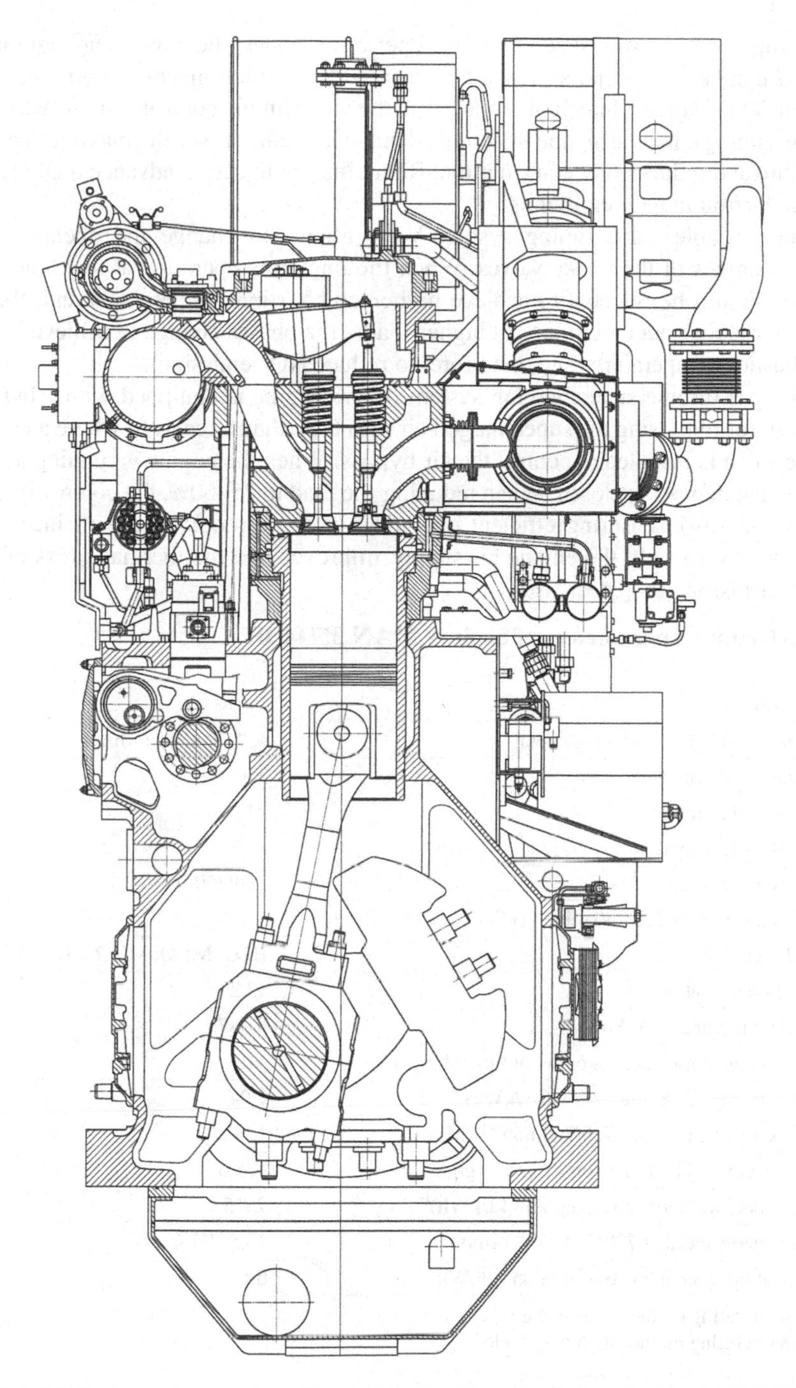

Fig. 2.4 Medium-speed gas engine with antechamber-flare electric flame ignition MAN 35/44DF [4]

To optimize the workflow in various operating modes, the engines are equipped with a number of special systems, which include the fuel injection timing control system VIT (Variable Injection Timing) and the valve timing control system (Variable Valve Timing). Increasing the injection advance angle increases the maximum cycle pressure and reduces fuel consumption. Reducing the injection advance angle leads to a reduction in NO_x emissions.

The variable valve timing system VVT allows to change the opening and closing angles of the intake valves. When the engine is operating at low load, the phases should be selected to achieve higher combustion temperatures and, therefore, to reduce soot emissions. At higher loads, the phases change to achieve lower combustion temperatures and, therefore, to reduce NO_x emissions.

The gas turbine supercharging system of the engine is equipped with a bypass channel for bypassing the supercharged air into the exhaust receiver, where an automatic valve is installed to control the air bypass. When the engine is running at full power, the damper is closed. When reducing the load to 25–60%, the control system opens the valve, ensuring efficient operation of the turbocharger with increased air flow. As a result, the engine air supply improves and the thermal stress of the cylinder-piston group decreases.

Main technical parameters of engines MAN 35/44DF

Parameter	Value
Number and cylinders arrangement	6, 7, 8, 9, 10 in-line
Cylinder bore (mm)	350
Piston stroke (mm)	440
Cylinder capacity (dm^3)	42.3
Rotation speed (min^{-1})	720/750
Cylinder power at 720/750 min^{-1} (kW)	510/530
Used fuels	HFO, MDO, MGO и LNG
Compression ratio	13.2
Air charging pressure (MPa)	0.347
Exhaust gas temperature at outlet of turbocharger (°C)	319
Brake specific air consumption (кg/kWh)	7.04
Mean effective pressure 720/750 min^{-1} (MPa)	2.0/2.01
Brake specific liquid fuel consumption (g/kWh)[a]	175.5
Brake specific energy consumption (kJ/kWh)[b]	7515
Mean piston speed at 720/750 min^{-1} (m/s)	10.56/11.0
Lubrication system oil consumption (g/kWh)	0.5

[a]When operating on the diesel cycle
[b]When operating on the gas-diesel cycle

Dimensions and weight engines series MAN 35/44DF

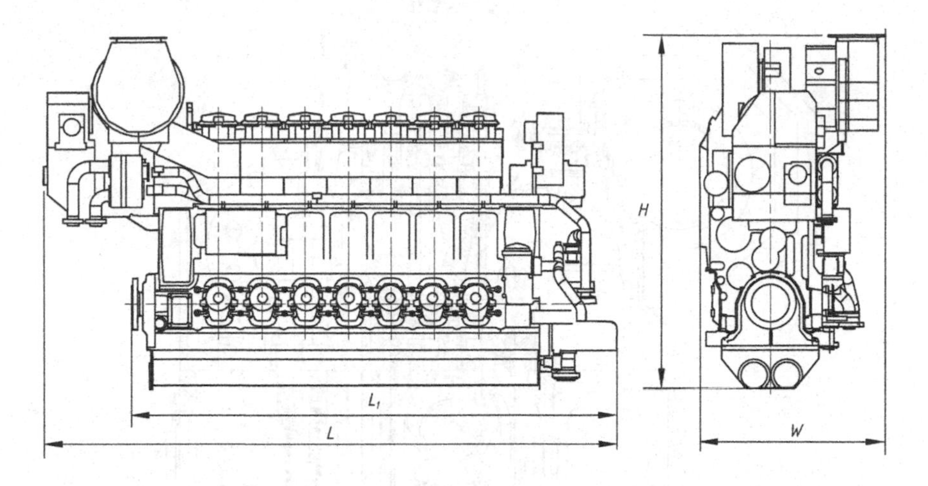

Engine version	L (mm)	L_1 (mm)	W (mm)	H (mm)	Weight (kg)
MAN 6L 35/44DF	6485	5265	2539	4163	40,500
MAN 7L 35/44DF	7015	5877	2678	4369	45,600
MAN 8L 35/44DF	7545	6407	2678	4369	50,700
MAN 9L 35/44DF	8075	6937	2678	4369	55,000
MAN 10L 35/44DF	8605	7556	2678	4369	59,700

2.2.2 *The Caterpillar MaK M46DF Engine*

The Caterpillar MaK M46DF Engine was developed on the basis of the MaK M43C diesel engine (Fig. 2.5). In gas diesel performance, the engine has 6, 7, 8 and 9 cylinders. The engines of this series are equipped with a standard fuel system for operation on liquid fuels, as well as a gas system with the dosing of gas fuel directly at the entrance to the working cylinder with the help of valves with an electromagnetic drive. A separate accumulator-type fuel system is installed on the engine to supply the pilot fuel. Gas supply systems and ignition are controlled by an electronic unit, which, in addition to the control, monitors and diagnoses the engine. Efficient selection of parameters for the operation of all systems allowed to obtain a stable engine operation in the entire load range on gas fuels with a methane number of 55 or more, which is the best indicator for gas-diesel engines of a similar dimension.

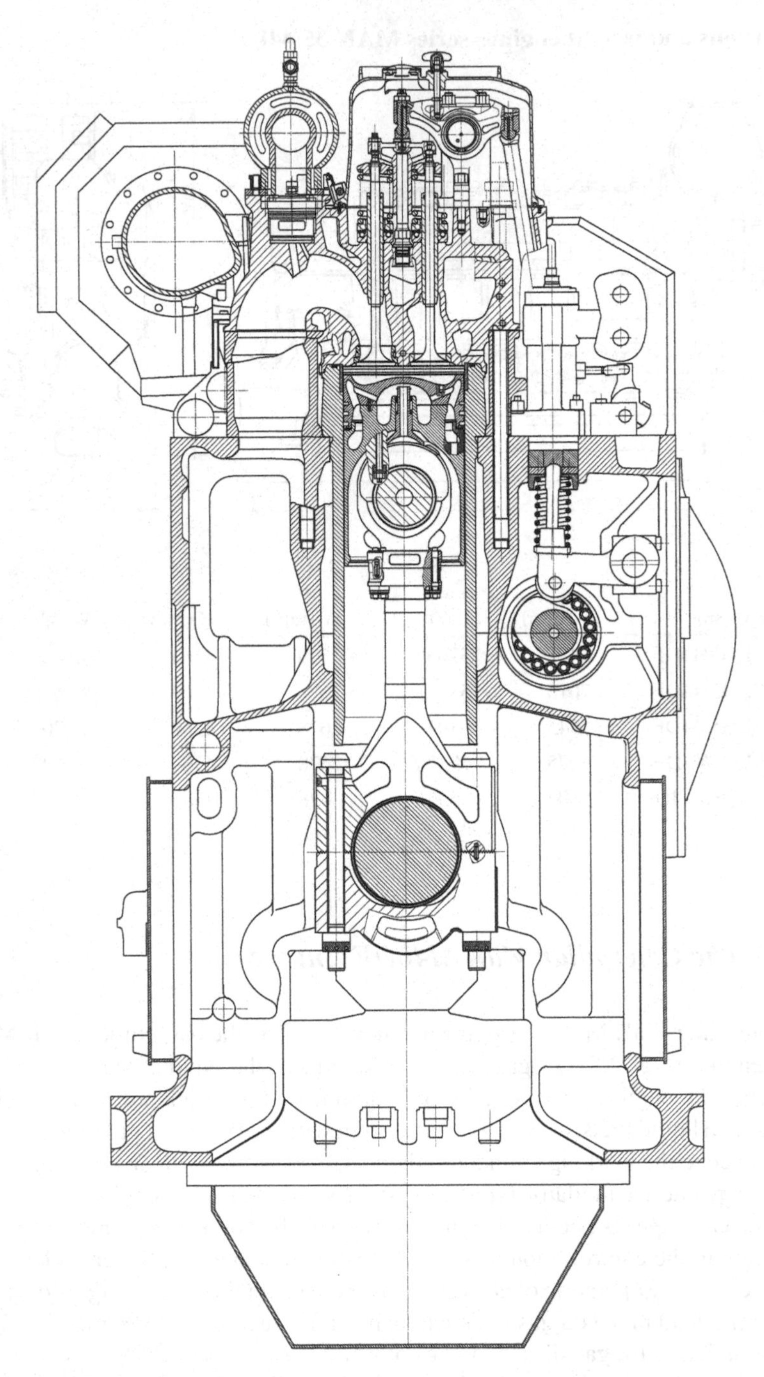

Fig. 2.5 Medium-speed gas engine with antechamber-flare electric flame ignition MaK M46DF [5]

Main technical parameters of engines MaK M46DF

Parameter	Value
Number and cylinders arrangement	6, 7, 8, 9 in-line
Cylinder bore (mm)	460
Piston stroke (mm)	610
Cylinder capacity (dm^3)	101.4
Rotation speed (min^{-1})	500/514
Cylinder power at 500/514 min^{-1} (kW)	900
Air charging pressure (MPa)	0.355
Maximum cycle pressure (MPa)	19.0
Exhaust gas temperature at outlet of turbocharger (°C)	345
Brake specific air consumption (кg/kWh)	5.94
Mean effective pressure (MPa)	2.13/2.07
Brake specific fuel oil consumption (g/kWh)[a]	186.0
Brake specific energy consumption (kJ/kWh)[b]	7200
Brake specific pilot oil consumption under load 100/15 (%)	2.0/6.9
Mean piston speed at 500/514 min^{-1} (m/s)	10.17/10.45

[a]When operating on the diesel cycle
[b]When operating on the gas-diesel cycle

Dimensions and weight engines series MaK M46DF

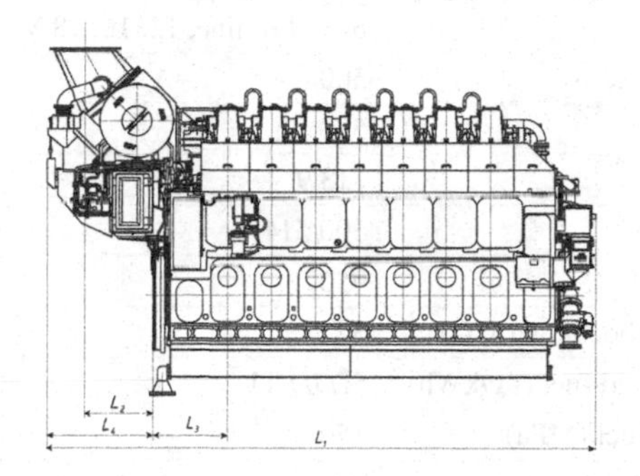

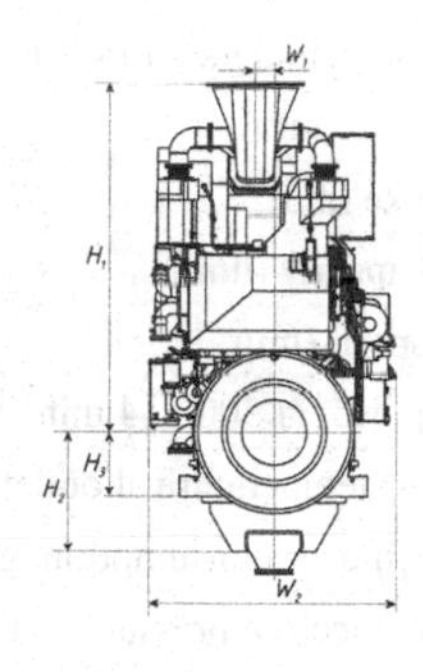

Engine version	L_1 (mm)	L_2 (mm)	L_3 (mm)	L_4 (mm)	H_1 (mm)
6 M 46 DF	8330	1086	1255	1723	3734
7 M 46 DF	9068	1119	1255	1740	4105
8 M 46 DF	9798	1119	1255	1740	4105
9 M 46 DF	10,768	1119	1255	1740	4105

(continued)

(continued)

Engine version	H_2 (mm)	H_3 (mm)	W_1 (mm)	W_2 (mm)	Weight (kg)
6 M 46 DF	1396	750	215	2961	94,000
7 M 46 DF	1396	750	232	2961	107,000
8 M 46 DF	1396	750	232	2961	114,000
9 M 46 DF	1396	750	232	2961	127,000

2.2.3 The L50DF Wärtsilä Engine

The L50DF Wärtsilä engine is based on 46 series diesel engines (Fig. 2.6). Engines of this series are manufactured in-line and V-shaped execution. For work on liquid fuels, the engine is equipped with a fuel system of volumetric action, which injects fuel into the working cylinder through an injector with a double nozzle tip. Through the same nozzle, the pilot fuel is injected while operating on gas fuel. Light fuel distillates are used as ignition fuels, for which the engine is equipped with a special accumulator fuel system. To prevent carbonization of the nozzles of the pilot fuel when operating on liquid fuel, the pilot injection system should be switched to the minimum supply mode.

Main technical parameters of engines Wärtsilä L50DF

Parameter	Value
Number and cylinders arrangement	6, 8, 9 in-line; 12, 16, 18 V-shaped
Cylinder bore (mm)	500
Piston stroke (mm)	580
Cylinder capacity (dm^3)	113.9
Rotation speed (min^{-1})	500/514
Cylinder power at 500/514 min^{-1} (kW)	950
Exhaust gas temperature at outlet of turbocharger (°C)	373
Brake specific air consumption, gas/liquid fuel (кg/kWh)	5.77/7.11
Maximum injection pressure of liquid fuel (MPa)	150
Injection pressure of pilot oil (MPa)	90
Mean effective pressure (MPa)	2.0/1.95
Brake specific fuel oil consumption (g/kWh)[a]	189
Brake specific energy consumption (kJ/kWh)[b]	7410
Brake specific pilot oil consumption (%)	1.0
Mean piston speed at 500/514 min^{-1} (m/s)	9.67/9.94

[a]When operating on the diesel cycle
[b]When operating on the gas-diesel cycle

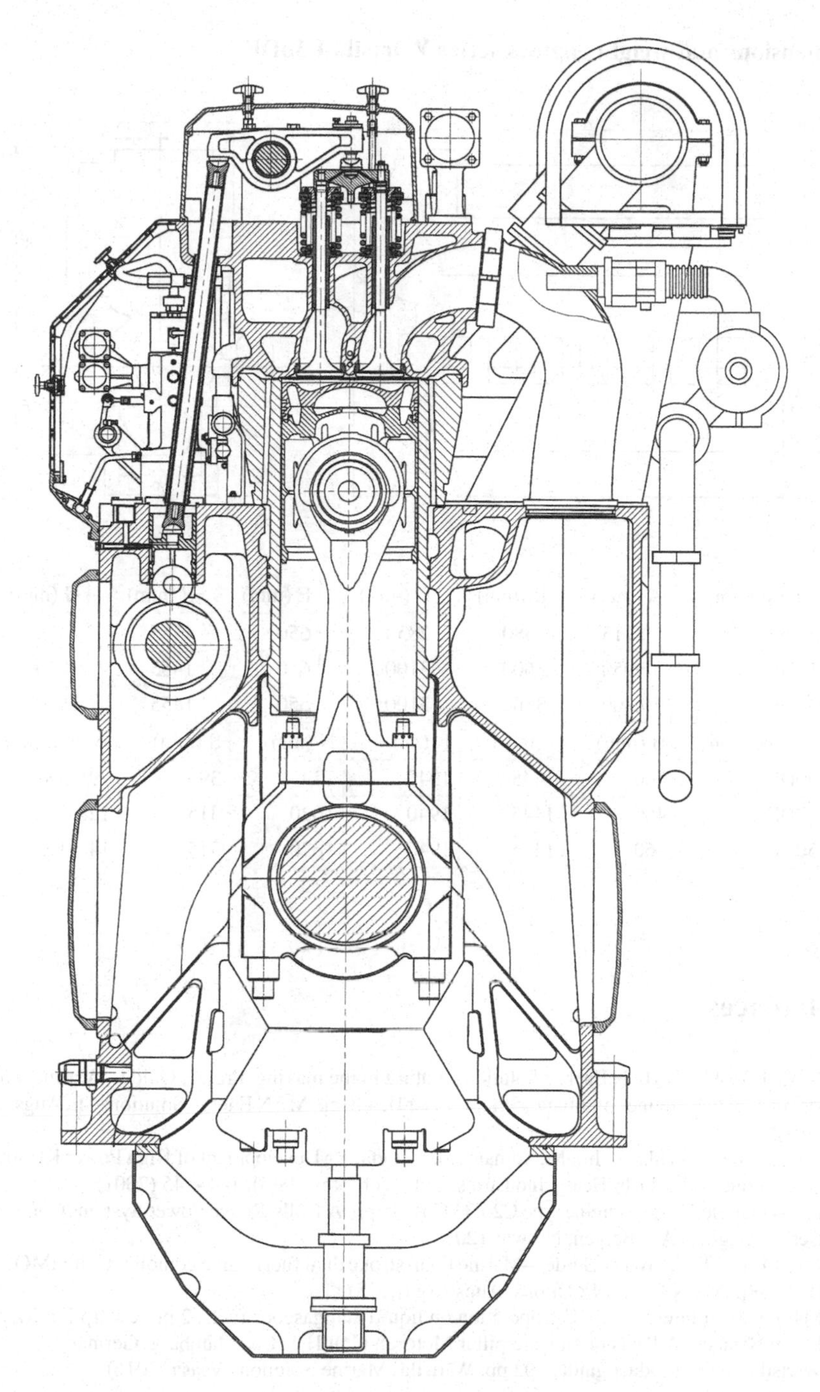

Fig. 2.6 Medium-speed gas engine with antechamber-flare electric flame ignition Wärtsilä L50DF [6]

Dimensions and weight engines series Wärtsilä L50DF

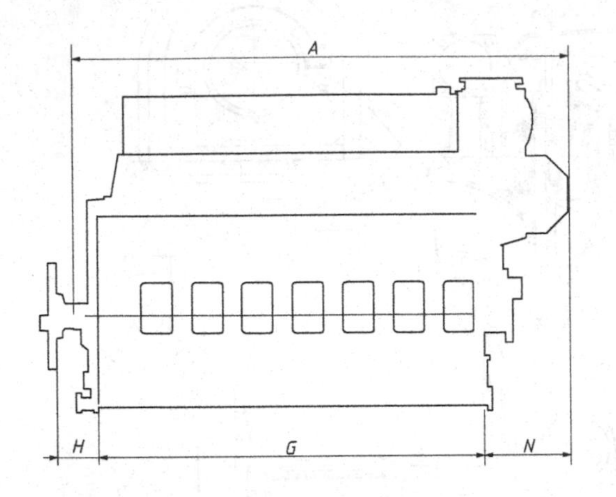

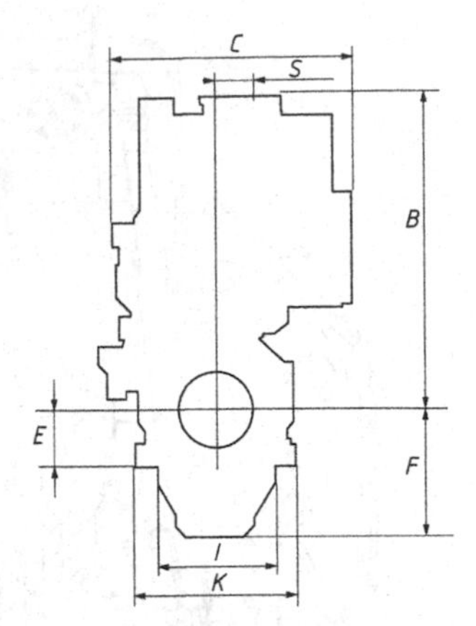

Engine version	A (mm)	B (mm)	C (mm)	E (mm)	F (mm)	G (mm)
6L50DF	8115	3580	2850	650	1455	6170
8L50DF	9950	3600	3100	650	1455	7810
9L50DF	10,800	3600	3100	650	1455	8630

Engine version	H (mm)	I (mm)	K (mm)	N (mm)	S (mm)	Weight (kg)
6L50DF	460	1445	1940	1295	395	96,000
8L50DF	460	1445	1940	1620	315	128,000
9L50DF	460	1445	1940	1620	315	148,000

References

1. MAN L35/44DF. MAN Energy Solutions. Future in the making. Project Guide—Marine. Four-stroke dual fuel engine compliant with IMO Tier II, 440 pp. MAN Energy Solutions SE, Augsburg (2018)
2. Nakano, R., Yasueda, S., Ito, K., Yamamoto, T., Oda, Y.: Development of High Power KU30GA Gas Engine. Mitsubishi Heavy Industries, Ltd. Tech. Rev. **38**(3), 141–145 (2001)
3. Project Guide. Bergen engine type C26:33 Gas, 85 pp. A Rolls-Royce Power Systems Company, Bergen Engines AS, Bergen, Norway (2016)
4. MAN L35/44DF. Project Guide—Marine Four-stroke dual fuel engine compliant with IMO Tier II, 440 pp. MAN Energy Solutions, Augsburg (2018)
5. M46DF Dual Fuel Engine. For operation on liquid and gaseous fuels, 2 pp. Caterpillar Marine Power Systems A Division of Caterpillar Motoren GmbH & Co., Hamburg, Germany
6. Wärtsilä 46 DF Product guide, 202 pp. Wärtsilä, Marine Solutions Vaasa (2016)

Chapter 3
Two-Stroke Ship Low-Speed Crosshead Engines

In 1879, the German inventor Karl Benz patented a two-stroke gas engine. English engineer Dugald Clerk significantly improved the 2-stroke cycle in the engines. He obtained a patent for them in 1881. The main advantage of a 2-stroke cycle over a four-stroke cycle is the ability to significantly (1.5–1.8 times) increase the power, taken from the same working volume. It was that advantage that forced the Swiss firm Sulzer Brothers Ltd., in 1905, to begin the development and construction of a ship low-speed crosshead diesel engine. The first two-stroke marine diesel engine was a four-cylinder engine with a cylinder diameter of 175 mm and a piston stroke of 250 mm. The engine developed a cylinder power of 25 horsepower at a rotational speed of 375 min^{-1}. An interesting fact is that this engine had a straight-flow-valve purge, which the company soon refused and returned to this scheme only in 1980. The only difference in the concept of a straight-flow-valve purge of 1905 from the modern one was, that the valves, installed in the cylinder head, were used to supply air, and the windows in the lower part of the cylinder sleeve for exhaust gas. Subsequently, similar purge schemes were used on their engines by other manufacturers, such as MAN, Krupp, Cards, AB Dicsel-Motorer, Fiat. The British company Doxford used a straight-flow-valve system through the purge and exhaust ports located at different ends of the cylinder liner, which were blocked by two working pistons located in one cylinder on both sides. Previously, a similar scheme was developed for aircraft diesel engines by German engineer Hugo Junkers.

Based on the intention to maximally simplify the design of the ship's engine, primarily to improve its reliability, Sulzer declined the straight-flow-valve scheme in favor of the transverse-slit purge, where both the inlet and outlet are made through two groups of windows in the cylinder liner. The first engine, operating according to this scheme, was built in 1909, and for this manufacturer this purge scheme for many years had become the main until 1956, when two-stroke low-speed RD series engines with loopback purge were introduced to the market.

I. Bilousov et al., *Modern Marine Internal Combustion Engines*, Springer Series on Naval Architecture, Marine Engineering, Shipbuilding and Shipping 8,
https://doi.org/10.1007/978-3-030-49749-1_3

The first passenger ship Monte Penedo, built in 1912, was the first large ocean vessel, equipped with two-stroke low-speed engines. The ship had two four-cylinder Sulzer 4S47 engines, installed with a cylinder diameter of 470 and a piston stroke of 680 mm. Each of the engines developed 850 horsepower at a rotational speed of 160 min^{-1}.

An important stage in the development of ship's low-speed engines was the replacement of the compressor injection system into the combustion chamber with direct-acting fuel injection systems, which began in 1929.

Features of the organization of gas exchange in two-stroke marine diesel engines for a long time did not allow to solve the problem of using gas turbine supercharging for them. In such engines, mounted piston pumps were used to purge the working cylinder, and subsequently it were piston pumps as additional charging set. The practical result on the use of gas turbine pressurization in two-stroke diesel engines was achieved only in 1952. This year the tanker Dorthe Maersk with a displacement of 18,000 tons, built at the shipyard A. P. Moller in Denmark. It was the first ship which had a Burmeister & Wain (B&W) two-stroke six-cylinder diesel engine with a gas turbine supercharger. Two turbochargers series VTR630, mounted on the engine, allowed to increase its power from 4070 to 5800 kW.

In the design of the crosshead engine, where the cylinder is separated from the crankcase by a diaphragm, the possibility of using the sub-piston space as sub-piston pumps was realized. An automatic air valve was used to control the flow of air between the sub-piston cavity, the working cylinder and the environment. Subsequently, with the appearance of gas turbine supercharging on low-speed diesel engines, sub-piston pumps were used as a second compression stage, which was connected in parallel or in a series with a turbo compressor, forcing air into a common second-stage receiver or into a special buffer space for each cylinder.

For a long time, the use of various types of gas exchange schemes dominated on low-speed two-stroke engines. This was due to the simplicity of the design, where there were no gas distribution valves with their actuator mechanisms, relatively simple solutions for the reversing systems and configuration of the combustion chamber with the central location of the injector, etc. For a long time these features overshadowed the disadvantages of the transverse-slit purge schemes, such as: poor cleaning of the working cylinders from combustion products; significant loss of the working stroke, associated with a large height of gas distribution windows; the need to have a sufficient height of the piston to avoid the simultaneous opening of the purge and discharge ports when the piston is close to the TDC; loss of charge due to the fact that from the moment of closing of the purge and until the moment of closing of the exhaust ports, the charge is displaced by the piston to the exhaust receiver. These disadvantages significantly limited the possibility of further increasing the cylinder power and the level of boost of low-speed engines. A special direction in the field of designing two-stroke low-speed diesel engines was an attempt to create double-acting engines, where both the over-piston and sub-piston cavities are workers. For example, the firm Harland & Wolf engines in the 30s of the last century produced dual-action crosshead engines with three pistons in the working cylinder, intake, exhaust and main.

By the end of the twentieth century, it had become clear, that all attempts to improve the transverse-slit purge schemes did not yield the expected result, and the desire to improve the performance of the engines led to an unjustified complication of the design. For this reason, since the 80s of the last century, some manufacturers refused to produce them, and the remaining manufacturers, after the reorganization of their production, switched to the production of engines with a straight-flow-valve purge scheme.

Thus, in all modern low-speed marine diesel engines (LSMDE), the working cylinder is purged through the purge windows, located in the lower part of the cylinder sleeve. Controls the opening of the windows working piston, simultaneously performing the functions of the spool. Exhaust gases are discharged through one exhaust valve, located in the cylinder head and having a hydraulic drive. This scheme allows to significantly simplify the design of the valve assembly and the drive mechanism, as well as to maximize the valve opening flow area. Its main disadvantage is the impossibility of the injector location along the axis of the working cylinder. In order to evenly distribute the fuel aerosol in the entire volume of the combustion chamber, on engines, made according to this scheme, two or three peripheral injectors are installed, which leads to a complication of the fuel system.

Today, only three manufacturers in the world together with their licensees produce low-speed engines, such companies as MAN Diesel & Turbo SE, Wärtsilä NSD and Mitsubishi Heavy Industries, Ltd.

3.1 MAN Diesel & Turbo SE

MAN Diesel & Turbo SE is by far the largest manufacturer of ship low-speed engines, which provides approximately 70% of all global needs in this market segment. The company has about 20 licensees, most of whom are concentrated in the Asia-Pacific region. Since the mid-60s of the twentieth century, the company introduced to the market its own line of low-speed crosshead engines such as KSZ, which were produced with cylinder diameters from 50 to 90 cm and had from 4 to 12 cylinders. In the engines of this series, the principle of loop-purge through the blowing-out and exhaust windows, located one above the other, cut through on one side of the cylinder bushing, was implemented. All modifications of engines of this series were marked with additional indexes KSZ-A, KSZ-B and KSZ-C. Starting with the KSZ-B modification engines, the company switched from the arc-form design of basic engine block to the box-form construction, which later become widespread, including engines from other manufacturers.

In the early 80s of the twentieth century, it became clear, that the gas exchange circuit, used by the company, was not promising for further forcing diesel engines, where the temperature of the bushings in the exhaust ports increased significantly. This resulted into their destruction and failure of the entire engine.

In 1980, MAN AG acquired a diesel engine plant from the Danish company Burmeister & Wain (B&W), inheriting the most promising developments in the field of two-stroke low-speed diesel engines. It should be noted, that the company

Burmeister & Wain at that time was one of the leaders in the design and manufacture of engines of this type, from its first models, releasing engines with a straight-flow-valve purge scheme.

Having bought the Burmeister & Wain production facilities located in the suburb of Copenhagen Christianshavn, MAN stopped developing and producing its own engine models, focused on bringing and launching the production of low-speed diesel engines of the MC type, which were manufactured under the MAN B&W brand for another 10 years. The first engine, put into production by the MC series, was the diesel L35MC, launched in 1982, and in September of the next year the first low-speed engine with a cylinder diameter of 600 mm L60MC was launched and tested. The first stage of the program included the development of engines of the L-MC group with diameters of working cylinders of 350, 500, 600, 700, 800 and 900 mm. In 1988, the industrial production of low-speed engines was moved from Christianshavn to another Copenhagen suburb, Teglholmen, and in 1992 a research center was created here.

In subsequent years, groups with the index K (short stroke, S/D = 2.45–2.88) were added to the already existing group of engines with the index L (Long stroke—long-stroke, S/D = 2.8–3.24) and with the S index (Super long-stroke, with S/D = 3.54–4.4). The range of cylinder diameters was expanded, dimensions 260, 420, 460 and 980 mm were added to the already existing dimensions, and work was also done to create an engine with a cylinder diameter of 1080 mm. Beginning in 1994, work began on the creation of compact versions on the basis of MC-type engines, which received an additional MC-C index (for example, K98MC-C) in the marking.

In 2002, on the basis of the MC-C type engines, a modification with electronic control of the fuel injection, valve timing, start-up and reverse processes, as well as the lubrication of the working cylinders was developed. Engines of this series began to be designated by an index of ME (electronically controlled, for example, K98ME-C). In 2006, another G group was added to the already existing size groups ("Green" super long with S/D = 4.4–4.7). The engines of this size group have a lower rotational speed, which allows to maximize the propulsive efficiency of the power plant of the vessel.

3.1.1 The Engines of the MC Series

The engines of the MC series have a large number of different sizes, that cover the size groups L, K and S. All models are adapted to operate on heavy fuels of viscosity up to 720 cSt with high efficiency and reliability.

The engine basic block forms the foundation frame, the crankcase box and the cylinder liner block. All crankcase elements are tightened with the help of long anchor studs, which pass through the entire structure in specially made wells.

The engines of the MC series have a straight-flow-valve purge scheme, which simplifies the design of the sleeves and the purge windows themselves. The central exhaust valves are actuated by hydraulic actuators from the camshaft fist washers. The camshaft is driven by roller chains, which improves the weight and size characteristics of the transmission and allows the camshaft to be positioned as close as possible to the cylinder covers to reduce the length of the hydraulic valve tubes and minimize the adverse effect of wave processes in high-pressure fuel tubes. The reverse of engines is made by a rerun of rollers of pushers of fuel pumps.

On each cylinder there are two or three uncooled injectors, through which constantly, except for the injection moment, the fuel, heated to a temperature of 120–140 °C, is pumped.

The cylinder cover is made of forged steel and has a central hole for installing the exhaust valve assembly, which is fastened with four studs. Other openings in the cover are provided for installing injectors, start valve, start air inlet, safety valve and indicator valve.

The cooling jacket is installed at the bottom of the cylinder cover, forming a cooling cavity. Another cooling cavity is formed around the exhaust valve seat after installation. These two cavities are communicated by a large number of inclined and radial drillings along the bottom of the cover for cooling it.

The sleeves and cylinder heads are cooled with water through the internal channels and drilling, and the cooling of the pistons is oil-based, which simplifies the design of the cooling system and reduces the risk of water entering the crankcase. The number of piston rings is limited to four.

The piston head on the side of the firing surface has a heat resistant layer. It is attached to the top of the rod with long bolts, and the piston skirt is attached with bolts to the piston head. Long bolts have some elasticity, which allows to compensate for the thermal expansion of the parts, of which consists the piston. The piston head has chrome-plated grooves for the four piston rings. The two upper rings may be of increased height.

From the purge air receiver, the crankcase is separated by a diaphragm where a piston rod gland seal is installed to prevent lubricant oil entering from the crankcase into the cavity of the purge air and purge air into the crankcase.

The sleeves of the working cylinders are installed in separate jacket, assembled with the help of side fasteners in a single unit, which is attached to the engine basic block by epy means of vertical anchor ties. The sleeves are mounted in the upper part through installation collar, which allows them to extend into the cavity of the jacket without mechanical stresses. The cylinder block has studs for fastening the cylinder head. The studs have o-rings that protect the threads from corrosion.

The crankshaft is made either semi-composite or with welded cranks. Cross-head device is made with one continuous support thin-walled bearing and double-sided sliders. The engines of the MC series are fitted with highly efficient, uncooled turbochargers, to which the exhaust gases are supplied with constant pressure. For starting and running at low speed engines are equipped with autonomous electric blowers.

Symbols for the following characteristics of engines, manufactured by MAN

Next will be given the characteristics and parameters of engines, manufactured by MAN, for which the manufacturer has established a number of symbols and concepts.

Field of permissible engine operating conditions
Since the same engine can be installed on different ships, for each diesel model there is a so-called contract capacity field of Pe, which in graphic form is determined by a specific quadrilateral L_1, L_2, L_3, L_4 (Fig. 3.1). The logarithmic coordinate system is used to build the field of permissible modes, therefore the screw characteristics and constant mean effective pressure curves are straight lines on the graph. The relative rotational speed in percent is indicated on the abscissa axis, and the relative power of the diesel engine is indicated on the vertical axis. The nominal maximum continuous power mode (Maximum continuous rating MCR) corresponds to the point L_1 (p_{enom} = 100%, n_{nom} = 100%) through which the nominal screw characteristic passes.

In the remaining points:
for the point L_2 ($p_e = 0.8\ p_{enom}$, $n = n_{nom}$);
for the point L_3 ($p_e = p_{enom}$, n = 0.75–0.82 n_{nom});
for the point L_4 ($p_e = 0.8p_{enom}$, n = 0.75–0.82 n_{nom}).

For all contractual capacities, the maximum combustion pressure is maintained at the level of the nominal mode.

When determining the weight and dimensions of diesel engines, the following notation is used:

L_{min}—is the minimum engine length;
A—is the distance between the axes of the working cylinders;
B—is the distance between the primed surfaces;
B_1—is the width of the base frame at the base;
B_2—is maximum width of the base frame;
C—is the distance from the axis of the crankshaft to the base plate;

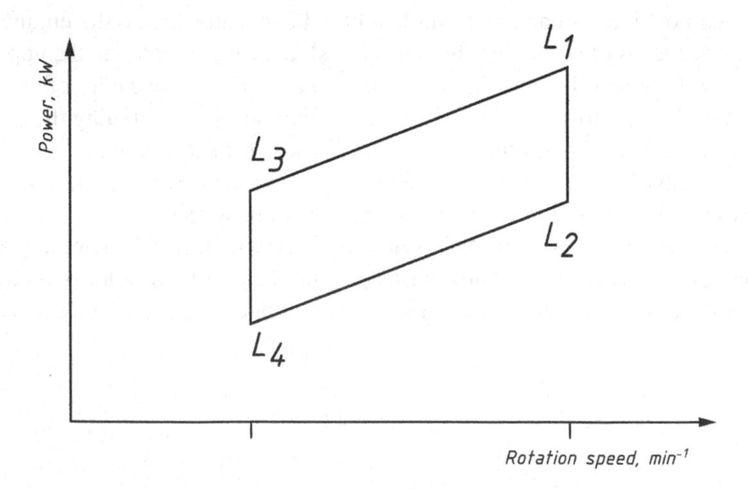

Fig. 3.1 Field of permissible engine operating conditions

H_1—is standard procedure for lifting parts of the cylinder-piston group;
H_2—is procedure with a reduced height of the cylinder-piston group;
H_3—is procedure for lifting with a MAN B&W Double-Jib electric crane;
H_4—is a normal lifting procedure with a MAN B&W Double-Jib crane.

Main technical parameters of engines MAN series S26MC (Fig. 3.2)

Parameter	Value
Number and cylinders arrangement	5, 6, 7, 8, 9, 10, 11, 12 in-line
Cylinder bore (mm)	260
Piston stroke (mm)	980
Cylinder capacity (dm^3)	52.03
Rotational speed (min^{-1})	250
Cylinder power, N_e (layout area)	
– maximum continuous rating at 250 min^{-1} (L_1) (kW)	400
– operational at 250 min^{-1} (L_2) (kW)	320
– maximum continuous rating at 212 min^{-1} (L_3) (kW)	340
– operational at 212 min^{-1} (L_4) (kW)	270
Air charging pressure (L_1) (MPa)	0.377
Compression pressure (L_1) (MPa)	15.2
Maximum cycle pressure (L_1) (MPa)	17.2
Exhaust gas temperature at inlet of turbocharger (L_1) (°C)	420
Exhaust gas temperature at outlet of turbocharger (L_1) (°C)	265
Mean effective pressure	
– at 250 min^{-1} (L_1, L_2) (MPa)	1.85
– at 212 min^{-1} (L_3, L_4) (MPa)	1.48
Brake specific fuel oil consumption (g/kWh)	
– at 250 min^{-1} (L_1/L_2) for N_e 100/80% (g/kWh)	179/177
– at 212 min^{-1} (L_3/L_4) for N_e 100/80% (g/kWh)	174/172
Mean piston speed at 250 min^{-1} (m/s)	8.17
Lubrication System oil consumption (g/kWh)	0.15
Cylinder oil consumption	
– mechanical lubricator (g/kWh)	1.0–1.5
– alpha lubricator (g/kWh)	0.7

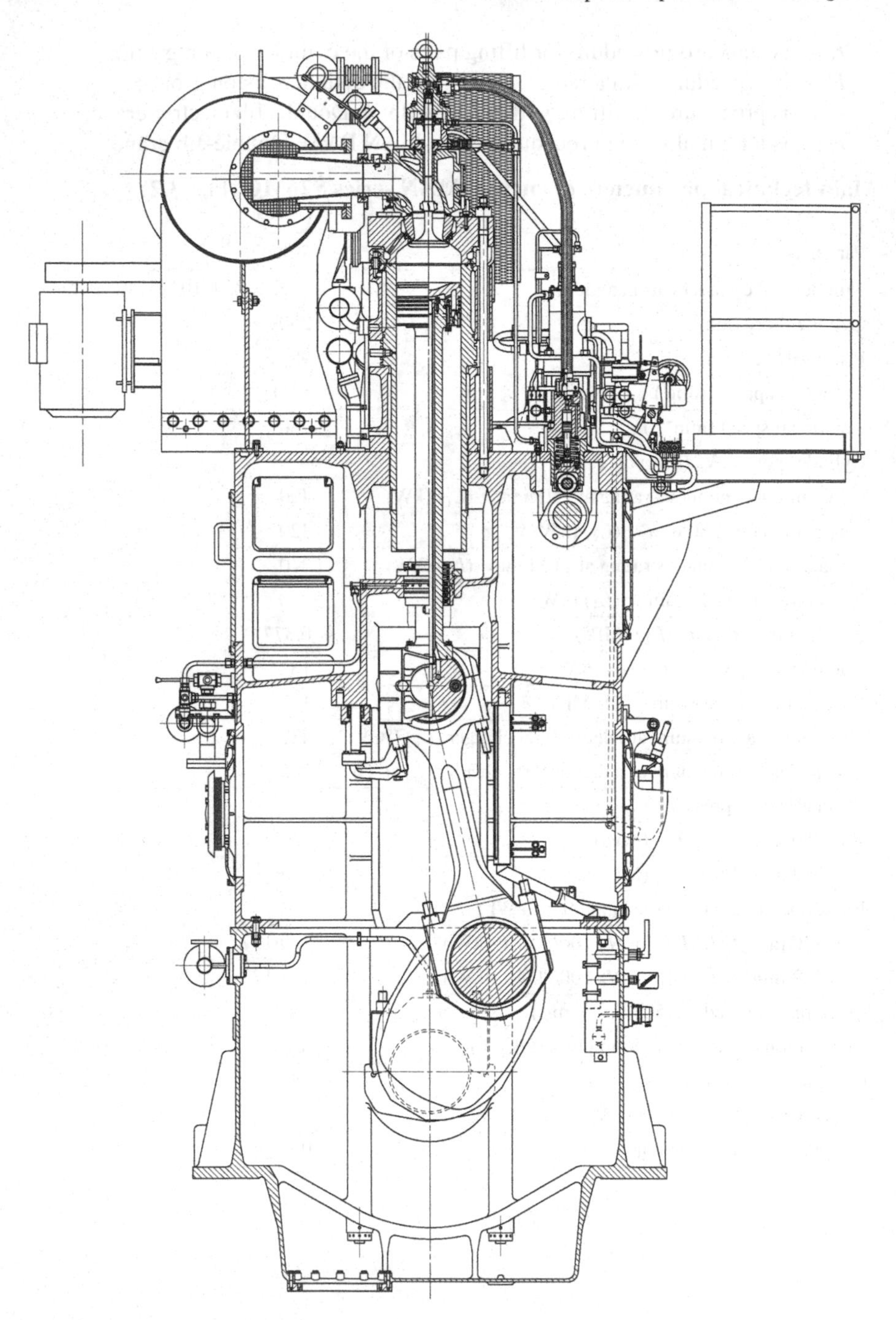

Fig. 3.2 Cross-section of the engine MAN series S26MC [1]

Dimensions and weight engines MAN series S26MC

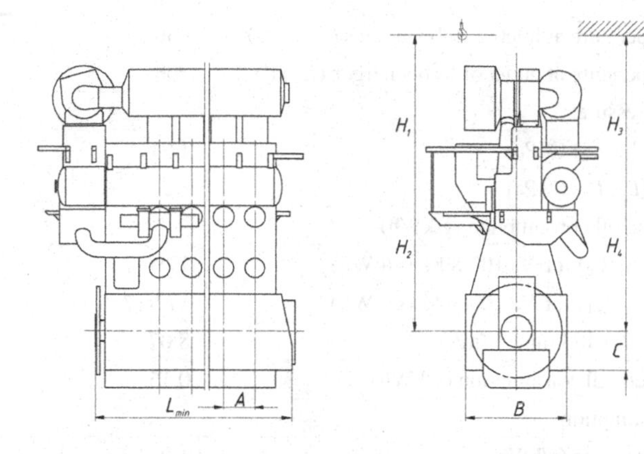

Number of cylinders	5	6	7	8	9	10	11	12
L_{min}	3637	4127	4617	5107	5597	6577	7067	7557
Weight (kg)	37,000	42,000	48,000	53,000	58,000	68,000	74,000	79,000

A (mm)	B (mm)	C (mm)	H_1 (mm)	H_2 (mm)	H_3 (mm)	H_4 (mm)
490	1880	420	4850	4750	4600	4525

Main technical parameters of engines MAN series S35MC (Fig. 3.3)

Parameter	Value
Number and cylinders arrangement	5, 6, 7, 8, 9, 10, 11, 12 in-line
Cylinder bore (mm)	350
Piston stroke (mm)	1400
Cylinder capacity (dm^3)	134.7
Rotational speed (min^{-1})	173
Cylinder power, N_e (layout area)	
– maximum continuous rating at 173 min^{-1} (L_1) (kW)	740
– operational at 173 min^{-1} (L_2) (kW)	595
– maximum continuous rating at 147 min^{-1} (L_3) (kW)	630
– operational at 147 min^{-1} (L_4) (kW)	505
Air charging pressure (L_1) (MPa)	0.36
Compression pressure (L_1) (MPa)	12.53
Maximum cycle pressure (L_1) (MPa)	14.52

(continued)

(continued)

Parameter	Value
Exhaust gas temperature at inlet of turbocharger (L_1) (°C)	430
Exhaust gas temperature at outlet of turbocharger (L_1) (°C)	268
Mean effective pressure:	
– at 173 min^{-1} (L_1, L_2) (MPa)	1.91
– at 147 min^{-1} (L_3, L_4) (MPa)	1.53
Brake specific fuel oil consumption (g/kWh)	
– at 173 min^{-1} (L_1/L_2) for N_e 100/80% (g/kWh)	178/176
– at 147 min^{-1} (L_3/L_4) for N_e 100/80% (g/kWh)	173/171
Mean piston speed at 173 min^{-1} (m/s)	8.07
Lubrication System oil consumption (g/kWh)	0.15
Cylinder oil consumption	
– mechanical lubricator (g/kWh)	1.0–1.5
– alpha lubricator (g/kWh)	0.7

Dimensions and weight engines MAN series S35MC

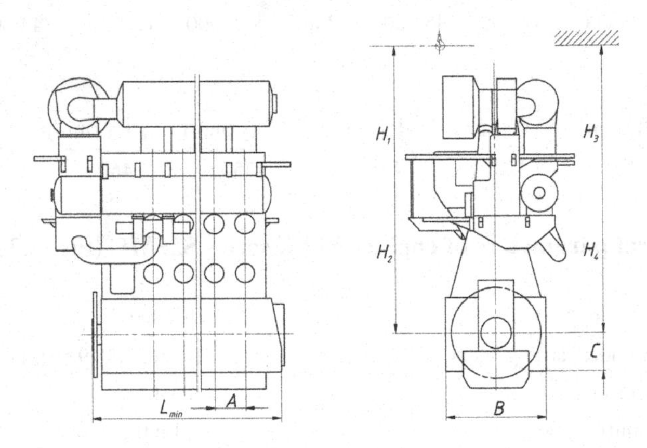

Number of cylinders	5	6	7	8	9	10	11	12
L_{min}	4209	4809	5409	6009	6609	7809	8409	9009
Weight (kg)	65,000	75,000	84,000	93,000	103,000	119,000	133,000	144,000

A (mm)	B (mm)	C (mm)	H_1 (mm)	H_2 (mm)	H_3 (mm)	H_4 (mm)
600	2200	650	6425	6275	6050	6075

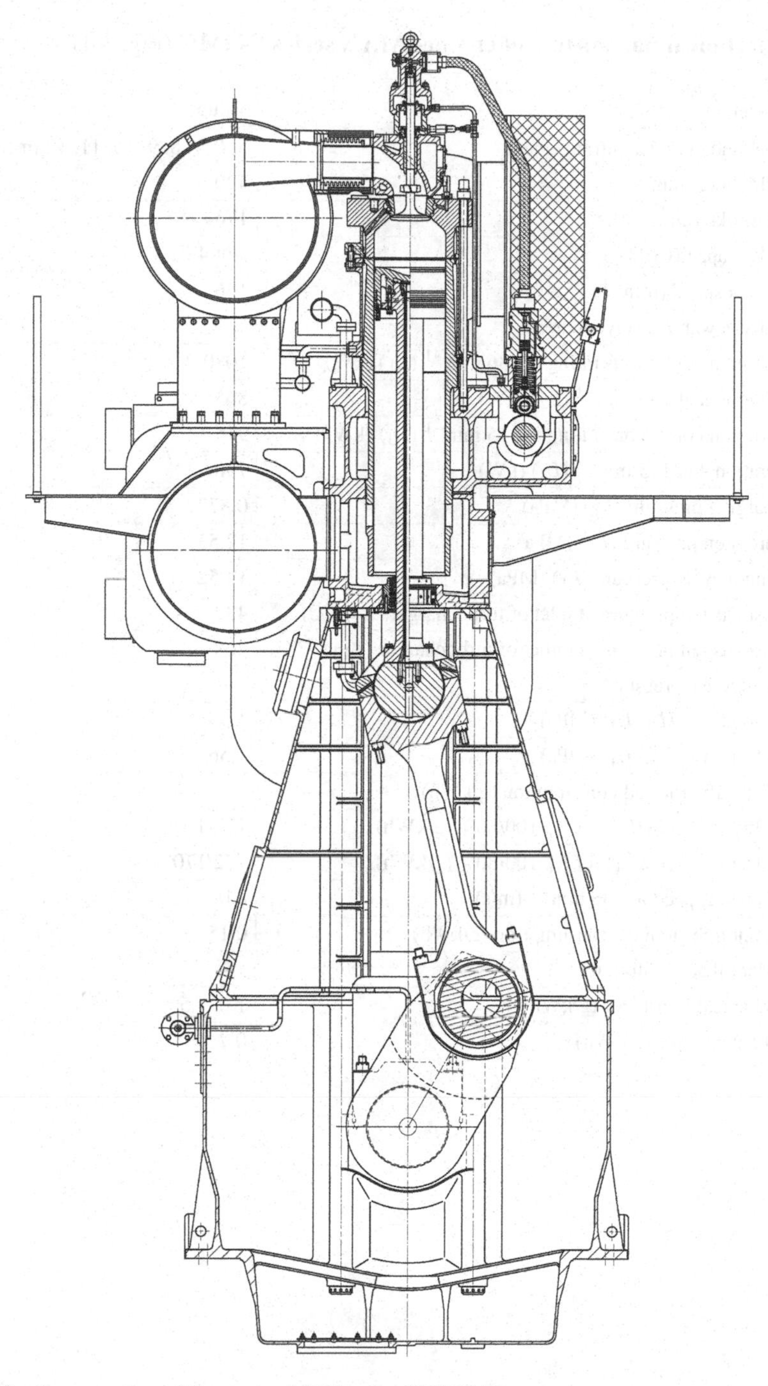

Fig. 3.3 Cross-section of the engine MAN series S35MC [2]

Main technical parameters of engines MAN series S42MC (Fig. 3.4)

Parameter	Value
Number and cylinders arrangement	5, 6, 7, 8, 9, 10, 11, 12 in-line
Cylinder bore (mm)	420
Piston stroke (mm)	1764
Cylinder capacity (dm^3)	244.4
Rotational speed (min^{-1})	136
Cylinder power, N_e (layout area)	
– maximum continuous rating at 136 min^{-1} (L_1) (kW)	1080
– operational at 136 min^{-1} (L_2) (kW)	865
– maximum continuous rating at 115 min^{-1} (L_3) (kW)	915
– operational at 115 min^{-1} (L_4) (kW)	730
Air charging pressure (L_1) (MPa)	0.372
Compression pressure (L_1) (MPa)	12.53
Maximum cycle pressure (L_1) (MPa)	14.52
Exhaust gas temperature at inlet of turbocharger (L_1) (°C)	432
Exhaust gas temperature at outlet of turbocharger (L_1) (°C)	268
Mean effective pressure:	
– at 136 min^{-1} (L_1, L_2) (MPa)	1.95
– at 115 min^{-1} (L_3, L_4) (MPa)	1.56
Brake specific fuel oil consumption (g/kWh)	
– at 136 min^{-1} (L_1/L_2) for N_e 100/80% (g/kWh)	177/175
– at 115 min^{-1} (L_3/L_4) for N_e 100/80% (g/kWh)	172/170
Mean piston speed at 136 min^{-1} (m/s)	8.0
Lubrication System oil consumption (g/kWh)	0.15
Cylinder oil consumption:	
– mechanical lubricator (g/kWh)	1.0–1.5
– alpha lubricator (g/kWh)	0.7

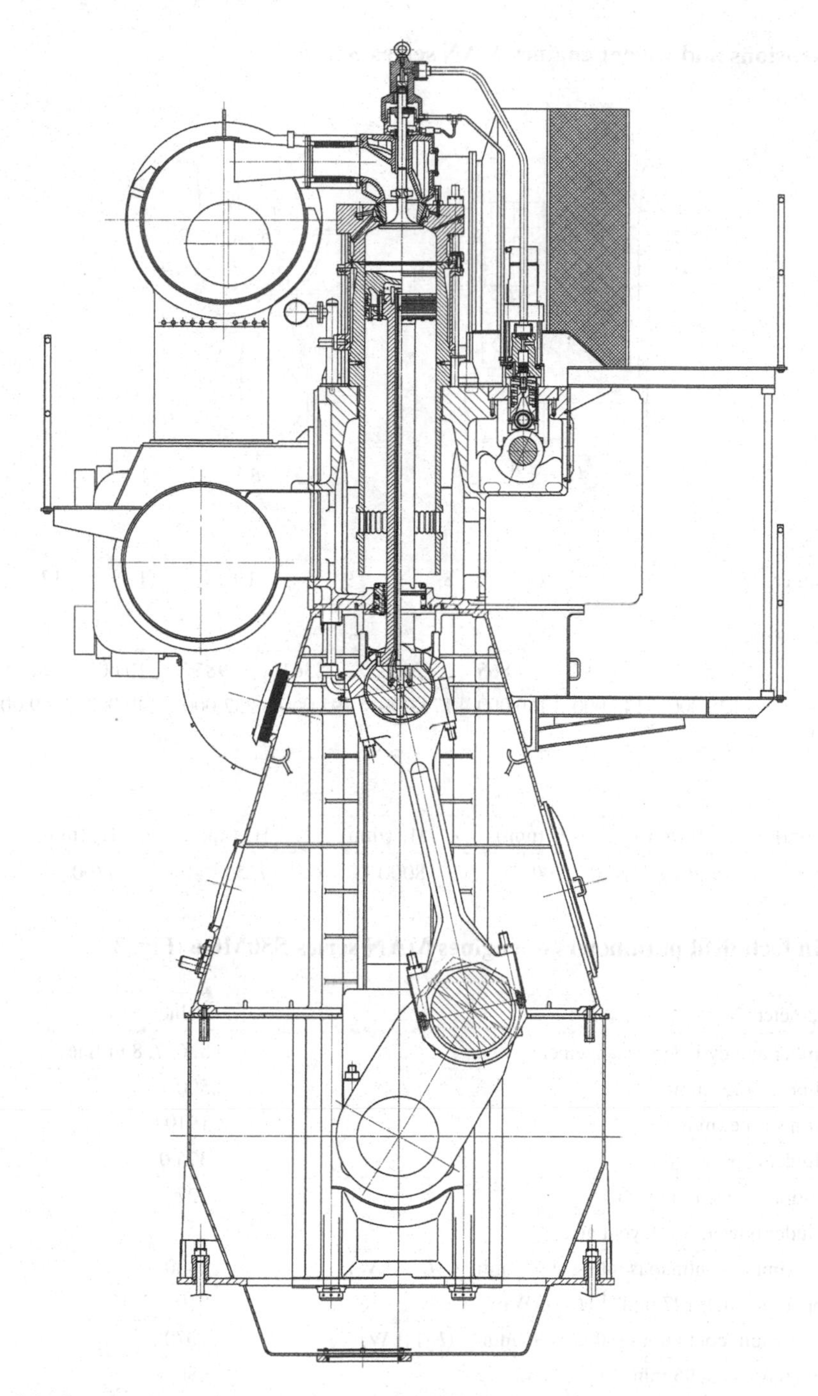

Fig. 3.4 Cross-section of the engine MAN series S42MC [3]

Dimensions and weight engines MAN series S42MC

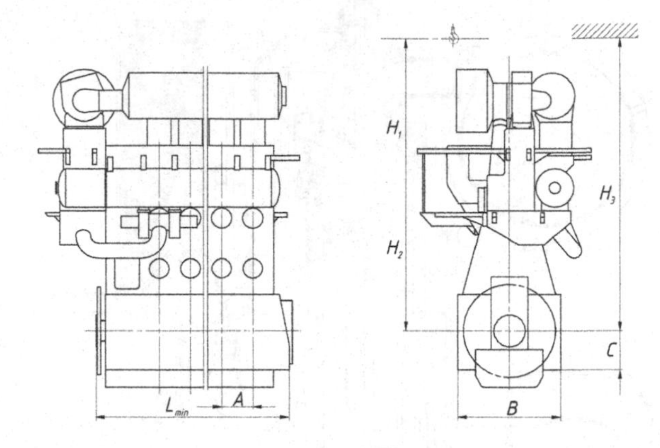

Number of cylinders	5	6	7	8	9	10	11	12
L_{min}	5369	6117	6865	7613	8361	9857	10,605	11,353
Weight (kg)	125,000	143,000	160,000	176,000	195,000	232,000	249,000	269,000

A (mm)	B (mm)	C (mm)	H_1 (mm)	H_2 (mm)	H_3 (mm)
748	2670	900	8000	7550	7300

Main technical parameters of engines MAN series S50MC6 (Fig. 3.5)

Parameter	Value
Number and cylinders arrangement	5, 6, 7, 8 in-line
Cylinder bore (mm)	500
Piston stroke (mm)	1910
Cylinder capacity (dm^3)	375.0
Rotational speed (min^{-1})	127
Cylinder power, N_e (layout area)	
– maximum continuous rating at 127 min^{-1} (L_1) (kW)	1430
– operational at 127 min^{-1} (L_2) (kW)	910
– maximum continuous rating at 95 min^{-1} (L_3) (kW)	1070
– operational at 95 min^{-1} (L_4) (kW)	680
Air charging pressure (L_1) (MPa)	0.352
Compression pressure (L_1) (MPa)	12.50
Maximum cycle pressure (L_1) (MPa)	14.20

(continued)

(continued)

Parameter	Value
Exhaust gas temperature at inlet of turbocharger (L_1) (°C)	410
Exhaust gas temperature at outlet of turbocharger (L_1) (°C)	252
Mean effective pressure	
– at 127 min^{-1} (L_1, L_2) (MPa)	1.80
– at 95 min^{-1} (L_3, L_4) (MPa)	1.15
Brake specific fuel oil consumption (g/kWh)	
– at 127 min^{-1} (L_1/L_2) for N_e 100/64% (g/kWh)	171/168
– at 95 min^{-1} (L_3/L_4) for N_e 100/64% (g/kWh)	159/157
Mean piston speed at 127 min^{-1} (m/s)	8.1
Lubrication System oil consumption (g/kWh)	0.15
Cylinder oil consumption	
– mechanical lubricator (g/kWh)	1.0–1.5
– alpha lubricator (g/kWh)	0.7

Dimensions and weight engines MAN series S50MC6

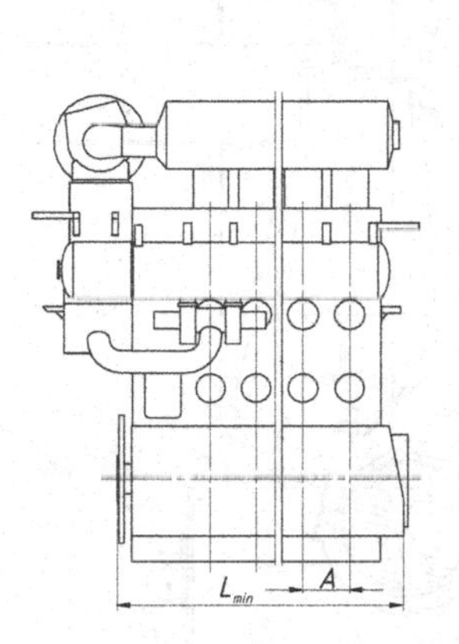

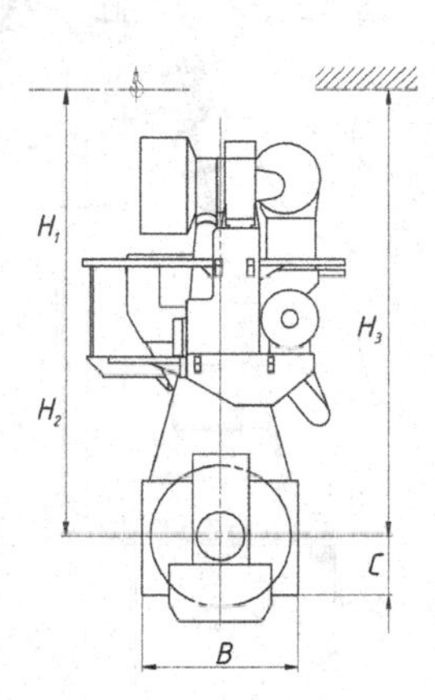

Number of cylinders	5	6	7	8
L_{min}	6602	7492	8382	9272
Weight (kg)	195,000	225,000	255,000	288,000

A (mm)	B (mm)	C (mm)	H_1 (mm)	H_2 (mm)	H_3 (mm)
890	2950	1085	8875	8300	8125

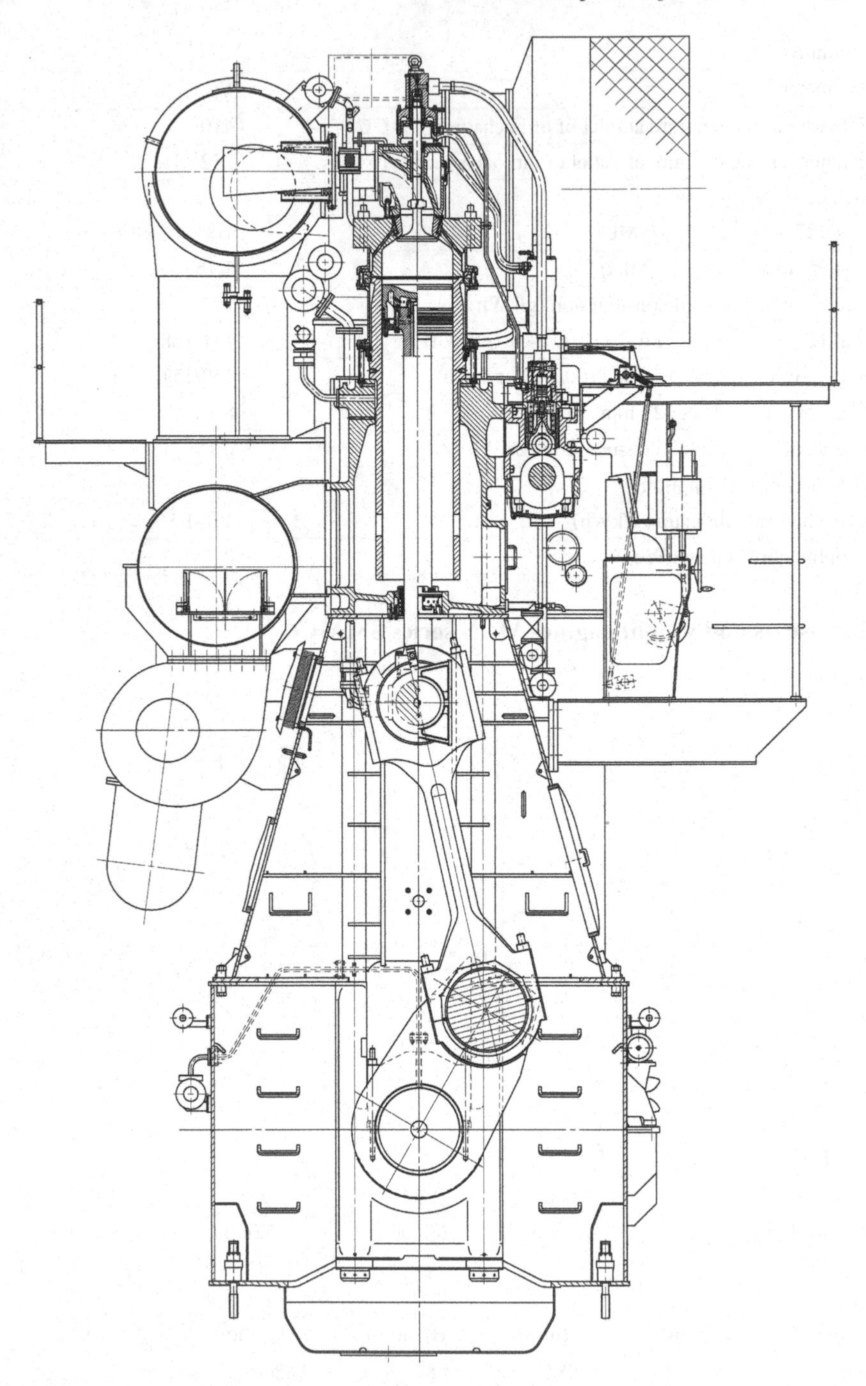

Fig. 3.5 Cross-section of the engine MAN series S50MC6 [4]

Main technical parameters of engines MAN series S60MC6 (Fig. 3.6)

Parameter	Value
Number and cylinders arrangement	5, 6, 7, 8 in-line
Cylinder bore (mm)	600
Piston stroke (mm)	2292
Cylinder capacity (dm^3)	648.03
Rotational speed (min^{-1})	105
Cylinder power, N_e (layout area)	
– maximum continuous rating at 105 min^{-1} (L_1) (kW)	2040
– operational at 105 min^{-1} (L_2) (kW)	1300
– maximum continuous rating at 79 min^{-1} (L_3) (kW)	1540
– operational at 79 min^{-1} (L_4) (kW)	980
Air charging pressure (L_1) (MPa)	0.352
Compression pressure (L_1) (MPa)	12.52
Maximum cycle pressure (L_1) (MPa)	14.20
Exhaust gas temperature at inlet of turbocharger (L_1) (°C)	392
Exhaust gas temperature at outlet of turbocharger (L_1) (°C)	246
Mean effective pressure	
– at 105 min^{-1} (L_1, L_2) (MPa)	1.80
– at 79 min^{-1} (L_3, L_4) (MPa)	1.15
Brake specific fuel oil consumption (g/kWh)	
– at 105 min^{-1} (L_1/L_2) for N_e 100/64% (g/kWh)	170/167
– at 79 min^{-1} (L_3/L_4) for N_e 100/64% (g/kWh)	158/156
Mean piston speed at 105 min^{-1} (m/s)	8.02
Lubrication System oil consumption (g/kWh)	0.15
Cylinder oil consumption	
– mechanical lubricator (g/kWh)	1.0–1.5
– alpha lubricator (g/kWh)	0.7

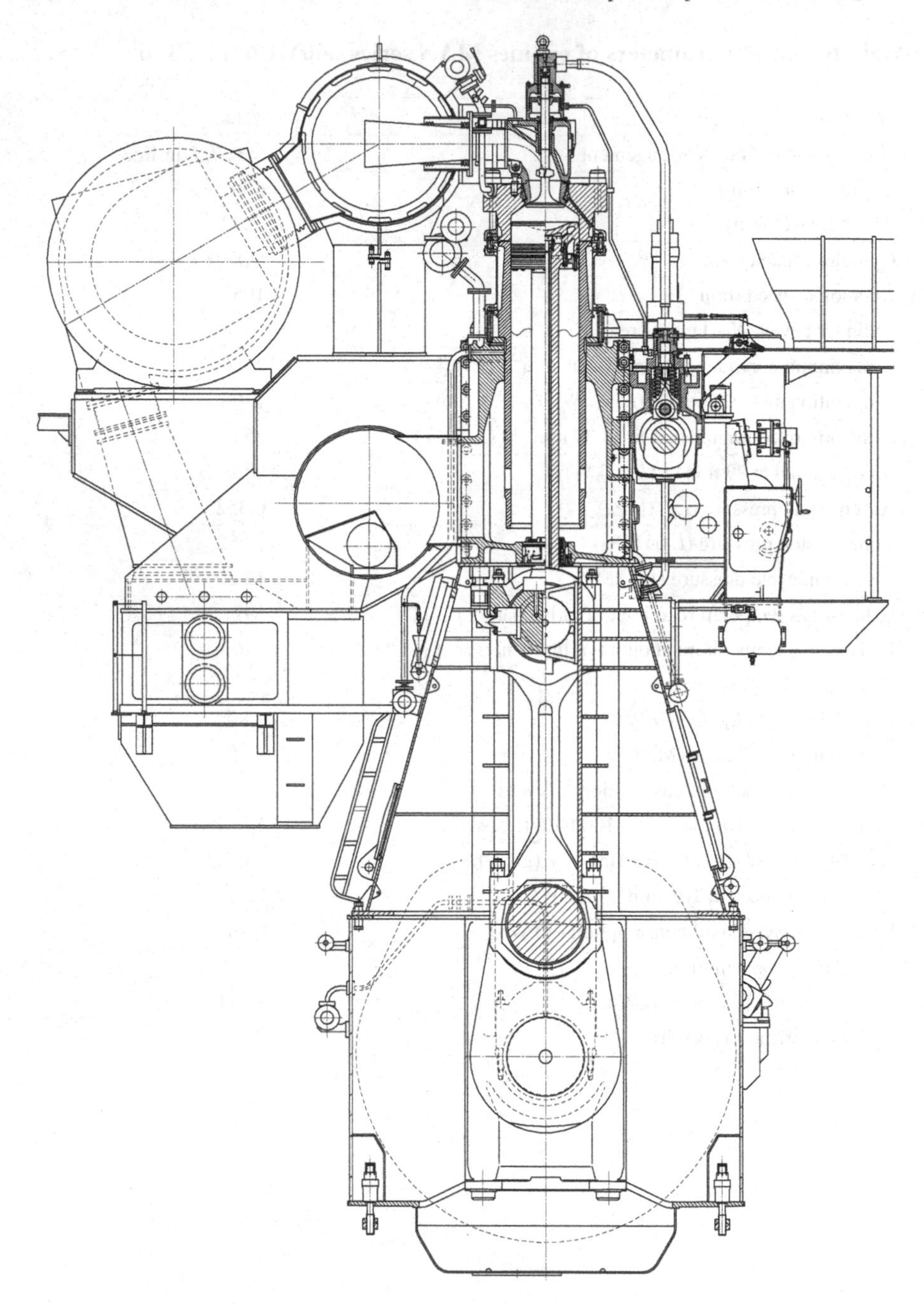

Fig. 3.6 Cross-section of the engine MAN series S60MC6 [5]

Dimensions and weight engines MAN series S60MC6

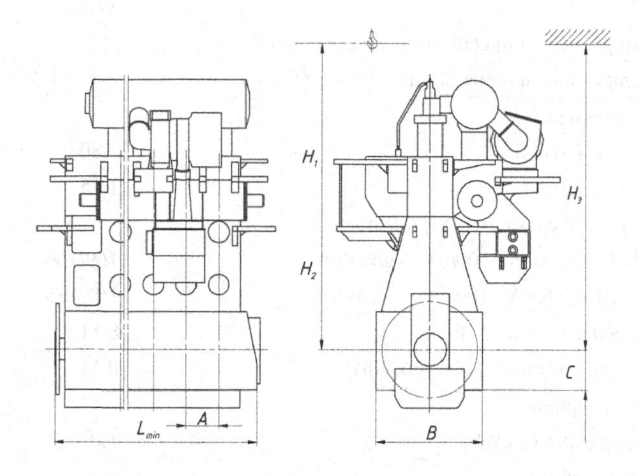

Number of cylinders	5	6	7	8
L_{min}	7655	8723	9791	10857
Weight (kg)	319,000	371,000	422,000	470,000

A (mm)	B (mm)	C (mm)	H_1 (mm)	H_2 (mm)	H_3 (mm)
1068	3478	1300	10,700	9800	9600

Main technical parameters of engines MAN series S70MC6 (Fig. 3.7)

Parameter	Value
Number and cylinders arrangement	5, 6, 7, 8 in-line
Cylinder bore (mm)	700
Piston stroke (mm)	2634
Cylinder capacity (dm^3)	1029.08
Rotational speed (min^{-1})	91
Cylinder power, N_e (layout area)	
– maximum continuous rating at 91 min^{-1} (L_1) (kW)	2810
– operational at 91 min^{-1} (L_2) (kW)	1790
– maximum continuous rating at 68 min^{-1} (L_3) (kW)	2100
– operational at 68 min^{-1} (L_4) (kW)	1340
Air charging pressure (L_1) (MPa)	0.353
Compression pressure (L_1) (MPa)	12.48
Maximum cycle pressure (L_1) (MPa)	14.20

(continued)

(continued)

Parameter	Value
Exhaust gas temperature at inlet of turbocharger (L_1) (°C)	392
Exhaust gas temperature at outlet of turbocharger (L_1) (°C)	246
Mean effective pressure	
– at 91 min^{-1} (L_1, L_2) (MPa)	1.80
– at 68 min^{-1} (L_3, L_4) (MPa)	1.15
Brake specific fuel oil consumption (g/kWh)	
– at 91 min^{-1} (L_1/L_2) for N_e 100/64% (g/kWh)	169/166
– at 68 min^{-1} (L_3/L_4) for N_e 100/64% (g/kWh)	157/155
Mean piston speed at 91 min^{-1} (m/s)	8.11
Lubrication system oil consumption (g/kWh)	0.15
Cylinder oil consumption	
– mechanical lubricator (g/kWh)	1.0–1.5
– alpha lubricator (g/kWh)	0.7

Dimensions and weight engines MAN series S70MC6

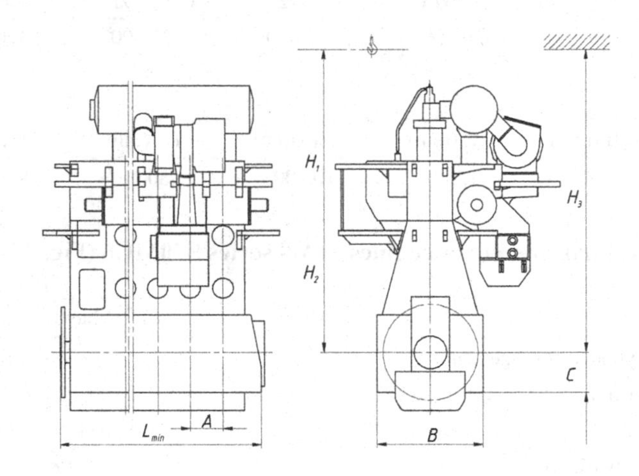

Number of cylinders	5	6	7	8
L_{min}	8981	10,227	10,688	11,878
Weight (kg)	492,000	562,000	648,000	722,000

A (mm)	B (mm)	C (mm)	H_1 (mm)	H_2 (mm)	H_3 (mm)
1246	4250	1520	12450	11475	11325

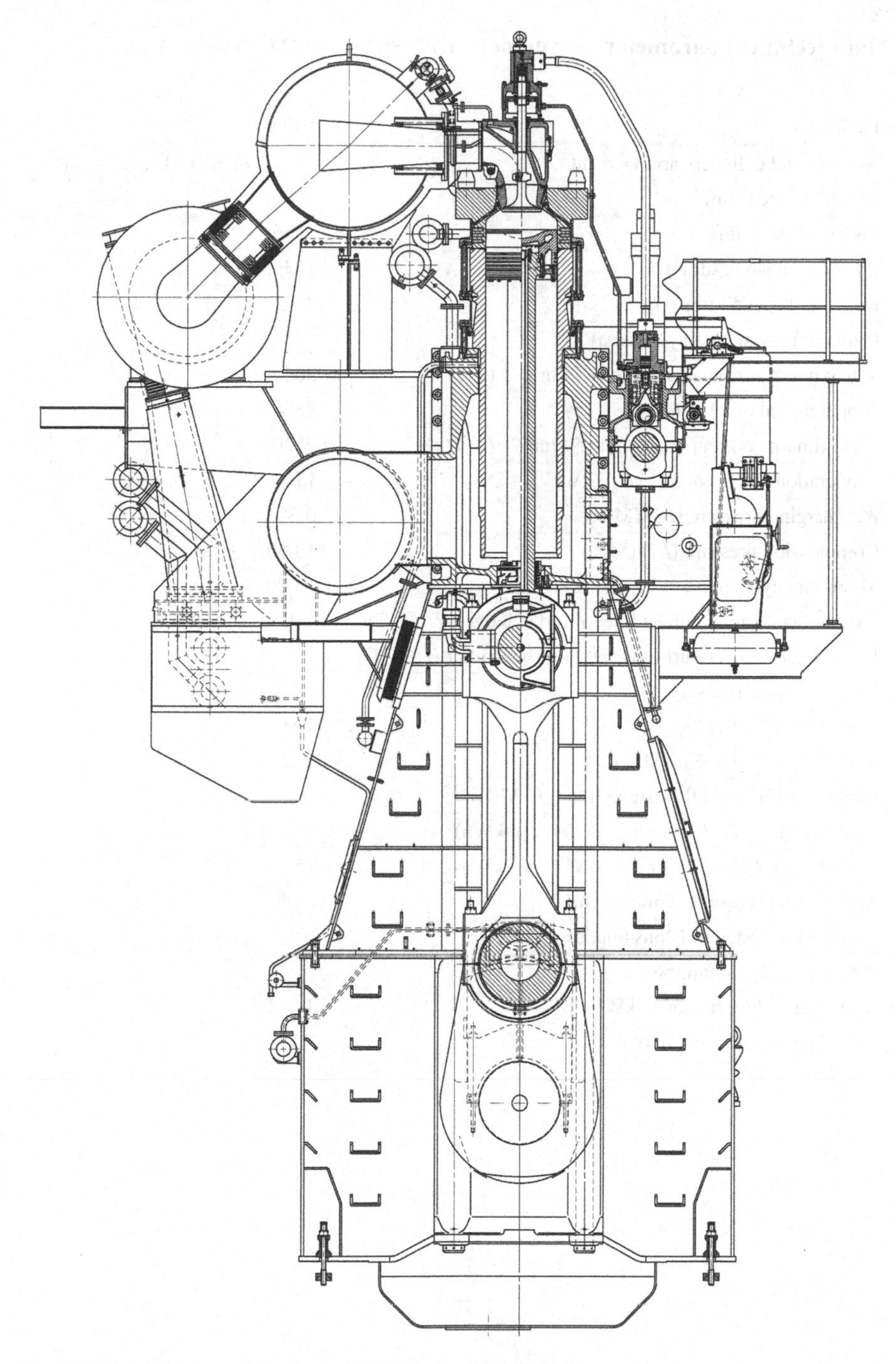

Fig. 3.7 Cross-section of the engine MAN series S70MC6 [6]

Main technical parameters of engines MAN series S80MC6 (Fig. 3.8)

Parameter	Value
Number and cylinders arrangement	5, 6, 7, 8, 9, 10, 11, 12 in-line
Cylinder bore (mm)	800
Piston stroke (mm)	3056
Cylinder capacity (dm^3)	1536.11
Rotational speed (min^{-1})	79
Cylinder power, N_e (layout area)	
– maximum continuous rating at 79 min^{-1} (L_1) (kW)	3640
– operational at 79 min^{-1} (L_2) (kW)	2330
– maximum continuous rating at 59 min^{-1} (L_3) (kW)	2720
– operational at 59 min^{-1} (L_4) (kW)	1740
Air charging pressure (L_1) (MPa)	0.352
Compression pressure (L_1) (MPa)	13.50
Maximum cycle pressure (L_1) (MPa)	15.20
Exhaust gas temperature at inlet of turbocharger (L_1) (°C)	410
Exhaust gas temperature at outlet of turbocharger (L_1) (°C)	255
Mean effective pressure	
– at 79 min^{-1} (L_1, L_2) (MPa)	1.80
– at 59 min^{-1} (L_3, L_4) (MPa)	1.15
Brake specific fuel oil consumption (g/kWh)	
– at 79 min^{-1} (L_1/L_2) for N_e 100/64% (g/kWh)	167/164
– at 59 min^{-1} (L_3/L_4) for N_e 100/64% (g/kWh)	155/153
Mean piston speed at 79 min^{-1} (m/s)	8.05
Lubrication system oil consumption (g/kWh)	0.15
Cylinder oil consumption	
– mechanical lubricator (g/kWh)	1.0–1.5
– alpha lubricator (g/kWh)	0.7

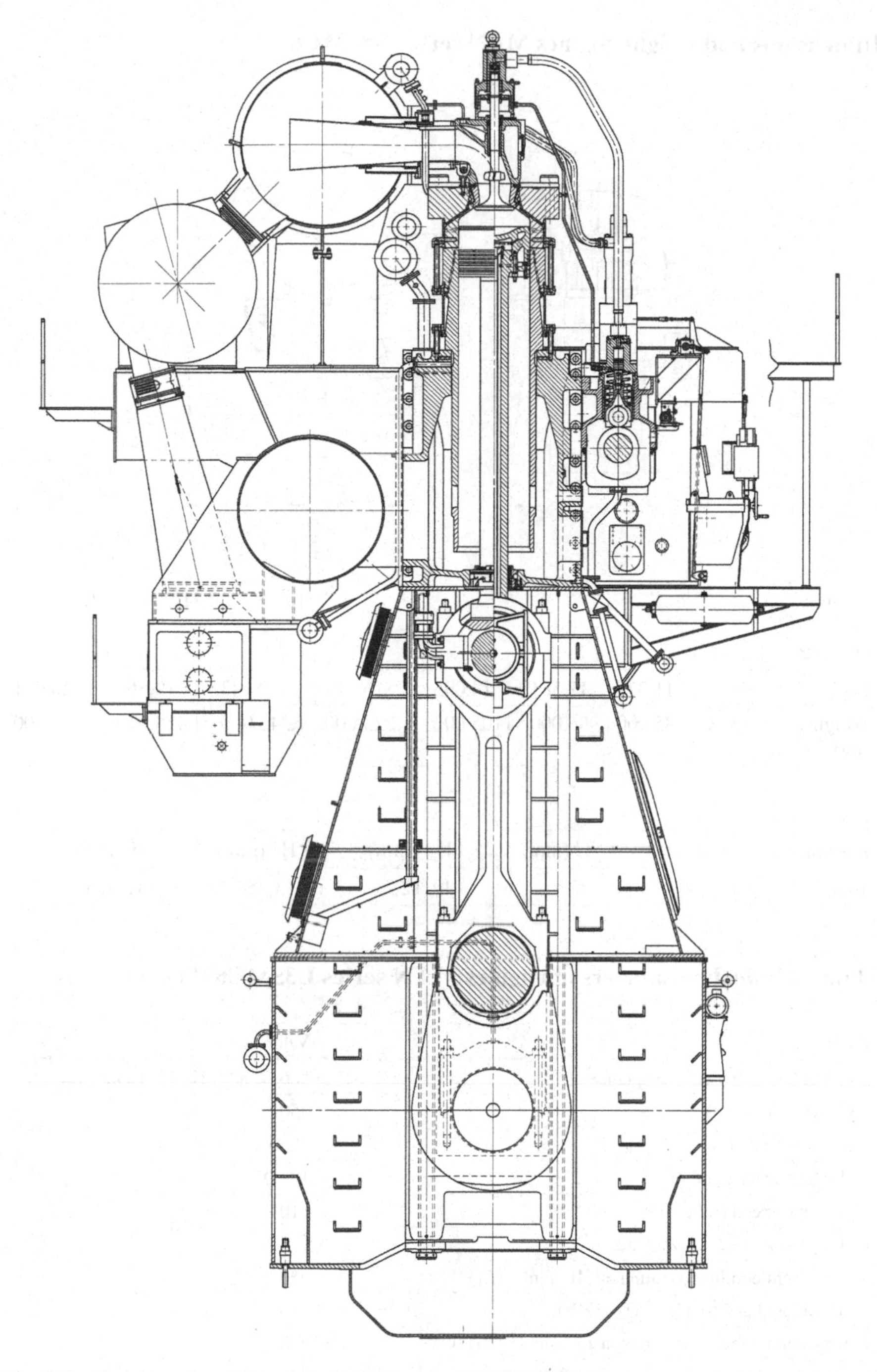

Fig. 3.8 Cross-section of the engine MAN series S80MC6 [7]

Dimensions and weight engines MAN series S80MC6

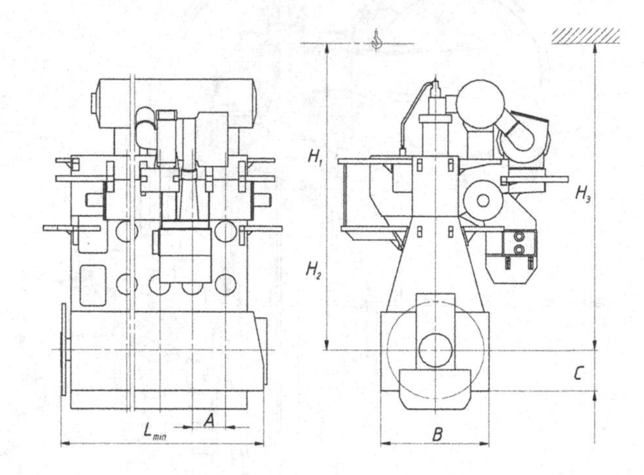

Number of cylinders	5	6	7	8	9	10	11	12
L_{min}	9953	11,377	12,581	14,005	16,719	18,143	19,567	20,991
Weight (kg)	777,000	885,000	996,000	1105,000	1,223,000	1,343,000	1,458,000	1,564,000

A (mm)	B (mm)	C (mm)	H_1 (mm)	H_2 (mm)	H_3 (mm)
1424	4824	1736	14,125	13,250	12,925

Main technical parameters of engines MAN series L35MC6 (Fig. 3.9)

Parameter	Value
Number and cylinders arrangement	5, 6, 7, 8, 9, 10, 11, 12 in-line
Cylinder bore (mm)	350
Piston stroke (mm)	1050
Cylinder capacity (dm^3)	101.02
Rotational speed (min^{-1})	210
Cylinder power, N_e (layout area)	
– maximum continuous rating at 210 min^{-1} (L_1) (kW)	650
– operational at 210 min^{-1} (L_2) (kW)	520
– maximum continuous rating at 178 min^{-1} (L_3) (kW)	550
– operational at 178 min^{-1} (L_4) (kW)	440
Air charging pressure (L_1) (MPa)	0.335

(continued)

(continued)

Parameter	Value
Compression pressure (L_1) (MPa)	12.50
Maximum cycle pressure (L_1) (MPa)	14.70
Exhaust gas temperature at inlet of turbocharger (L_1) (°C)	425
Exhaust gas temperature at outlet of turbocharger (L_1) (°C)	270
Mean effective pressure	
– at 210 min^{-1} (L_1, L_2) (MPa)	1.84
– at 178 min^{-1} (L_3, L_4) (MPa)	1.47
Brake specific fuel oil consumption (g/kWh)	
– at 210 min^{-1} (L_1/L_2) for N_e 100/80% (g/kWh)	177/175
– at 178 min^{-1} (L_3/L_4) for N_e 100/80% (g/kWh)	172/170
Mean piston speed at 210 min^{-1} (m/s)	7.35
System oil consumption per cylinder and per day (kg)	2.0–3.0
Cylinder oil consumption (g/kWh)	0.8–1.2

Dimensions and weight engines MAN series L35MC6

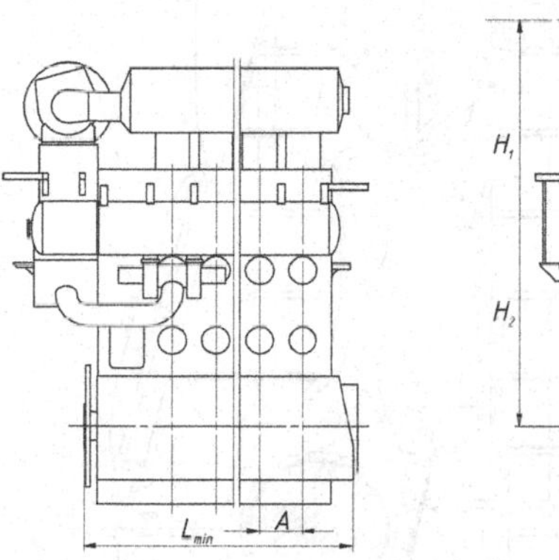

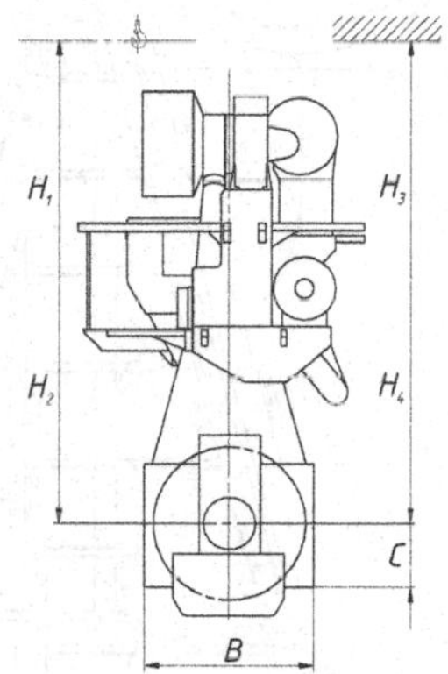

Number of cylinders	5	6	7	8	9	10	11	12
L_{min}	4174	4774	5374	5974	6574	7774	8374	8974
Weight (kg)	58,000	67,000	75,000	83,000	92,000	111,000	120,000	128,000

A (mm)	B (mm)	C (mm)	H_1 (mm)	H_2 (mm)	H_3 (mm)	H_4 (mm)
600	1980	550	5400	5350	5125	5150

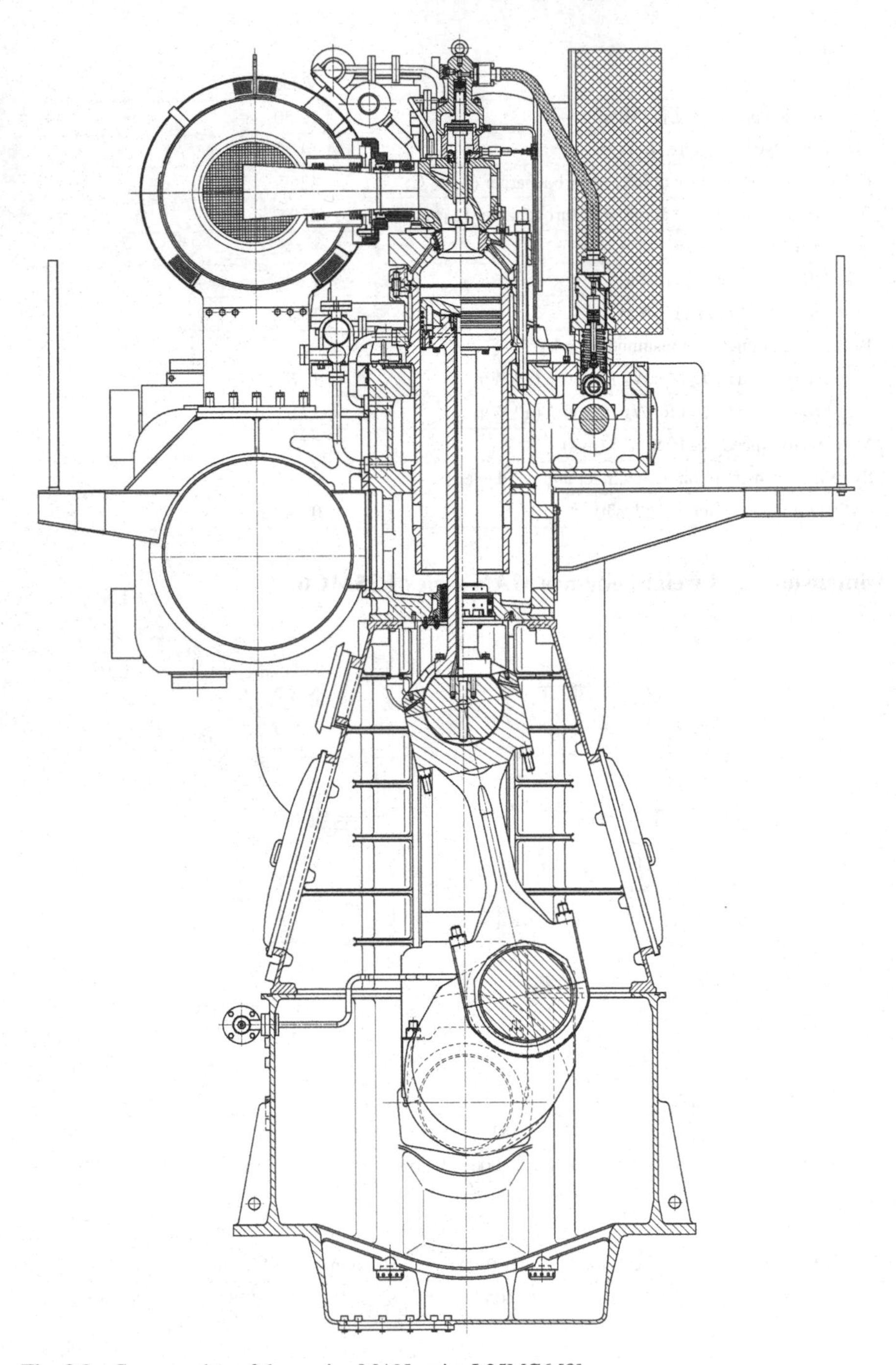

Fig. 3.9 Cross-section of the engine MAN series L35MC6 [8]

Main technical parameters of engines MAN series L42MC6 (Fig. 3.10)

Parameter	Value
Number and cylinders arrangement	4, 5, 6, 7, 8, 9, 10, 11, 12 in-line
Cylinder bore (mm)	420
Piston stroke (mm)	1360
Cylinder capacity (dm^3)	188.42
Rotational speed (min^{-1})	176
Cylinder power, N_e (layout area)	
– maximum continuous rating at 176 min^{-1} (L_1) (kW)	995
– operational at 176 min^{-1} (L_2) (kW)	635
– maximum continuous rating at 132 min^{-1} (L_3) (kW)	745
– operational at 132 min^{-1} (L_4) (kW)	480
Air charging pressure (L_1) (MPa)	0.36
Compression pressure (L_1) (MPa)	12.55
Maximum cycle pressure (L_1) (MPa)	14.20
Exhaust gas temperature at inlet of turbocharger (L_1) (°C)	420
Exhaust gas temperature at outlet of turbocharger (L_1) (°C)	253
Mean effective pressure	
– at 176 min^{-1} (L_1, L_2) (MPa)	1.80
– at 132 min^{-1} (L_3, L_4) (MPa)	1.15
Brake specific fuel oil consumption (g/kWh)	
– at 176 min^{-1} (L_1/L_2) for N_e 100/80% (g/kWh)	177/174
– at 132 min^{-1} (L_3/L_4) for N_e 100/80% (g/kWh)	165/165
Mean piston speed at 176 min^{-1} (m/s)	7.98
System oil consumption per cylinder and per day (kg)	3.0
Cylinder oil consumption (g/kWh)	0.9 1.4

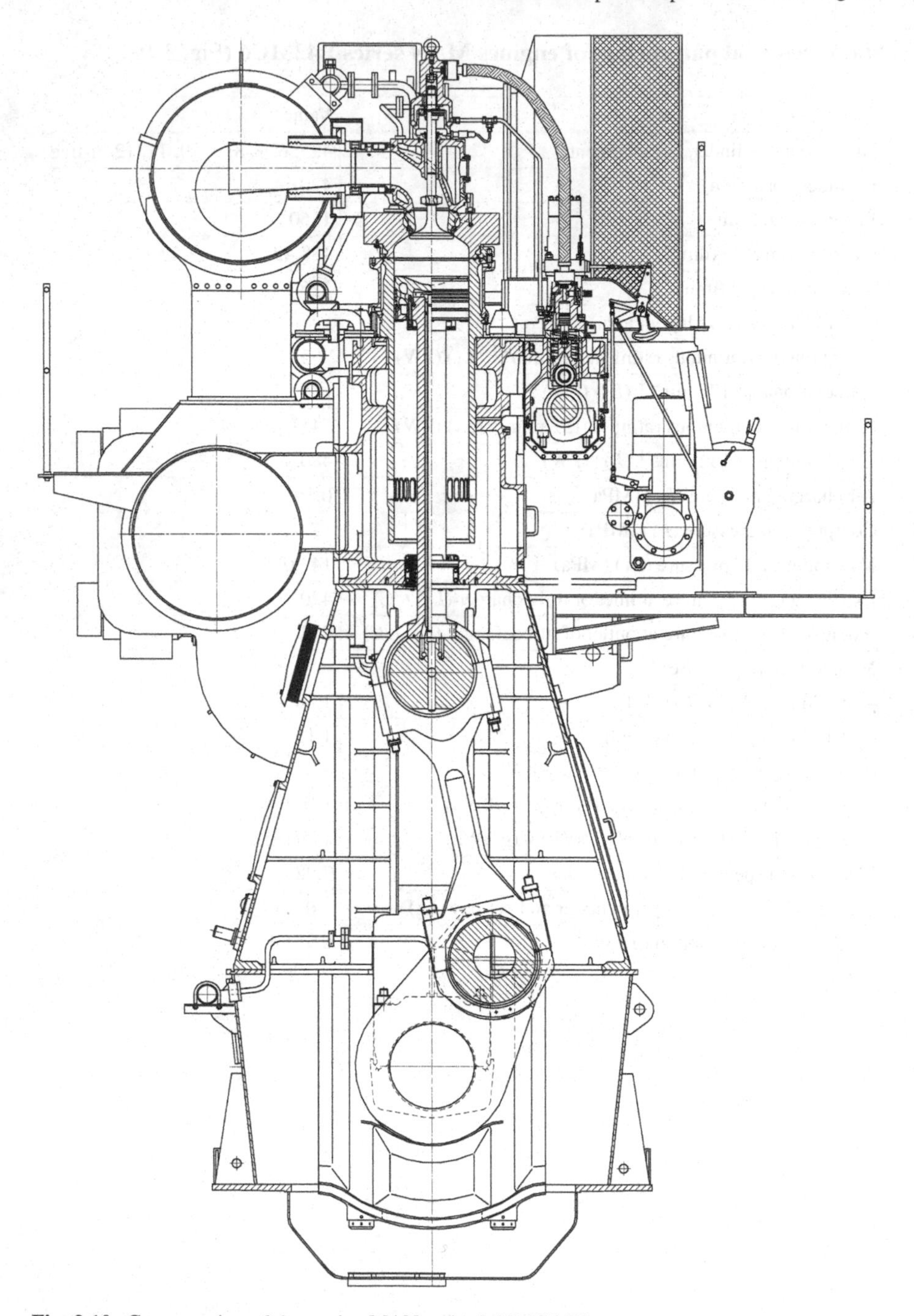

Fig. 3.10 Cross-section of the engine MAN series L42MC6 [9]

Dimensions and weight engines MAN series L42MC6

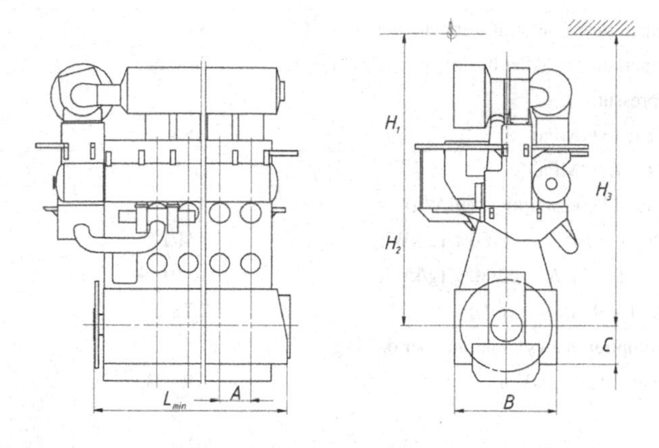

Number of cylinders	4	5	6	7	8
L_{min}	4661	5409	6157	6905	7653
Weight (kg)	95,000	110,000	125,000	14,3000	15,8000

Number of cylinders	9	10	11	12
L_{min}	8401	9897	10,645	11,393
Weight (kg)	176,000	210,000	229,000	244,000

A (mm)	B (mm)	C (mm)	H_1 (mm)	H_2 (mm)	H_3 (mm)
748	2460	690	6700	6250	6350

Main technical parameters of engines MAN series L80MC6 (Fig. 3.11)

Parameter	Value
Number and cylinders arrangement	4, 5, 6, 7, 8, 9, 10, 11, 12 in-line
Cylinder bore (mm)	800
Piston stroke (mm)	2592
Cylinder capacity (dm^3)	1302.88
Rotational speed (min^{-1})	93
Cylinder power, N_e (layout area)	
– maximum continuous rating at 93 min^{-1} (L_1) (kW)	3640
– operational at 93 min^{-1} (L_2) (kW)	2330
– maximum continuous rating at 70 min^{-1} (L_3) (kW)	2740
– operational at 70 min^{-1} (L_4) (kW)	1750
Air charging pressure (L_1) (MPa)	0.355
Compression pressure (L_1) (MPa)	12.50
Maximum cycle pressure (L_1) (MPa)	14.20

(continued)

(continued)

Parameter	Value
Exhaust gas temperature at inlet of turbocharger (L_1) (°C)	390
Exhaust gas temperature at outlet of turbocharger (L_1) (°C)	243
Mean effective pressure	
– at 93 min^{-1} (L_1, L_2) (MPa)	1.80
– at 70 min^{-1} (L_3, L_4) (MPa)	1.15
Brake specific fuel oil consumption (g/kWh)	
– at 93 min^{-1} (L_1/L_2) for N_e 100/80% (g/kWh)	174/171
– at 70 min^{-1} (L_3/L_4) for N_e 100/80% (g/kWh)	162/159
Mean piston speed at 93 min^{-1} (m/s)	7.78
System oil consumption per cylinder and per day (kg)	12
Cylinder oil consumption (g/kWh)	0.9–1.4

Dimensions and weight engines MAN series L80MC6

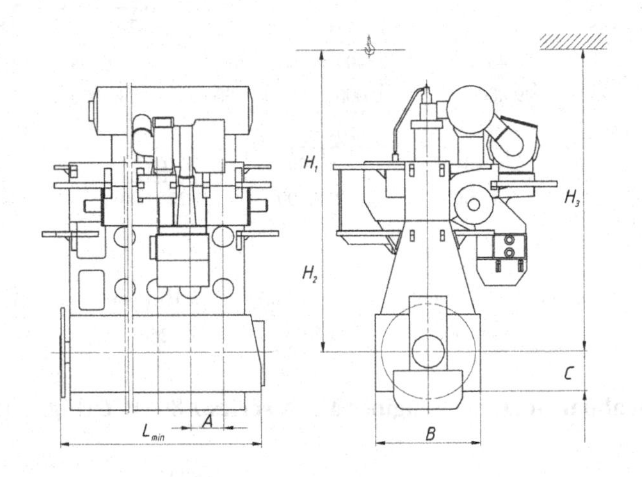

Number of cylinders	4	5	6	7	8
L_{min}	8386	9810	11,234	12,658	14,082
Weight (kg)	580,000	681,000	791,000	864,000	974,000

Number of cylinders	9	10	11	12
L_{min}	16,786	18,210	19,634	21,058
Weight (kg)	110,000	1,218,000	1,339,000	1,440,000

A (mm)	B (mm)	C (mm)	H_1 (mm)	H_2 (mm)	H_3 (mm)
1424	4388	1510	11,575	12,400	11,775

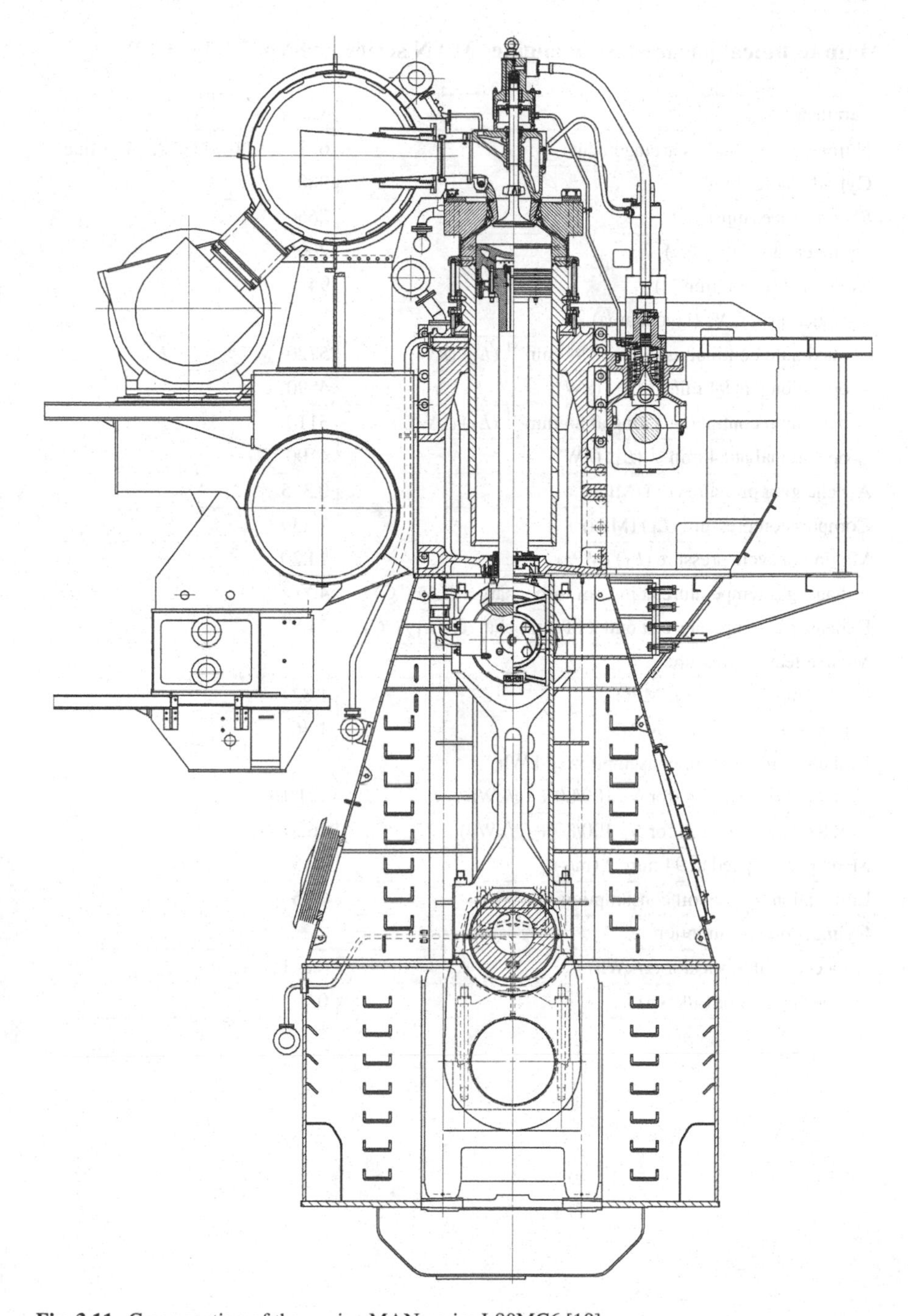

Fig. 3.11 Cross-section of the engine MAN series L80MC6 [10]

Main technical parameters of engines MAN series K98MC7 (Fig. 3.12)

Parameter	Value
Number and cylinders arrangement	6, 7, 8, 9, 10, 11, 12, 14 in-line
Cylinder bore (mm)	980
Piston stroke (mm)	2660
Cylinder capacity (dm^3)	2006.43
Rotational speed (min^{-1})	94
Cylinder power, N_e (layout area)	
– maximum continuous rating at 94 min^{-1} (L_1) (kW)	5720
– operational at 94 min^{-1} (L_2) (kW)	4590
– maximum continuous rating at 84 min^{-1} (L_3) (kW)	5110
– operational at 84 min^{-1} (L_4) (kW)	4100
Air charging pressure (L_1) (MPa)	0.355
Compression pressure (L_1) (MPa)	12.30
Maximum cycle pressure (L_1) (MPa)	14.20
Exhaust gas temperature at inlet of turbocharger (L_1) (°C)	400
Exhaust gas temperature at outlet of turbocharger (L_1) (°C)	242
Mean effective pressure	
– at 94 min^{-1} (L_1, L_2) (MPa)	1.82
– at 84 min^{-1} (L_3, L_4) (MPa)	1.46
Brake specific fuel oil consumption (g/kWh)	
– at 94 min^{-1} (L_1/L_2) for N_e 100/80% (g/kWh)	171/165
– at 84 min^{-1} (L_3/L_4) for N_e 100/80% (g/kWh)	162/158
Mean piston speed at 94 min^{-1} (m/s)	8.33
Lubrication System oil consumption (g/kWh)	0.15
Cylinder oil consumption	
– mechanical lubricator (g/kWh)	0.8–1.2
– alpha lubricator (g/kWh)	0.7

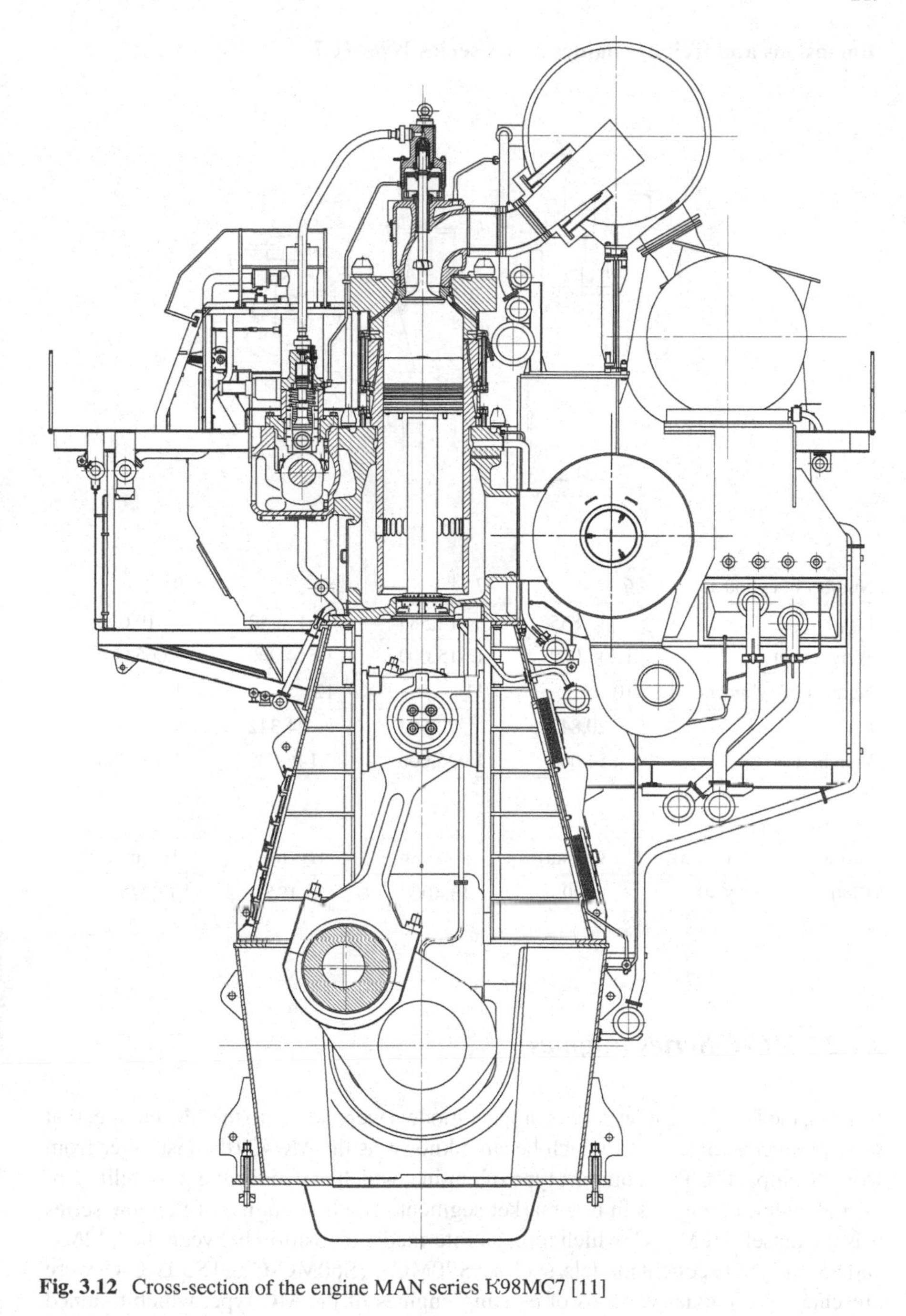

Fig. 3.12 Cross-section of the engine MAN series K98MC7 [11]

Dimensions and weight engines MAN series K98MC7

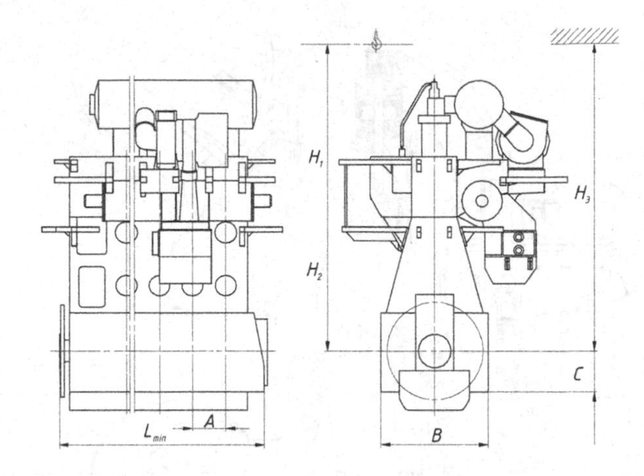

Number of cylinders	6	7	8	9
L_{min}	12,835	14,585	17,342	19,092
Weight (kg)	1,143,000	1,315,000	1,514,000	1,666,000
Number of cylinders	10	11	12	14
L_{min}	20,842	22,592	24,342	26,092
Weight (kg)	1,854,000	1,996,000	2,146,000	2,446,000

A (mm)	B (mm)	C (mm)	H_1 (mm)	H_2 (mm)	H_3 (mm)
1750	4640	1700	13,400	13,125	13,275

3.1.2 MC-C Series Engines

In 1996, the following new, more compact models were added to the MC engine, that were produced since 1982, which became known as the MC-C (the last letter from word "Compact"). This compact type of engine, which expanded the possibilities of using low-speed engines in this market segment. The first engine of the new series was the diesel S46MC-C, which took an intermediate position between the S42MC and S50MC. Subsequent models, such as: S70MC-C, S60MC-C and S50MC-C, were presented as compact versions of existing engines of the MC type, which retained cylinder diameters, but increased the piston stroke and increased to 1.9 MPa mean effective pressure. The length and weight of the engines have been reduced. So, for example, the 6S50MC-C engine is about 1000 mm shorter and 25 tons lighter compared to the same 6S50MC. Also, managed to reduce the height, required for repairs. Due to the increase in the average effective pressure, a number of design changes were made. Thus, the design of the foundation frame was changed, from rectangular to trapezoidal. This made it possible to shift the mounting bolts to the foundation of the vessel beyond the foundation frame, which reduced the effect of foundation deformations on the frame itself and on the crankshaft frame bearings. In addition, access to the mounting bolts themselves was facilitated for their control and maintenance. The transverse seat used to install the thrust bearing, as a result of giving it a trapezoidal shape, acquired additional rigidity, which made it possible to improve the working conditions of the thrust bearing. As frame bearings on all series MC-C engines, thin-walled steel strip, coated with antifriction alloy, are used.

The increase in operating pressures in the cylinders, when the engines were forced at the average effective pressure, led to the fact that the required tightening forces of the solid anchor links have resulted in the deformation of the engine core and, above all, the foundation of the crankshaft. This required an increase in the rigidity of the basic elements of the skeleton and, as a consequence, an increase in their mass. For this reason, when creating compact versions of MC-C engines, MAN switched to the use of short anchor ties, where the stud is screwed into the threaded hole on the bottom of the foundation frame shelf with a lower end. For a more uniform distribution of efforts, the number of anchor ties was doubled (paired anchor ties).

To protect the piston rings from high temperatures, the height of the piston head was increased, which made it possible to transfer the gas joint to the region of lower temperatures. As a result, it was possible to reduce the temperature in the location of the upper piston ring by about 100 °C.

A deeper mounting collar on the cylinder head can significantly reduce the thermal load on the sleeve, since at the beginning of expansion the hot combustion products contact only with the fire surface of the cover, and by the time the piston begins to open the working surface of the cylinder liner, the temperature is already much less than at the end of combustion. At the bottom of the cover, at the junction with the end surface of the cylinder liner, an anti-polishing ring is installed, which removes carbon deposits from the cylindrical part of the piston head. The fire surface of the piston head and cylinder head has a chemically and corrosion-resistant coating "Inconel" which is an austenitic alloy, based on nickel and chromium. The grooves of the piston rings in the lower part have a wear-resistant coating, based on a chromium alloy of increased thickness. The piston head itself is made of forgings made of heat-resistant steel.

The fuel system has undergone significant changes. In high-pressure fuel pumps, a new system of seals against fuel leakage has been applied. The design of the nozzle allowed to minimize the volume under needle chamber.

Main technical parameters of engines MAN series S46MC-C (Fig. 3.13)

Parameter	Value
Number and cylinders arrangement	5, 6, 7, 8 in-line
Cylinder bore (mm)	460
Piston stroke (mm)	1932
Cylinder capacity, dm^3	321.1
Rotational speed, min^{-1}	129
Cylinder power, N_e (layout area)	
– maximum continuous rating at 129 min^{-1} (L_1) (kW)	1310
– operational at 129 min^{-1} (L_2) (kW)	1050
– maximum continuous rating at 108 min^{-1} (L_3) (kW)	1100
– operational at 108 min^{-1} (L_4) (kW)	880
Air charging pressure (L_1) (MPa)	0.362
Compression pressure (L_1) (MPa)	13.2
Maximum cycle pressure (L_1) (MPa)	15.2
Exhaust gas temperature at inlet of turbocharger (L_1) (°C)	430
Exhaust gas temperature at outlet of turbocharger (L_1) (°C)	265
Mean effective pressure	
– at 129 min^{-1} (L_1, L_2) (MPa)	1.90
– at 108 min^{-1} (L_3, L_4) (MPa)	1.52
Brake specific fuel oil consumption (g/kWh)	
– at 129 min^{-1} (L_1/L_2) for N_e 100/80% (g/kWh)	174/172
– at 108 min^{-1} (L_3/L_4) for N_e 100/80% (g/kWh)	169/167
Mean piston speed at 129 min^{-1} (m/s)	8.31
Lubrication System oil consumption (g/kWh)	0.15
Cylinder oil consumption	
– mechanical lubricator (g/kWh)	1.0–1.5
– alpha lubricator (g/kWh)	0.7

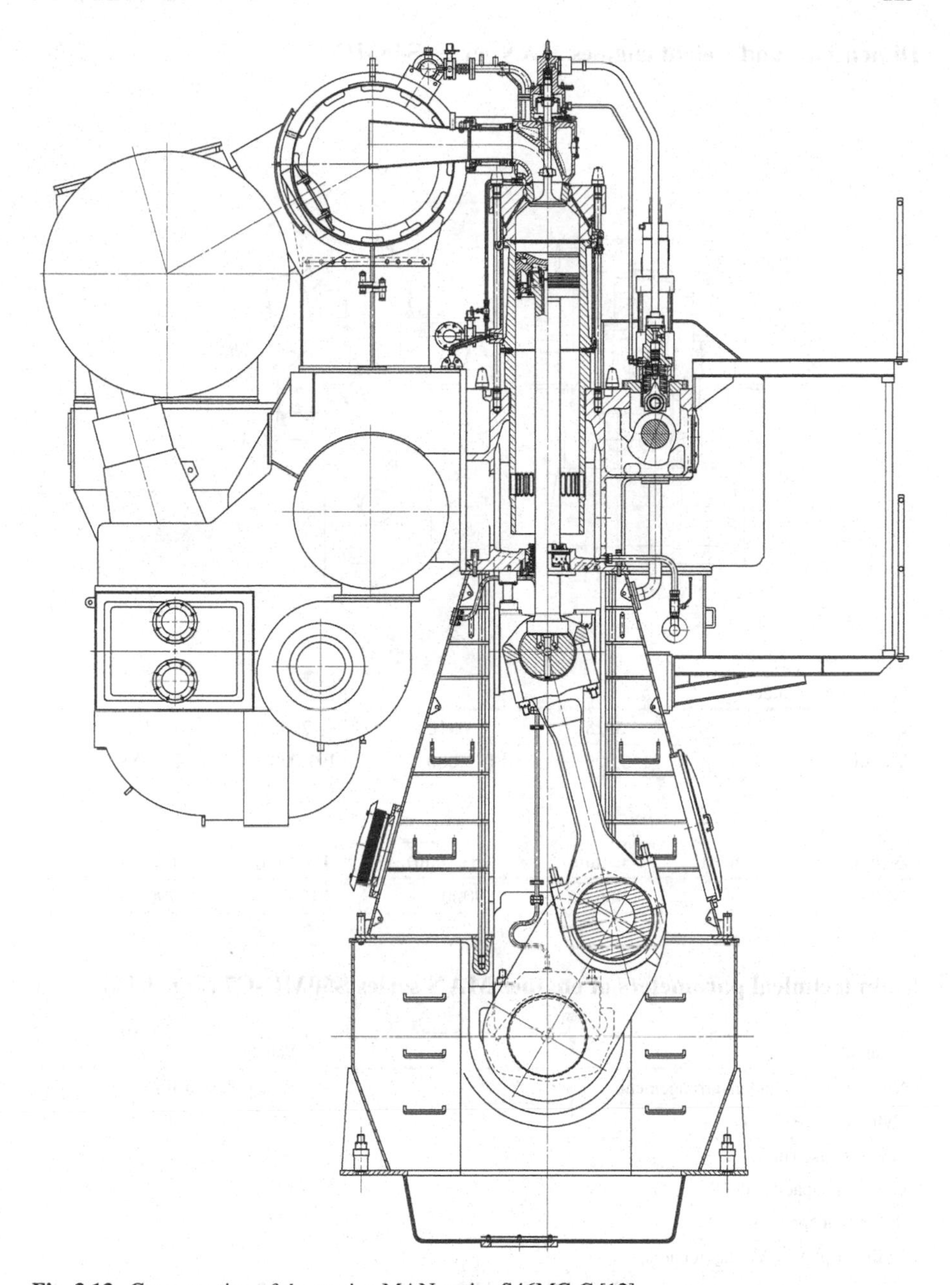

Fig. 3.13 Cross-section of the engine MAN series S46MC-C [12]

Dimensions and weight engines MAN series S46MC-C

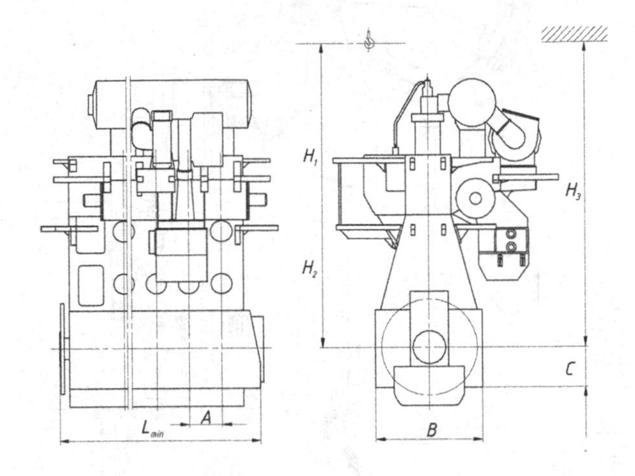

Number of cylinders	5	6	7	8
L_{min}	5528	6310	7092	7874
Weight (kg)	153,000	171,000	197,000	217,000

A (mm)	B (mm)	C (mm)	H_1 (mm)	H_2 (mm)	H_3 (mm)
782	2924	986	9000	8175	7900

Main technical parameters of engines MAN series S50MC-C7 (Fig. 3.14)

Parameter	Value
Number and cylinders arrangement	5, 6, 7, 8, 9 in-line
Cylinder bore (mm)	500
Piston stroke (mm)	2000
Cylinder capacity (dm^3)	392.7
Rotational speed (min^{-1})	127
Cylinder power, N_e (layout area)	
– maximum continuous rating at 127 min^{-1} (L_1) (kW)	1660
– operational at 127 min^{-1} (L_2) (kW)	1330
– maximum continuous rating at 108 min^{-1} (L_3) (kW)	1410
– operational at 108 min^{-1} (L_4) (kW)	1130
Air charging pressure (L_1) (MPa)	0.375
Compression pressure (L_1) (MPa)	14.2

(continued)

(continued)

Parameter	Value
Maximum cycle pressure (L_1) (MPa)	16.2
Exhaust gas temperature at inlet of turbocharger (L_1) (°C)	405
Exhaust gas temperature at outlet of turbocharger (L_1) (°C)	242
Mean effective pressure	
– at 127 min^{-1} (L_1, L_2) (MPa)	2.0
– at 108 min^{-1} (L_3, L_4) (MPa)	1.6
Brake specific fuel oil consumption (g/kWh)	
– at 127 min^{-1} (L_1/L_2) for N_e 100/80% (g/kWh)	171/168
– at 108 min^{-1} (L_3/L_4) for N_e 100/80% (g/kWh)	164/161
Mean piston speed at 127 min^{-1} (m/s)	8.47
Lubrication System oil consumption (g/kWh)	0.15
Cylinder oil consumption	
– mechanical lubricator (g/kWh)	1.0–1.5
– alpha lubricator (g/kWh)	0.7

Dimensions and weight engines MAN series S50MC-C7

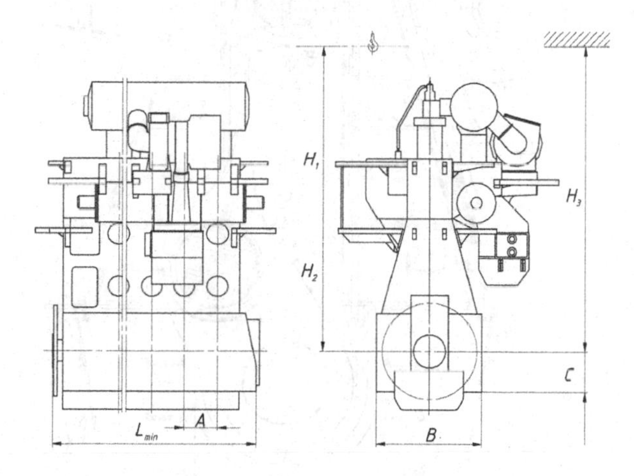

Number of cylinders	5	6	7	8	9
L_{min}	5924	6774	7624	8474	9324
Weight (kg)	181,000	207,000	238,000	270,000	300,000

A (mm)	B (mm)	C (mm)	H_1 (mm)	H_2 (mm)	H_3 (mm)
850	3150	1085	9000	8475	8250

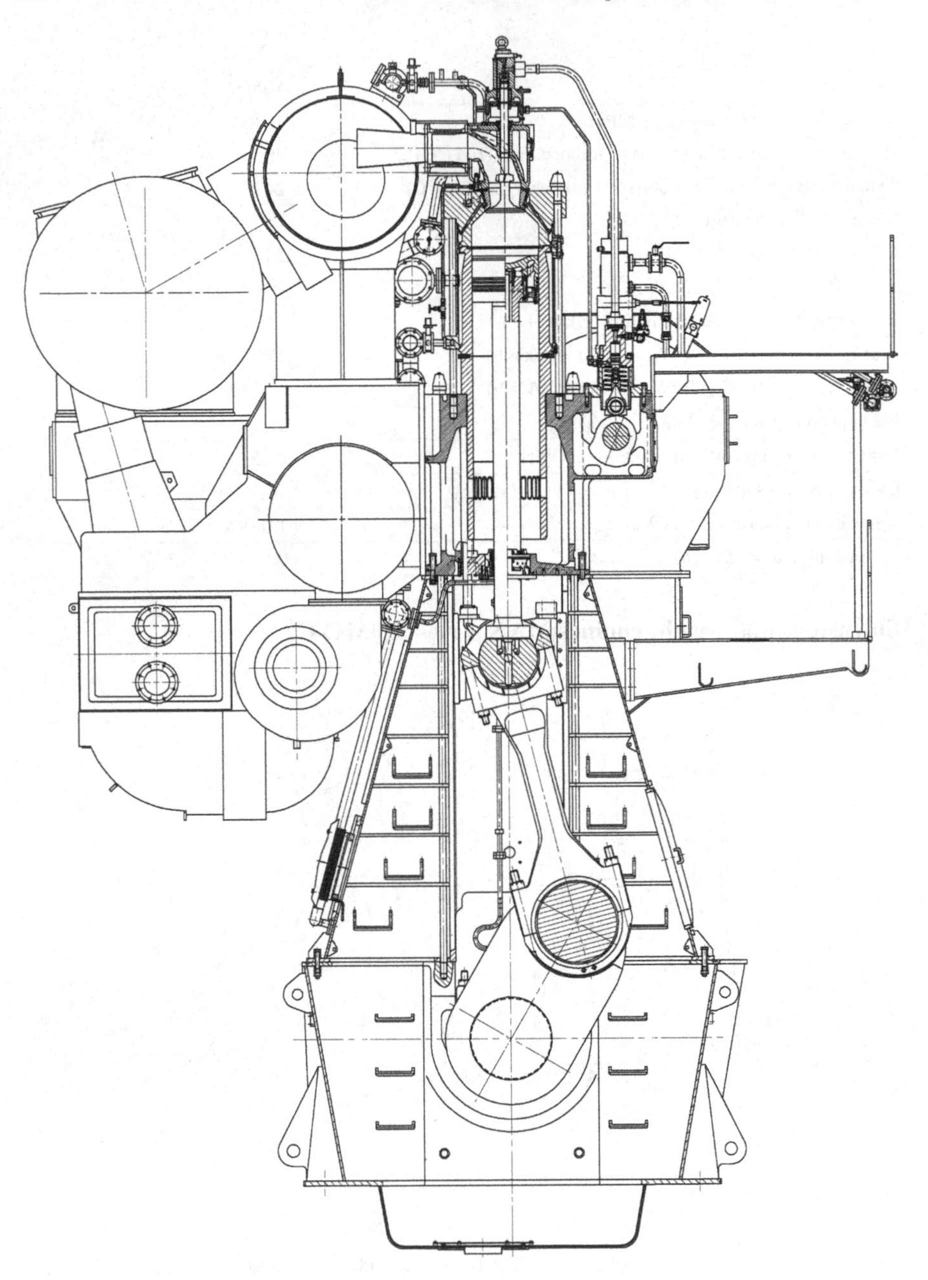

Fig. 3.14 Cross-section of the engine MAN series S50MC-C7 [13]

Main technical parameters of engines MAN series S60MC-C7 (Fig. 3.15)

Parameter	Value
Number and cylinders arrangement	5, 6, 7, 8 in-line
Cylinder bore (mm)	600
Piston stroke (mm)	2400
Cylinder capacity (dm^3)	678.58
Rotational speed (min^{-1})	105
Cylinder power, N_e (layout area)	
– maximum continuous rating at 105 min^{-1} (L_1) (kW)	2260
– operational at 105 min^{-1} (L_2) (kW)	1450
– maximum continuous rating at 79 min^{-1} (L_3) (kW)	1700
– operational at 79 min^{-1} (L_4) (kW)	1090
Air charging pressure (L_1) (MPa)	0.362
Compression pressure (L_1) (MPa)	13.25
Maximum cycle pressure (L_1) (MPa)	15.20
Exhaust gas temperature at inlet of turbocharger (L_1) (°C)	395
Exhaust gas temperature at outlet of turbocharger (L_1) (°C)	244
Mean effective pressure	
– at 105 min^{-1} (L_1, L_2) (MPa)	1.90
– at 79 min^{-1} (L_3, L_4) (MPa)	1.22
Brake specific fuel oil consumption (g/kWh)	
– at 105 min^{-1} (L_1/L_2) for N_e 100/80% (g/kWh)	170/167
– at 79 min^{-1} (L_3/L_4) for N_e 100/80% (g/kWh)	158/156
Mean piston speed at 105 min^{-1} (m/s)	8.40
Lubrication system oil consumption (g/kWh)	0.15
Cylinder oil consumption	
– mechanical lubricator (g/kWh)	1.0–1.5
– alpha lubricator (g/kWh)	0.7

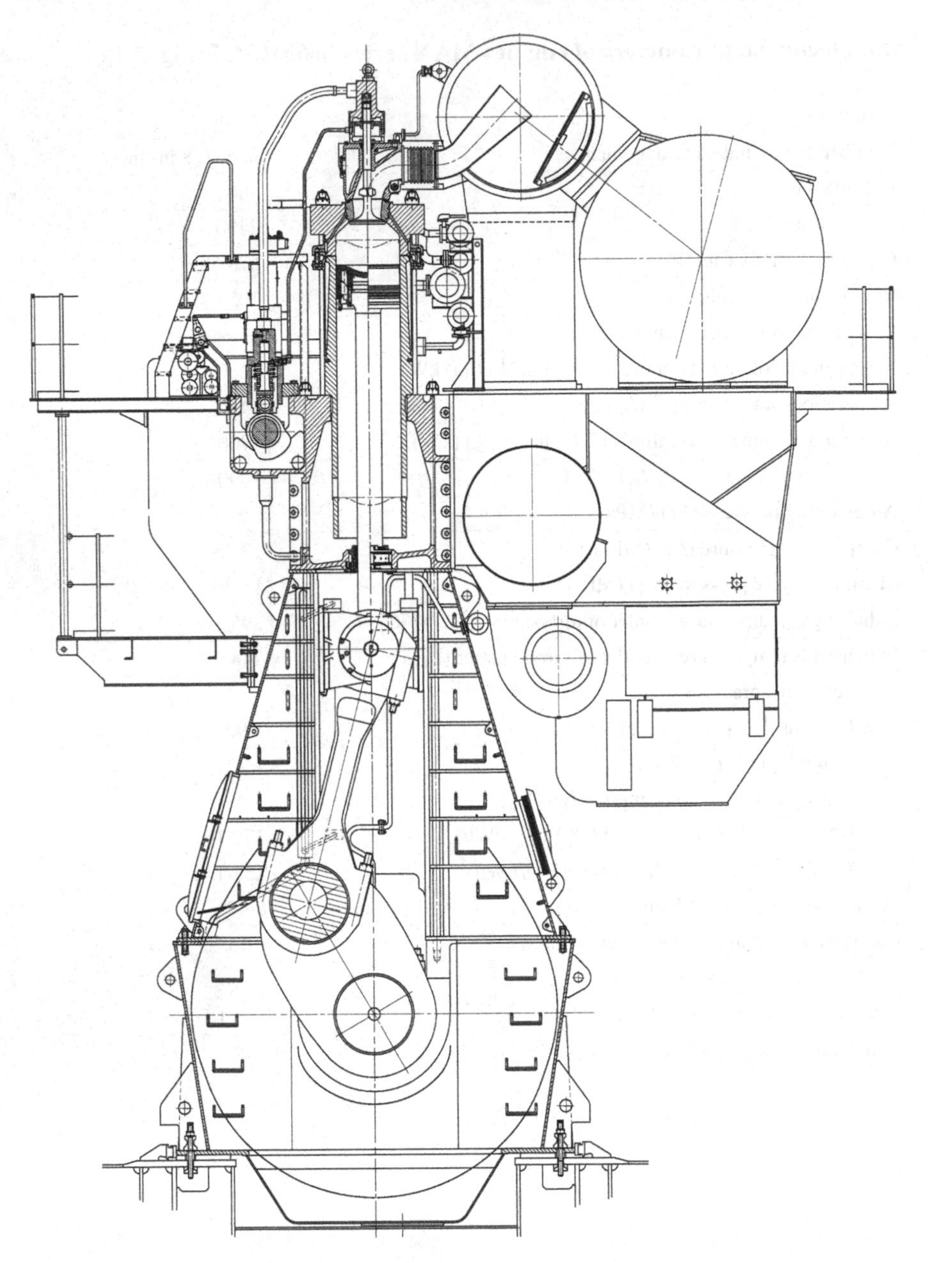

Fig. 3.15 Cross-section of the engine MAN series S60MC-C7 [14]

Dimensions and weight engines MAN series S60MC-C7

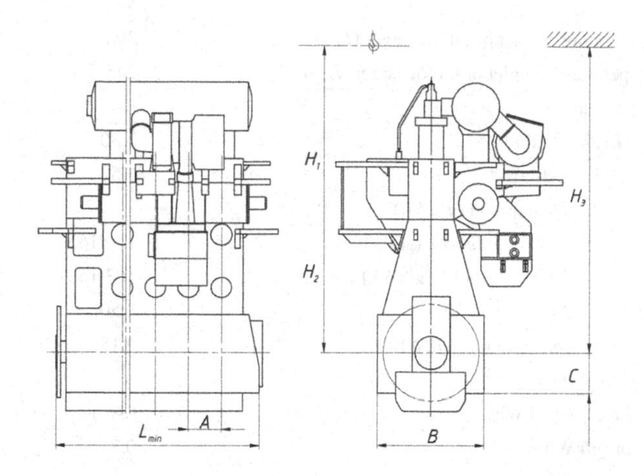

Number of cylinders	5	6	7	8
L_{min}	7251	8271	9162	10,182
Weight (kg)	314,000	358,000	410,000	467,000

A (mm)	B (mm)	C (mm)	H_1 (mm)	H_2 (mm)	H_3 (mm)
1020	3770	1300	10,700	10,050	9800

Main technical parameters of engines MAN series S70MC-C7 (Fig. 3.16)

Parameter	Value
Number and cylinders arrangement	5, 6, 7, 8 in-line
Cylinder bore (mm)	700
Piston stroke (mm)	2360
Cylinder capacity (dm^3)	908.23
Rotational speed (min^{-1})	108
Cylinder power, N_e (layout area)	
– maximum continuous rating at 108 min^{-1} (L_1) (kW)	3110
– operational at 108 min^{-1} (L_2) (kW)	2480
– maximum continuous rating at 91 min^{-1} (L_3) (kW)	2620
– operational at 91 min^{-1} (L_4) (kW)	2090
Air charging pressure (L_1) (MPa)	0.370
Compression pressure (L_1) (MPa)	13.40
Maximum cycle pressure (L_1) (MPa)	15.20

(continued)

(continued)

Parameter	Value
Exhaust gas temperature at inlet of turbocharger (L_1) (°C)	398
Exhaust gas temperature at outlet of turbocharger (L_1) (°C)	245
Mean effective pressure	
– at 108 min^{-1} (L_1, L_2) (MPa)	1.90
– at 91 min^{-1} (L_3, L_4) (MPa)	1.52
Brake specific fuel oil consumption (g/kWh)	
– at 108 min^{-1} (L_1/L_2) for N_e 100/80% (g/kWh)	170/167
– at 91 min^{-1} (L_3/L_4) for N_e 100/80% (g/kWh)	163/160
Mean piston speed at 108 min^{-1} (m/s)	8.50
Lubrication system oil consumption (g/kWh)	0.15
Cylinder oil consumption	
– mechanical lubricator (g/kWh)	0.8–1.2
– alpha lubricator (g/kWh)	0.7

Dimensions and weight engines MAN series S70MC-C7

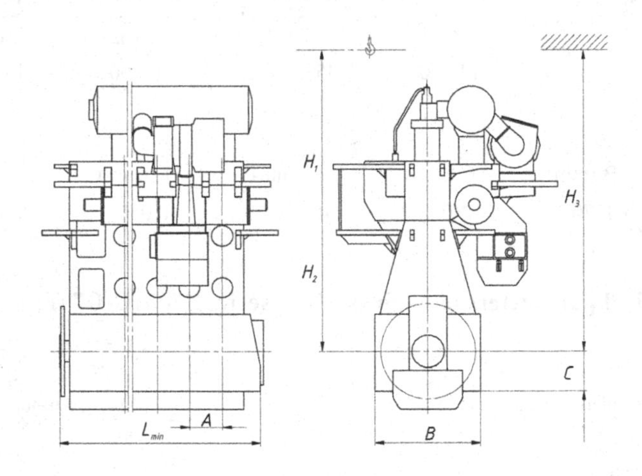

Number of cylinders	5	6	7	8
L_{min}	8308	9498	10,688	11,878
Weight (kg)	480,000	555,000	624,000	704,000

A (mm)	B (mm)	C (mm)	H_1 (mm)	H_2 (mm)	H_3 (mm)
1190	4390	1520	12475	11675	11425

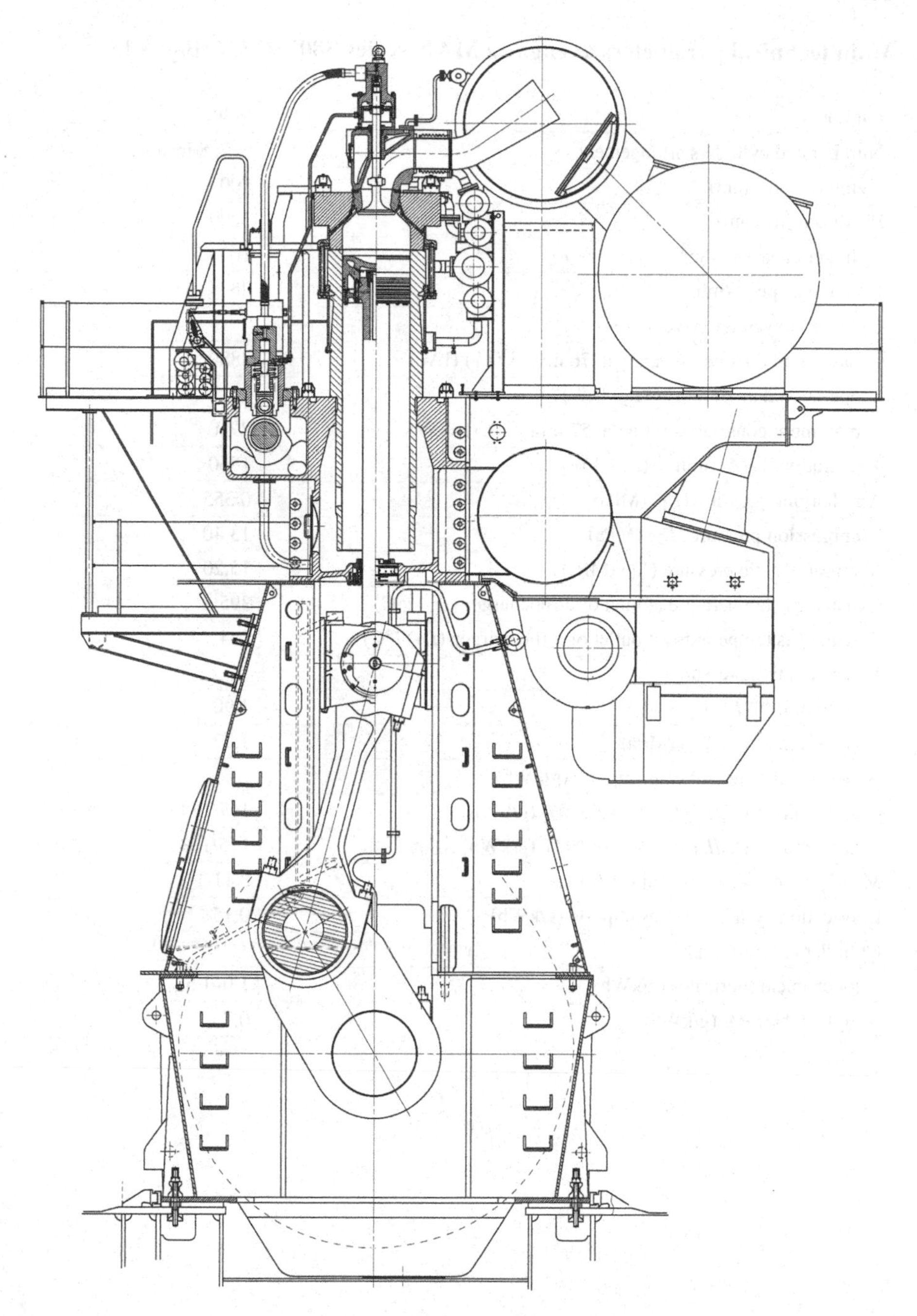

Fig. 3.16 Cross-section of the engine MAN series S70MC-C7 [15]

Main technical parameters of engines MAN series S80MC-C7 (Fig. 3.17)

Parameter	Value
Number and cylinders arrangement	6, 7, 8 in-line
Cylinder bore (mm)	800
Piston stroke (mm)	3200
Cylinder capacity (dm^3)	1608.5
Rotational speed (min^{-1})	76
Cylinder power, N_e (layout area)	
– maximum continuous rating at 76 min^{-1} (L_1) (kW)	3880
– operational at 76 min^{-1} (L_2) (kW)	2480
– maximum continuous rating at 57 min^{-1} (L_3) (kW)	2910
– operational at 57 min^{-1} (L_4) (kW)	1860
Air charging pressure (L_1) (MPa)	0.355
Compression pressure (L_1) (MPa)	13.40
Maximum cycle pressure (L_1) (MPa)	15.20
Exhaust gas temperature at inlet of turbocharger (L_1) (°C)	395
Exhaust gas temperature at outlet of turbocharger (L_1) (°C)	244
Mean effective pressure	
– at 76 min^{-1} (L_1, L_2) (MPa)	1.90
– at 57 min^{-1} (L_3, L_4) (MPa)	1.22
Brake specific fuel oil consumption (g/kWh)	
– at 76 min^{-1} (L_1/L_2) for N_e 100/80% (g/kWh)	167/164
– at 57 min^{-1} (L_3/L_4) for N_e 100/80% (g/kWh)	155/153
Mean piston speed at 76 min^{-1} (m/s)	8.11
Lubrication system oil consumption (g/kWh)	0.15
Cylinder oil consumption	
– mechanical lubricator (g/kWh)	1.0–1.5
– alpha lubricator (g/kWh)	0.7

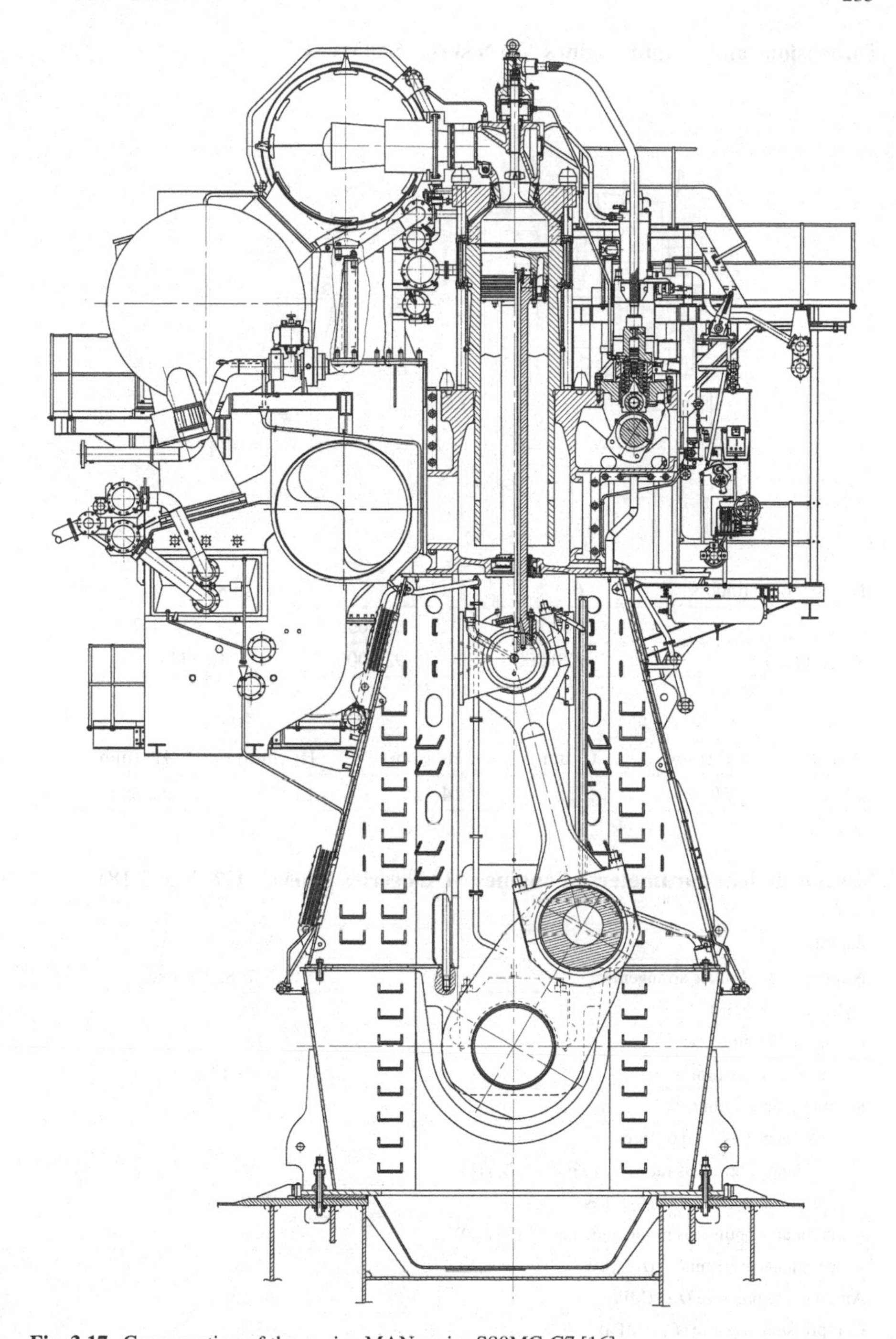

Fig. 3.17 Cross-section of the engine MAN series S80MC-C7 [16]

Dimensions and weight engines MAN series S80MC-C7

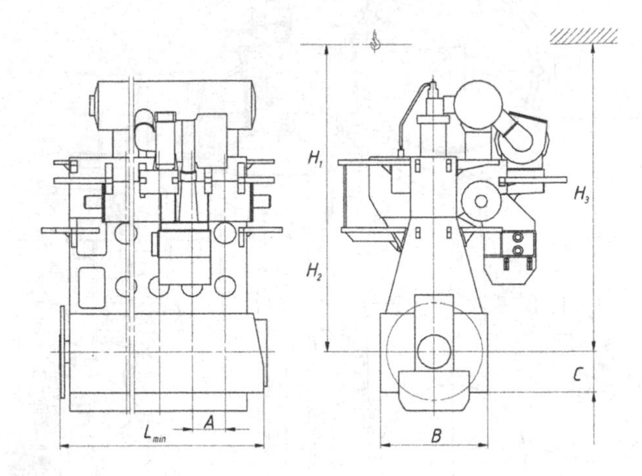

Number of cylinders	6	7	8
L_{min}	11,229	12,653	14,077
Weight (kg)	820,000	922,000	10,234,000

A (mm)	B (mm)	C (mm)	H_1 (mm)	H_2 (mm)	H_3 (mm)
1424	5000	1736	14,325	13,175	12,850

Main technical parameters of engines MAN series S90MC-C7 (Fig. 3.18)

Parameter	Value
Number and cylinders arrangement	6, 7, 8, 9 in-line
Cylinder bore (mm)	900
Piston stroke (mm)	3188
Cylinder capacity (dm^3)	2028.12
Rotational speed (min^{-1})	76
Cylinder power, N_e (layout area)	
– maximum continuous rating at 76 min^{-1} (L_1) (kW)	4890
– operational at 76 min^{-1} (L_2) (kW)	3920
– maximum continuous rating at 61 min^{-1} (L_3) (kW)	3930
– operational at 61 min^{-1} (L_4) (kW)	3140
Air charging pressure (L_1) (MPa)	0.355
Compression pressure (L_1) (MPa)	13.40
Maximum cycle pressure (L_1) (MPa)	15.20

(continued)

(continued)

Parameter	Value
Exhaust gas temperature at inlet of turbocharger (L_1) (°C)	395
Exhaust gas temperature at outlet of turbocharger (L_1) (°C)	244
Mean effective pressure	
– at 76 min^{-1} (L_1, L_2) (MPa)	1.90
– at 61 min^{-1} (L_3, L_4) (MPa)	1.52
Brake specific fuel oil consumption (g/kWh)	
– at 76 min^{-1} (L_1/L_2) for N_e 100/80% (g/kWh)	167/164
– at 61 min^{-1} (L_3/L_4) for N_e 100/80% (g/kWh)	160/157
Mean piston speed at 76 min^{-1} (m/s)	8.08
Lubrication system oil consumption (g/kWh)	0.15
Cylinder oil consumption	
– mechanical lubricator (g/kWh)	1.0–1.5
– alpha lubricator (g/kWh)	0.7

Dimensions and weight engines MAN series S90MC-C7

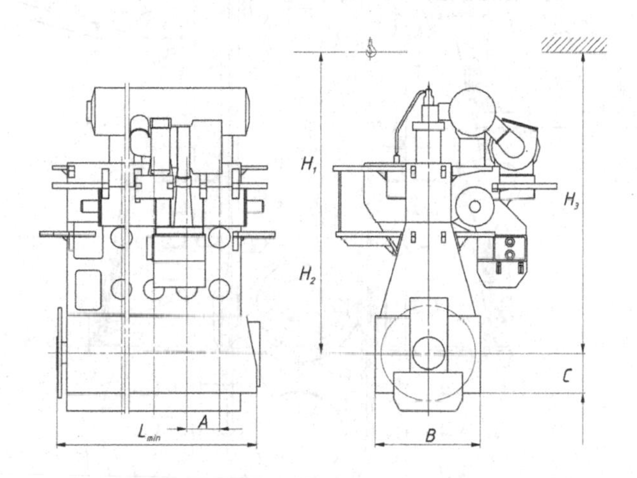

Number of cylinders	6	7	8	9
L_{min}	12,802	14,404	16,006	17,608
Weight (kg)	1,074,000	1,203,000	1,372,000	1,543,000

A (mm)	B (mm)	C (mm)	H_1 (mm)	H_2 (mm)	H_3 (mm)
1602	5000	1800	14500	13525	14,275

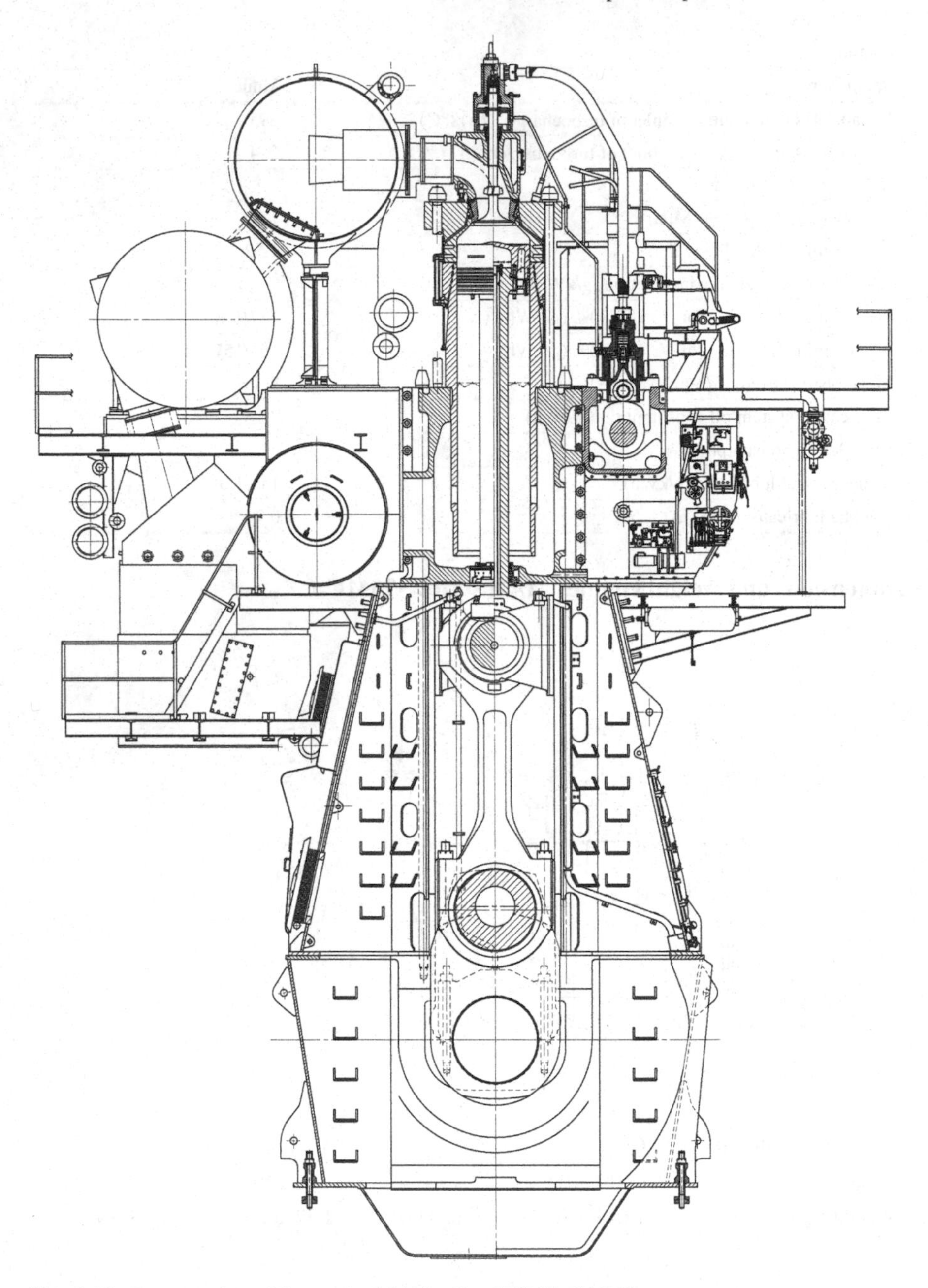

Fig. 3.18 Cross-section of the engine MAN series S90MC-C7 [17]

Main technical parameters of engines MAN series L60MC-C7 (Fig. 3.19)

Parameter	Value
Number and cylinders arrangement	5, 6, 7, 8, 9 in-line
Cylinder bore (mm)	600
Piston stroke (mm)	2022
Cylinder capacity (dm^3)	571.7
Rotational speed (min^{-1})	123
Cylinder power, N_e (layout area)	
– maximum continuous rating at 123 min^{-1} (L_1) (kW)	2230
– operational at 123 min^{-1} (L_2) (kW)	1780
– maximum continuous rating at 105 min^{-1} (L_3) (kW)	1900
– operational at 105 min^{-1} (L_4) (kW)	1520
Air charging pressure (L_1) (MPa)	0.365
Compression pressure (L_1) (MPa)	13.20
Maximum cycle pressure (L_1) (MPa)	15.20
Exhaust gas temperature at inlet of turbocharger (L_1) (°C)	405
Exhaust gas temperature at outlet of turbocharger (L_1) (°C)	244
Mean effective pressure	
– at 123 min^{-1} (L_1, L_2) (MPa)	1.90
– at 105 min^{-1} (L_3, L_4) (MPa)	1.52
Brake specific fuel oil consumption (g/kWh)	
– at 123 min^{-1} (L_1/L_2) for N_e 100/80% (g/kWh)	171/168
– at 105 min^{-1} (L_3/L_4) for N_e 100/80% (g/kWh)	164/161
Mean piston speed at 123 min^{-1} (m/s)	8.29
Lubrication system oil consumption (g/kWh)	0.15
Cylinder oil consumption	
– mechanical lubricator (g/kWh)	0.8–1.2
– alpha lubricator (g/kWh)	0.7

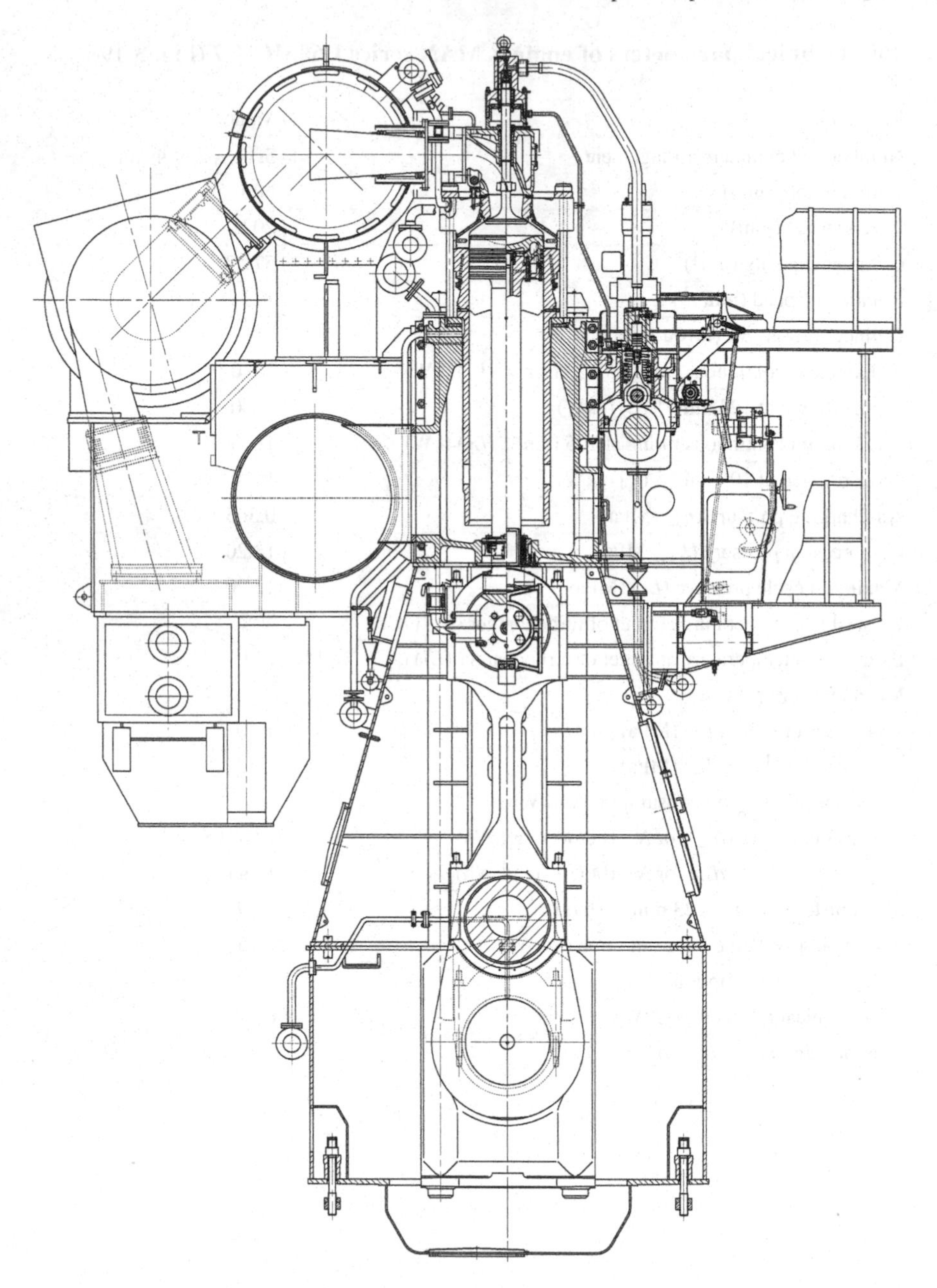

Fig. 3.19 Cross-section of the engine MAN series L60MC-C7 [18]

Dimensions and weight engines MAN series L60MC-C7

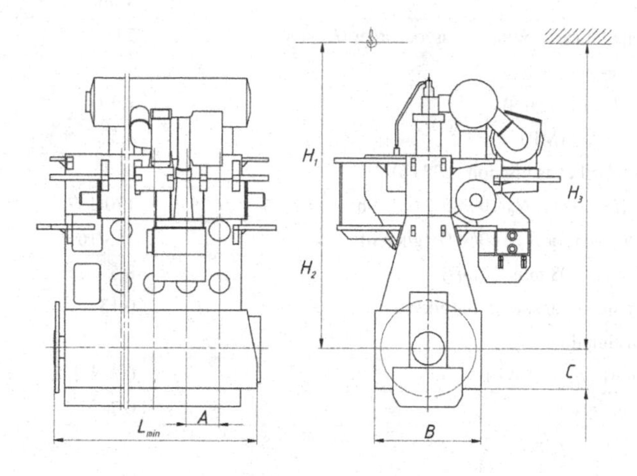

Number of cylinders	5	6	7	8	9
L_{min}	7122	8142	9162	10,182	11,202
Weight (kg)	304,000	347,000	397,000	453,000	510,000

A (mm)	B (mm)	C (mm)	H_1 (mm)	H_2 (mm)	H_3 (mm)
1020	3228	1134	9675	9125	8925

Main technical parameters of engines MAN series L70MC-C7 (Fig. 3.20)

Parameter	Value
Number and cylinders arrangement	5, 6, 7, 8 in-line
Cylinder bore (mm)	700
Piston stroke (mm)	2360
Cylinder capacity (dm^3)	908.23
Rotational speed (min^{-1})	108
Cylinder power, N_e (layout area)	
– maximum continuous rating at 108 min^{-1} (L_1) (kW)	3110
– operational at 108 min^{-1} (L_2) (kW)	2480
– maximum continuous rating at 91 min^{-1} (L_3) (kW)	2620
– operational at 91 min^{-1} (L_4) (kW)	2090
Air charging pressure (L_1) (MPa)	0.362
Compression pressure (L_1) (MPa)	13.25
Maximum cycle pressure (L_1) (MPa)	15.20
Exhaust gas temperature at inlet of turbocharger (L_1) (°C)	408

(continued)

(continued)

Parameter	Value
Exhaust gas temperature at outlet of turbocharger (L_1) (°C)	248
Mean effective pressure	
– at 108 min^{-1} (L_1, L_2) (MPa)	1.90
– at 91 min^{-1} (L_3, L_4) (MPa)	1.52
Brake specific fuel oil consumption (g/kWh)	
– at 108 min^{-1} (L_1/L_2) for N_e 100/80% (g/kWh)	170/167
– at 91 min^{-1} (L_3/L_4) for N_e 100/80% (g/kWh)	163/160
Mean piston speed at 108 min^{-1} (m/s)	8.5
Lubrication system oil consumption (g/kWh)	0.15
Cylinder oil consumption	
– mechanical lubricator (g/kWh)	0.8–1.2
– alpha lubricator (g/kWh)	0.7

Dimensions and weight engines MAN series L70MC-C7

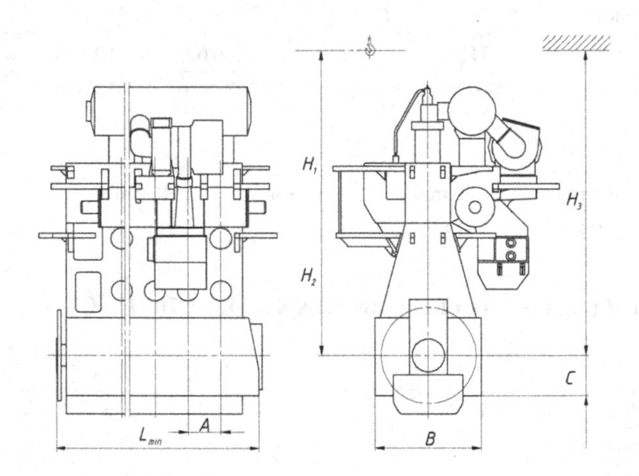

Number of cylinders	5	6	7	8
L_{min}	7781	8971	10,161	11,351
Weight (kg)	465,000	538,000	605,000	683,000

A (mm)	B (mm)	C (mm)	H_1 (mm)	H_2 (mm)	H_3 (mm)
1190	3980	1262	11,250	10,475	10,475

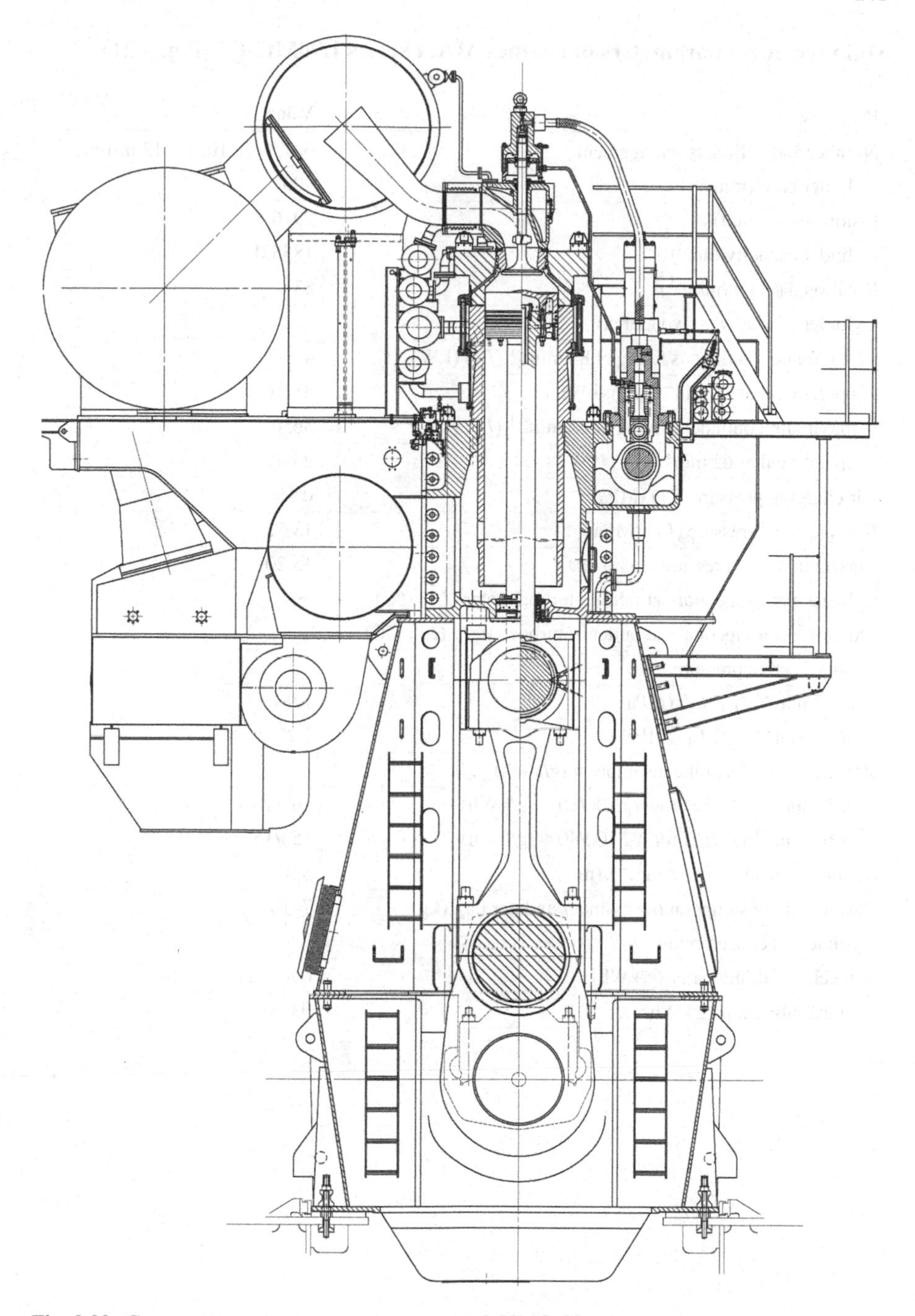

Fig. 3.20 Cross-section of the engine MAN series L70MC-C7 [19]

Main technical parameters of engines MAN series L90MC-C7 (Fig. 3.21)

Parameter	Value
Number and cylinders arrangement	6, 7, 8, 9, 10, 11, 12 in-line
Cylinder bore (mm)	900
Piston stroke (mm)	2916
Cylinder capacity (dm^3)	1855.0
Rotational speed (min^{-1})	83
Cylinder power, N_e (layout area)	
– maximum continuous rating at 83 min^{-1} (L_1) (kW)	4890
– operational at 83 min^{-1} (L_2) (kW)	3130
– maximum continuous rating at 62 min^{-1} (L_3) (kW)	3650
– operational at 62 min^{-1} (L_4) (kW)	2340
Air charging pressure (L_1) (MPa)	0.350
Compression pressure (L_1) (MPa)	13.52
Maximum cycle pressure (L_1) (MPa)	15.20
Exhaust gas temperature at inlet of turbocharger (L_1) (°C)	392
Exhaust gas temperature at outlet of turbocharger (L_1) (°C)	245
Mean effective pressure	
– at 83 min^{-1} (L_1, L_2) (MPa)	1.90
– at 62 min^{-1} (L_3, L_4) (MPa)	1.22
Brake specific fuel oil consumption (g/kWh)	
– at 83 min^{-1} (L_1/L_2) for N_e 100/80% (g/kWh)	167/165
– at 62 min^{-1} (L_3/L_4) for N_e 100/80% (g/kWh)	155/154
Mean piston speed at 83 min^{-1} (m/s)	8.07
System oil consumption per cylinder and per day (kg)	7–10
Cylinder oil consumption	
– mechanical lubricator (g/kWh)	0.8–1.2
– alpha lubricator (g/kWh)	0.6–0.8

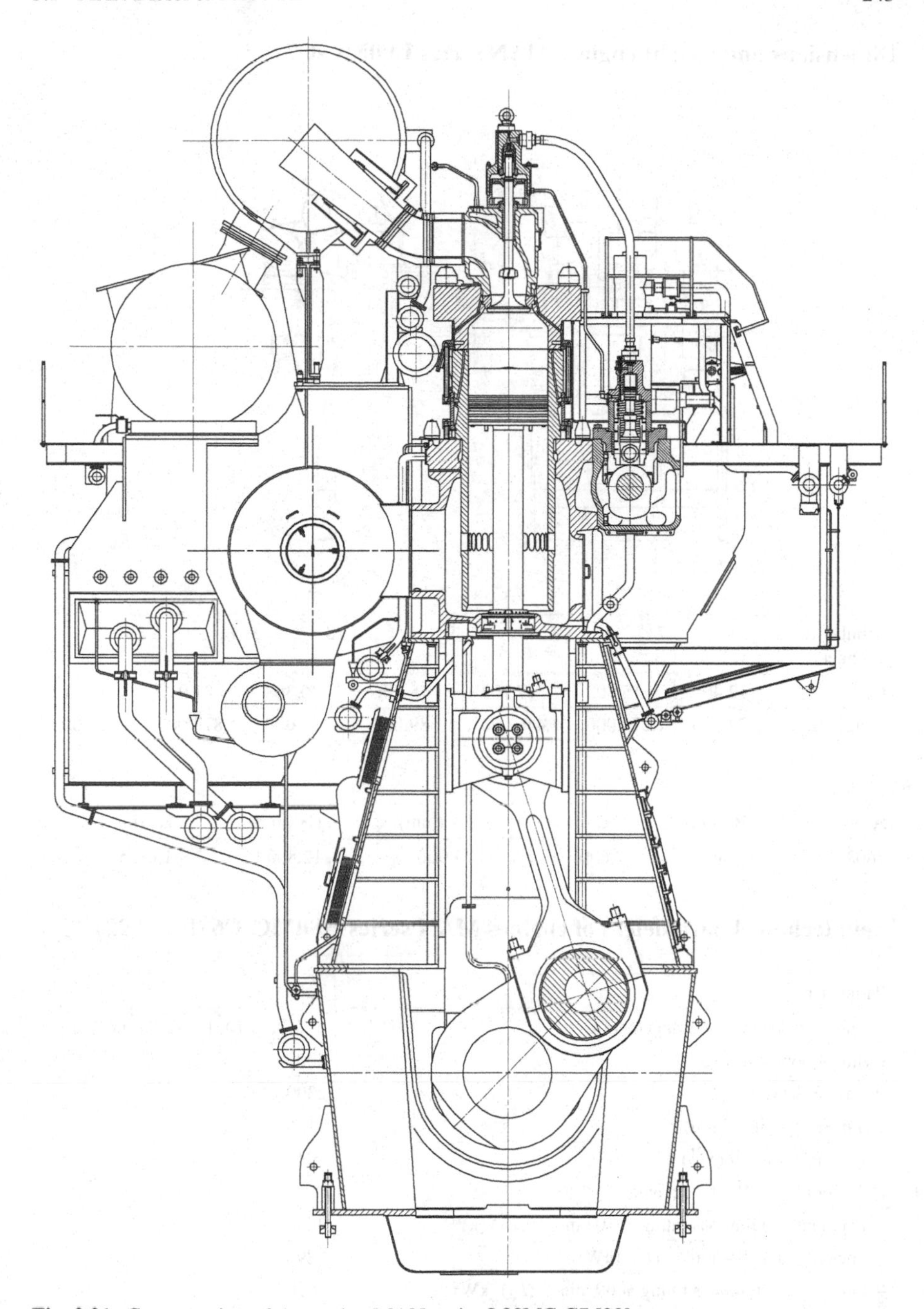

Fig. 3.21 Cross-section of the engine MAN series L90MC-C7 [20]

Dimensions and weight engines MAN series L90MC-C7

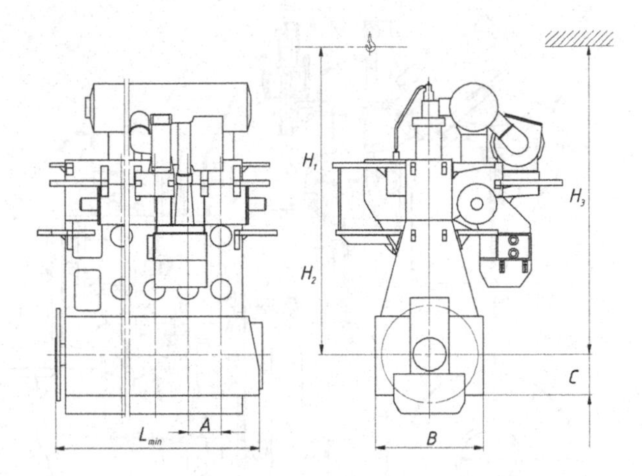

Number of cylinders	6	7	8	9	10	11	12
L_{min}	12,400	15,502	17,102	18,706	20,308	21,910	23,512
Weight (kg)	1,077,000	1,279,000	1,446,000	1,589,000	1,734,000	1,877,000	2,038,000

A (mm)	B (mm)	C (mm)	H_1 (mm)	H_2 (mm)	H_3 (mm)
1602	4936	699	13,900	12,800	13,125

Main technical parameters of engines MAN series K80MC-C6 (Fig. 3.22)

Parameter	Value
Number and cylinders arrangement	6, 7, 8, 9, 10, 11, 12 in-line
Cylinder bore (mm)	800
Piston stroke (mm)	2300
Cylinder capacity (dm^3)	1156.1
Rotational speed (min^{-1})	104
Cylinder power, N_e (layout area)	
– maximum continuous rating at 104 min^{-1} (L_1) (kW)	3610
– operational at 104 min^{-1} (L_2) (kW)	2890
– maximum continuous rating at 69 min^{-1} (L_3) (kW)	3090
– operational at 69 min^{-1} (L_4) (kW)	2470
Air charging pressure (L_1) (MPa)	0.355
Compression pressure (L_1) (MPa)	12.50
Maximum cycle pressure (L_1) (MPa)	14.20
Exhaust gas temperature at inlet of turbocharger (L_1) (°C)	385

(continued)

(continued)

Parameter	Value
Exhaust gas temperature at outlet of turbocharger (L_1) (°C)	244
Mean effective pressure	
– at 104 min^{-1} (L_1, L_2) (MPa)	1.80
– at 69 min^{-1} (L_3, L_4) (MPa)	1.44
Brake specific fuel oil consumption (g/kWh)	
– at 104 min^{-1} (L_1/L_2) for N_e 100/80% (g/kWh)	171/169
– at 69 min^{-1} (L_3/L_4) for N_e 100/80% (g/kWh)	165/162
Mean piston speed at 104 min^{-1} (m/s)	7.97
System oil consumption per cylinder and per day (kg)	6–9
Cylinder oil consumption	
– mechanical lubricator (g/kWh)	0.8–1.2
– alpha lubricator (g/kWh)	0.6–0.9

Dimensions and weight engines MAN series K80MC-C6

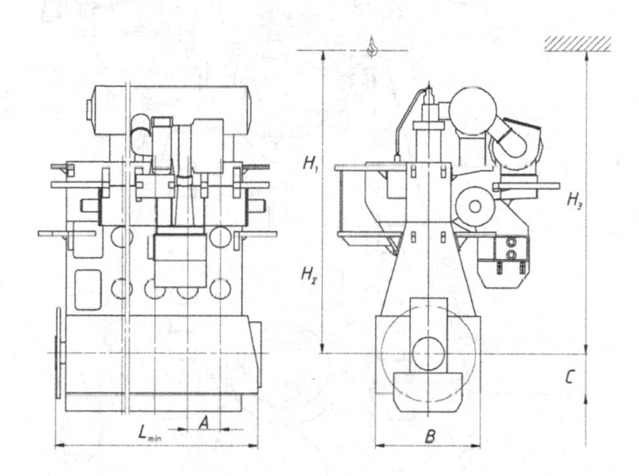

Number of cylinders	6	7	8	9	10	11	12
L_{min}	11,104	12,528	13,952	16,526	17,950	19,374	20,798
Weight (kg)	737,000	830,000	926,000	1,062,000	1,178,000	1,276,000	1,374,000

A (mm)	B (mm)	C (mm)	H_1 (mm)	H_2 (mm)	H_3 (mm)
1424	4088	1510	11,900	11,500	11,300

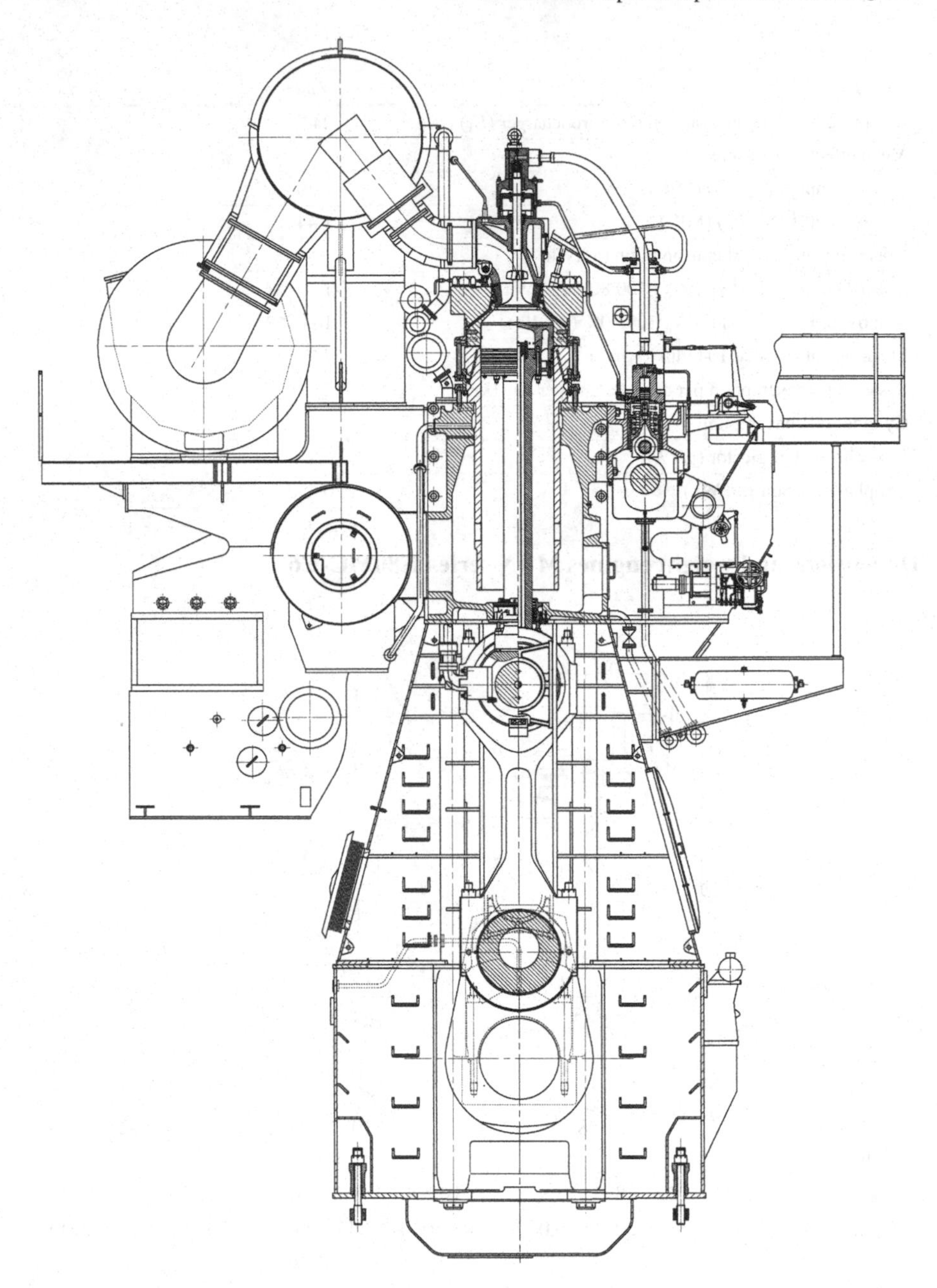

Fig. 3.22 Cross-section of the engine MAN series K80MC-C6 [21]

Main technical parameters of engines MAN series K90MC-C6 (Fig. 3.23)

Parameter	Value
Number and cylinders arrangement	6, 7, 8, 9, 10, 11, 12 in-line
Cylinder bore (mm)	900
Piston stroke (mm)	2300
Cylinder capacity (dm^3)	1463.2
Rotational speed (min^{-1})	104
Cylinder power, N_e (layout area)	
– maximum continuous rating at 104 min^{-1} (L_1) (kW)	4570
– operational at 104 min^{-1} (L_2) (kW)	3650
– maximum continuous rating at 89 min^{-1} (L_3) (kW)	3910
– operational at 89 min^{-1} (L_4) (kW)	3130
Air charging pressure (L_1) (MPa)	0.354
Compression pressure (L_1) (MPa)	12.50
Maximum cycle pressure (L_1) (MPa)	14.20
Exhaust gas temperature at inlet of turbocharger (L_1) (°C)	385
Exhaust gas temperature at outlet of turbocharger (L_1) (°C)	240
Mean effective pressure	
– at 104 min^{-1} (L_1, L_2) (MPa)	1.80
– at 89 min^{-1} (L_3, L_4) (MPa)	1.44
Brake specific fuel oil consumption (g/kWh)	
– at 104 min^{-1} (L_1/L_2) for N_e 100/80% (g/kWh)	177/174
– at 89 min^{-1} (L_3/L_4) for N_e 100/80% (g/kWh)	171/168
Mean piston speed at 104 min^{-1} (m/s)	7.97
Lubrication system oil consumption (g/kWh)	0.1
Cylinder oil consumption	
– mechanical lubricator (g/kWh)	0.8–1.8
– alpha lubricator (g/kWh)	0.6–1.35

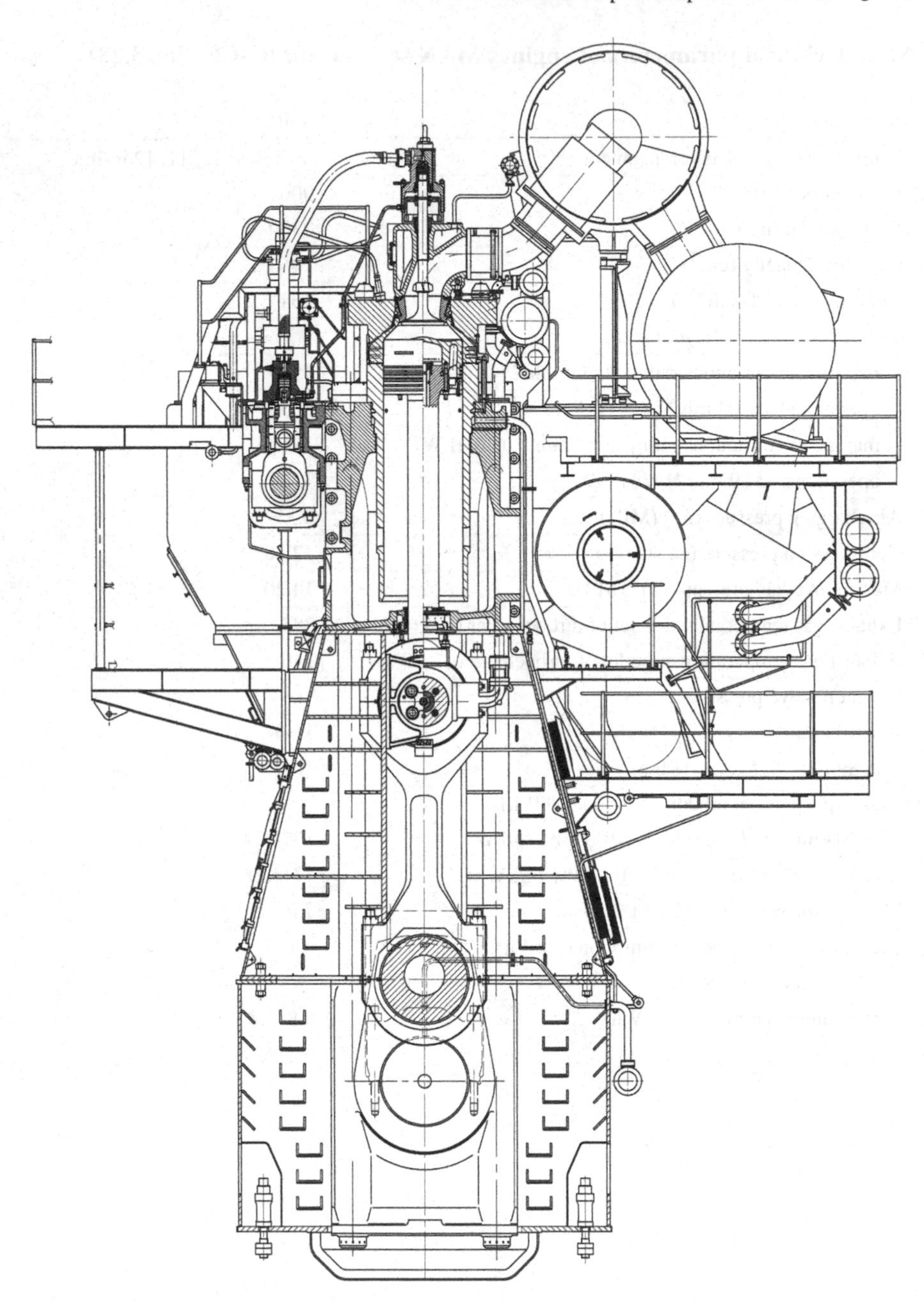

Fig. 3.23 Cross-section of the engine MAN series K90MC-C6 [22]

Dimensions and weight engines MAN series K90MC-C6

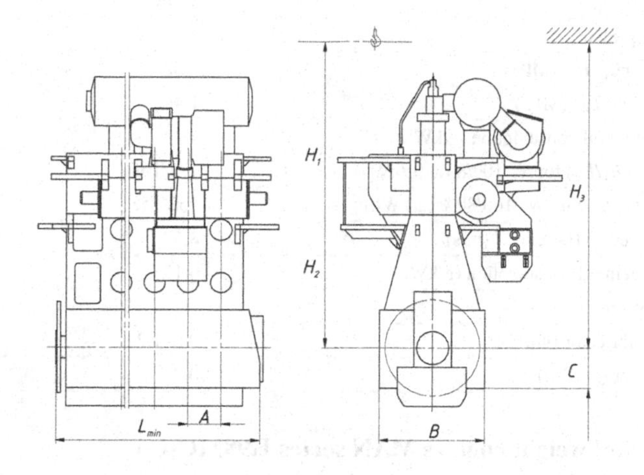

Number of cylinders	6	7	8	9	10	11	12
L_{min}	12,502	14,104	15,706	18,458	20,060	21,662	23,267
Weight (kg)	986,000	1,106,000	1,253,000	1,415,000	1,561,000	1,686,000	1,826,000

A (mm)	B (mm)	C (mm)	H_1 (mm)	H_2 (mm)	H_3 (mm)
1602	4286	1699	12,800	12,600	12,375

Main technical parameters of engines MAN series K98MC-C6 (Fig. 3.24)

Parameter	Value
Number and cylinders arrangement	6, 7, 8, 9, 10, 11, 12, 14 in-line
Cylinder bore (mm)	980
Piston stroke (mm)	2300
Cylinder capacity (dm^3)	1810.3
Rotational speed (min^{-1})	104
Cylinder power, N_e (layout area)	
– maximum continuous rating at 104 min^{-1} (L_1) (kW)	5710
– operational at 104 min^{-1} (L_2) (kW)	4580
– maximum continuous rating at 94 min^{-1} (L_3) (kW)	5160
– operational at 94 min^{-1} (L_4) (kW)	4140
Air charging pressure (L_1) (MPa)	0.355
Compression pressure (L_1) (MPa)	12.52
Maximum cycle pressure (L_1) (MPa)	14.20
Exhaust gas temperature at inlet of turbocharger (L_1) (°C)	402
Exhaust gas temperature at outlet of turbocharger (L_1) (°C)	242

(continued)

(continued)

Parameter	Value
Mean effective pressure	
– at 104 min^{-1} (L_1, L_2) (MPa)	1.82
– at 94 min^{-1} (L_3, L_4) (MPa)	1.46
Brake specific fuel oil consumption (g/kWh)	
– at 104 min^{-1} (L_1/L_2) for N_e 100/80% (g/kWh)	171/165
– at 94 min^{-1} (L_3/L_4) for N_e 100/80% (g/kWh)	162/158
Mean piston speed at 104 min^{-1} (m/s)	8.32
Lubrication system oil consumption (g/kWh)	0.15
Cylinder oil consumption	
– mechanical lubricator (g/kWh)	0.8–1.2
– alpha lubricator (g/kWh)	0.7

Dimensions and weight engines MAN series K98MC-C6

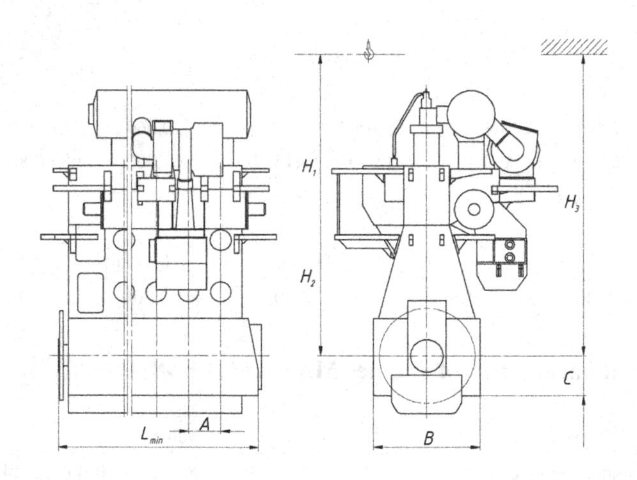

Number of cylinders	6	7	8	9
L_{min}	12,835	14,585	17,342	19,092
Weight (kg)	1,102,000	1,277,000	1,470,000	1,618,000
Number of cylinders	10	11	12	14
L_{min}	20,842	22,592	24,342	26,092
Weight (kg)	1,789,000	1,932,000	2,075,000	2,361,000

A (mm)	B (mm)	C (mm)	H_1 (mm)	H_2 (mm)	H_3 (mm)
1750	4370	1700	12,825	12,875	12,825

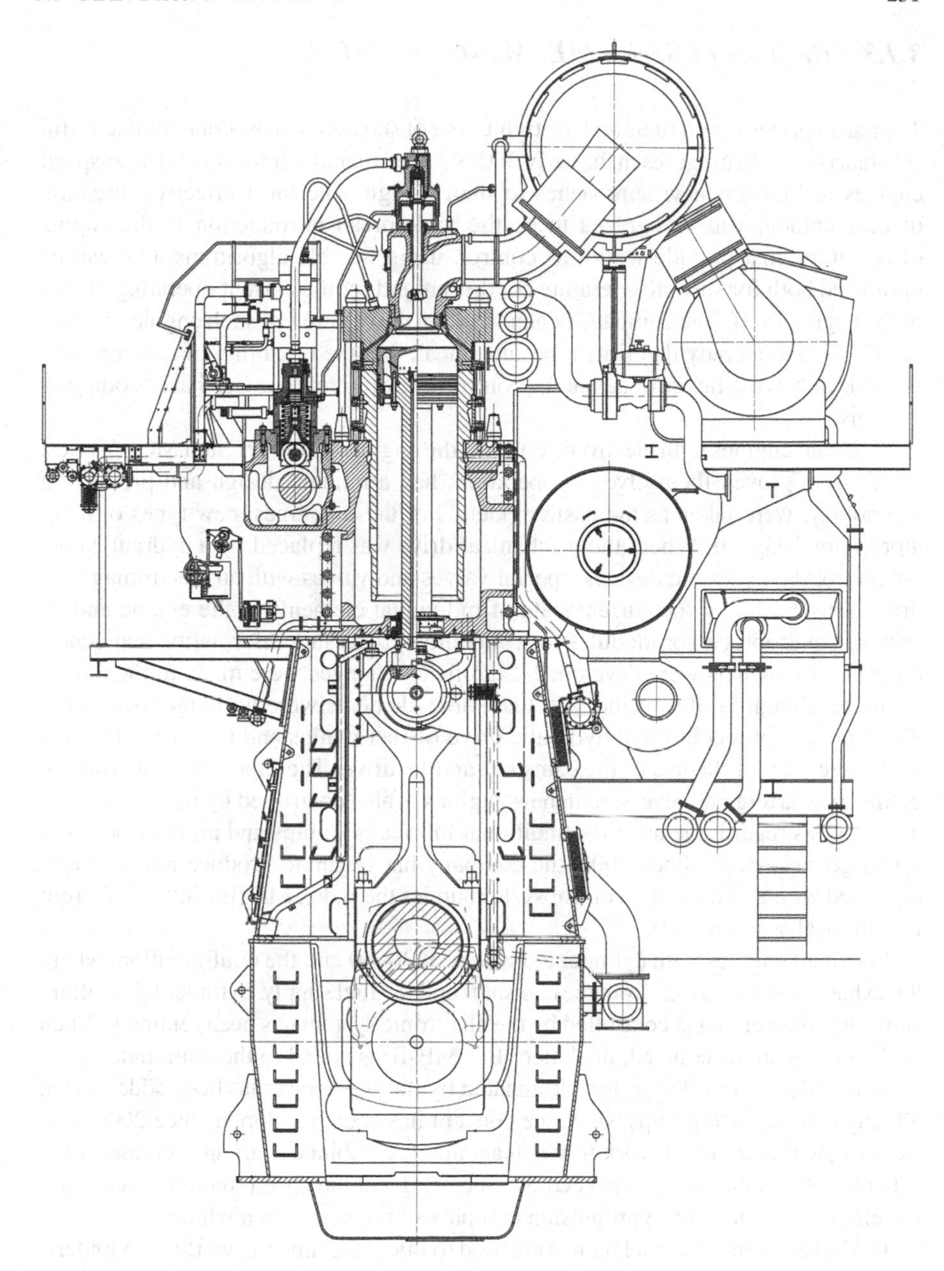

Fig. 3.24 Cross-section of the engine MAN series K98MC-C6 [23]

3.1.3 Engines of Series ME, ME-C and ME-B

The rise in prices for oil fuels and the tightening of standards in the content of harmful substances in exhaust gases in the early 2000s forced manufacturers of ship low-speed engines to look for new approaches to their design. The most effective direction of development was recognized to be the widespread introduction to the engine of actuators, that provide workflow control, using flexible algorithms, that can be optimized both for specific operating conditions and for the current operating modes of the engine itself. The company began work on the creation of such engines back in the 1990s, and already that time it became clear, that the flexibility of algorithms is possible only with the use of actuators, controlled by microprocessor-based computer systems.

To create engines with electronic control, the engines of the MC and MC-C series, which have proven themselves in operation, had a reliable design and production technology, were taken as the basic models. For these engines, new types of actuators were designed, Where the mechanical drive was replaced by a hydraulic one. To control the hydraulic devices, special valves and valves with an electromagnetic drive, sensors, fixing the current state of individual elements of the engine and its systems, microprocessor modules for collecting, processing information and generating control signals were developed. Significant changes were made to the design of the base engines, due to the fact, that some elements were no longer necessary. Thus, the use of a controlled hydraulic drive of fuel pumps and exhaust valves led to the rejection of the use of the camshaft and its drive. The lubrication of working cylinders, starting and reversing in new engines is also controlled by hydraulic actuators, in this regard, the need for traditional lubricator pumps and air distributors is no longer necessary. Since 2003, the company has begun to produce new engines, equipped with an electronic control system under the index ME (the letter "E" from Electronically controlled).

For small engines with cylinder diameters of 30–50 cm, the configuration, where the exhaust valve is driven from the camshaft and the fuel supply, cylinder lubrication, starting and reversing is controlled by the electronic system has been retained. When such engines are designated, the letter "B" (ME-B) is added to the main index.

Since 2010, a new dimension, designated by the "G" index, has been added to the already existing size groups, since the concept has been developed since 2006. The use of new principles of workflow management, combined with an increase in the S/D ratio, has significantly reduced the operating engine speed, thereby increasing the efficiency of the entire propulsion complex of the vessel as a whole.

In ME series engines, fuel pumps are used to inject fuel into the working cylinders, which, in essence, are hydraulic pressure multipliers. Structurally, they are simpler, than their mechanical counterparts. Plungers do not have regulating edges, which significantly simplifies their manufacturing technology and increases resource. Each pump unit is equipped with gas pressure accumulators with a membrane separator, which provide a steady supply of oil and eliminate the occurrence of dangerous fluctuations in the control oil system. The oil goes to the membrane pressure accumulators from the control line and from them to the distribution devices of the fuel pump, the exhaust valve actuator, and the lubricator pump. In the hydraulic drive system, oil is taken from the circulating engine lubrication system. Before entering the control line, it undergoes additional cleaning in an automatic self-cleaning filter that retains mechanical impurities up to 5 μm in size. Next, the oil is fed to the pump unit. The main pumps are driven by the engine and provide oil to the system under a pressure of 20 MPa during diesel operation. Starting pumps are electrically driven and provide the system with an oil under pressure of 17.5 MPa during engine start-up or when operating in emergency situations. In case of failure of one or several main pumps, the engine can continue to be used, reducing its power to 50–80%.

After the pumps, the oil enters into the accumulating cavity, which allows to reduce pressure pulsations. Next, through high-pressure pipelines, oil is supplied to each block of the engine pumps.

To supply oil to the control line, axial-plunger pumps with adjustable capacity are used. All oil lines on the engine cover with protective covers. The internal space between the working cavity and the casing is used to collect leaks.

The moment of oil supply to the actuators, as well as the beginning of its discharge is determined by the flow of the control pulse from the electronic unit, with which each engine cylinder is equipped. The electrical signals are sent to two-stage electrically controlled distribution mechanisms of the valve spindle type, which have an identical design, both in the drive system of the fuel pumps and in the exhaust valve actuators. The main valve with chopped distribution grooves is located in the switchgear housing.

The main valve spindle is moved by means of a protrusion, which works as hydraulic piston. In the cavities, located on both sides of the protrusion, the oil enters through the so-called "pilot" spindle valve with an electromagnetic actuator, to which the control signal enters.

The fuel injection pump consists of a hydraulic cylinder and a plunger bushing. In the hydraulic cylinder is located a piston, that is mechanically connected to the plunger of the fuel pump. The diameter of the hydraulic cylinder is several times greater, than that of the plunger. This makes it possible to increase the fuel pressure from 20 MPa in the control oil line to 60–100 MPa at the outlet of the fuel pump. At the same time, unlike the mechanically driven fuel injection pump, the pressure, under which the fuel is supplied to the injector, does not depend on the engine speed, thus, significantly improves the quality of the mixture formation throughout the entire engine load range.

In the course of research work on a laboratory engine, it was found that the nature of fuel injection significantly affects on the Brake specific fuel oil consumption, as well as the content of harmful emissions from the exhaust gases. The nature of the injection, which gives the best result in terms of fuel economy, leads to an increase in the content of nitrogen oxides (NO_x) in the exhaust gases, and an attempt to improve the environmental performance of the engine on the contrary leads to an increase in fuel consumption. In this regard, for engines of type ME, changing the fuel injection law is an effective tool that can be used to find a compromise solution between reducing consumption while maintaining harmful emissions within specified limits.

The electronic control system, based on microprocessor technology, in combination with the hydraulic system of the injection pump drive, allows implementing various methods of fuel supply into the cylinders, single injection of the entire injection rate, preliminary injection of the so-called "pilot portion of fuel", prior to the main injection, multistage injection, etc. The corresponding law of fuel supply can be optimized for a specific mode of engine operation and can be applied when entering this mode.

The transition from one fuel control algorithm to another can be performed in the time interval between two consecutive injections.

Main technical parameters of engines MAN series S30ME-B 9.5 (Fig. 3.25)

Parameter	Value
Number and cylinders arrangement	5, 6, 7, 8 in-line
Cylinder bore (mm)	300
Piston stroke (mm)	1328
Cylinder capacity (dm^3)	93.87
Rotational speed (min^{-1})	195
Cylinder power, N_e (layout area)	
– maximum continuous rating at 195 min^{-1} (L_1) (kW)	640
– operational at 195 min^{-1} (L_2) (kW)	510
– maximum continuous rating at 148 min^{-1} (L_3) (kW)	485
– operational at 148 min^{-1} (L_4) (kW)	390
Air charging pressure (L_1) (MPa)	0.375
Compression pressure (L_1) (MPa)	14.25
Maximum cycle pressure (L_1) (MPa)	16.20
Exhaust gas temperature at inlet of turbocharger (L_1) (°C)	442
Exhaust gas temperature at outlet of turbocharger (L_1) (°C)	260
Mean effective pressure	
– at 195 min^{-1} (L_1, L_2) (MPa)	2.10
– at 148 min^{-1} (L_3, L_4) (MPa)	1.69
Brake specific fuel oil consumption (g/kWh)	
– at 195 min^{-1} (L_1/L_2) for N_e 100/75% (g/kWh)	176/173
– at 148 min^{-1} (L_3/L_4) for N_e 100/75% (g/kWh)	172/169
Brake specific air consumption (kg/kWh)	7.60
Brake specific exhaust gas flow (kg/kWh)	7.78
Cylinder oil consumption (g/kWh)	0.8
Mean piston speed at 195 min^{-1} (m/s)	8.63

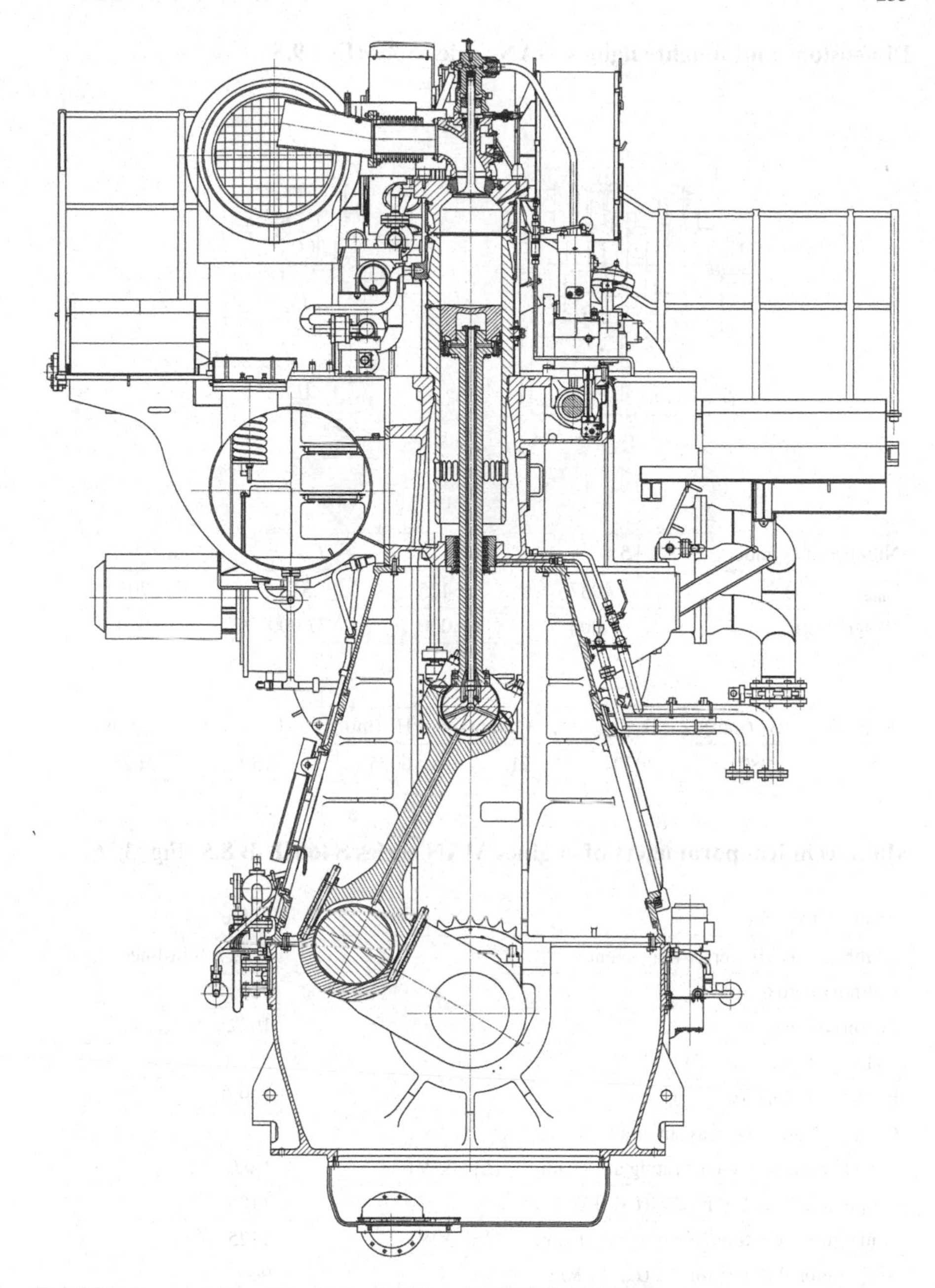

Fig. 3.25 Cross-section of the engine MAN series S30ME-B 9.5 [24]

Dimensions and weight engines MAN series S30ME-B 9.5

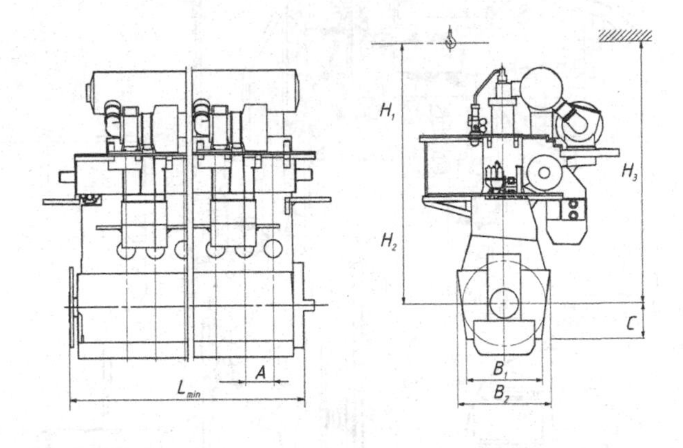

Number of cylinders	5	6	7	8
L_{min}	4087	6425	5163	5701
Weight (kg)	61,000	69,000	77,000	86,000

A (mm)	B_1 (mm)	B_2 (mm)	C (mm)	H_1 (mm)	H_2 (mm)	H_3 (mm)
538	1980	2020	712	6025	5950	5625

Main technical parameters of engines MAN series S46ME-B 8.5 (Fig. 3.26)

Parameter	Value
Number and cylinders arrangement	5, 6, 7, 8 in-line
Cylinder bore (mm)	460
Piston stroke (mm)	1932
Cylinder capacity (dm^3)	321.1
Rotational speed (min^{-1})	129.0
Cylinder power, N_e (layout area)	
– maximum continuous rating at 129 min^{-1} (L_1) (kW)	1380
– operational at 129 min^{-1} (L_2) (kW)	1105
– maximum continuous rating at 105 min^{-1} (L_3) (kW)	1125
– operational at 105 min^{-1} (L_4) (kW)	900
Air charging pressure (L_1) (MPa)	0.375
Compression pressure (L_1) (MPa)	14.25
Maximum cycle pressure (L_1) (MPa)	16.20
Exhaust gas temperature at inlet of turbocharger (L_1) (°C)	440

(continued)

(continued)

Parameter	Value
Exhaust gas temperature at outlet of turbocharger (L_1) (°C)	235
Mean effective pressure	
– at 129 min^{-1} (L_1, L_2) (MPa)	2.00
– at 105 min^{-1} (L_3, L_4) (MPa)	1.60
Brake specific fuel oil consumption (g/kWh)	
– at 129 min^{-1} (L_1/L_2) for N_e 100/75% (g/kWh)	172/167
– at 105 min^{-1} (L_3/L_4) for N_e 100/75% (g/kWh)	167/163
Brake specific air consumption (kg/kWh)	8.30
Brake specific exhaust gas flow (kg/kWh)	8.47
Cylinder oil consumption (g/kWh)	0.6
Mean piston speed at 129 min^{-1} (m/s)	8.31

Dimensions and weight engines MAN series S46ME-B 8.5

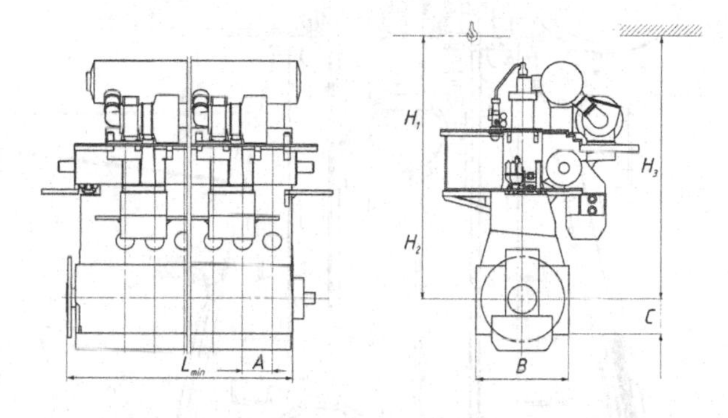

Number of cylinders	5	6	7	8
L_{min}	5528	6310	7092	7874
Weight (kg)	159,000	177,000	199,000	219,000

A (mm)	B (mm)	C (mm)	H_1 (mm)	H_2 (mm)	H_3 (mm)
782	2924	986	9000	8175	7900

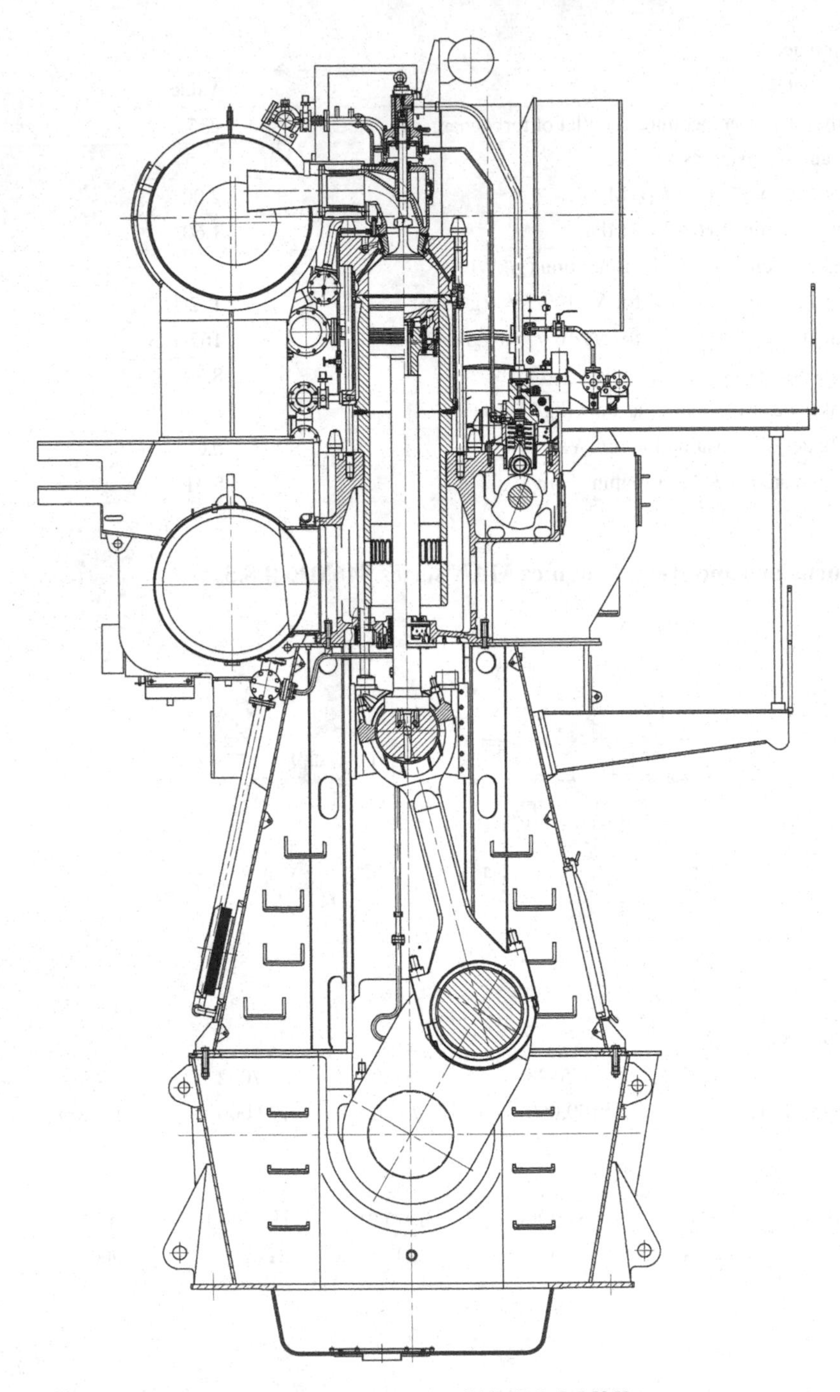

Fig. 3.26 Cross-section of the engine MAN series S46ME-B 8.5 [25]

Main technical parameters of engines MAN series S50ME-B 9.5 (Fig. 3.27)

Parameter	Value
Number and cylinders arrangement	5, 6, 7, 8, 9 in-line
Cylinder bore (mm)	500
Piston stroke (mm)	2214
Cylinder capacity, dm^3	434.72
Rotational speed, min^{-1}	117.0
Cylinder power, N_e (layout area)	
– maximum continuous rating at 117 min^{-1} (L_1) (kW)	1780
– operational at 117 min^{-1} (L_2) (kW)	1420
– maximum continuous rating at 89 min^{-1} (L_3) (kW)	1350
– operational at 89 min^{-1} (L_4) (kW)	1080
Air charging pressure (L_1) (MPa)	0.375
Compression pressure (L_1) (MPa)	14.40
Maximum cycle pressure (L_1) (MPa)	16.20
Exhaust gas temperature at inlet of turbocharger (L_1) (°C)	405
Exhaust gas temperature at outlet of turbocharger (L_1) (°C)	235
Mean effective pressure	
– at 117 min^{-1} (L_1, L_2) (MPa)	2.10
– at 89 min^{-1} (L_3, L_4) (MPa)	1.68
Brake specific fuel oil consumption (g/kWh)	
– at 117 min^{-1} (L_1/L_2) for N_e 100/75% (g/kWh)	168/165
– at 89 min^{-1} (L_3/L_4) for N_e 100/75% (g/kWh)	162/159,5
Brake specific air consumption (kg/kWh)	7.52
Brake specific exhaust gas flow (kg/kWh)	7.67
Cylinder oil consumption (g/kWh)	0.6
Mean piston speed at 117 min^{-1} (m/s)	8.63

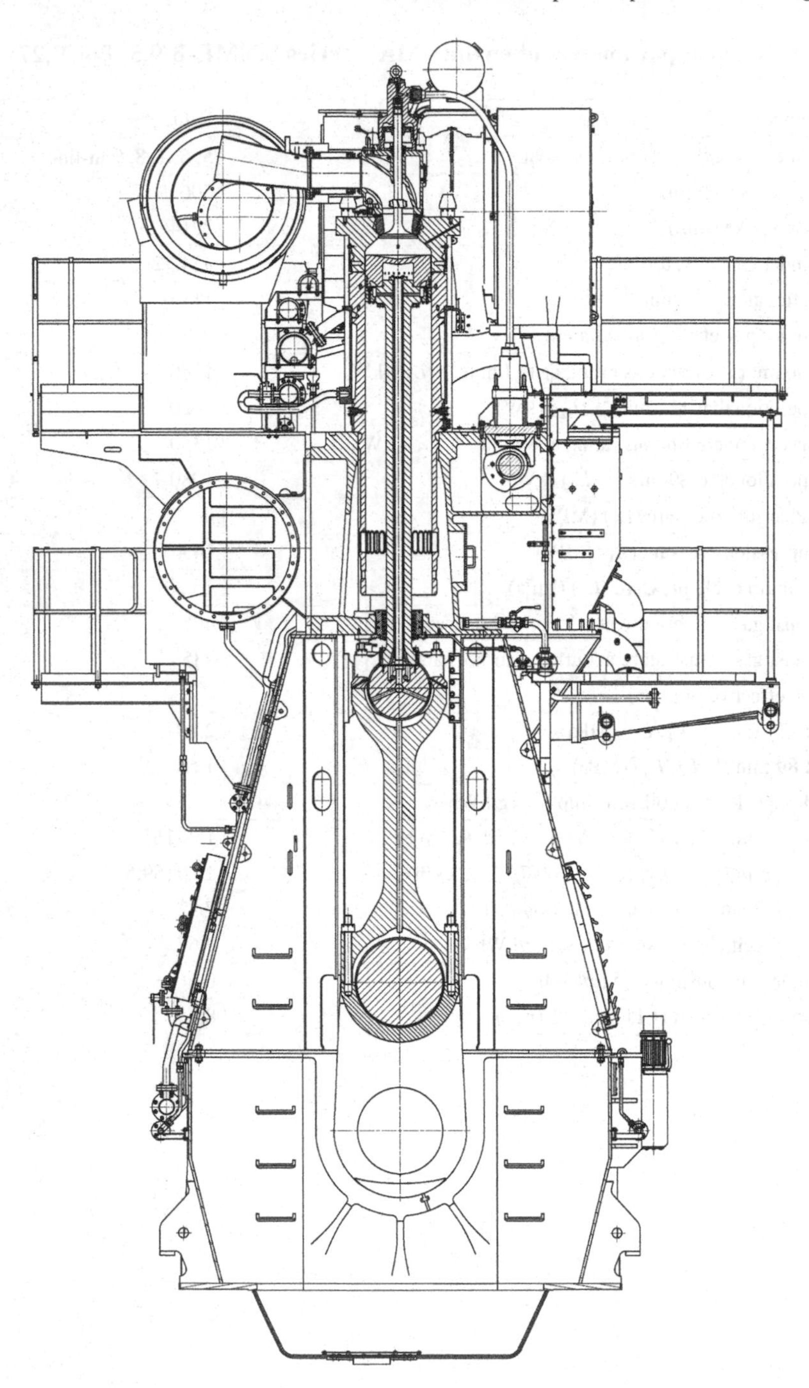

Fig. 3.27 Cross-section of the engine MAN series S50ME-B 9.5 [26]

Dimensions and weight engines MAN series S50ME-B 9.5

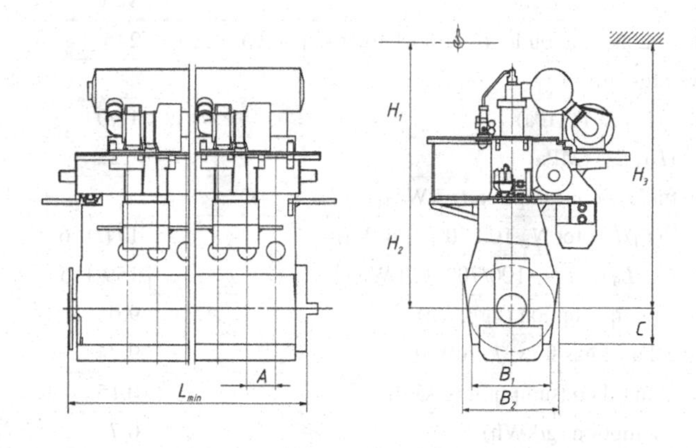

Number of cylinders	5	6	7	8	9
L_{min}	6073	6948	7823	8698	9573
Weight (kg)	190,000	220,000	255,000	258,000	315,000

A (mm)	B_1 (mm)	B_2 (mm)	C (mm)	H_1 (mm)	H_2 (mm)	H_3 (mm)
875	3350	3290	1190	9775	9200	8900

Main technical parameters of engines MAN series S50ME-C7 (Fig. 3.28)

Parameter	Value
Number and cylinders arrangement	5, 6, 7, 8, 9 in-line
Cylinder bore (mm)	500
Piston stroke (mm)	2000
Cylinder capacity (dm^3)	392.7
Rotational speed (min^{-1})	127.0
Cylinder power, N_e (layout area)	
– maximum continuous rating at 127 min^{-1} (L_1) (kW)	2230
– operational at 127 min^{-1} (L_2) (kW)	1880
– maximum continuous rating at 95 min^{-1} (L_3) (kW)	1900
– operational at 95 min^{-1} (L_4) (kW)	1520
Air charging pressure (L_1) (MPa)	0.360
Compression pressure (L_1) (MPa)	13.40
Maximum cycle pressure (L_1) (MPa)	15.20
Exhaust gas temperature at inlet of turbocharger (L_1) (°C)	400

(continued)

(continued)

Parameter	Value
Exhaust gas temperature at outlet of turbocharger (L_1) (°C)	245
Mean effective pressure	
– at 127 min^{-1} (L_1, L_2) (MPa)	1.90
– at 95 min^{-1} (L_3, L_4) (MPa)	1.22
Brake specific fuel oil consumption (g/kWh)	
– at 127 min^{-1} (L_1/L_2) for N_e 100/70% (g/kWh)	171/166
– at 95 min^{-1} (L_3/L_4) for N_e 100/70% (g/kWh)	159/155
Brake specific air consumption (kg/kWh)	9.06
Brake specific exhaust gas flow (kg/kWh)	9.24
Lubrication system oil consumption (g/kWh)	0.15
Cylinder oil consumption (g/kWh)	0.7
Mean piston speed at 127 min^{-1} (m/s)	8.47

Dimensions and weight engines MAN series S50ME-C7

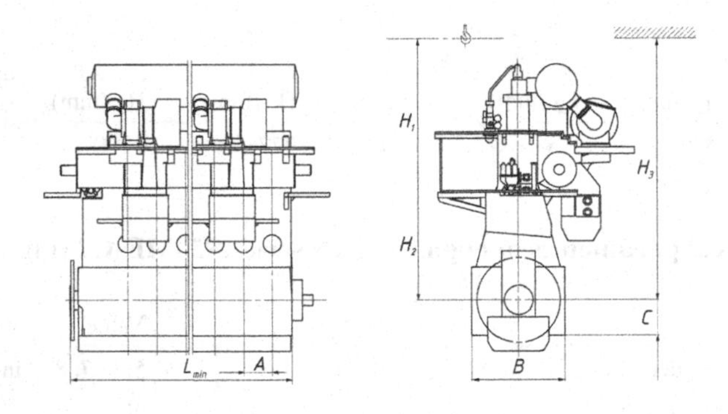

Number of cylinders	5	6	7	8	9
L_{min}	7122	8142	9162	10,182	11,202
Weight (kg)	286,000	326,000	354,000	426,000	479,000

A (mm)	B (mm)	C (mm)	H_1 (mm)	H_2 (mm)	H_3 (mm)
1020	3228	1134	9675	9125	8925

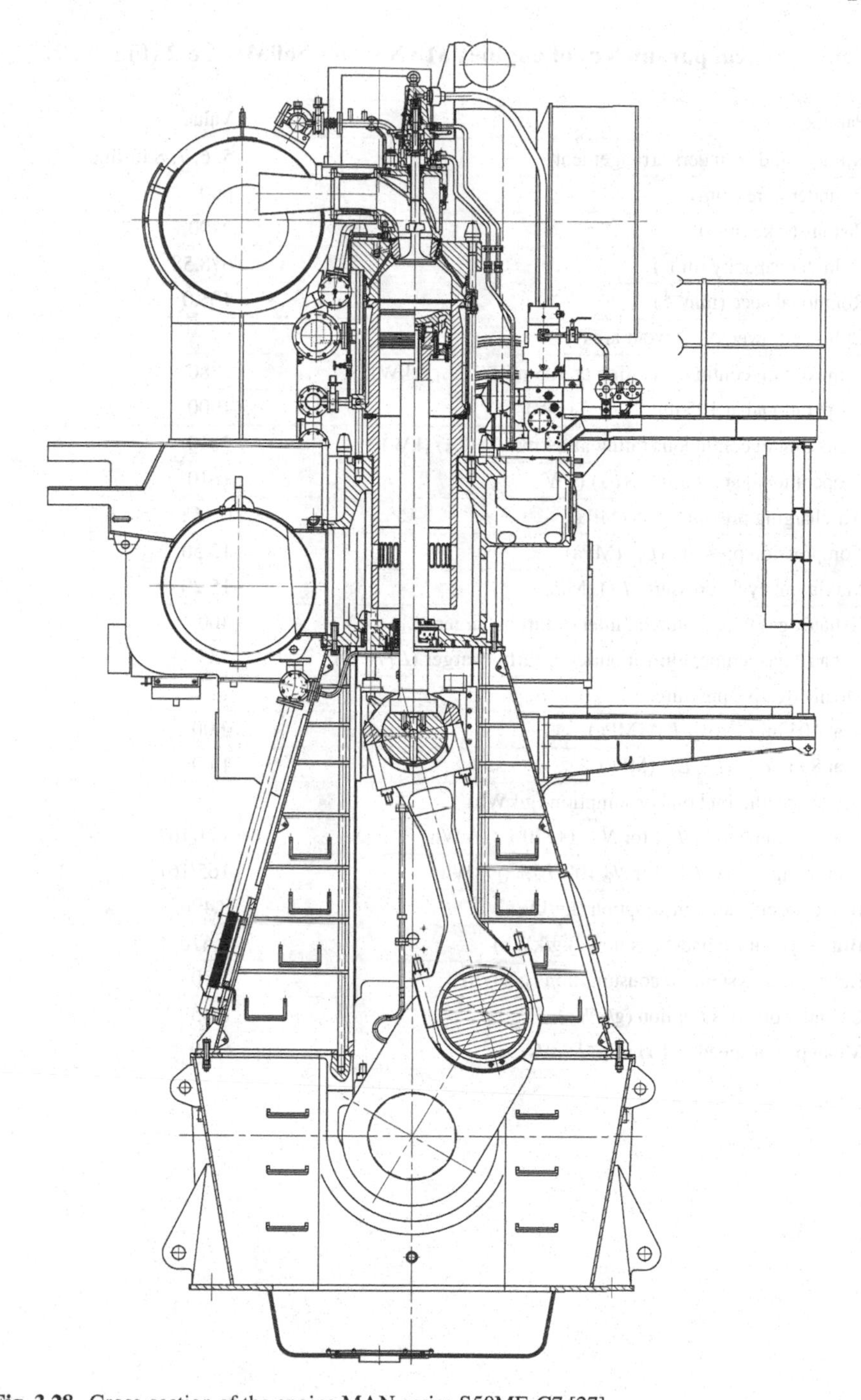

Fig. 3.28 Cross-section of the engine MAN series S50ME-C7 [27]

Main technical parameters of engines MAN series S60ME-C8.2 (Fig. 3.29)

Parameter	Value
Number and cylinders arrangement	5, 6, 7, 8 in-line
Cylinder bore (mm)	600
Piston stroke (mm)	2400
Cylinder capacity (dm^3)	678.58
Rotational spee (min^{-1})	105.0
Cylinder power, N_e (layout area)	
– maximum continuous rating at 105 min^{-1} (L_1) (kW)	2380
– operational at 105 min^{-1} (L_2) (kW)	1900
– maximum continuous rating at 89 min^{-1} (L_3) (kW)	2010
– operational at 89 min^{-1} (L_4) (kW)	1610
Air charging pressure (L_1) (MPa)	0.362
Compression pressure (L_1) (MPa)	13.50
Maximum cycle pressure (L_1) (MPa)	15.20
Exhaust gas temperature at inlet of turbocharger (L_1) (°C)	400
Exhaust gas temperature at outlet of turbocharger (L_1) (°C)	235
Mean effective pressure	
– at 105 min^{-1} (L_1, L_2) (MPa)	2.00
– at 89 min^{-1} (L_3, L_4) (MPa)	1.60
Brake specific fuel oil consumption (g/kWh)	
– at 105 min^{-1} (L_1/L_2) for N_e 100/70% (g/kWh)	171/167
– at 89 min^{-1} (L_3/L_4) for N_e 100/70% (g/kWh)	165/161
Brake specific air consumption (kg/kWh)	8.496
Brake specific exhaust gas flow (kg/kWh)	8.670
Lubrication system oil consumption (g/kWh)	0.10
Cylinder oil consumption (g/kWh)	0.65
Mean piston speed at 105 min^{-1} (m/s)	8.40

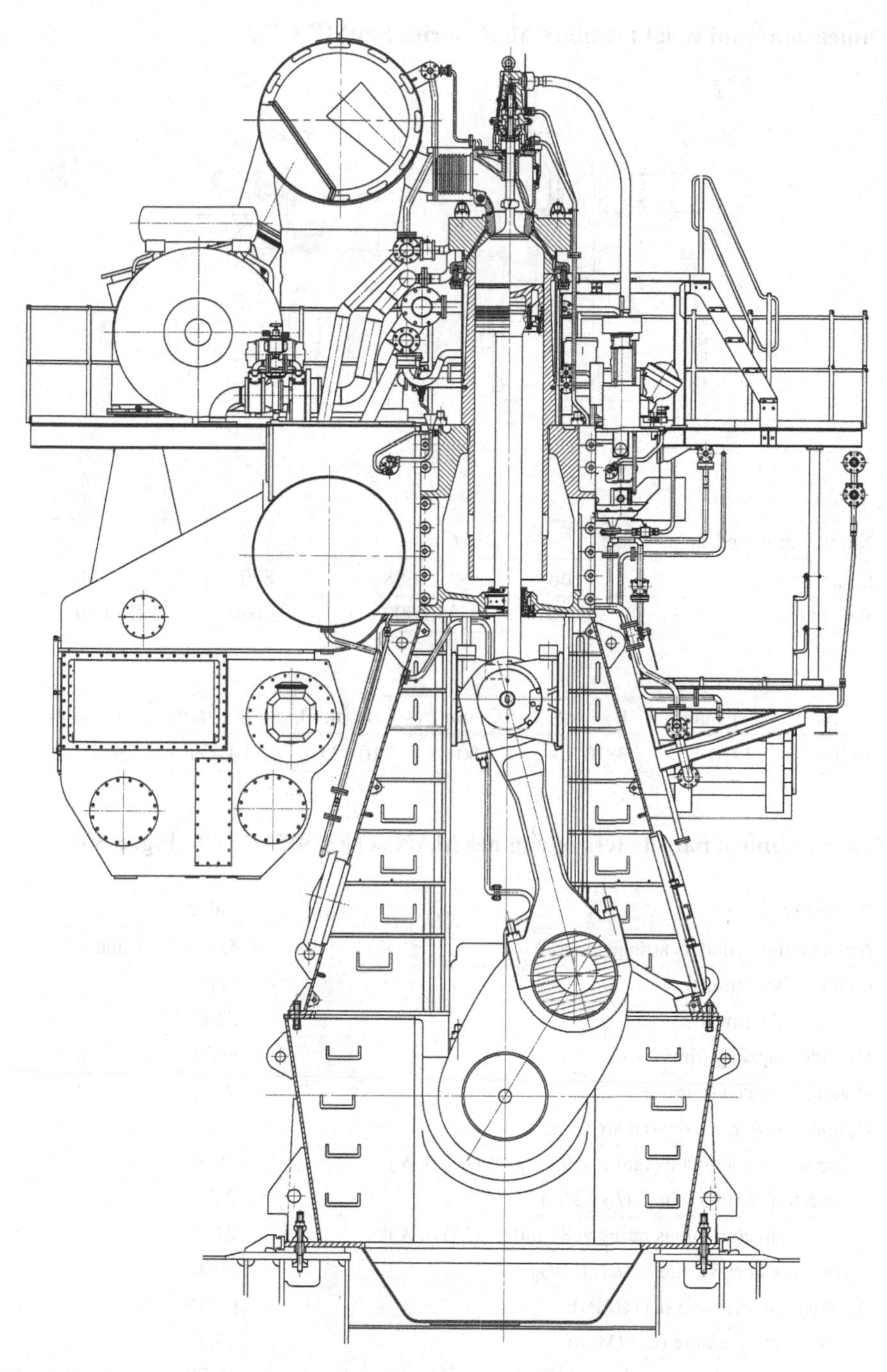

Fig. 3.29 Cross-section of the engine MAN series S60ME-C8.2 [28]

Dimensions and weight engines MAN series S60ME-C8.2

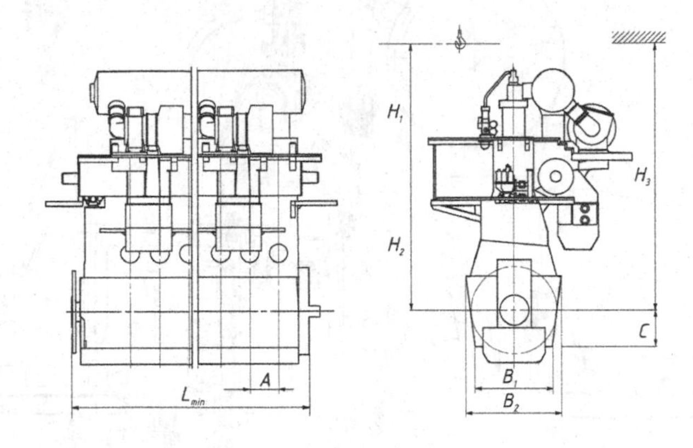

Number of cylinders	5	6	7	8
L_{min}	6668	7688	8708	9728
Weight (kg)	308,000	350,000	393,000	452,000

A (mm)	B_1 (mm)	B_2 (mm)	C (mm)	H_1 (mm)	H_2 (mm)	H_3 (mm)
1020	3770	3840	1300	10,825	10,000	9775

Main technical parameters of engines MAN series S65ME-C8 (Fig. 3.30)

Parameter	Value
Number and cylinders arrangement	5, 6, 7, 8 in-line
Cylinder bore (mm)	650
Piston stroke (mm)	2730
Cylinder capacity (dm^3)	905.9
Rotational speed (min^{-1})	95
Cylinder power, N_e (layout area)	
– maximum continuous rating at 95 min^{-1} (L_1) (kW)	2870
– operational at 95 min^{-1} (L_2) (kW)	2290
– maximum continuous rating at 81 min^{-1} (L_3) (kW)	2450
– operational at 81 min^{-1} (L_4) (kW)	1960
Air charging pressure (L_1) (MPa)	0.375
Compression pressure (L_1) (MPa)	14.40
Maximum cycle pressure (L_1) (MPa)	16.20
Exhaust gas temperature at inlet of turbocharger (L_1) (°C)	410

(continued)

(continued)

Parameter	Value
Exhaust gas temperature at outlet of turbocharger (L_1) (°C)	244
Mean effective pressure	
– at 95 min^{-1} (L_1, L_2) (MPa)	2.00
– at 81 min^{-1} (L_3, L_4) (MPa)	1.60
Brake specific fuel oil consumption (g/kWh)	
– at 95 min^{-1} (L_1/L_2) for N_e 100/75% (g/kWh)	170/166
– at 81 min^{-1} (L_3/L_4) for N_e 100/75% (g/kWh)	164/160
Brake specific air consumption (kg/kWh)	9.11
Brake specific exhaust gas flow (kg/kWh)	9.27
Lubrication system oil consumption (g/kWh)	0.15
Cylinder oil consumption (g/kWh)	0.70
Mean piston speed at 95 min^{-1} (m/s)	8.65

Dimensions and weight engines MAN series S65ME-C8

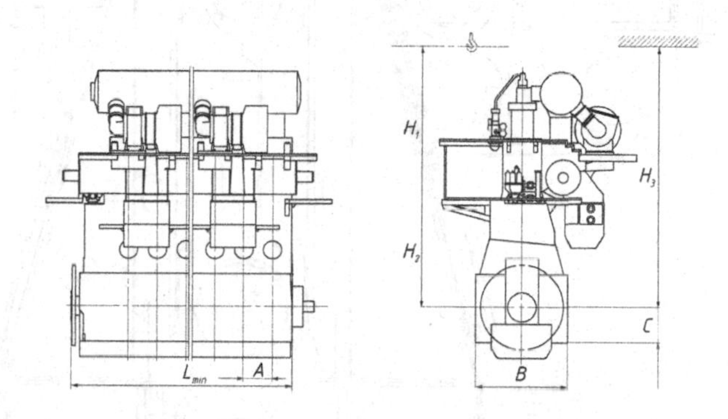

Number of cylinders	5	6	7	8
L_{min}	6914	7998	9062	10,138
Weight (kg)	382,000	451,000	512,000	575,000

A (mm)	B MM	C (mm)	H_1 (mm)	H_2 (mm)	H_3 (mm)
1084	4124	1410	11,950	11,225	11,025

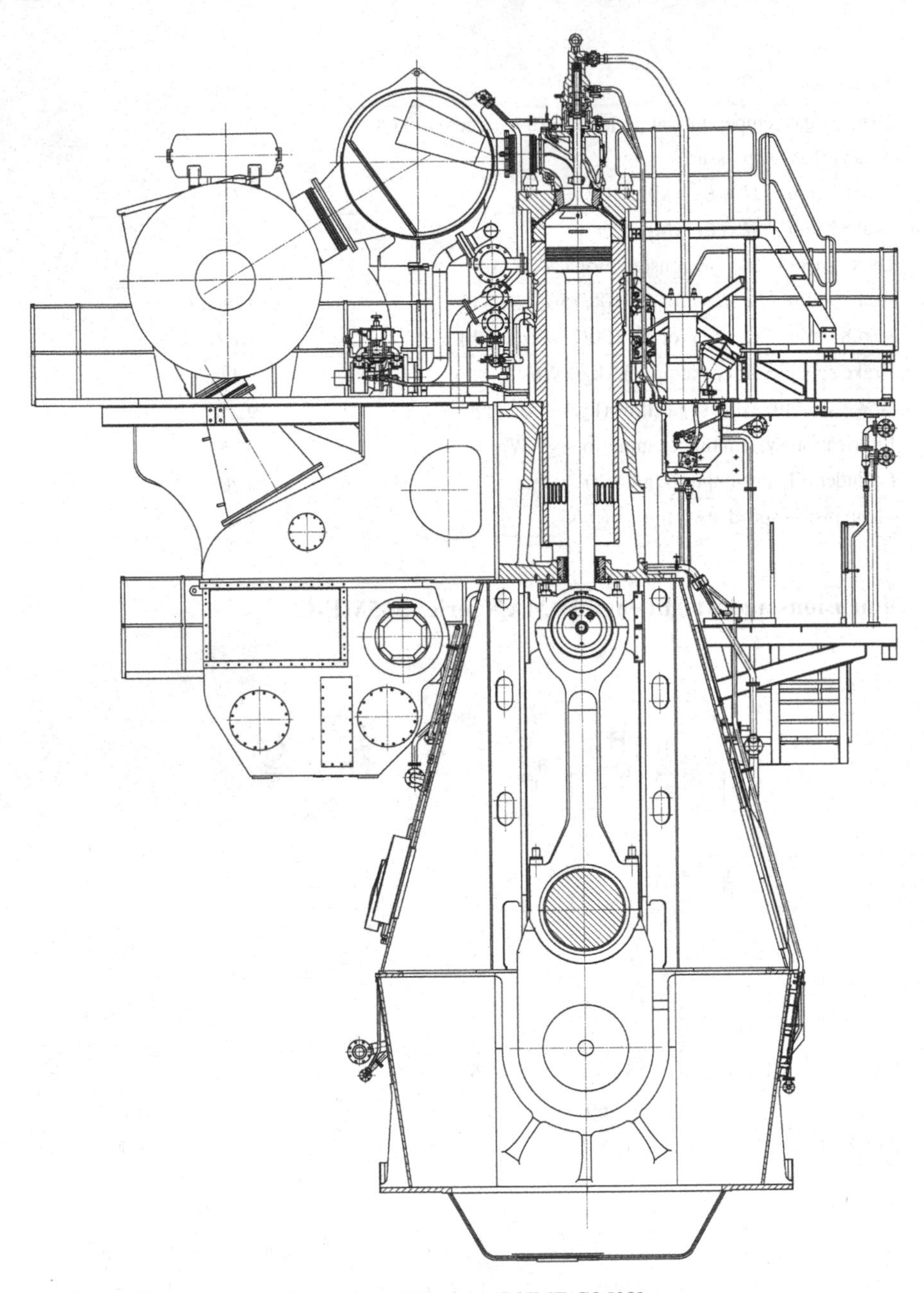

Fig. 3.30 Cross-section of the engine MAN series S65ME-C8 [29]

Main technical parameters of engines MAN series S70ME-C8 (Fig. 3.31)

Parameter	Value
Number and cylinders arrangement	5, 6, 7, 8 in-line
Cylinder bore (mm)	700
Piston stroke (mm)	2800
Cylinder capacity (dm^3)	1077.57
Rotational speed (min^{-1})	91
Cylinder power, N_e (layout area)	
– maximum continuous rating at 91 min^{-1} (L_1) (kW)	3270
– operational at 91 min^{-1} (L_2) (kW)	2610
– maximum continuous rating at 73 min^{-1} (L_3) (kW)	2620
– operational at 73 min^{-1} (L_4) (kW)	2100
Air charging pressure (L_1) (MPa)	0.365
Compression pressure (L_1) (MPa)	13.50
Maximum cycle pressure (L_1) (MPa)	15.20
Exhaust gas temperature at inlet of turbocharger (L_1) (°C)	395
Exhaust gas temperature at outlet of turbocharger (L_1) (°C)	235
Mean effective pressure	
– at 91 min^{-1} (L_1, L_2) (MPa)	2.00
– at 73 min^{-1} (L_3, L_4) (MPa)	1.60
Brake specific fuel oil consumption (g/kWh)	
– at 91 min^{-1} (L_1/L_2) for N_e 100/75% (g/kWh)	169/165
– at 73 min^{-1} (L_3/L_4) for N_e 100/75% (g/kWh)	163/159
Brake specific air consumption (kg/kWh)	8.09
Brake specific exhaust gas flow (kg/kWh)	8.27
Lubrication system oil consumption (g/kWh)	0.15
Cylinder oil consumption (g/kWh)	0.70
Mean piston speed at 91 min^{-1} (m/s)	8.49

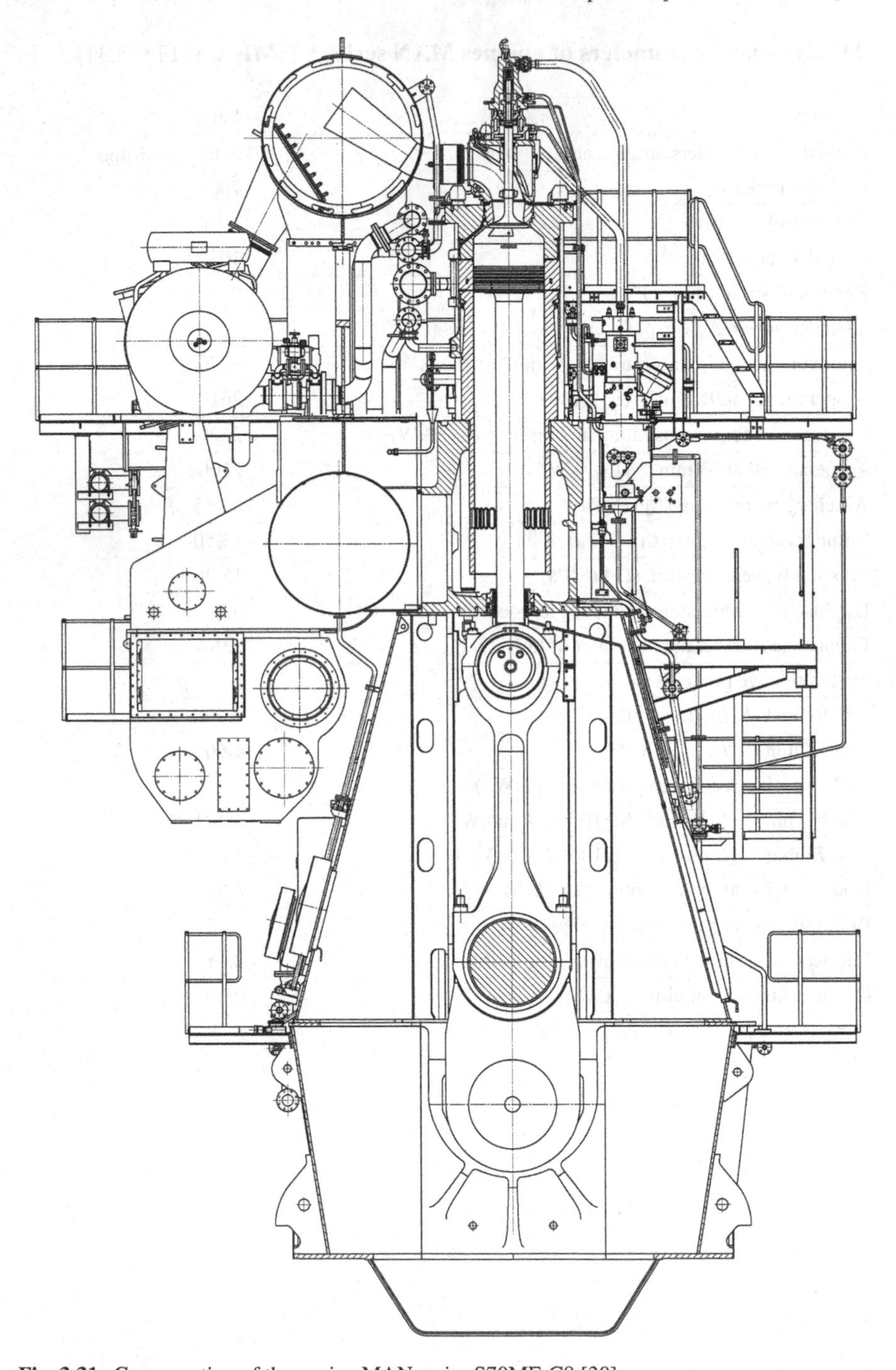

Fig. 3.31 Cross-section of the engine MAN series S70ME-C8 [30]

Dimensions and weight engines MAN series S70ME-C8

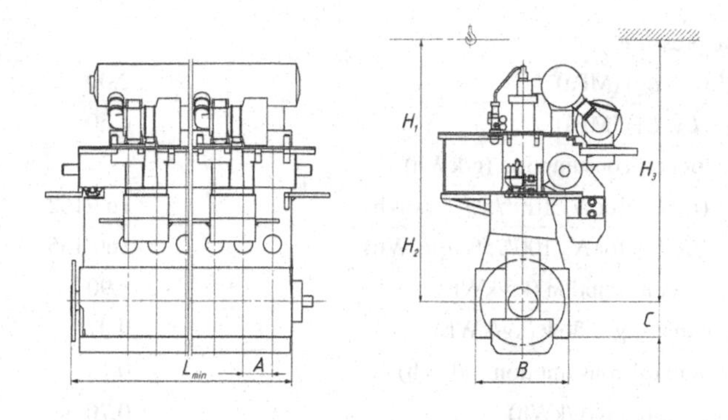

Number of cylinders	5	6	7	8
L_{min}	7514	8704	9894	11,084
Weight (kg)	451,000	534,000	605,000	681,000

A (mm)	B MM	C (mm)	H_1 (mm)	H_2 (mm)	H_3 (mm)
1190	4390	1520	12,550	11,725	11,500

Main technical parameters of engines MAN series S80ME-C9 (Fig. 3.32)

Parameter	Value
Number and cylinders arrangement	6, 7, 8, 9 in-line
Cylinder bore (mm)	800
Piston stroke (mm)	3450
Cylinder capacity (dm^3)	1734.16
Rotational speed (min^{-1})	78
Cylinder power, N_e (layout area)	
– maximum continuous rating at 78 min^{-1} (L_1) (kW)	4510
– operational at 78 min^{-1} (L_2) (kW)	3610
– maximum continuous rating at 66 min^{-1} (L_3) (kW)	3820
– operational at 66 min^{-1} (L_4) (kW)	3050
Air charging pressure (L_1) (MPa)	0.375
Compression pressure (L_1) (MPa)	14.30
Maximum cycle pressure (L_1) (MPa)	16.20
Exhaust gas temperature at inlet of turbocharger (L_1) (°C)	410
Exhaust gas temperature at outlet of turbocharger (L_1) (°C)	250

(continued)

(continued)

Parameter	Value
Mean effective pressure	
– at 78 min^{-1} (L_1, L_2) (MPa)	2.00
– at 66 min^{-1} (L_3, L_4) (MPa)	1.60
Brake specific fuel oil consumption (g/kWh)	
– at 78 min^{-1} (L_1/L_2) for N_e 100/75% (g/kWh)	167/162
– at 66 min^{-1} (L_3/L_4) for N_e 100/75% (g/kWh)	160/155
Brake specific air consumption (kg/kWh)	8.90
Brake specific exhaust gas flow (kg/kWh)	9.07
Lubrication system oil consumption (g/kWh)	0.15
Cylinder oil consumption (g/kWh)	0.70
Mean piston speed at 78 min^{-1} (m/s)	8.97

Dimensions and weight engines MAN series S80ME-C9

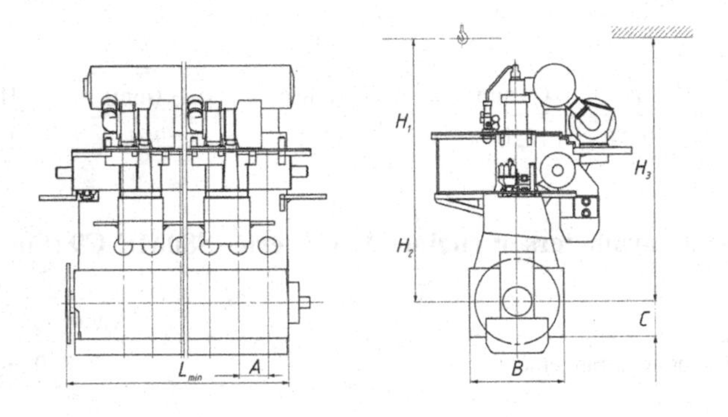

Number of cylinders	6	7	8	9
L_{min}	10,100	11,434	12,768	14,102
Weight (kg)	833,000	933,000	1,043,000	1,153,000

A (mm)	B MM	C (mm)	H_1 (mm)	H_2 (mm)	H_3 (mm)
1334	5280	1890	15,050	13,925	13,500

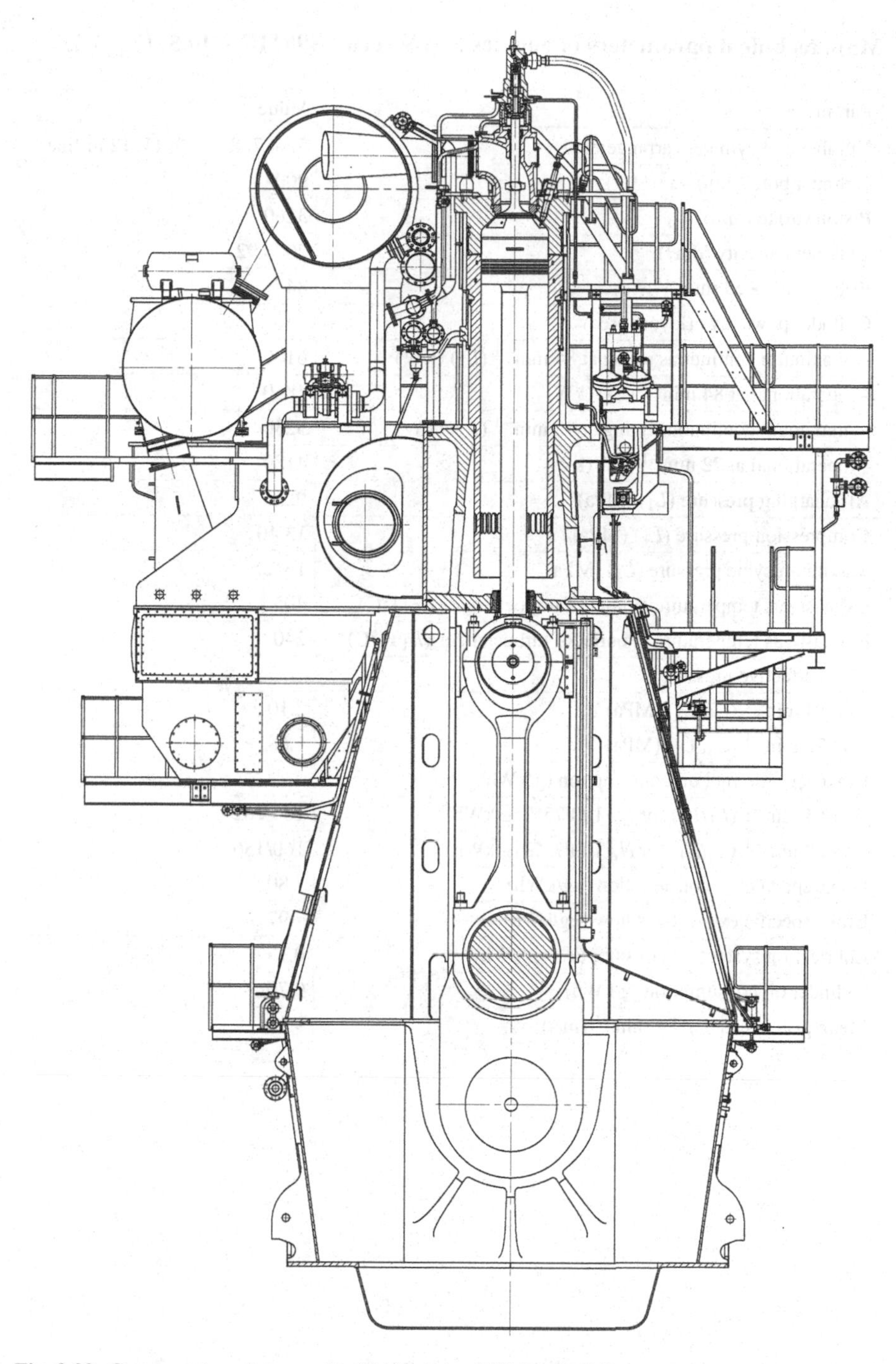

Fig. 3.32 Cross-section of the engine MAN series S80ME-C9 [31]

Main technical parameters of engines MAN series S90ME-C10.5 (Fig. 3.33)

Parameter	Value
Number and cylinders arrangement	5, 6, 7, 8, 9, 10, 11, 12 in-line
Cylinder bore (mm)	900
Piston stroke (mm)	3260
Cylinder capacity (dm^3)	2073.92
Rotational speed (min^{-1})	84
Cylinder power, N_e (layout area)	
– maximum continuous rating at 84 min^{-1} (L_1) (kW)	6100
– operational at 84 min^{-1} (L_2) (kW)	4860
– maximum continuous rating at 72 min^{-1} (L_3) (kW)	5230
– operational at 72 min^{-1} (L_4) (kW)	4180
Air charging pressure (L_1) (MPa)	0.375
Compression pressure (L_1) (MPa)	13.40
Maximum cycle pressure (L_1) (MPa)	15.20
Exhaust gas temperature at inlet of turbocharger (L_1) (°C)	405
Exhaust gas temperature at outlet of turbocharger (L_1) (°C)	240
Mean effective pressure	
– at 84 min^{-1} (L_1, L_2) (MPa)	2.10
– at 72 min^{-1} (L_3, L_4) (MPa)	1.68
Brake specific fuel oil consumption (g/kWh)	
– at 84 min^{-1} (L_1/L_2) for N_e 100/75% (g/kWh)	166/162
– at 72 min^{-1} (L_3/L_4) for N_e 100/75% (g/kWh)	160/156
Brake specific air consumption (kg/kWh)	7.80
Brake specific exhaust gas flow (kg/kWh)	7.97
Lubrication system oil consumption (g/kWh)	0.15
Cylinder oil consumption (g/kWh)	0.70
Mean piston speed at 84 min^{-1} (m/s)	9.13

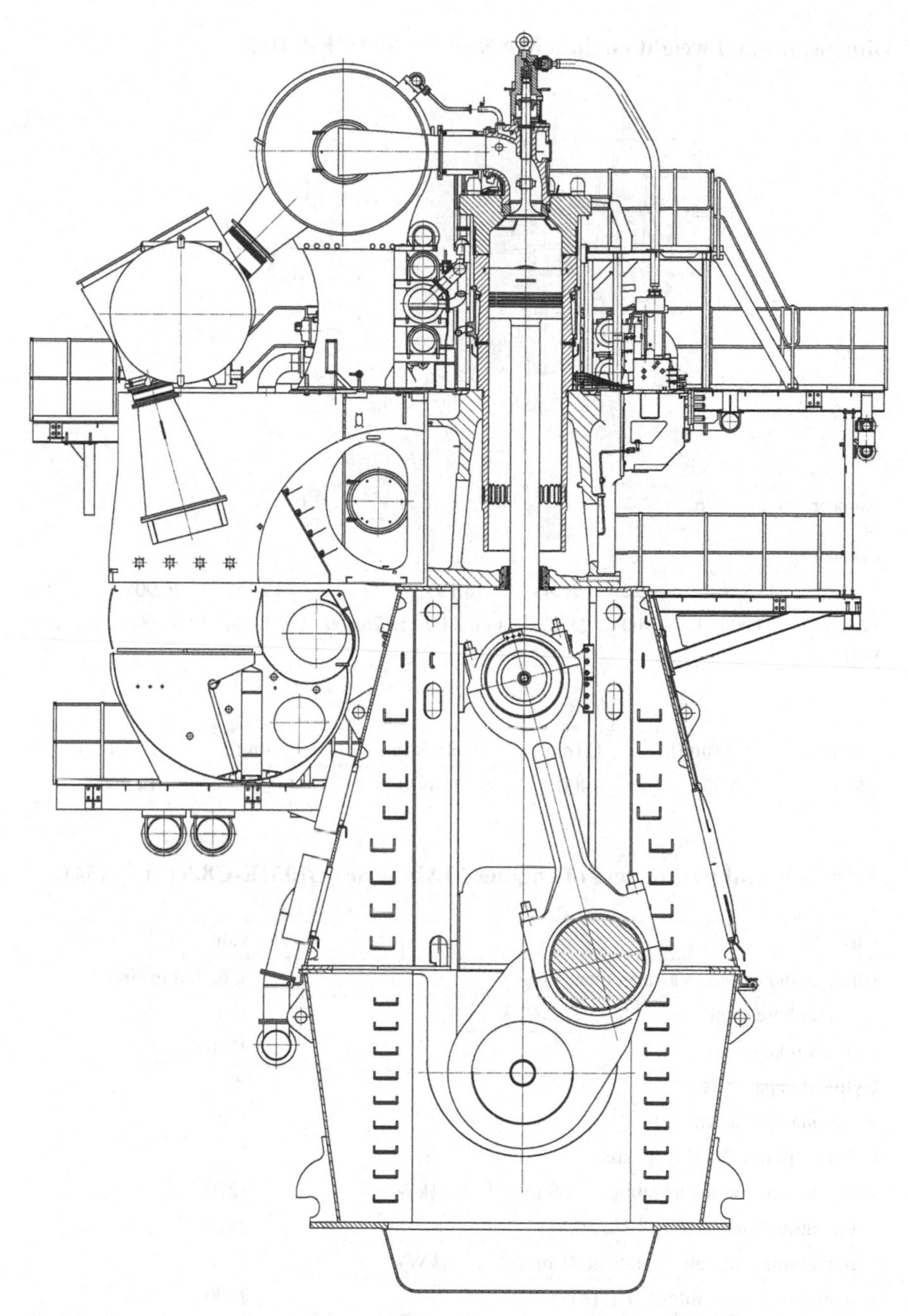

Fig. 3.33 Cross-section of the engine MAN series S90ME-C10.5 [32]

Dimensions and weight engines MAN series S90ME-C10.5

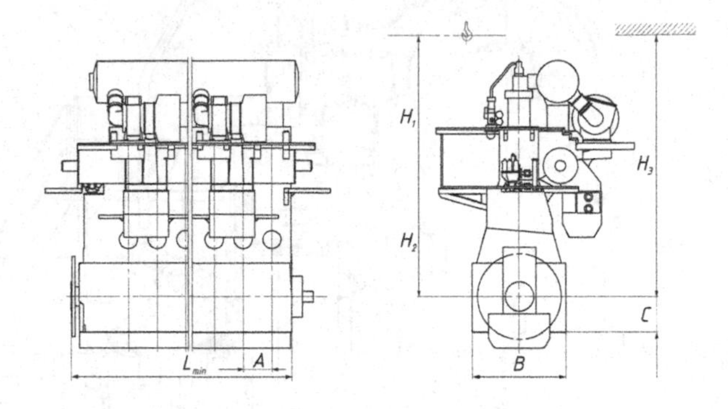

Number of cylinders	5	6	7	8	9	10	11	12
L_{min}	11,232	128,22	13,485	16,135	17,725	19,315	20,905	22,495
Weight (kg)	953,000	1,104,000	1,255,000	144,6000	1,626,000	1,771,000	1,942,000	2,088,000

A (mm)	B (mm)	C (mm)	H_1 (mm)	H_2 (mm)	H_3 (mm)
1590	5160	1900	15,000	14,025	14,500

Main technical parameters of engines MAN series L70ME-C8.5 (Fig. 3.34)

Parameter	Value
Number and cylinders arrangement	5, 6, 7, 8 in-line
Cylinder bore (mm)	700
Piston stroke (mm)	2360
Cylinder capacity (dm^3)	908.23
Rotational speed (min^{-1})	108
Cylinder power, N_e (layout area)	
– maximum continuous rating at 108 min^{-1} (L_1) (kW)	3270
– operational at 108 min^{-1} (L_2) (kW)	2620
– maximum continuous rating at 91 min^{-1} (L_3) (kW)	2750
– operational at 91 min^{-1} (L_4) (kW)	2200
Air charging pressure (L_1) (MPa)	0.378
Compression pressure (L_1) (MPa)	14.40
Maximum cycle pressure (L_1) (MPa)	16.20

(continued)

(continued)

Parameter	Value
Exhaust gas temperature at inlet of turbocharger (L_1) (°C)	422
Exhaust gas temperature at outlet of turbocharger (L_1) (°C)	235
Mean effective pressure	
– at 108 min^{-1} (L_1, L_2) (MPa)	2.00
– at 91 min^{-1} (L_3, L_4) (MPa)	1.60
Brake specific fuel oil consumption (g/kWh)	
– at 108 min^{-1} (L_1/L_2) for N_e 100/75% (g/kWh)	170/166
– at 91 min^{-1} (L_3/L_4) for N_e 100/75% (g/kWh)	164/160
Brake specific air consumption (kg/kWh)	8.50
Brake specific exhaust gas flow (kg/kWh)	8.67
Lubrication system oil consumption (g/kWh)	0.15
Cylinder oil consumption (g/kWh)	0.70
Mean piston speed at 108 min^{-1} (m/s)	8.5

Dimensions and weight engines MAN series L70ME-C8.5

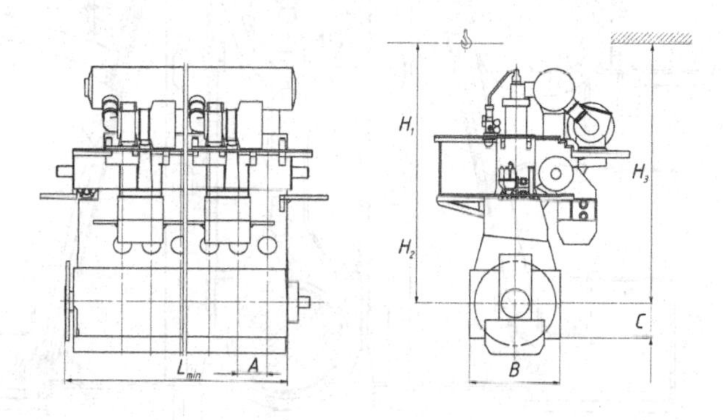

Number of cylinders	5	6	7	8
L_{min}	7639	8829	10,019	11,209
Weight (kg)	437,000	506,000	569,000	642,000

A (mm)	B (mm)	C (mm)	H_1 (mm)	H_2 (mm)	H_3 (mm)
1190	3980	1262	11,250	10,650	10,625

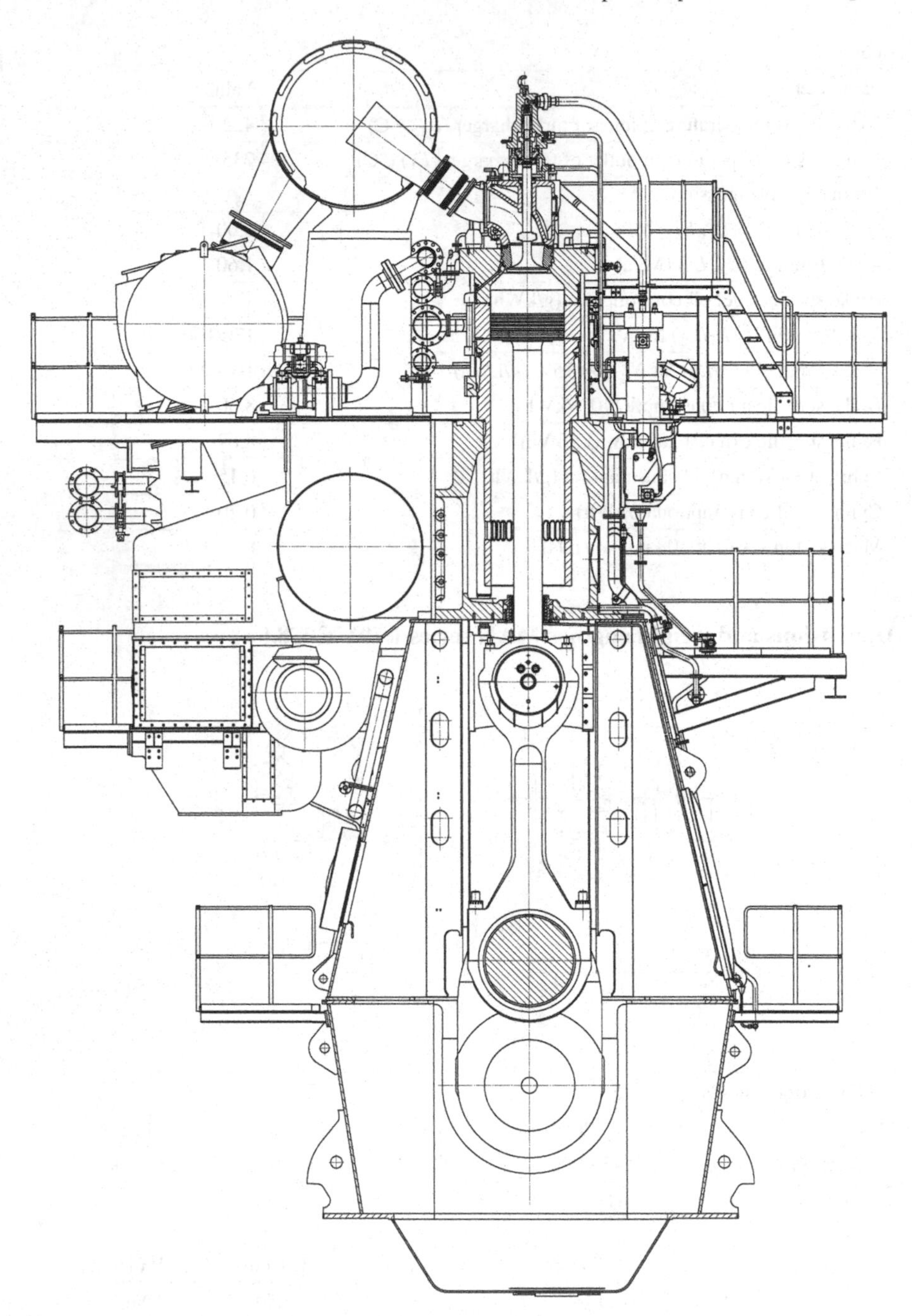

Fig. 3.34 Cross-section of the engine MAN series L70ME-C8.5 [33]

Main technical parameters of engines MAN series K98ME-C6 (Fig. 3.35)

Parameter	Value
Number and cylinders arrangement	6, 7, 8, 9, 10, 11, 12, 14 in-line
Cylinder bore (mm)	980
Piston stroke (mm)	2400
Cylinder capacity (dm^3)	1810.31
Rotational speed (min^{-1})	104
Cylinder power, N_e (layout area)	
– maximum continuous rating at 104 min^{-1} (L_1) (kW)	5710
– operational at 104 min^{-1} (L_2) (kW)	4830
– maximum continuous rating at 94 min^{-1} (L_3) (kW)	4580
– operational at 94 min^{-1} (L_4) (kW)	4140
Air charging pressure (L_1) (MPa)	0.360
Compression pressure (L_1) (MPa)	12.70
Maximum cycle pressure (L_1) (MPa)	14.20
Exhaust gas temperature at inlet of turbocharger (L_1) (°C)	400
Exhaust gas temperature at outlet of turbocharger (L_1) (°C)	245
Mean effective pressure	
– at 104 min^{-1} (L_1, L_2) (MPa)	1.82
– at 94 min^{-1} (L_3, L_4) (MPa)	1.46
Brake specific fuel oil consumption (g/kWh)	
– at 104 min^{-1} (L_1/L_2) for N_e 100/70% (g/kWh)	171/163
– at 94 min^{-1} (L_3/L_4) for N_e 100/70% (g/kWh)	162/156
Brake specific air consumption (kg/kWh)	9.50
Brake specific exhaust gas flow (kg/kWh)	9.67
Lubrication system oil consumption (g/kWh)	0.15
Cylinder oil consumption (g/kWh)	0.70
Mean piston speed at 104 min^{-1} (m/s)	8.64

Dimensions and weight engines MAN series K98ME-C6

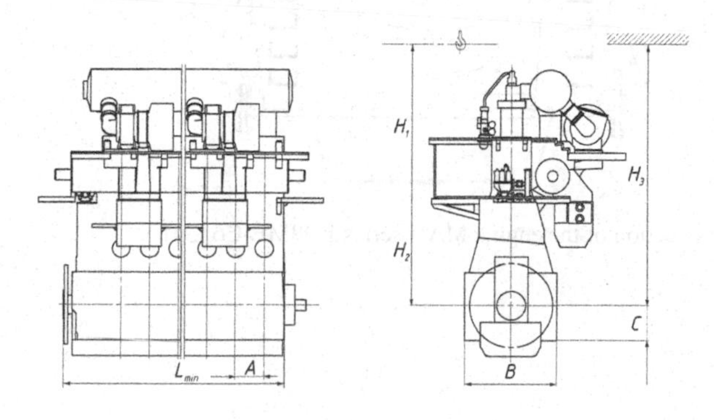

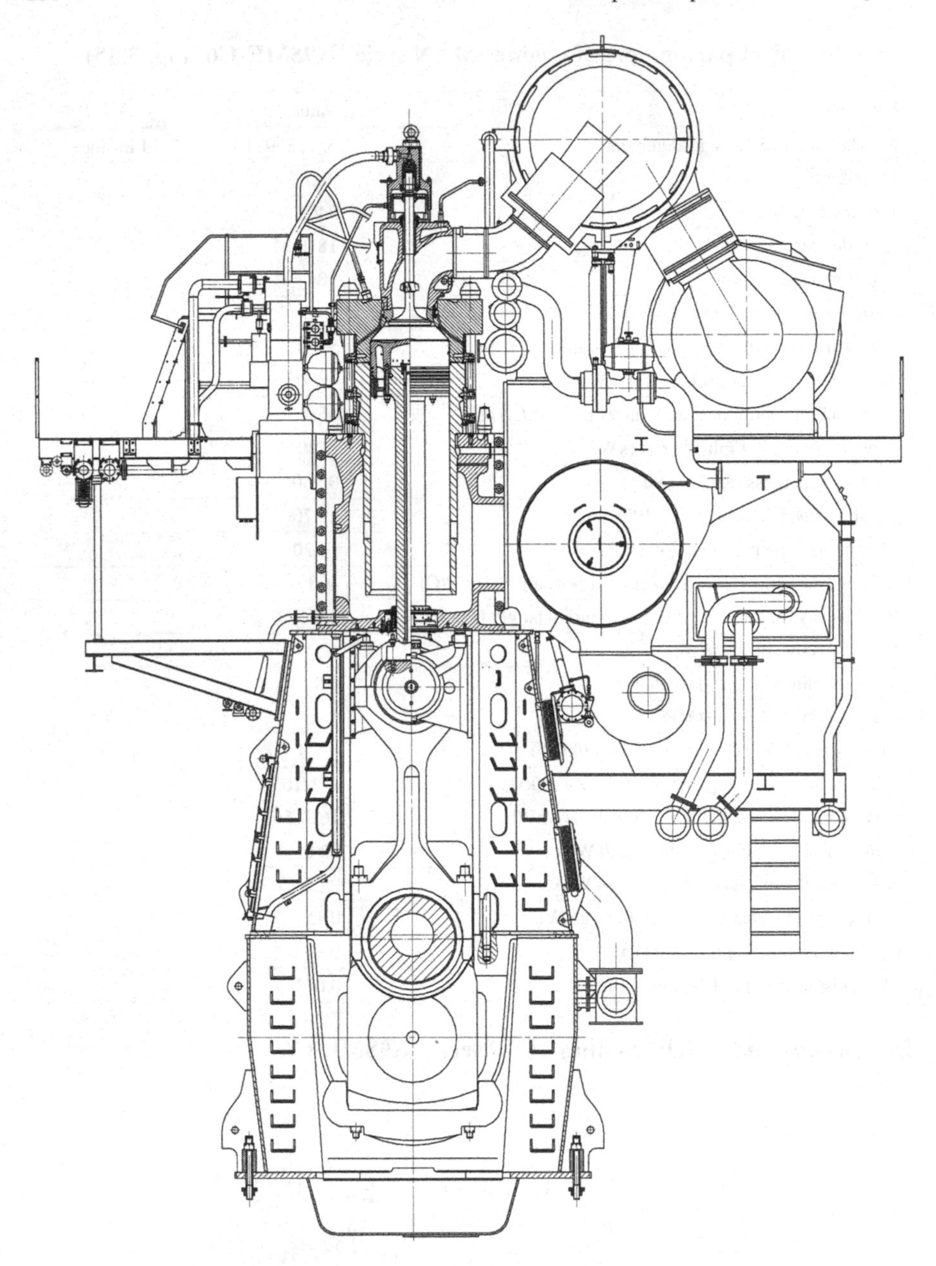

Fig. 3.35 Cross-section of the engine MAN series K98ME-C6 [34]

Number of cylinders	6	7	8	9
L_{min}	14,360	17,430	19,370	21,310
Weight (kg)	1,326,000	1,536,000	1,769,000	1,945,000
Number of cylinders	10	11	12	14
L_{min}	23,370	25,310	27,250	32,650
Weight (kg)	2,150,000	2,320,000	2,489,000	2,828,000

A (mm)	B (mm)	C (mm)	H_1 (mm)	H_2 (mm)	H_3 (mm)
1940	4640	1800	13,800	13,525	13,475

Main technical parameters of engines MAN series G50ME-B9 (Fig. 3.36)

Parameter	Value
Number and cylinders arrangement	5, 6, 7, 8, 9 in-line
Cylinder bore (mm)	500
Piston stroke (mm)	2500
Cylinder capacity (dm^3)	490.87
Rotational speed (min^{-1})	100
Cylinder power, N_e (layout area)	
– maximum continuous rating at 100 min^{-1} (L_1) (kW)	1720
– operational at 100 min^{-1} (L_2) (kW)	1370
– maximum continuous rating at 79 min^{-1} (L_3) (kW)	1360
– operational at 79 min^{-1} (L_4) (kW)	1090
Air charging pressure (L_1) (MPa)	0,370
Compression pressure (L_1) (MPa)	14.40
Maximum cycle pressure (L_1) (MPa)	16.20
Exhaust gas temperature at inlet of turbocharger (L_1) (°C)	405
Exhaust gas temperature at outlet of turbocharger (L_1) (°C)	235
Mean effective pressure:	
– at 100 min^{-1} (L_1, L_2) (MPa)	2.10
– at 79 min^{-1} (L_3, L_4) (MPa)	1.67
Brake specific fuel oil consumption (g/kWh)	
– at 100 min^{-1} (L_1/L_2) for N_e 100/75% (g/kWh)	167.0/164.0
– at 79 min^{-1} (L_3/L_4) for N_e 100/75% (g/kWh)	161.0/158.5
Brake specific air consumption (kg/kWh)	7.60
Brake specific exhaust gas flow (kg/kWh)	7.77
Lubrication System oil consumption (g/kWh)	0.15
Cylinder oil consumption (g/kWh)	0.60
Mean piston speed at 100 min^{-1} (m/s)	8.33

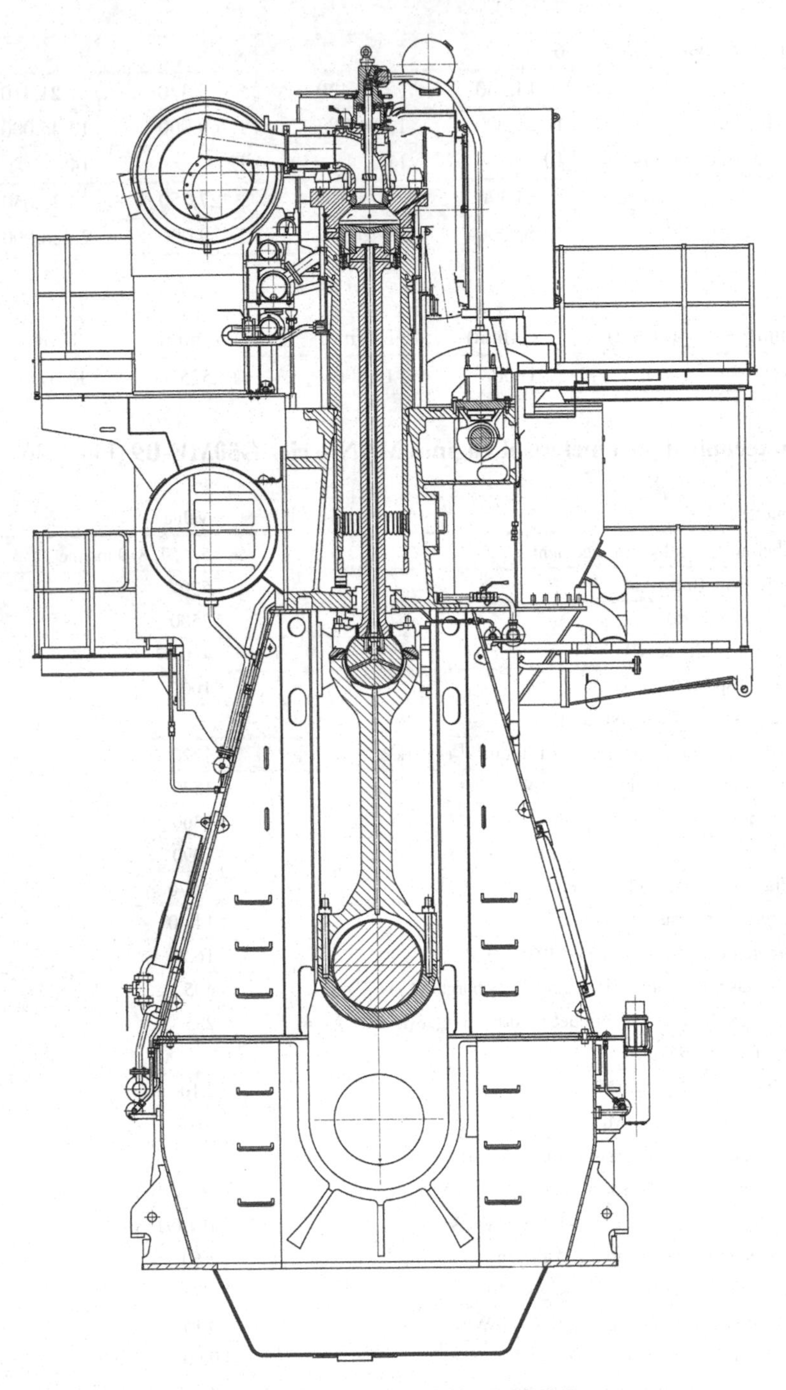

Fig. 3.36 Cross-section of the engine MAN series G50ME-B9 [35]

Dimensions and weight engines MAN series G50ME-B9

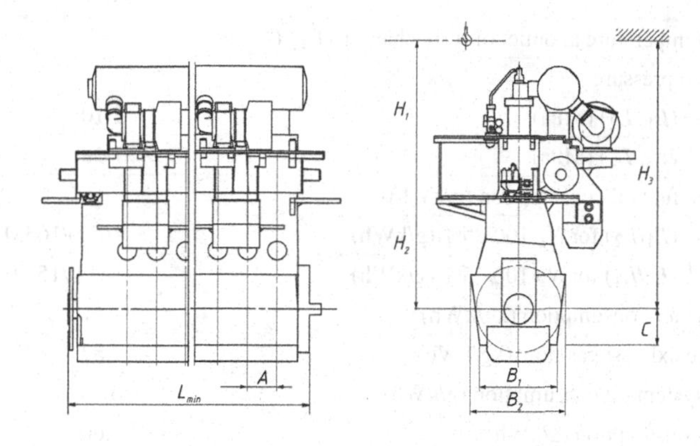

Number of cylinders	5	6	7	8	9
L_{min}	6325	7219	8113	9007	9901
Weight (kg)	225,000	260,000	295,000	330,000	365,000

A (mm)	B_1 (mm)	B_2 (mm)	C (mm)	H_1 (mm)	H_2 (mm)	H_3 (mm)
894	3896	3672	1205	10,750	10,175	9825

Main technical parameters of engines MAN series G60ME-C9.2 (Fig. 3.37)

Parameter	Value
Number and cylinders arrangement	5, 6, 7, 8 in-line
Cylinder bore (mm)	600
Piston stroke (mm)	2790
Cylinder capacity (dm^3)	788.85
Rotational speed (min^{-1})	97
Cylinder power, N_e (layout area)	
– maximum continuous rating at 97 min^{-1} (L_1) (kW)	2680
– operational at 97 min^{-1} (L_2) (kW)	2140
– maximum continuous rating at 77 min^{-1} (L_3) (kW)	2130
– operational at 77 min^{-1} (L_4) (kW)	1700
Air charging pressure (L_1) (MPa)	0.372
Compression pressure (L_1) (MPa)	16.50
Maximum cycle pressure (L_1) (MPa)	17.20
Exhaust gas temperature at inlet of turbocharger (L_1) (°C)	400

(continued)

(continued)

Parameter	Value
Exhaust gas temperature at outlet of turbocharger (L_1) (°C)	235
Mean effective pressure	
– at 97 min^{-1} (L_1, L_2) (MPa)	2.10
– at 77 min^{-1} (L_3, L_4) (MPa)	1.68
Brake specific fuel oil consumption (g/kWh)	
– at 97 min^{-1} (L_1/L_2) for N_e 100/75% (g/kWh)	167.0/163.0
– at 77 min^{-1} (L_3/L_4) for N_e 100/75% (g/kWh)	161.0/157.0
Brake specific air consumption (kg/kWh)	7.61
Brake specific exhaust gas flow (kg/kWh)	7.87
Lubrication system oil consumption (g/kWh)	0.15
Cylinder oil consumption (g/kWh)	0.60
Mean piston speed at 97 min^{-1} (m/s)	9.02

Dimensions and weight engines MAN series G60ME-C9.2

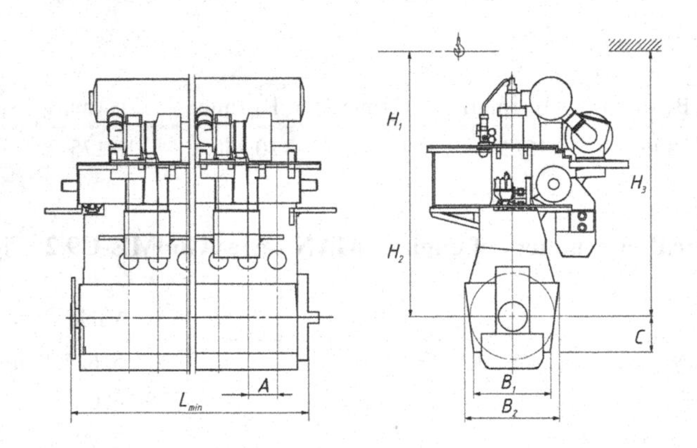

Number of cylinders	5	6	7	8
L_{min}	7390	8470	9550	10,630
Weight (kg)	395,000	439,000	491,000	543,000

A (mm)	B_1 (mm)	B_2 (mm)	C (mm)	H_1 (mm)	H_2 (mm)	H_3 (mm)
1080	4090	4220	1500	12,175	11,400	11,075

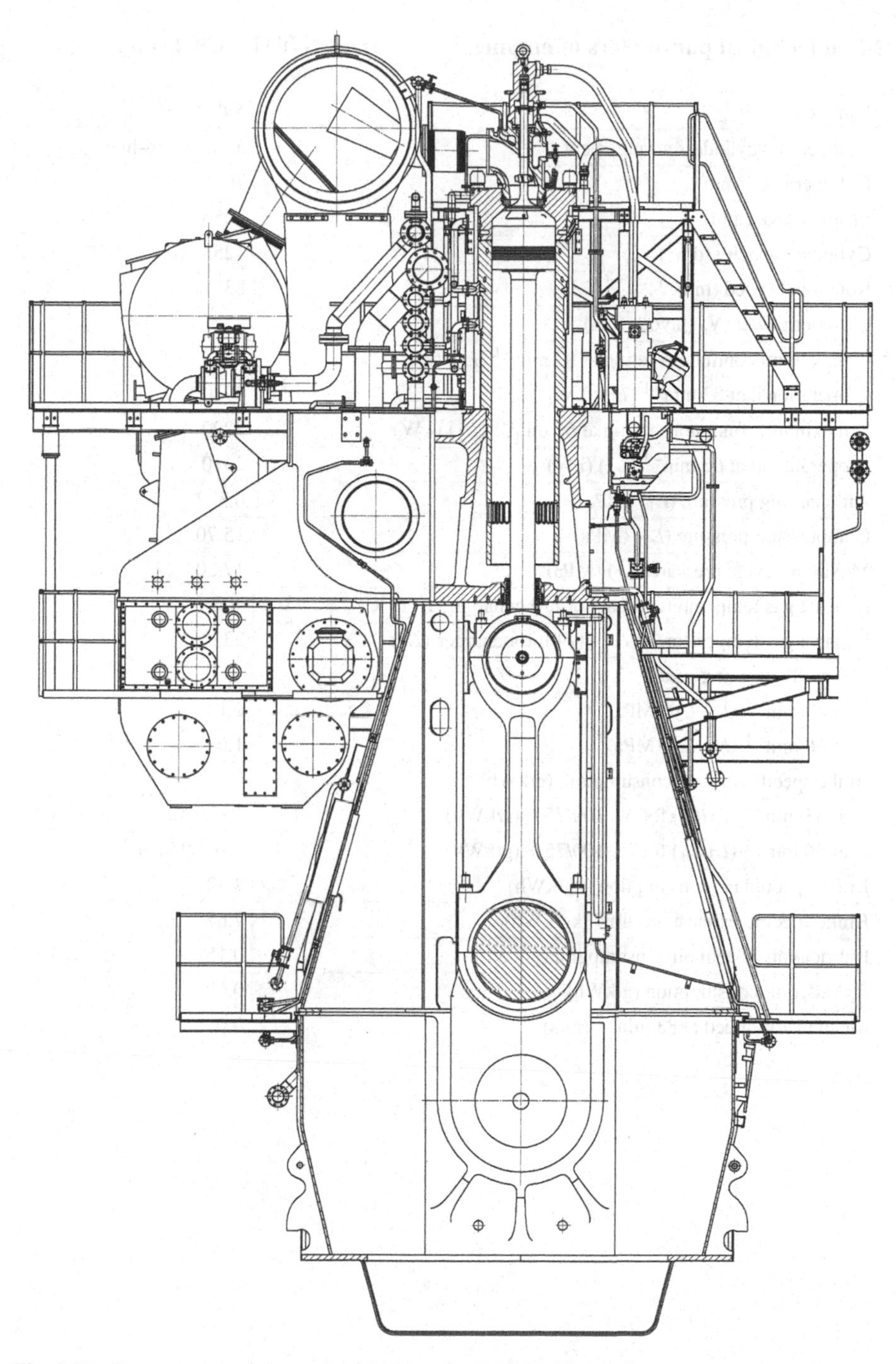

Fig. 3.37 Cross-section of the engine MAN series G60ME-C9.2 [36]

Main technical parameters of engines MAN series G70ME-C9.2 (Fig. 3.38)

Parameter	Value
Number and cylinders arrangement	5, 6, 7, 8 in-line
Cylinder bore (mm)	700
Piston stroke (mm)	3256
Cylinder capacity (dm^3)	1253.06
Rotational speed (min^{-1})	83
Cylinder power, N_e (layout area)	
– maximum continuous rating at 83 min^{-1} (L_1) (kW)	3640
– operational at 83 min^{-1} (L_2) (kW)	2910
– maximum continuous rating at 66 min^{-1} (L_3) (kW)	2890
– operational at 66 min^{-1} (L_4) (kW)	2310
Air charging pressure (L_1) (MPa)	0.372
Compression pressure (L_1) (MPa)	15.70
Maximum cycle pressure (L_1) (MPa)	17.20
Exhaust gas temperature at inlet of turbocharger (L_1) (°C)	400
Exhaust gas temperature at outlet of turbocharger (L_1) (°C)	235
Mean effective pressure	
– at 83 min^{-1} (L_1, L_2) (MPa)	2.10
– at 66 min^{-1} (L_3, L_4) (MPa)	1.68
Brake specific fuel oil consumption (g/kWh)	
– at 83 min^{-1} (L_1/L_2) for N_e 100/75% (g/kWh)	167.0/163.0
– at 66 min^{-1} (L_3/L_4) for N_e 100/75% (g/kWh)	161.0/157.0
Brake specific air consumption (kg/kWh)	7.42
Brake specific exhaust gas flow (kg/kWh)	7.67
Lubrication system oil consumption (g/kWh)	0.15
Cylinder oil consumption (g/kWh)	0.60
Mean piston speed at 83 min^{-1} (m/s)	9.01

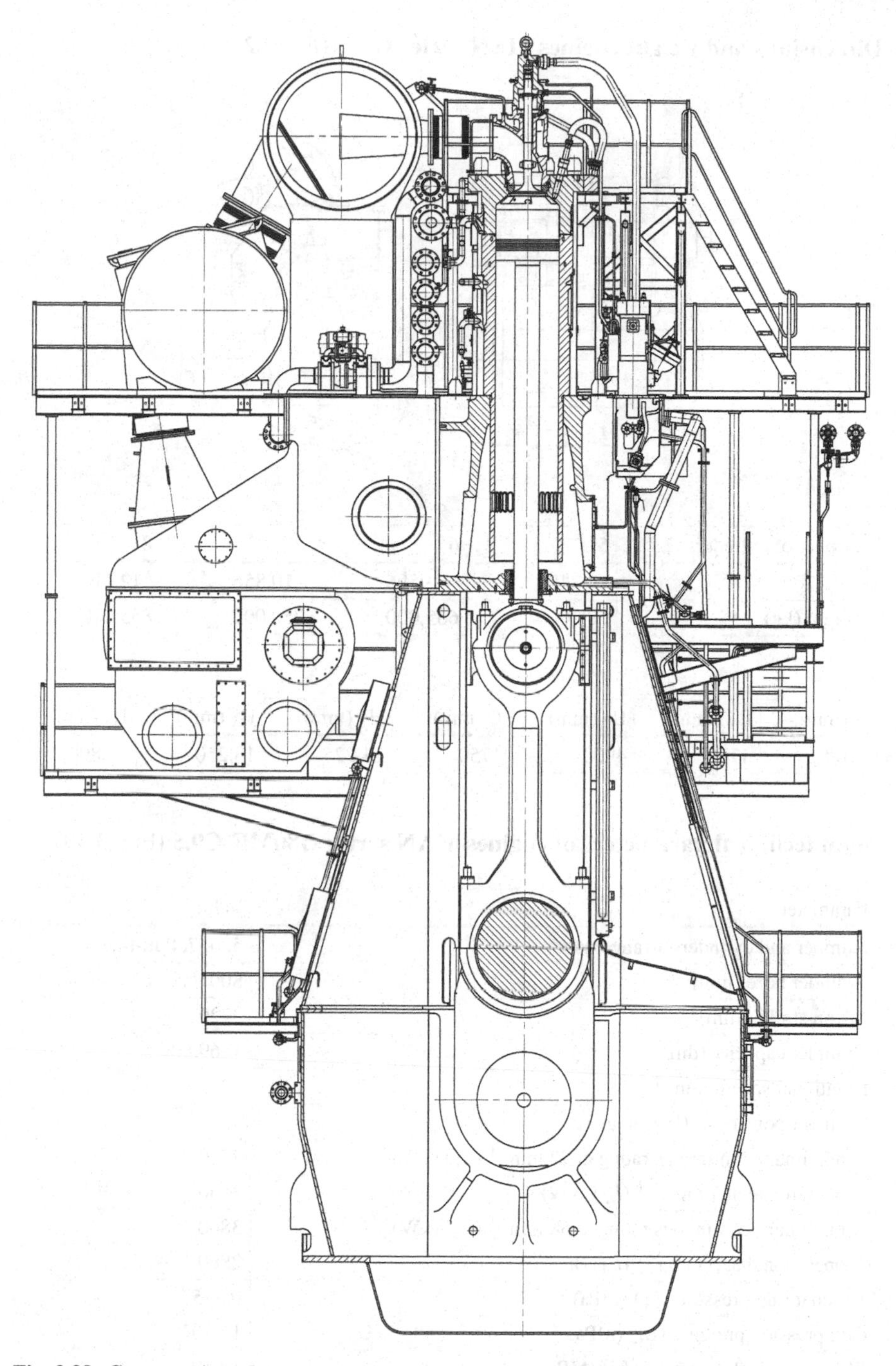

Fig. 3.38 Cross-section of the engine MAN series G70ME-C9.2 [37]

Dimensions and weight engines MAN series G70ME-C9.2

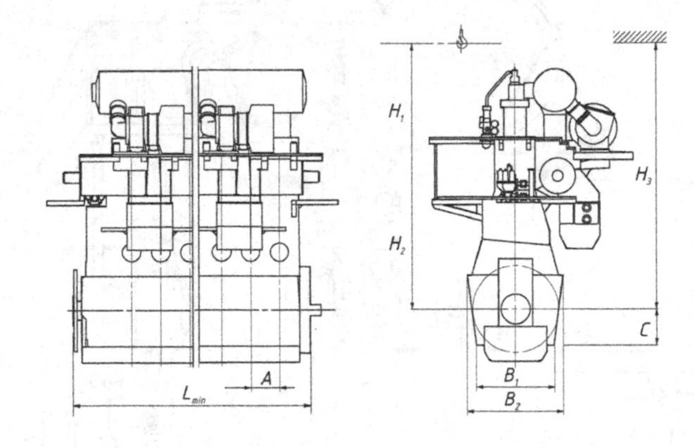

Number of cylinders	5	6	7	8
L_{min}	8486	9596	10,856	12,116
Weight (kg)	585,000	665,000	750,000	855,000

A (mm)	B_1 (mm)	B_2 (mm)	C (mm)	H_1 (mm)	H_2 (mm)	H_3 (mm)
1260	4760	4900	1750	14,225	13,250	12,800

Main technical parameters of engines MAN series G80ME-C9.5 (Fig. 3.39)

Parameter	Value
Number and cylinders arrangement	5, 6, 7, 8 in-line
Cylinder bore (mm)	800
Piston stroke (mm)	3720
Cylinder capacity (dm^3)	1869.88
Rotational speed, min^{-1}	72
Cylinder power, N_e (layout area)	
– maximum continuous rating at 72 min^{-1} (L_1) (kW)	4710
– operational at 72 min^{-1} (L_2) (kW)	3550
– maximum continuous rating at 58 min^{-1} (L_3) (kW)	3800
– operational at 58 min^{-1} (L_4) (kW)	2860
Air charging pressure (L_1) (MPa)	0.375
Compression pressure (L_1) (MPa)	16.80
Maximum cycle pressure (L_1) (MPa)	18.50
Exhaust gas temperature at inlet of turbocharger (L_1) (°C)	400

(continued)

(continued)

Parameter	Value
Exhaust gas temperature at outlet of turbocharger (L_1) (°C)	235
Mean effective pressure	
– at 72 min^{-1} (L_1, L_2) (MPa)	2.10
– at 58 min^{-1} (L_3, L_4) (MPa)	1.58
Brake specific fuel oil consumption (g/kWh)	
– at 72 min^{-1} (L_1/L_2) for N_e 100/75% (g/kWh)	166.0/162.0
– at 58 min^{-1} (L_3/L_4) for N_e 100/75% (g/kWh)	159.0/156.0
Brake specific air consumption (kg/kWh)	7.60
Brake specific exhaust gas flow (kg/kWh)	7.77
Lubrication system oil consumption (g/kWh)	0.15
Cylinder oil consumption (g/kWh)	0.60
Mean piston speed at 72 min^{-1} (m/s)	8.93

Dimensions and weight engines MAN series G80ME-C9.5

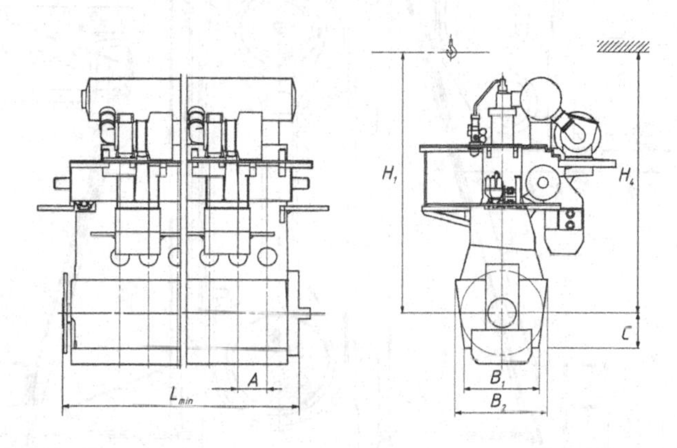

Number of cylinders	6	7	8	9
L_{min}	10,735	12,135	13,535	15,825
Weight (kg)	945,000	1,055,000	1,175,000	1,350,000

A (mm)	B_1 (mm)	B_2 (mm)	C (mm)	H_1 (mm)	H_4 (mm)
1400	5320	5680	1960	16,100	15,825

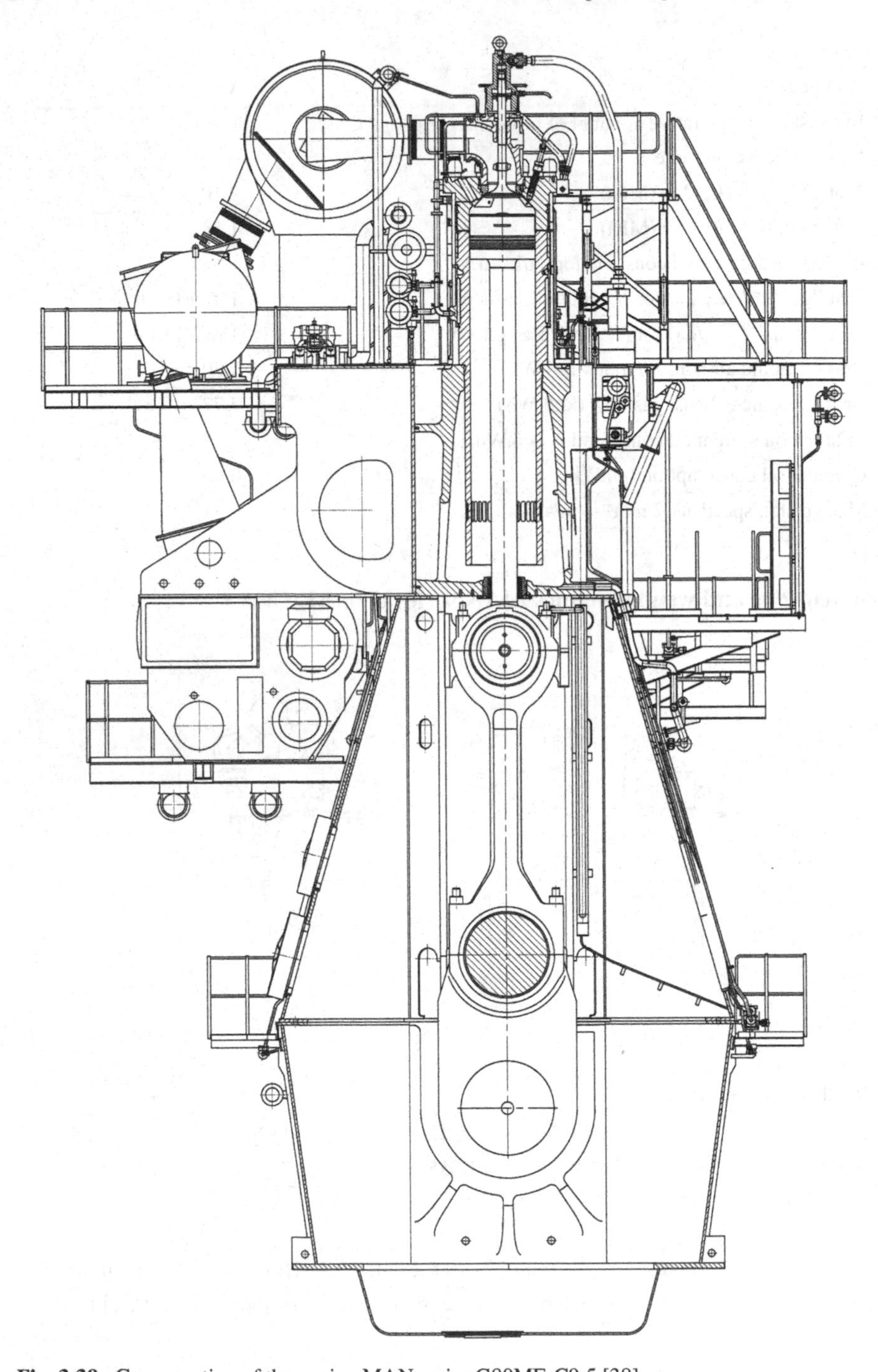

Fig. 3.39 Cross-section of the engine MAN series G80ME-C9.5 [38]

3.2 Mitsubishi Heavy Industries

Mitsubishi Heavy Industries is one of the largest Japanese corporations, specializing, among other things, in the production of ship low-speed two-stroke diesel engines. Initially, the company was founded under the name "Tsukomo Shokai" by the son of a provincial farmer from the impoverished ancient samurai family Yataro Iwasaki. In 1868, Yataro found a job at a shipowning trading company, that belonged to the samurai clan Tosa, headed by Tsukumo Shokai. In 1873, he bought the assets of the company and renamed it Mitsubishi. The name Mitsubishi is a combination of the words " mitsu" which means "three" and "hishi" water chestnut, which in Japan is associated with a rhomb-shaped diamond. The official translation sounds like " three diamonds". The Mitsubishi company logo is a combination of the family crest of the Iwasaki family in the form of three rhomb-shaped diamonds, arranged as on the three-faced crest of the Tosa clan. Initially, the company developed as a shipbuilding and ship repair company, however, even during the life of its founder, it extended its interests to all sorts of technical innovations of the time. In 1875, after several changes of names, it became known as the Mitsubishi Mail Steamship Company.

20 years after the foundation, the company's president's nephew, Kayota Iwasaki, takes the presidency of the company. By this time, the company was engaged not only in shipping, but also in the construction of ships, oil production and metallurgy. The young talented manager brought Mitsubishi to the level of a giant corporation, which later became known as Mitsubishi Motors.

In 1912, the company's engineers inspected the world's first ocean-going vessel "Selandia", equipped with diesel engines, which triggered the beginning of its own program of creating a ship low-speed engine. As a result, by 1927, the development of the first in Japan low-speed two-stroke engine of the MS type was completed. The engine differed from many of its European counterparts because of instead of the cumbersome compressor fuel injection system, it used a direct fuel injection system and a number of innovative solutions. The production of engines type MS began in 1932. A total of 84 engines of this series were produced, and on each subsequent engine the identified design flaws were eliminated and various kinds of improvements were made. In parallel with the release of its own engines, Mitsubishi manufactured two-stroke low-speed engines under license from B&W. The accumulation of industrial experience led to the fact, that in 1955 a new series was developed on the basis of the MS engine, which was called UE. Under this program, a laboratory test base was created, where all new solutions were tested on laboratory engines. Thus, the first engine of the new series was a laboratory engine of the 3UEC72/150 type, and the first commercial engine was the diesel 9UEC75/150 with a power of 8832 kW. Before the 1973 oil crisis, the main focus in designing ship engines was to increase power. By this time, five series of UE engines were developed, denoted by letters from A to E, on which a gas turbine supercharged impulse system was applied. To increase the cylinder capacity in 1975, the company, the first in the world, developed and introduced a two-stage turbocharging system, applying it to the 8UEC52E engine. As a result, the engine power was increased by 30%, and dimensions were reduced by 25%.

Rising oil prices forced the company to focus on improving the efficiency of its engines. As a result, since 1990, the production of UEC-LSII engines has begun. This series covered a wide power range, including models ranging from the smallest UEC33LSII to the largest UE85LSII. In 2001, the UEC-LSII series of diesel engines was launched, and in 2006, the LSE series with an electronic control system (E—Economy, Easy operation and maintenance. Environmentally friendly).

3.2.1 Engines Series UEC LSH-Eco and LSE-Eco

The UEC-LSE series engines have been developed, based on the UEC-LSII series. The main distinctive feature of these engines is the use of an electronic control system for them in the main processes that affect for their economic and environmental performance.

The concept of the UEC-Eco and LSE diesel engine control system is the same as that of an electronically controlled diesel engine, manufactured by MAN Diesel & Turbo. The processes of fuel supply, opening and closing of exhaust valves and supply of oil to the wall surface are carried out by means of hydraulic servomotors, electronically controlled from microprocessor units. Flexible control allows reducing fuel consumption on partial modes by 1–2% with the same NO_x emissions as a traditional diesel engine or reducing NO_x emissions by 10–15% while maintaining constant consumption.

The fuel supply to the cylinders is being made, using a high-pressure pump, having a hydraulic plunger drive, which is controlled by two valves with an electromagnetic drive. Consecutive, parallel or mixed opening of these valves allows to control the process of fuel injection by changing the fuel pressure and the feed rate (feed law). Solenoids are controlled by a program, incorporated into the microprocessor control module.

The exhaust valve is driven by a hydraulic transmission, where the servo piston moves not from the camshaft cam, but by means of a hydraulic actuator from the same control hydraulic system. The pressure, by analogy with the hydraulic drive of the injection pump, is regulated by two valves with an electromagnetic drive. These valves are controlled by a program, integrated in the microprocessor unit and providing for changing the valve opening and closing phases depending on the engine operating mode. This allows to reduce fuel consumption by another 1–2%.

In 2015, another LSH-Eco Engine version was added to already existing engines. The first engine of this model was the 6UEC50LSH-EcoC2, built by Kobe Diesel Co., Ltd. and delivered to the customer in March 2015. When creating these engines, three-dimensional design methods were applied, which made it possible to significantly optimize the design parameters and determine the optimal algorithms for managing work processes. In engines of this type, a large proportion of the piston stroke to the diameter of the cylinder (S/D = 4.4–4.7) is used, which, with a reduced rotational speed, allows to maximize the efficiency of propulsion of the vessel. The next stage in the development was the production of UEC35LSE-Eco-B2-MGO engines, which use a two-stage gas turbine boost, exhaust gas recirculation system and water injection into the engine combustion chamber along with the fuel. The use of a two-stage supercharging allowed in these engines to realize the Miller cycle, where the exhaust valve closes with a lag at the compression stroke, thereby reducing the geometric compression ratio, and during the expansion the cylinder volume is used completely, allowing to get maximum work per cycle. Water injection reduces the temperature in the combustion chamber and improves the quality of fuel combustion by improving the mixture formation, associated with intensive evaporation of water. Exhaust gas recirculation reduces the oxygen concentration in the combustion chamber, therefore inhibiting nitrogen oxidation processes at high temperatures typical of the combustion process in the working cylinder. The listed innovative solutions allowed the engines of this series to fulfill the existing standards for NO_x emissions, while even reducing the specific fuel consumption by about 5% compared to other engines of the UE series. According to the company, their engines have a consumption of 2–3% less than those of other manufacturers.

To achieve NO_x emissions according to Tier III, engines can be equipped with catalytic reactors.

Main technical parameters of engines series UEC35LSE-Eco-B2 (Fig. 3.40)

Parameter	Value
Number and cylinders arrangement	5, 6, 7, 8 in-line
Cylinder bore (mm)	350
Piston stroke (mm)	1550
Cylinder capacity (dm^3)	149.13
Rotational speed (min^{-1})	167
Cylinder power, N_e (layout area)	
– maximum continuous rating at 167 min^{-1} (L_1) (kW)	870
– operational at 167 min^{-1} (L_2) (kW)	695
– maximum continuous rating at 142 min^{-1} (L_3) (kW)	740
– operational at 142 min^{-1} (L_4) (kW)	590
Air charging pressure (L_1) (MPa)	0.377
Compression pressure (L_1) (MPa)	17.2
Maximum cycle pressure (L_1) (MPa)	19.2
Exhaust gas temperature at inlet of turbocharger (L_1) (°C)	420
Exhaust gas temperature at outlet of turbocharger (L_1) (°C)	236
Mean effective pressure	
– at 167 min^{-1} (L_1, L_2) (MPa)	2.10
– at 142 min^{-1} (L_3, L_4) (MPa)	1.67
Brake specific fuel oil consumption (g/kWh)	
– at 167 min^{-1} (L_1/L_2) for N_e 100/75% (g/kWh)	167/163
– at 142 min^{-1} (L_3/L_4) for N_e 100/75% (g/kWh)	161/157

(continued)

(continued)

Parameter	Value
Mean piston speed at 167 min^{-1} (m/s)	8.63
Brake specific air consumption (kg/kWh)	7.76
Brake specific exhaust gas flow (kg/kWh)	7.93
Cylinder oil consumption (g/kWh)	1.2–1.4

Dimensions and weight engines UEC35LSE-Eco-B2

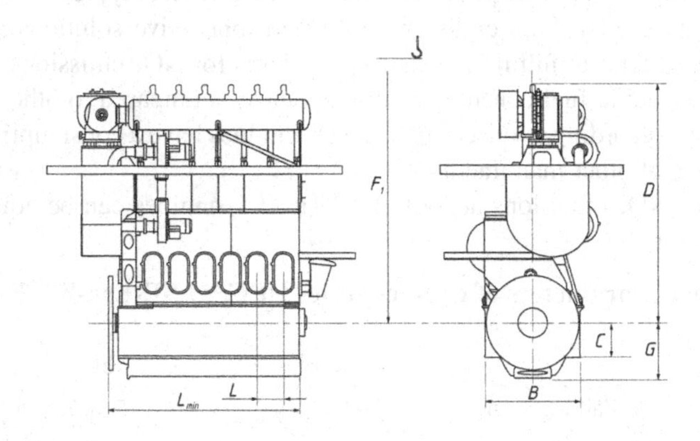

Number of cylinders	5	6	7	8
L_{min}	4398	5010	5622	6234
Weight (kg)	70,000	82,000	92,000	100,000

L (mm)	B (mm)	C (mm)	D (mm)	F_1 (mm)	G (mm)
612	2284	830	5622,5	6725	1326

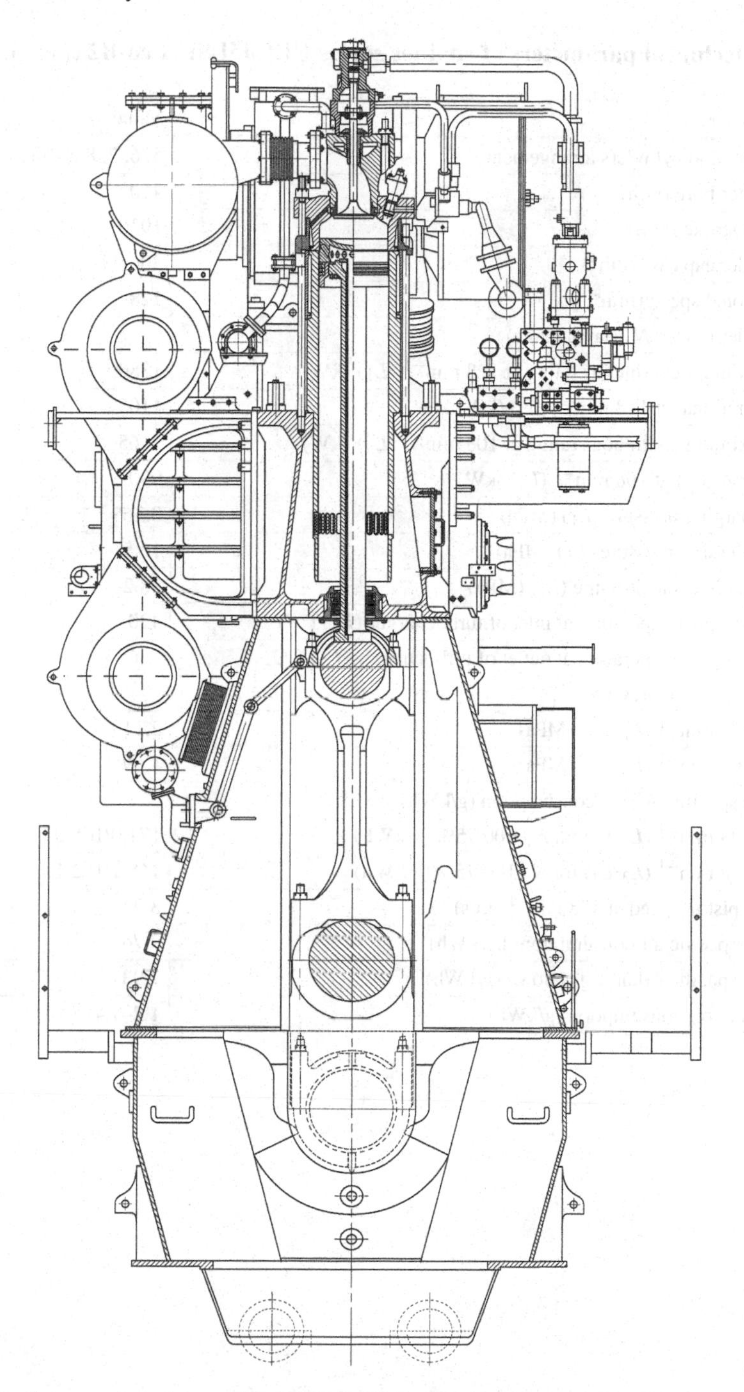

Fig. 3.40 Cross-section of the engine UEC35LSE-Eco-B2 by Mitsubishi firm [39]

Main technical parameters of engines series UEC45LSE-Eco-B2 (Fig. 3.41)

Parameter	Value
Number and cylinders arrangement	5, 6, 7, 8 in-line
Cylinder bore (mm)	450
Piston stroke (mm)	1930
Cylinder capacity (dm^3)	306.95
Rotational speed (min^{-1})	128
Cylinder power, N_e (layout area)	
– maximum continuous rating at 128 min^{-1} (L_1) (kW)	1380
– operational at 128 min^{-1} (L_2) (kW)	1105
– maximum continuous rating at 108 min^{-1} (L_3) (kW)	1165
– operational at 108 min^{-1} (L_4) (kW)	930
Air charging pressure (L_1) (MPa)	0.365
Compression pressure (L_1) (MPa)	16.5
Maximum cycle pressure (L_1) (MPa)	18.2
Exhaust gas temperature at inlet of turbocharger (L_1) (°C)	420
Exhaust gas temperature at outlet of turbocharger (L_1) (°C)	236
Mean effective pressure	
– at 128 min^{-1} (L_1, L_2) (MPa)	2.11
– at 108 min^{-1} (L_3, L_4) (MPa)	1.69
Brake specific fuel oil consumption (g/kWh)	
– at 128 min^{-1} (L_1/L_2) for N_e 100/75% (g/kWh)	171.0/165.0
– at 108 min^{-1} (L_3/L_4) for N_e 100/75% (g/kWh)	168.1/162.1
Mean piston speed at 128 min^{-1} (m/s)	8.23
Brake specific air consumption (kg/kWh)	7.76
Brake specific exhaust gas flow (kg/kWh)	7.93
Cylinder oil consumption (g/kWh)	1.2–1.4

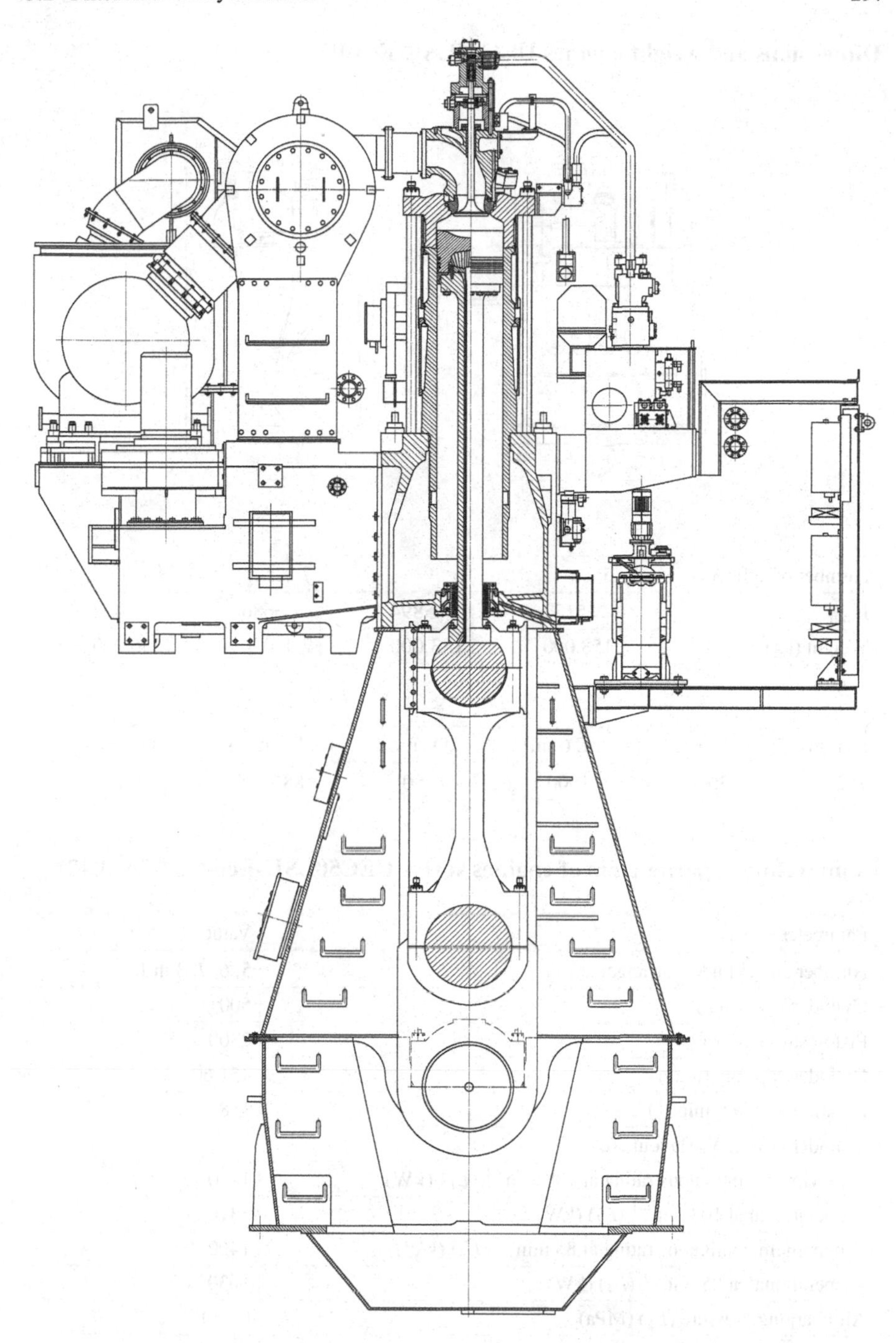

Fig. 3.41 Cross-section of the engine UEC45LSE-Eco-B2 by Mitsubishi firm [40]

Dimensions and weight engines UEC45LSE-Eco-B2

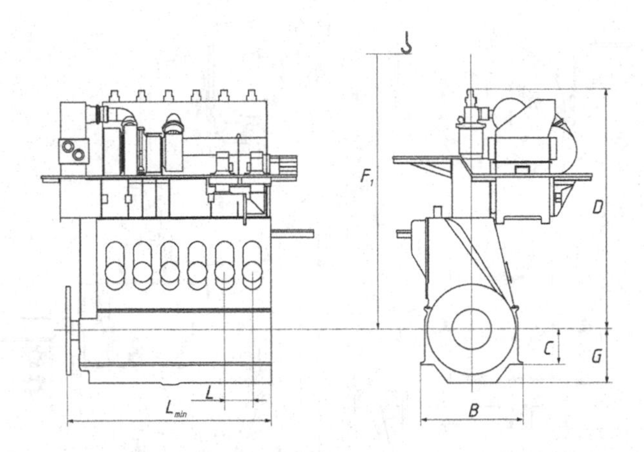

Number of cylinders	5	6	7	8
L_{min}	5102	5894	6686	7478
Weight (kg)	158,000	183,000	208,000	236,000

L (mm)	B (mm)	C (mm)	D (mm)	F_1 (mm)	G (mm)
792	3000	1000	7080	8860	1540

Main technical parameters of engines series UEC50LSH-Eco-C2 (Fig. 3.42)

Parameter	Value
Number and cylinders arrangement	5, 6, 7, 8 in-line
Cylinder bore (mm)	500
Piston stroke (mm)	2300
Cylinder capacity (dm^3)	451.60
Rotational speed (min^{-1})	108
Cylinder power, N_e (layout area)	
– maximum continuous rating at 108 min^{-1} (L_1) (kW)	1780
– operational at 108 min^{-1} (L_2) (kW)	1310
– maximum continuous rating at 85 min^{-1} (L_3) (kW)	1400
– operational at 85 min^{-1} (L_4) (kW)	1030
Air charging pressure (L_1) (MPa)	0.340
Compression pressure (L_1) (MPa)	19.5
Maximum cycle pressure (L_1) (MPa)	20.0

(continued)

(continued)

Parameter	Value
Exhaust gas temperature at inlet of turbocharger (L_1) (°C)	430
Exhaust gas temperature at outlet of turbocharger (L_1) (°C)	248
Mean effective pressure	
– at 108 min^{-1} (L_1, L_2) (MPa)	2.19
– at 85 min^{-1} (L_3, L_4) (MPa)	1.61
Brake specific fuel oil consumption (g/kWh)	
– at 108 min^{-1} (L_1/L_2) for N_e 100/75% (g/kWh)	164.0/158.0
– at 85 min^{-1} (L_3/L_4) for N_e 100/75% (g/kWh)	161.0/155.1
Mean piston speed at 108 min^{-1} (m/s)	8.28
Brake specific air consumption (kg/kWh)	7.37
Brake specific exhaust gas flow (kg/kWh)	7.55
Cylinder oil consumption (g/kWh)	0.6–1.2

Dimensions and weight engines UEC50LSH-Eco-C2

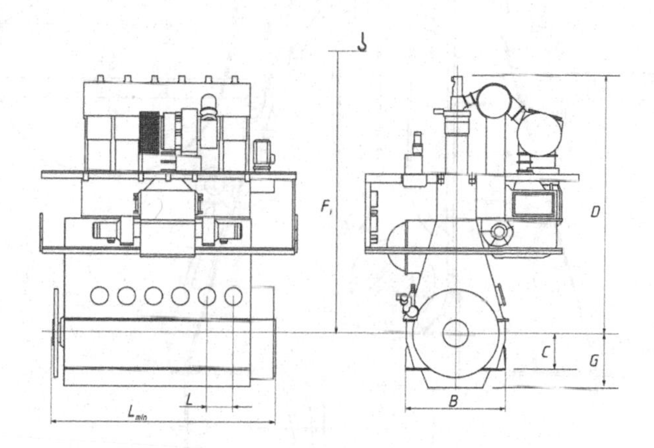

Number of cylinders	5	6	7	8
L_{min}	5560	6430	7300	8170
Weight (kg)	194,000	225,000	257,000	289,000

L (mm)	B (mm)	C (mm)	D (mm)	F_1 (mm)	G (mm)
870	3350	1190	8448	10,050	1743

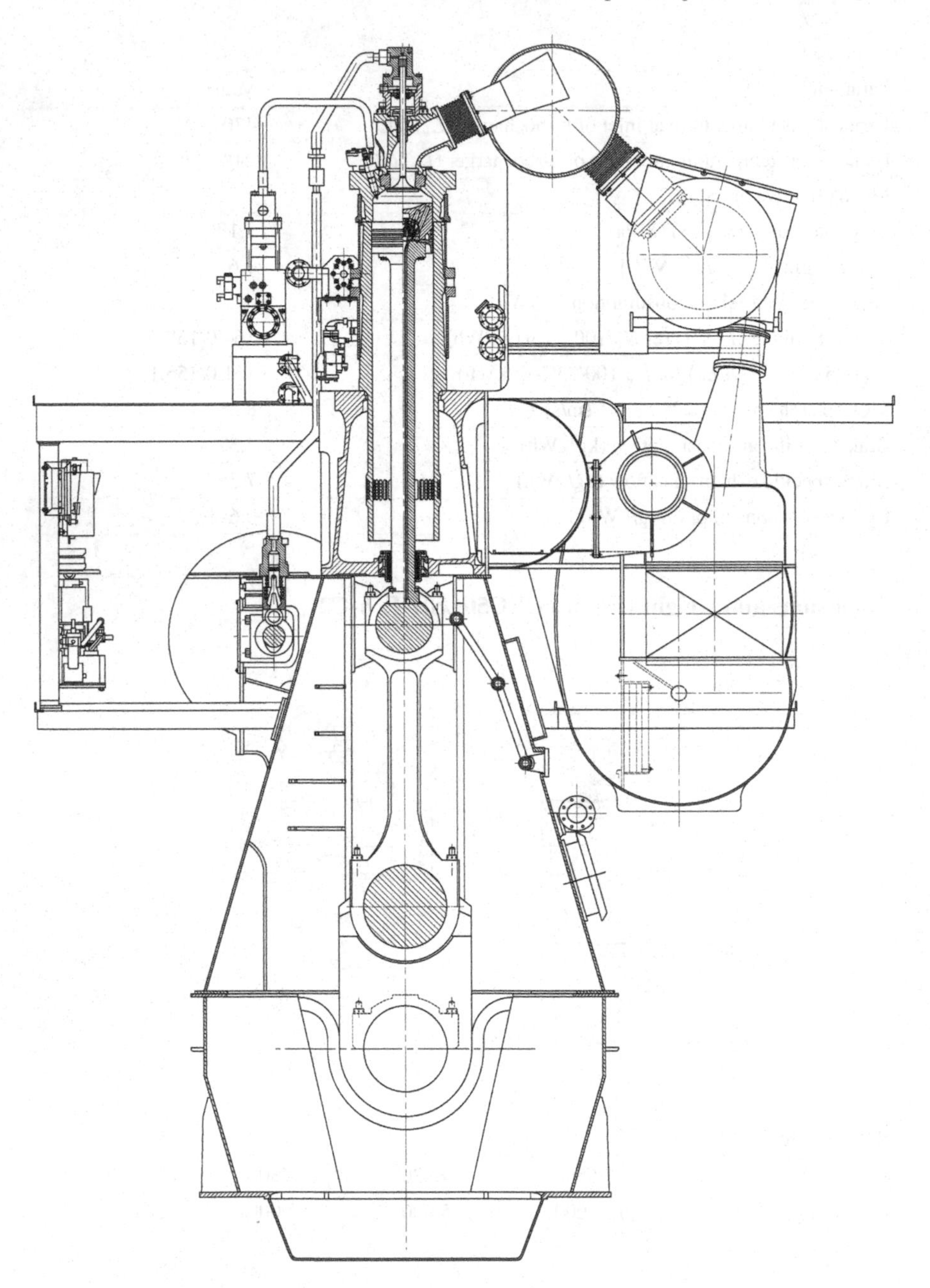

Fig. 3.42 Cross-section of the engine UEC50LSH-Eco-C2 by Mitsubishi firm [41]

Parameter	Value
Number and cylinders arrangement	5, 6, 7, 8 in-line
Cylinder bore (mm)	600
Piston stroke (mm)	2400
Cylinder capacity (dm^3)	678.58
Rotational speed (min^{-1})	105
Cylinder power, N_e (layout area)	
– maximum continuous rating at 105 min^{-1} (L_1) (kW)	2490
– operational at 105 min^{-1} (L_2) (kW)	1990
– maximum continuous rating at 79 min^{-1} (L_3) (kW)	1875
– operational at 79 min^{-1} (L_4) (kW)	1500
Air charging pressure (L_1) (MPa)	0.340
Compression pressure (L_1) (MPa)	17.5
Maximum cycle pressure (L_1) (MPa)	19.0
Exhaust gas temperature at inlet of turbocharger (L_1) (°C)	425
Exhaust gas temperature at outlet of turbocharger (L_1) (°C)	245
Mean effective pressure	
– at 105 min^{-1} (L_1, L_2) (MPa)	2.10
– at 79 min^{-1} (L_3, L_4) (MPa)	1.68
Brake specific fuel oil consumption (g/kWh)	
– at 105 min^{-1} (L_1/L_2) for N_e 100/75% (g/kWh)	168.0/162.7
– at 79 min^{-1} (L_3/L_4) for N_e 100/75% (g/kWh)	164.9/160.8
Mean piston speed at 105 min^{-1} (m/s)	8.40
Brake specific air consumption (kg/kWh)	7.67
Brake specific exhaust gas flow (kg/kWh)	7.84
Cylinder oil consumption (g/kWh)	0.6–1.2

Main technical parameters of engines series UEC60LSE-Eco-B1 (Fig. 3.43)

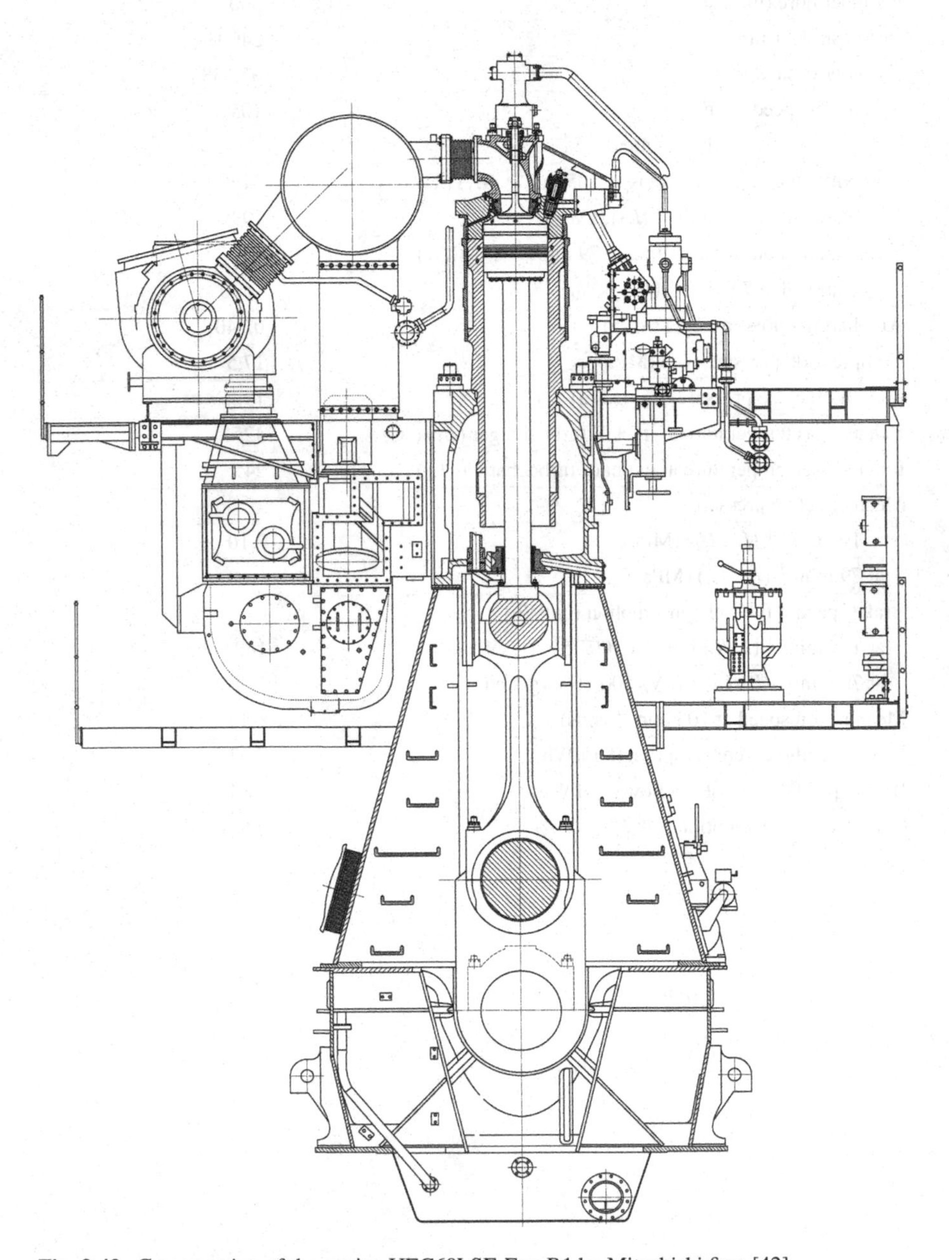

Fig. 3.43 Cross-section of the engine UEC60LSE-Eco-B1 by Mitsubishi firm [42]

Dimensions and weight engines UEC60LSE-Eco-B1

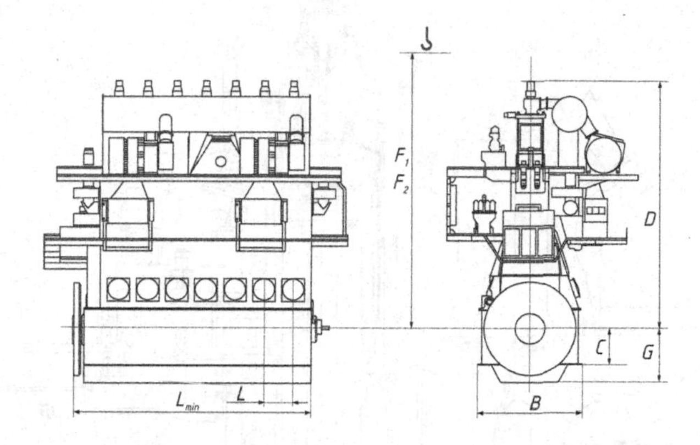

Number of cylinders	5	6	7	8
L_{min}	6780	7866	8952	10,038
Weight (kg)	306,000	356,000	407,000	456,000

L (mm)	B (mm)	C (mm)	D (mm)	F_1 (mm)	F_2 (mm)	G (mm)
1086	3770	1300	8903	10,800	10,040	1944

Main technical parameters of engines series UEC80LSE-Eco-B1 (Fig. 3.44)

Parameter	Value
Number and cylinders arrangement	5, 6, 7, 8 in-line
Cylinder bore (mm)	800
Piston stroke (mm)	3150
Cylinder capacity (dm^3)	1583.36
Rotational speed (min^{-1})	80
Cylinder power, N_e (layout area)	
– maximum continuous rating at 80 min^{-1} (L_1) (kW)	4440
– operational at 80 min^{-1} (L_2) (kW)	3550
– maximum continuous rating at 68 min^{-1} (L_3) (kW)	3775
– operational at 68 min^{-1} (L_4) (kW)	3020
Air charging pressure (L_1) (MPa)	0.325
Compression pressure (L_1) (MPa)	11.8
Maximum cycle pressure (L_1) (MPa)	14.8
Exhaust gas temperature at inlet of turbocharger (L_1) (°C)	390
Exhaust gas temperature at outlet of turbocharger (L_1) (°C)	250
Mean effective pressure	
– at 80 min^{-1} (L_1, L_2) (MPa)	2.10
– at 68 min^{-1} (L_3, L_4) (MPa)	1.68
Brake specific fuel oil consumption (g/kWh)	
– at 80 min^{-1} (L_1/L_2) for N_e 100/75% (g/kWh)	166.0/160.7
– at 68 min^{-1} (L_3/L_4) for N_e 100/75% (g/kWh)	163.4/158.8
Mean piston speed at 108 min^{-1} (m/s)	8.40
Brake specific air consumption (kg/kWh)	7.46
Brake specific exhaust gas flow (kg/kWh)	7.63
Cylinder oil consumption (g/kWh)	0.6–1.2

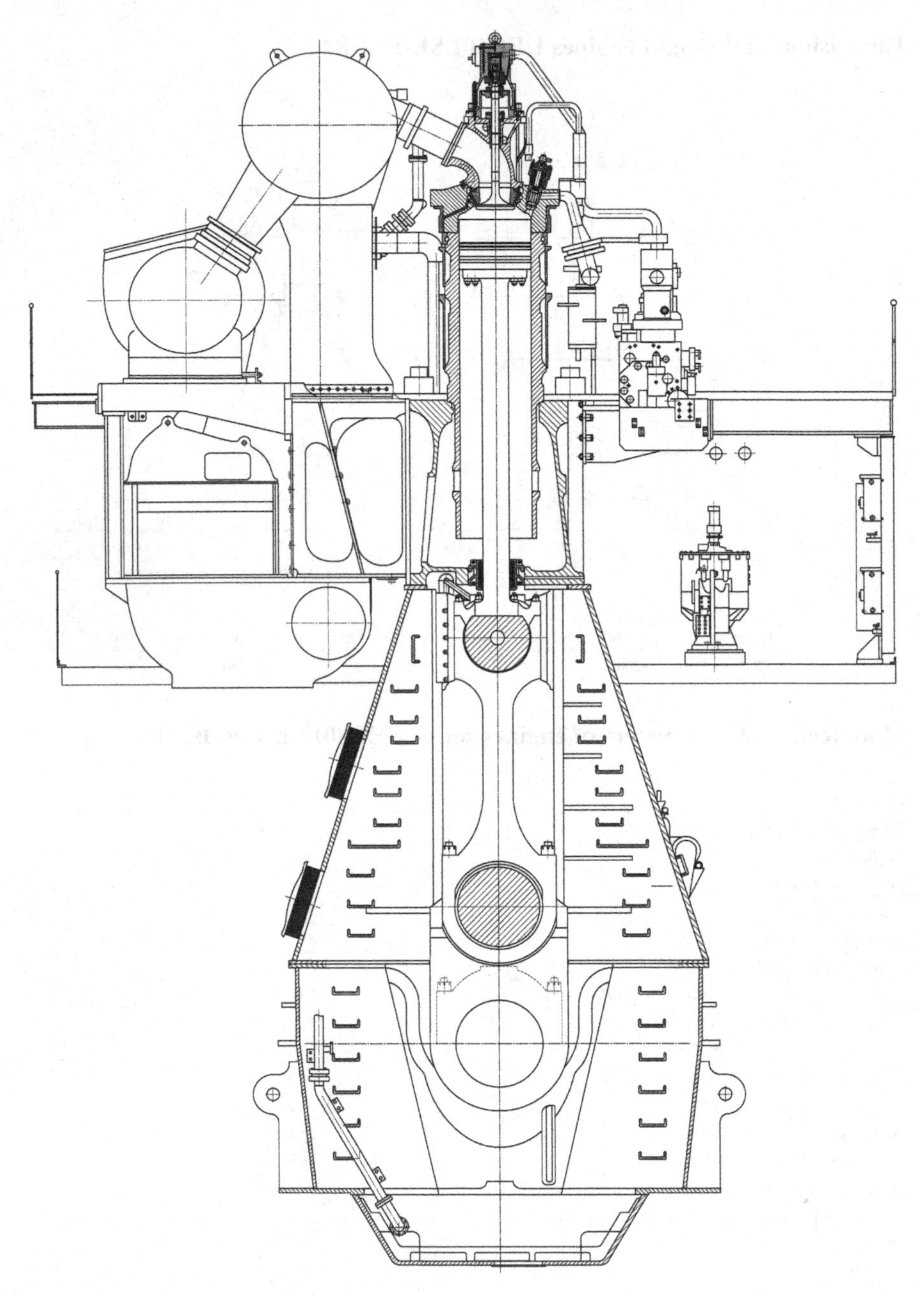

Fig. 3.44 Cross-section of the engine UEC80LSE-Eco-B1by Mitsubishi firm [43]

Dimensions and weight engines UEC80LSE-Eco-B1

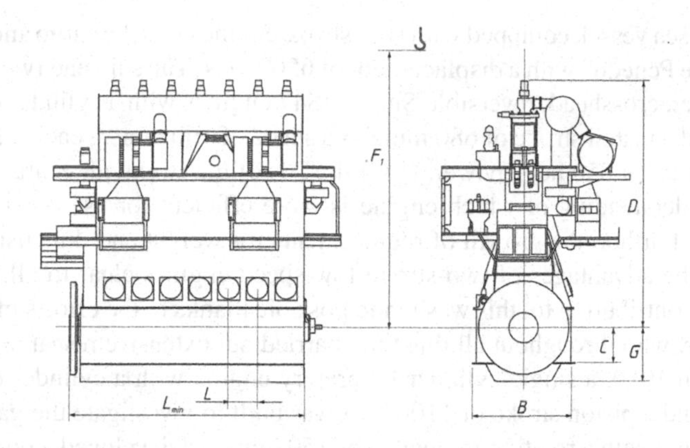

Number of cylinders	5	6	7	8
L_{min}	8658	10,038	11,418	12,798
Weight (kg)	693,000	794,000	895,000	996,000

L (mm)	B (mm)	C (mm)	D (mm)	F_1 (mm)	G (mm)
1380	5000	1736	11,725	14,247	2524

3.3 Wärtsilä-Sulzer

The Swiss company Sulzer, which previously specialized in the production of ship steam engines, was one of the first in the world to restructure its work to produce marine diesel engines. Having acquired in 1893 the rights to use the invention of R. Diesel, the company began to develop its own structures for use on ships. In an effort to improve the weight and dimensions of the engines, the engineers of the company, which at that time was led by the son of one of the founders, Joann Jakob Sulzer, focused on using the two-stroke cycle. The first engine of this type was built in 1905. It was a four-cylinder engine having a straight-flow valve purge with the location of the intake valves in the cylinder head. In the future, to facilitate the design, the efforts of the designers were directed to the use of valve-free gas distribution. In 1909, a cross-flow scavenging engine was tested, where the exhaust ports were located on one side of the cylinder liner and the purge ports on the other. The purge windows had a large height, which made it possible to charge the cylinder after the piston overlapped the outlet ports. To prevent the exhaust gas from being thrown into the intake receiver, a special motorized valve was installed at its outlet. Subsequently, this valve was replaced with a rotating spool valve, and later, on an automatic non-return valve.

The first sea vessel, equipped with two-stroke engines was the cargo and passenger ship "Monte Penedo" with a displacement of 6500 tons. The ship had two low-speed, four-cylinder, crosshead, reversible Sulzer 4S47 engines with a cylinder diameter of 470 mm and a piston stroke of 680 mm. At a speed of 160 min^{-1}, each engine developed power up to 850 horsepower. Until the 30s of the twentieth century there was no clear understanding of which engine is more efficient for use on ships, two or four stroke. Further, the growth of requirements for weight and dimensions indicators made the advantages of two-stroke low-speed engines clear to all. To make a significant contribution to, this was made possible thanks to the efforts of specialists from Sulzer, who throughout all this time carried out extensive research and design work. So, in 1910, a single-cylinder laboratory engine with a cylinder diameter of 1000 mm and a piston stroke of 1100 mm was built to investigate the gas exchange processes, and with a rotation frequency of 150 min^{-1}, it developed a power of 2000 horsepower.

In 1915, Sulzer specialists have been successful in work on the transfer of two-stroke engines with air-atomizing fuel systems to uncompressed systems, using high-pressure fuel pumps and closed injectors. In the same period, the first studies on the use of accumulator fuel injection systems were also conducted; however, the technological level of that time did not allow obtaining satisfactory commercial results; therefore, these developments did not find practical application.

In total, in the period from 1908 to 1930, Sulzer and its licensees built 640 two-stroke S-type engines with a total capacity of 1353010 horsepower. The S-series engines were manufactured with cylinder diameters of 600, 680, 760 and 820 mm. The largest engine of this series was the S90 Sulzer, released in 1929. The five-cylinder engine with a cylinder diameter of 900 mm developed a power of 4650 horsepower at a speed of 80 min^{-1}. The most powerful engine of this series was built in 1939 for the "Oranje" vessel with a displacement of 20,000 tons. It was a 12-cylinder SDT76 engine with a cylinder diameter of 760 mm which developed power at 37500 horsepower. For many years, this engine remained the most powerful in the world until this record was broken in 1977 with the construction of the Sulzer RND90M engine for the "Thames Maru" container ship by Mitsubishi. This 12-cylinder engine developed a power of 48,000 horsepower.

The first low-speed engines were created by Sulzer, using a lot of experience, gained during the construction of piston steam engines, which had a significant impact on their design. The basis of the design of the first diesel engines were A-shaped transverse arches, mounted on a cast-iron iron base frame, forming an open engine frame. Higher pressures in the working cylinders led to the need to increase the rigidity of the base frame and lubrication of the main friction pairs under pressure, which required closing the space between the arches, thereby creating an oil-tight crankcase. Subsequently, on diesel engines began to use cylinder liners, made separately from other structural elements. Phased engine design has changed, A-shaped arches were transformed into a box-shaped structure, made by casting from cast iron for one or several cylinders. Separate elements of the crankcase of the engine were connected to each other with the help of bolted joints, forming a rigid structure. During the period from 1932 to 1952, the company focused on the production of SD engines with cylinder diameters of 600 and 720 mm, which varied power from 1500 to 8400 horsepower by installing different numbers of working cylinders from 4 to 12. In this period, the company and its licensees were produced 568 engines with a total capacity of 2,795,180 horsepower. Starting from the 50s of the twentieth century, the company is gradually moving to the use of welded structures of the engine core elements, helped by the development of electric welding technology in the 30–40s. On the one hand, this made it possible to reduce the metal consumption of the structure by about 15%, and on the other hand, it removed the restrictions on the dimensions, which cast iron founding were limited to. The first engine, manufactured by the method of welded construction, was the RS series engine, put into production in 1950–52. The engines of this series were first used long anchor studs, which tightened the entire structure, unloading it from the forces, resulting from the pressure of gases. In addition, a diaphragm with a piston rod stem packing, separating the crankcase space from the piston cavity, was first used on engines of this series. This made it possible to ensure the lubrication of the working cylinders with certain types of oils, which was a necessary condition for the conversion of low-speed engines to heavy fuels. A rotary valve was installed in the exhaust channel, which blocked the exhaust ports when the piston was in the TDC area. This made it possible to significantly reduce both the height of the piston and the entire engine as a whole. Subsequently, a similar valve was used in the design of RD-type engines, the production of which was started in 1956. This series served as the basis for all subsequent RND, RND-M and RL series, including the modern series RTA and RT-flex. Since 1953, fuel systems have been introduced on series RS engines, designed to ensure their efficient operation on heavy fuels. To this end, the system provides continuous circulation of heated fuel to maintain the thermal conditions of all its elements. To prevent blocking of the plunger pairs and needle valves of the injectors, the gaps were increased in them. It was organized effective cooling nozzle tips to prevent the formation of carbon on them. As a result of the measures taken, it was possible to achieve the fact that engines could operate on heavy fuels in all modes, including starting and maneuvering. The introduction of supercharged two-stroke engines was a technological breakthrough in the development of this type of engine. The highly efficient turbo-compressor units, developed in the 40s–50s, allowed to successfully solve all the problems, preventing the use of supercharging for this type of engine. For the first time, a gas turbine boost was applied on the Sulzer SD72 engine in 1954, which made it possible to raise the cylinder power from 700 to 900 horsepower. Initially, this engine was not designed to use gas turbine supercharging; therefore, a more simple supercharging scheme was used with a constant exhaust gas pressure in front of the turbine. In 1956, on the series RD engines, the previously used longitudinal purge scheme was replaced with a loop scavenging one, which, in the presence of pressurization, made it possible to significantly improve the cleaning of cylinders from exhaust gases. In 1955–56, engines of the Sulzer RSD76 series were designed and tested, where pulsed supercharging was applied. The presence of the separation diaphragm between the crankcase and the piston cavity allowed to use this factor for the organization of the second stage of compression of the purge air. The use of pulsed supercharging allowed to increase in engine power of about 170% compared to their naturally aspirated clones. In addition, the design of the engine itself has been simplified. The refusal of the purge cylinder made it possible to place the turbocharger, the moisture separator and the charge air cooler directly on the engine, arranging them in a separate unit. This made it possible to significantly reduce the length of the intake and exhaust manifolds, ensuring minimal pressure and heat losses.

The RD engine series included engines with cylinder diameters of 440, 560, 680, 760 and 900 mm and covered a range of cylinder capacities from 500 to 2300 horsepower. The biggest engine of the series was the 12-cylinder RD90, which developed a power of 27,600 horsepower.

Other design features of the RD-type engines include the use of water-cooled pistons, the average arrangement of fuel pumps, which simplified their drive through gears, the use of a rotary non-return valve with a mechanical drive in the exhaust manifold.

In the 1960s, there was a period when the company conducted a series of studies on the use of one and two-stage gas turbine supercharging on two-stroke engines of various systems. For these studies the engine 5RD68 was used. In the framework of these studies systems with a pulsed supercharging and supercharging with constant pressure were compared, new design solutions were worked out.

Return to supercharging at constant pressure took place in 1968 with the start of production of engines of the RND type. This has significantly reduced the specific fuel consumption and increased mean effective pressure without a significant increase in thermal stresses in the engine parts. In addition, it made it possible to abandon the use of rotary valves in the exhaust system, while simplifying the design of the engine. The first engine in the new RND series had a cylinder diameter of 1050 mm, which, combined with a higher average effective pressure of 1,05 MPa, made it possible to increase the cylinder power to 4000 horsepower.

An important step in the design of powerful two-stroke engines was the control of the thermal tension of their main elements. On series RND engines, cylinder liner cooling was first used with inclined bores, that are as close as possible to the combustion surface of the combustion chamber. Starting in 1976, in the RND-M series engines, this method of cooling was also applied to the cylinder head, and in series RL engines, the launch of which was started in 1979, jet cooling of the piston heads was used for the first time. Cooling was through dead-end drilled holes, made in them, as close as possible to the bottom fire surface. Subsequently, all these solutions were used on the engines of the RTA and RT-flex series.

In the early 1980s, it became clear, that the spool valve gas exchange systems had exhausted their capabilities. In this regard, the company is developing a new range of engines with direct-flow valve purge the RTA series.

In 1997, Sulzer sells all of its assets, related to the production of low-speed engines, to the Finnish company Wärtsilä, which continues to produce two-stroke engines under the Sulzer brand. In 2001, the RT-flex series of electronically controlled engines was added to the already existing RTA series engines, and in 2016, Wärtsilä merged with the China State Shipbuilding Corporation (CSSC) and introduced the W-X series of engines to the market.

3.3.1 *RTA Series Engines*

The experience of designing engines with spool valve timing schemes, which Sulzer had been producing for more than 70 years, by the beginning of the 1980s has shown, that further forcing up them is either impossible or leads to unjustified design complexity. In this regard, in 1982, a new line of RTA-type engines was launched on the market, in which the company switched to a straight-flow-valve purge scheme. For exhaust emissions in engines of this series uses one central valve poppet type, installed in the cylinder head, and the fuel injectors have a peripheral arrangement. Their number depends on the diameter of the cylinder, on engines with a cylinder diameter up to 600 mm there are two of them, and with a larger diameter there are three.

Initially, the RTA series included six engine models with cylinder diameters of 380, 480, 580, 680, 760, and 840 mm and which had a piston stroke/diameter ratio of S/D = 2.86. In the designation of the engine after the indication of the series followed the figure for the diameter of the cylinder in centimeters (for example, RTA 48).

When developing engines of the RTA type, a large amount of research and design work was carried out aimed at reducing mechanical losses and wear of the main elements of the cylinder-piston group. This complex of constructive and technological measures was called TriboPack and included: a two-level supply of oil to the cylinder wall; drawing on the working surface cylinder liners of the microrelief by honing them; careful selection of the cooling modes of the cylinder liners through the inclined drilling in their upper collar; profiling of piston rings with a chrome-ceramic coating on their working surface; installation of anti-polishing ring; chrome plating on the surface of the grooves of the piston rings. In 1984, models with cylinder diameters of 520, 620 and 840 mm were added to already existing engines, and in 1986 with a diameter of 720 mm. In the designation of which the letter "M" was added (for example, RTA 52M). Since 1988, an improved version of the engine with a cylinder diameter of 840 mm, which received the designation RTA84C, was launched. In the "M" and "C" models, the S/D ratio was increased to 3.47, which made it possible to significantly reduce the rotational speed and increase the propulsive efficiency of the entire power plant. A number of improvements, made to the design of engines in the period from 1987 to 1992, allowed to increase the affective power without increasing the diameter of the cylinder by about 9%. Engines of improved models received the letter "U" in the designation. The first such engine was the RTA-2U, released in 1988. Starting from 1991, the company began developing engines with an even greater S/D ratio, which was 3.75. Initially it was assumed, that the main area of use of such engines would be the tanker fleet, so the engines began to be denoted by the index "T" (from the English Tanker). The first engine of this modification was the RTA84T, which had a rotational speed of 54 min^{-1}.

In the period 1995–1996, three more versions with cylinder diameters of 480, 580 and 680 mm with an S/D ratio added to 4.17 were added to the "T" engines. These engines were intended to be used not only on tankers but also on Suezmax class bulk carriers. The engines of the models RTA48T, RTA58T and RTA68T have more compact dimensions with high specific and mass-dimensional parameters.

Intensive development of the container fleetat the end of 1990 and the beginning of the 2000 was dictated by the need to create super-power engines, capable of developing an aggregate capacity of 40–80 thousand kW. In response to this market demand in 1994, Sulzer launches the RTA96C engine with a cylinder diameter of 960 mm, which becomes the largest series engine. This engine, designed for high-speed vessels, is made relatively short-stroke with $S/D = 2.6$. With a piston stroke of 2500 mm and a rotational speed of 108 min^{-1}, it develops a cylinder power of 7770 horsepower.

Main technical parameters of engines series RTA 48T-B (Fig. 3.45)

Parameter	Value
Number and cylinders arrangement	5, 6, 7, 8 in-line
Cylinder bore (mm)	480
Piston stroke (mm)	2000
Cylinder capacity (dm^3)	361.9
Rotational speed (min^{-1})	127
Cylinder power, N_e (layout area)	
– maximum continuous rating at 127 min^{-1} (L_1) (kW)	1455
– operational at 127 min^{-1} (L_2) (kW)	1020
– maximum continuous rating at 102 min^{-1} (L_3) (kW)	1165
– operational at 102 min^{-1} (L_4) (kW)	1020
Air charging pressure (L_1) (MPa)	0.374
Compression pressure (L_1) (MPa)	13.2
Maximum cycle pressure (L_1) (MPa)	15.0
Exhaust gas temperature at inlet of turbocharger (L_1) (°C)	440.0
Exhaust gas temperature at outlet of turbocharger (L_1) (°C)	248.0
Mean effective pressure	
– at 127 min^{-1} (L_1/L_2) (MPa)	1.90/1.33
– at 102 min^{-1} (L_3/L_4) (MPa)	1.90/1.66
Brake specific fuel oil consumption (g/kWh)	
– at 127 min^{-1} (L_1/L_2) for N_e 100/75% (g/kWh)	173/167
– at 102 min^{-1} (L_3/L_4) for N_e 100/75% (g/kWh)	173/169
Mean piston speed at 127 min^{-1} (m/s)	8.47
Brake specific air consumption (kg/kWh)	8.47
Brake specific exhaust gas flow (kg/kWh)	8.55
Cylinder oil consumption (g/kWh)	1.10

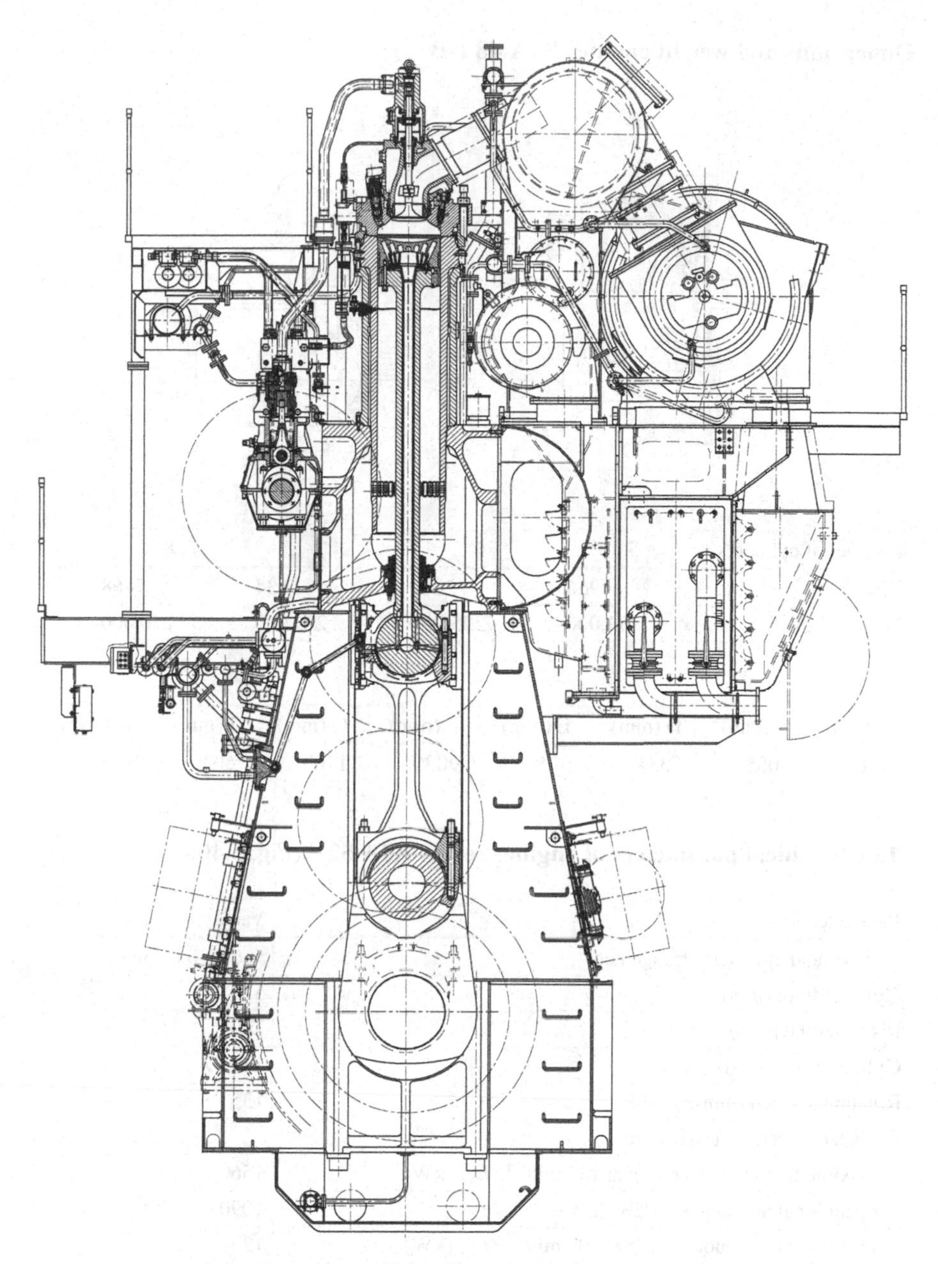

Fig. 3.45 Cross-section of the engine RTA 48T-B by Sulzer firm [44]

Dimensions and weight engines RTA 48T-B

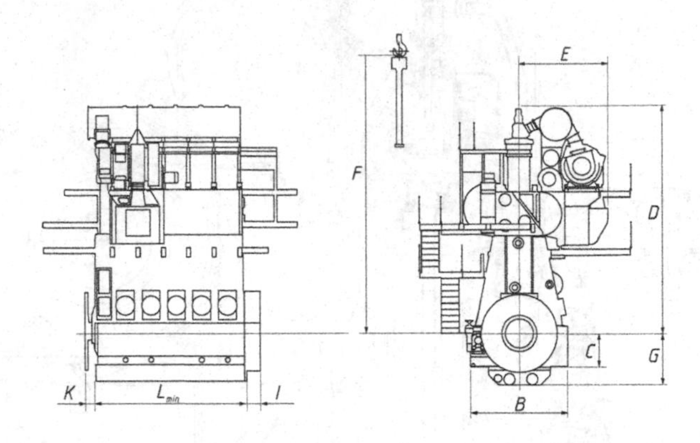

Number of cylinders	5	6	7	8
L_{min}	4966	5800	6634	7468
Weight (kg)	171,000	205,000	225,000	250,000

B (mm)	C (mm)	D (mm)	E (mm)	F (mm)	G (mm)	I (mm)	K (mm)
3170	1085	7334	3253	9030	1700	603	348

Main technical parameters of engines series RTA 52U (Fig. 3.46)

Parameter	Value
Number and cylinders arrangement	5, 6, 7, 8 in-line
Cylinder bore (mm)	520
Piston stroke (mm)	1800
Cylinder capacity (dm^3)	382.27
Rotational speed (min^{-1})	135
Cylinder power, N_e (layout area)	
– maximum continuous rating at 135 min^{-1} (L_1) (kW)	1560
– operational at 135 min^{-1} (L_2) (kW)	1090
– maximum continuous rating at 108 min^{-1} (L_3) (kW)	1250
– operational at 108 min^{-1} (L_4) (kW)	1090
Air charging pressure (L_1) (MPa)	0.254
Compression pressure (L_1) (MPa)	12.70
Maximum cycle pressure (L_1) (MPa)	14.50
Exhaust gas temperature at inlet of turbocharger (L_1) (°C)	390.0

(continued)

(continued)

Parameter	Value
Exhaust gas temperature at outlet of turbocharger (L_1) (°C)	240.0
Mean effective pressure	
– at 135 min^{-1} (L_1/L_2) (MPa)	1.81/1.27
– at 108 min^{-1} (L_3/L_4) (MPa)	1.81/1.58
Brake specific fuel oil consumption (g/kWh)	
– at 135 min^{-1} (L_1/L_2) for N_e 100/75% (g/kWh)	174.0/166.0
– at 108 min^{-1} (L_3/L_4) for N_e 100/75% (g/kWh)	173.0/169.0
Mean piston speed at 135 min^{-1} (m/s)	8.10
Brake specific air consumption (kg/kWh)	8.51
Brake specific exhaust gas flow (kg/kWh)	8.70
Cylinder oil consumption (g/kWh)	0.8–1.5

Dimensions and weight engines RTA 52U

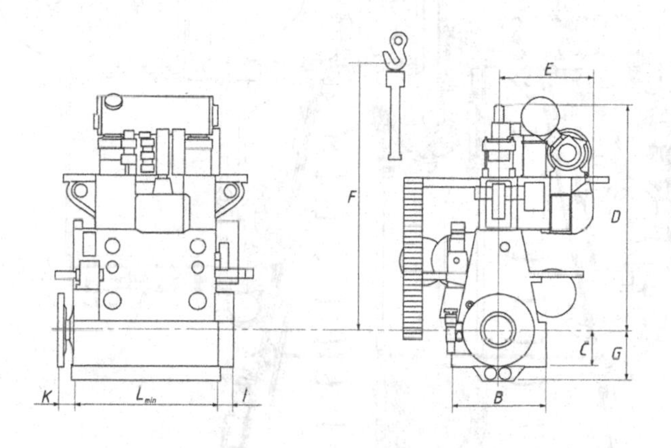

Number of cylinders	5	6	7	8
L_{min}	5605	6525	7445	8365
Weight (kg)	210,000	240,000	270,000	300,000

B (mm)	C (mm)	D (mm)	E (mm)	F (mm)	G (mm)	I (mm)	K (mm)
3030	1150	7480	3540	8745	1595	570	480

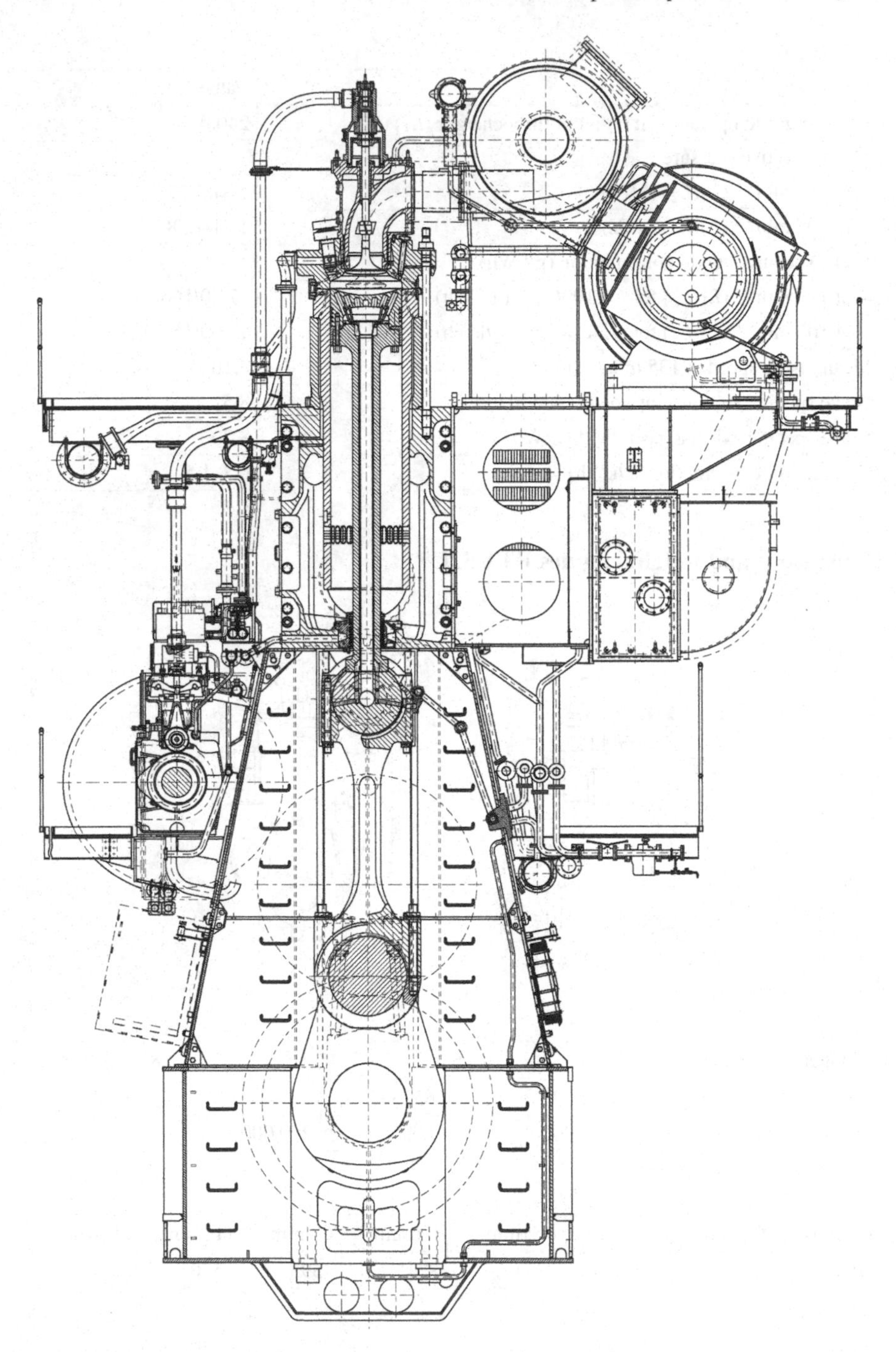

Fig. 3.46 Cross-section of the engine RTA 52U by Sulzer firm [45]

Main technical parameters of engines series RTA 58T (Fig. 3.47)

Parameter	Value
Number and cylinders arrangement	5, 6, 7, 8 in-line
Cylinder bore (mm)	580
Piston stroke (mm)	2416
Cylinder capacity (dm^3)	638.33
Rotational speed (min^{-1})	105
Cylinder power, N_e (layout area)	
– maximum continuous rating at 105 min^{-1} (L_1) (kW)	2260
– operational at 105 min^{-1} (L_2) (kW)	1580
– maximum continuous rating at 84 min^{-1} (L_3) (kW)	1810
– operational at 84 min^{-1} (L_4) (kW)	1580
Air charging pressure (L_1) (MPa)	0.32
Compression pressure (L_1) (MPa)	13.5
Maximum cycle pressure (L_1) (MPa)	15.0
Exhaust gas temperature at inlet of turbocharger (L_1) (°C)	400
Exhaust gas temperature at outlet of turbocharger (L_1) (°C)	240
Mean effective pressure	
– at 105 min^{-1} (L_1/L_2) (MPa)	2.02/1.41
– at 84 min^{-1} (L_3/L_4) (MPa)	2.02/1.77
Brake specific fuel oil consumption (g/kWh)	
– at 105 min^{-1} (L_1/L_2) for N_e 100/75% (g/kWh)	170.0/162.0
– at 84 min^{-1} (L_3/L_4) for N_e 100/75% (g/kWh)	170.0/166.0
Mean piston speed at 105 min^{-1} (m/s)	8.46
Brake specific air consumption (kg/kWh)	8.28
Brake specific exhaust gas flow (kg/kWh)	8.45
Cylinder oil consumption (g/kWh)	1.0–1.4

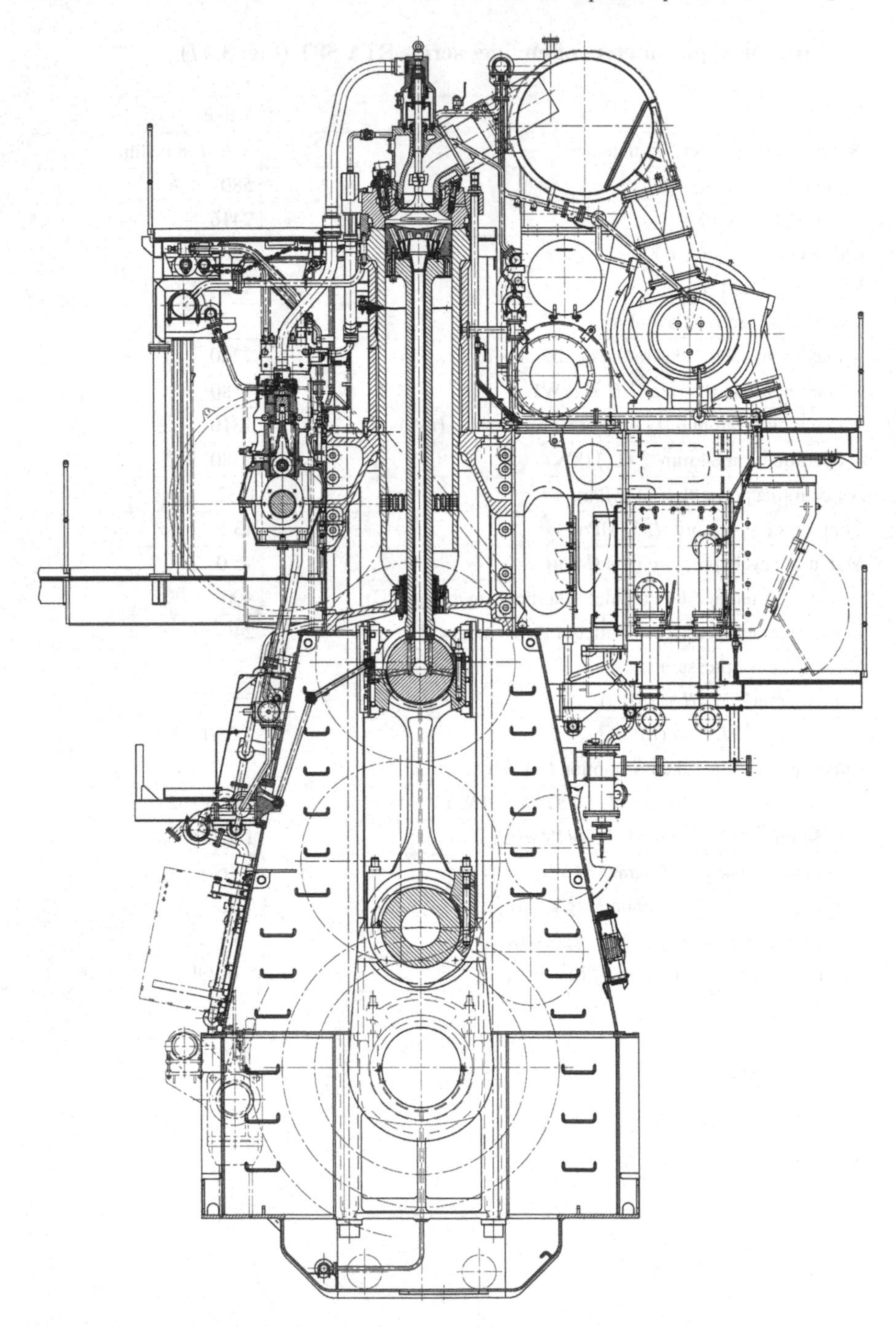

Fig. 3.47 Cross-section of the engine RTA 58T by Sulzer firm [46]

Dimensions and weight engines RTA 58T

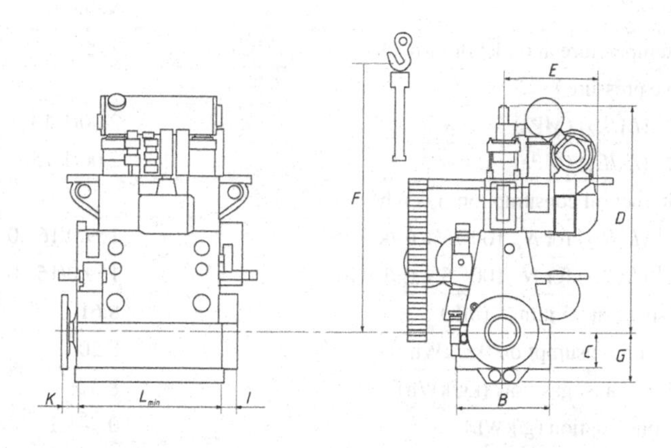

Number of cylinders	5	6	7	8
L_{min}	5981	6987	7993	8999
Weight (kg)	281,000	322,000	377,000	418,000

B (mm)	C (mm)	D (mm)	E (mm)	F (mm)	G (mm)	I (mm)	K (mm)
3820	1300	8810	3475	10,880	2000	604	400

Main technical parameters of engines series RTA 68T-B (Fig. 3.48)

Parameter	Value
Number and cylinders arrangement	5, 6, 7, 8 in-line
Cylinder bore (mm)	680
Piston stroke (mm)	2720
Cylinder capacity (dm^3)	987.82
Rotational speed (min^{-1})	95
Cylinder power, N_e (layout area)	
– maximum continuous rating at 95 min^{-1} (L_1) (kW)	3130
– operational at 95 min^{-1} (L_2) (kW)	2190
– maximum continuous rating at 76 min^{-1} (L_3) (kW)	2500
– operational at 76 min^{-1} (L_4) (kW)	2190
Air charging pressure (L_1) (MPa)	0.24
Compression pressure (L_1) (MPa)	12.2
Maximum cycle pressure (L_1) (MPa)	14.5
Exhaust gas temperature at inlet of turbocharger (L_1) (°C)	430

(continued)

(continued)

Parameter	Value
Exhaust gas temperature at outlet of turbocharger (L_1) (°C)	238
Mean effective pressure	
– at 95 min^{-1} (L_1/L_2) (MPa)	2.00/1.40
– at 76 min^{-1} (L_3/L_4) (MPa)	2.00/1.75
Brake specific fuel oil consumption (g/kWh)	
– at 95 min^{-1} (L_1/L_2) for N_e 100/75% (g/kWh)	169.0/161.0
– at 76 min^{-1} (L_3/L_4) for N_e 100/75% (g/kWh)	169.0/165.0
Mean piston speed at 95 min^{-1} (m/s)	8.61
Brake specific air consumption (kg/kWh)	8.20
Brake specific exhaust gas flow (kg/kWh)	8.37
Cylinder oil consumption (g/kWh)	0.7–1.2

Dimensions and weight engines RTA 68T-B

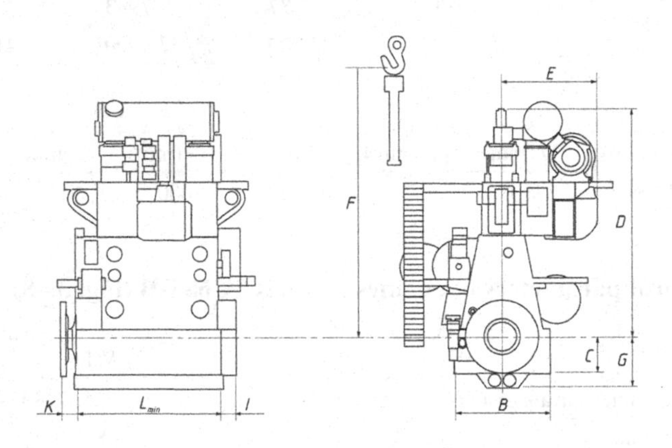

Number of cylinders	5	6	7	8
L_{min}	7005	8185	9365	10,545
Weight (kg)	412,000	472,000	533,000	593,000

B (mm)	C (mm)	D (mm)	E (mm)	F (mm)	G (mm)	I (mm)	K (mm)
4300	1520	10,400	4490	12,545	2340	789	525

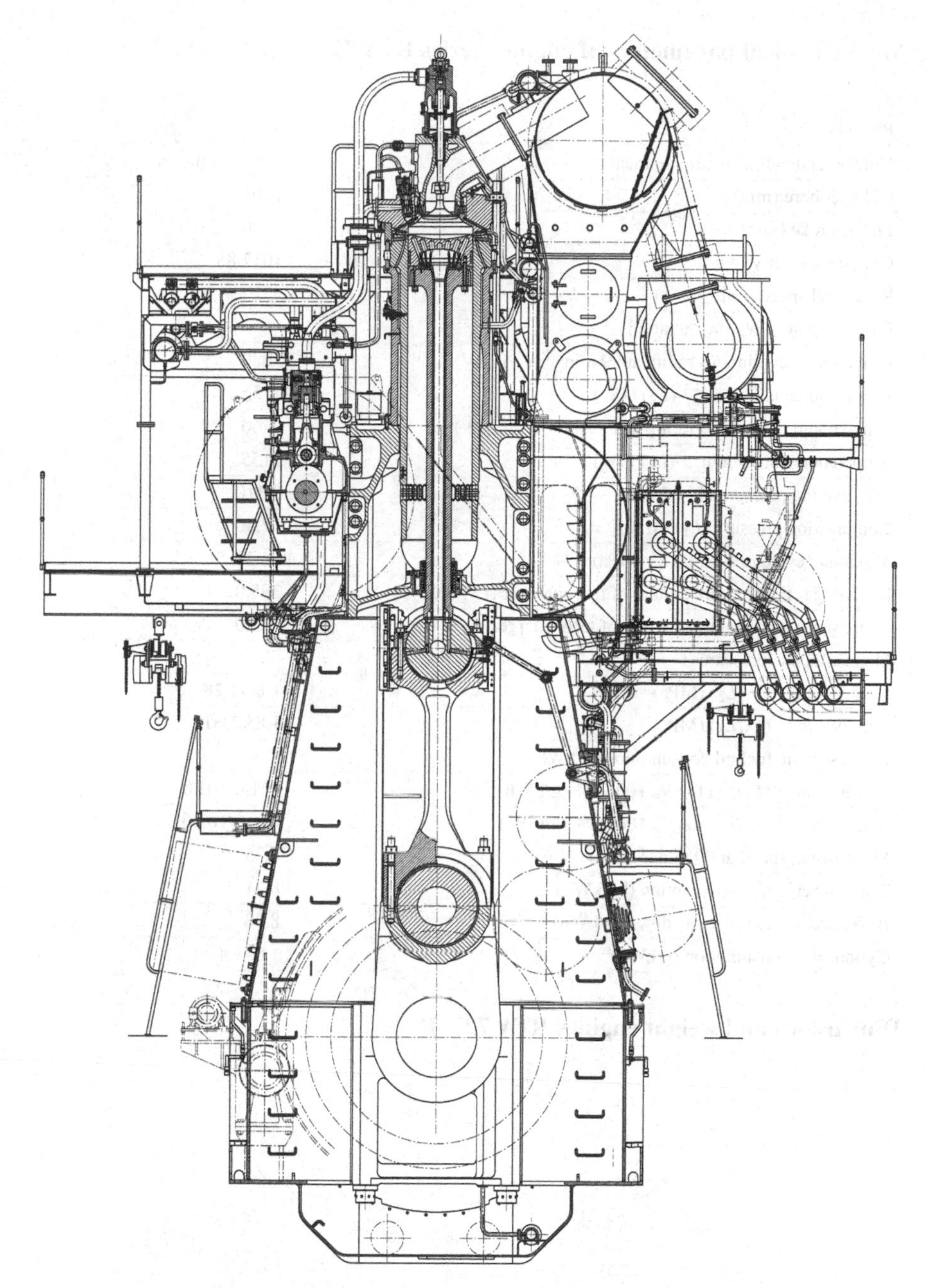

Fig. 3.48 Cross-section of the engine RTA 68T-B by Sulzer firm [47]

Main technical parameters of engines series RTA 72U-B (Fig. 3.49)

Parameter	Value
Number and cylinders arrangement	5, 6, 7, 8 in-line
Cylinder bore (mm)	720
Piston stroke (mm)	2500
Cylinder capacity (dm^3)	1017.88
Rotational speed (min^{-1})	99
Cylinder power, N_e (layout area)	
– maximum continuous rating at 99 min^{-1} (L_1) (kW)	3080
– operational at 99 min^{-1} (L_2) (kW)	2155
– maximum continuous rating at 79 min^{-1} (L_3) (kW)	2460
– operational at 79 min^{-1} (L_4) (kW)	2155
Air charging pressure (L_1) (MPa)	0.310
Compression pressure (L_1) (MPa)	12.50
Maximum cycle pressure (L_1) (MPa)	14.20
Exhaust gas temperature at inlet of turbocharger (L_1) (°C)	395.0
Exhaust gas temperature at outlet of turbocharger (L_1) (°C)	242.0
Mean effective pressure	
– at 99 min^{-1} (L_1/L_2) (MPa)	1.83/1.28
– at 79 min^{-1} (L_3/L_4) (MPa)	1.83/1.61
Brake specific fuel oil consumption (g/kWh)	
– at 99 min^{-1} (L_1/L_2) for N_e 100/75% (g/kWh)	171.0/165.0
– at 79 min^{-1} (L_3/L_4) for N_e 100/75% (g/kWh)	171.0/167.0
Mean piston speed at 99 min^{-1} (m/s)	8.25
Brake specific air consumption (kg/kWh)	8.21
Brake specific exhaust gas flow (kg/kWh)	8.38
Cylinder oil consumption (g/kWh)	1.0–1.4

Dimensions and weight engines RTA 72U-B

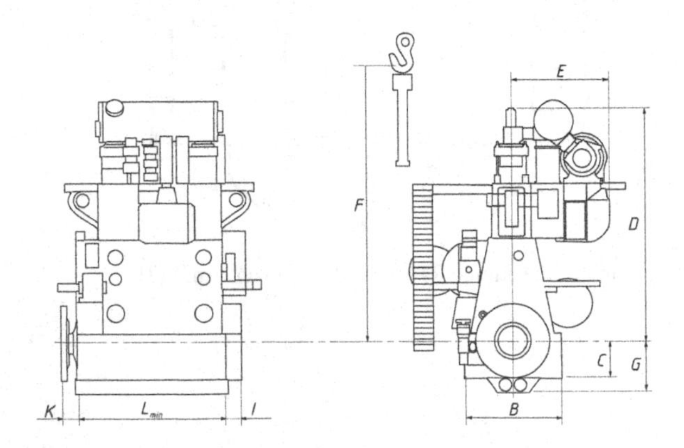

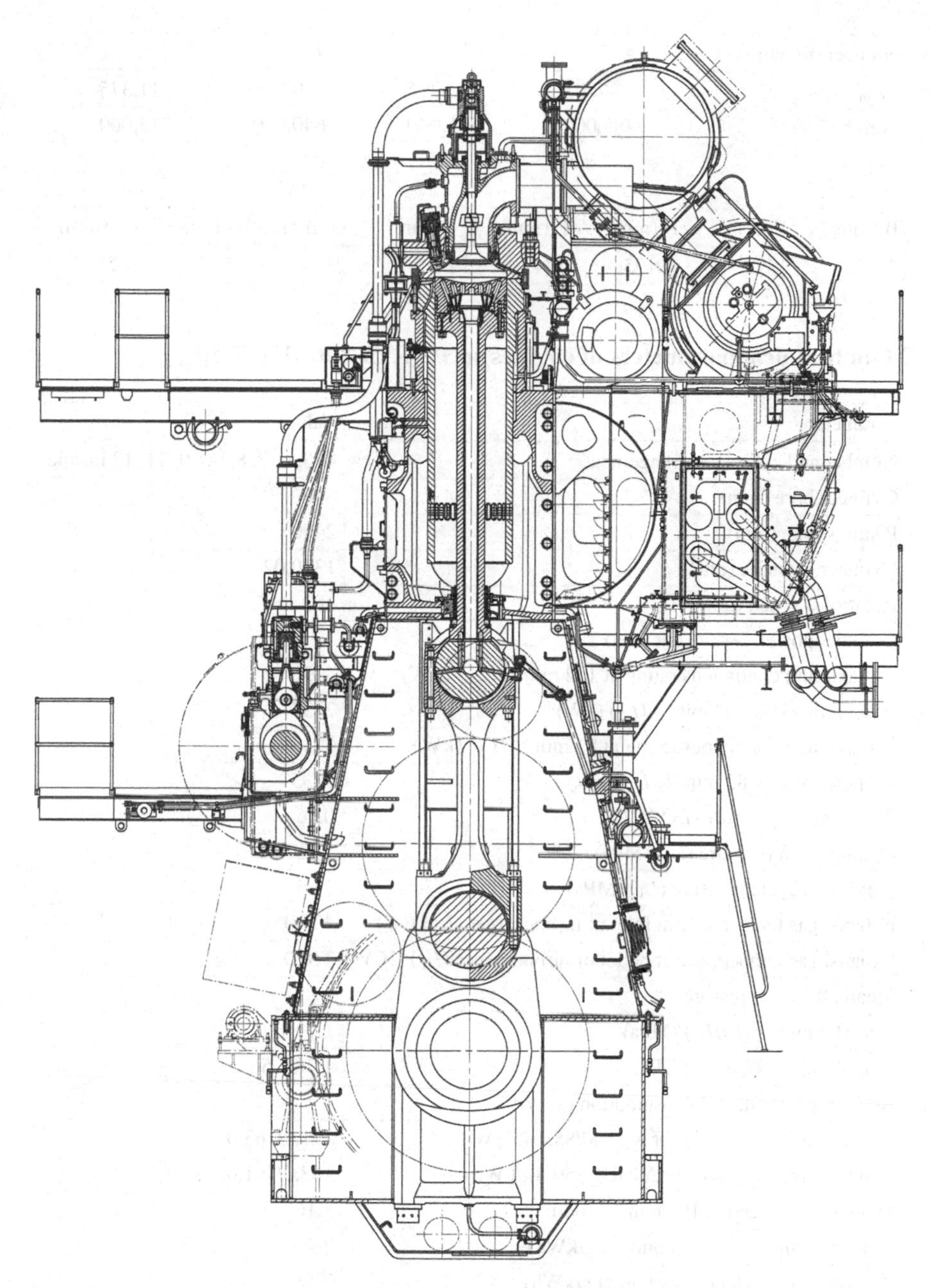

Fig. 3.49 Cross-section of the engine RTA 72U-B by Sulzer firm [48]

Number of cylinders	5	6	7	8
L_{min}	7505	8795	10,085	11,375
Weight (kg)	485,000	565,000	640,000	715,000

B (mm)	C (mm)	D (mm)	E (mm)	F (mm)	G (mm)	I (mm)	K (mm)
4070	1570	10195	3843	11875	2155	715	475

Main technical parameters of engines series RTA 84C (Fig. 3.50)

Parameter	Value
Number and cylinders arrangement	4, 5, 6, 7, 8, 9, 10, 11, 12 in-line
Cylinder bore (mm)	840
Piston stroke (mm)	2400
Cylinder capacity (dm^3)	1330.02
Rotational speed (min^{-1})	102
Cylinder power, N_e (layout area)	
– maximum continuous rating at 102 min^{-1} (L_1) (kW)	4050
– operational at 102 min^{-1} (L_2) (kW)	2835
– maximum continuous rating at 82 min^{-1} (L_3) (kW)	3240
– operational at 82 min^{-1} (L_4) (kW)	2835
Air charging pressure (L_1) (MPa)	0.26
Compression pressure (L_1) (MPa)	11.6
Maximum cycle pressure (L_1) (MPa)	13.5
Exhaust gas temperature at inlet of turbocharger (L_1) (°C)	455.0
Exhaust gas temperature at outlet of turbocharger (L_1) (°C)	318.0
Mean effective pressure	
– at 102 min^{-1} (L_1/L_2) (MPa)	1.79/1.25
– at 82 min^{-1} (L_3/L_4) (MPa)	1.78/1.56
Brake specific fuel oil consumption (g/kWh)	
– at 102 min^{-1} (L_1/L_2) for N_e 100/85% (g/kWh)	171.0/163.0
– at 82 min^{-1} (L_3/L_4) for N_e 100/85% (g/kWh)	168.0/161.0
Mean piston speed at 102 min^{-1} (m/s)	8.16
Brake specific air consumption (kg/kWh)	7.40
Brake specific exhaust gas flow (kg/kWh)	7.58
Cylinder oil consumption (g/kWh)	0.9–1.3

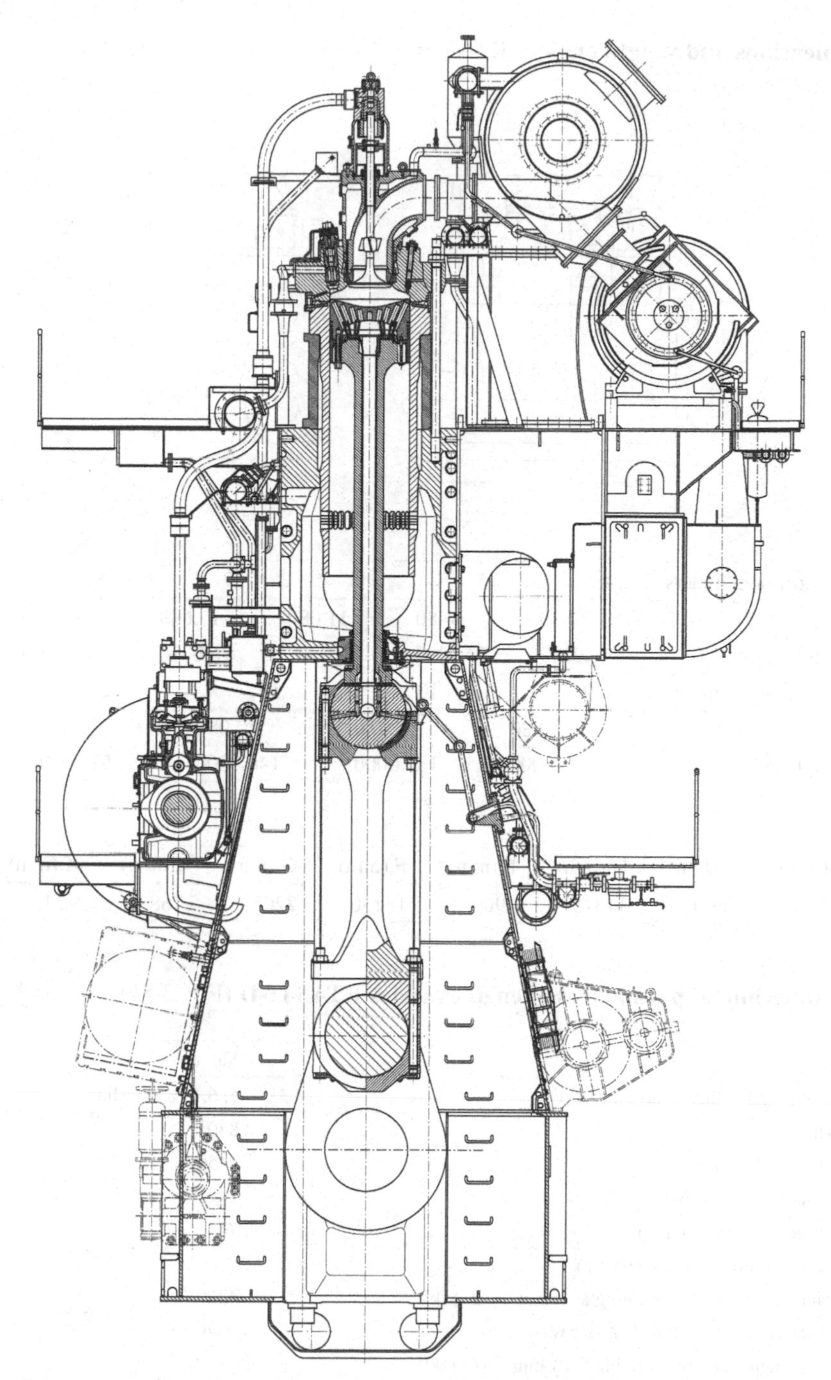

Fig. 3.50 Cross-section of the engine RTA 84C by Sulzer firm [49]

Dimensions and weight engines RTA 84C

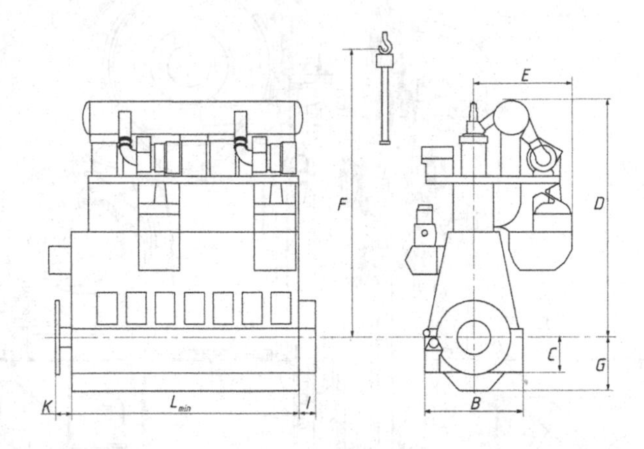

Number of cylinders	4	5	6	7	8
L_{min}	7880	9480	11,080	12,680	15,280
Weight (kg)	63,0000	74,0000	85,0000	96,0000	111,0000

Number of cylinders	9	10	11	12
L_{min}	16,880	18,480	20,080	21,680
Weight (kg)	123,0000	135,0000	146,0000	157,0000

B (mm)	C (mm)	D (mm)	E (mm)	F (mm)	G (mm)	I (mm)	K (mm)
4320	1600	11315	4900	13130	2205	768	920

Main technical parameters of engines series RTA84T-D (Fig. 3.51)

Parameter	Value
Number and cylinders arrangement	5, 6, 7, 8, 9 in-line
Cylinder bore (mm)	840
Piston stroke (mm)	3150
Cylinder capacity (dm^3)	1745.66
Rotational speed (min^{-1})	76
Cylinder power, N_e (layout area)	
– maximum continuous rating at 76 min^{-1} (L_1) (kW)	4200
– operational at 76 min^{-1} (L_2) (kW)	2940
– maximum continuous rating at 61 min^{-1} (L_3) (kW)	3370
– operational at 61 min^{-1} (L_4) (kW)	2940
Air charging pressure (L_1) (MPa)	0.23

(continued)

(continued)

Parameter	Value
Compression pressure (L_1) (MPa)	12.0
Maximum cycle pressure (L_1) (MPa)	13.8
Exhaust gas temperature at inlet of turbocharger (L_1) (°C)	430.0
Exhaust gas temperature at outlet of turbocharger (L_1) (°C)	296.0
Mean effective pressure	
– at 76 min^{-1} (L_1/L_2) (MPa)	1.90/1.33
– at 61 min^{-1} (L_3/L_4) (MPa)	1.90/1.66
Brake specific fuel oil consumption (g/kWh)	
– at 76 min^{-1} (L_1/L_2) for N_e 100/85% (g/kWh)	173.0/167.0
– at 61 min^{-1} (L_3/L_4) for N_e 100/85% (g/kWh)	173.0/169.0
Mean piston speed at 76 min^{-1} (m/s)	7.98
Brake specific air consumption (kg/kWh)	7.60
Brake specific exhaust gas flow (kg/kWh)	7.77
Cylinder oil consumption (g/kWh)	0.9–1.3

Dimensions and weight engines RTA84T-D

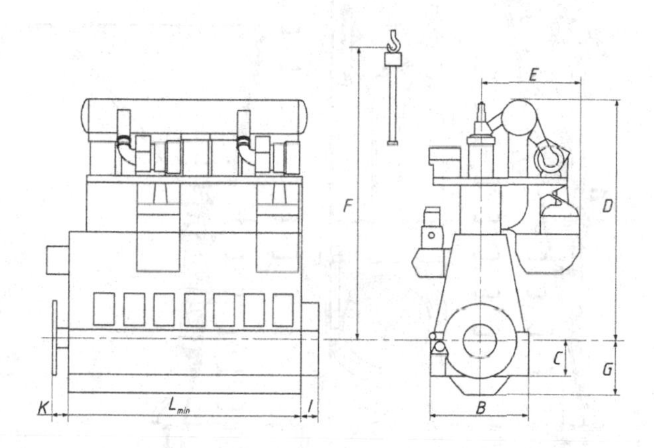

Number of cylinders	5	6	7	8	9
L_{min}	9695	11,195	12,695	15,195	16,695
Weight (kg)	740,000	870,000	990,000	1,140,000	1,260,000

B (mm)	C (mm)	D (mm)	E (mm)	F (mm)	G (mm)	I (mm)	K (mm)
5000	1800	12,150	5105	14,500	2700	760	805

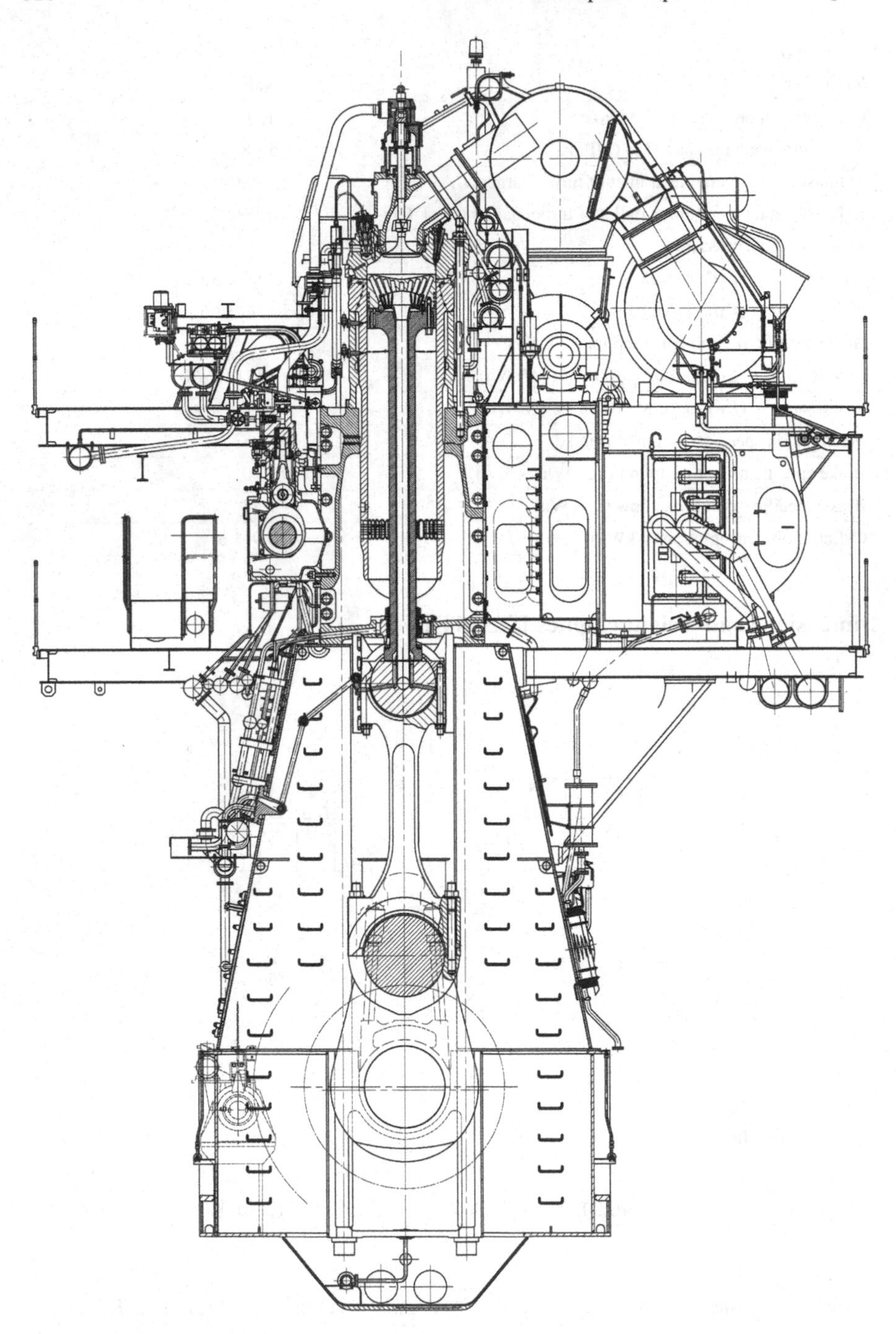

Fig. 3.51 Cross-section of the engine RTA84T-D by Sulzer firm [50]

Main technical parameters of engines series RTA 96C-B (Fig. 3.52)

Parameter	Value
Number and cylinders arrangement	6, 7, 8, 9, 10, 11, 12 in-line
Cylinder bore (mm)	960
Piston stroke (mm)	2500
Cylinder capacity (dm^3)	1809.56
Rotational speed (min^{-1})	102
Cylinder power, N_e (layout area)	
– maximum continuous rating at 102 min^{-1} (L_1) (kW)	5720
– operational at 102 min^{-1} (L_2) (kW)	4000
– maximum continuous rating at 92 min^{-1} (L_3) (kW)	5160
– operational at 92 min^{-1} (L_4) (kW)	4000
Air charging pressure (L_1) (MPa)	0.280
Compression pressure (L_1) (MPa)	13.50
Maximum cycle pressure (L_1) (MPa)	14.25
Exhaust gas temperature at inlet of turbocharger (L_1) (°C)	400.0
Exhaust gas temperature at outlet of turbocharger (L_1) (°C)	255.0
Mean effective pressure	
– at 102 min^{-1} (L_1/L_2) (MPa)	1.86/1.30
– at 92 min^{-1} (L_3/L_4) (MPa)	1.86/1.44
Brake specific fuel oil consumption (g/kWh)	
– at 102 min^{-1} (L_1/L_2) for N_e 100/75% (g/kWh)	177.0/171.0
– at 92 min^{-1} (L_3/L_4) for N_e 100/75% (g/kWh)	177.0/171.0
Mean piston speed at 102 min^{-1} (m/s)	8.50
Brake specific air consumption (kg/kWh)	7.32
Brake specific exhaust gas flow (kg/kWh)	7.48
Cylinder oil consumption (g/kWh)	0.8–1.6

Dimensions and weight engines RTA 96C-B

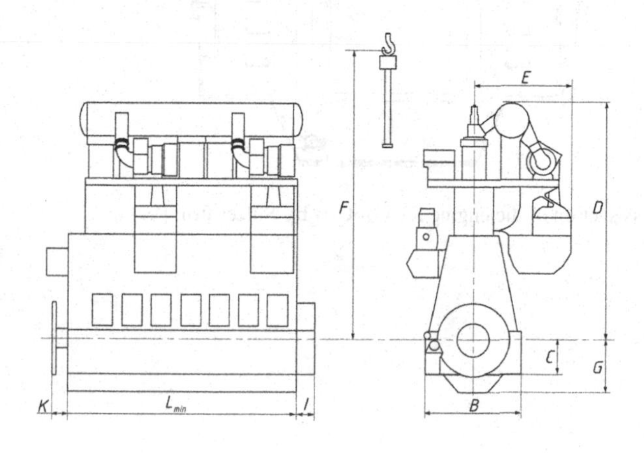

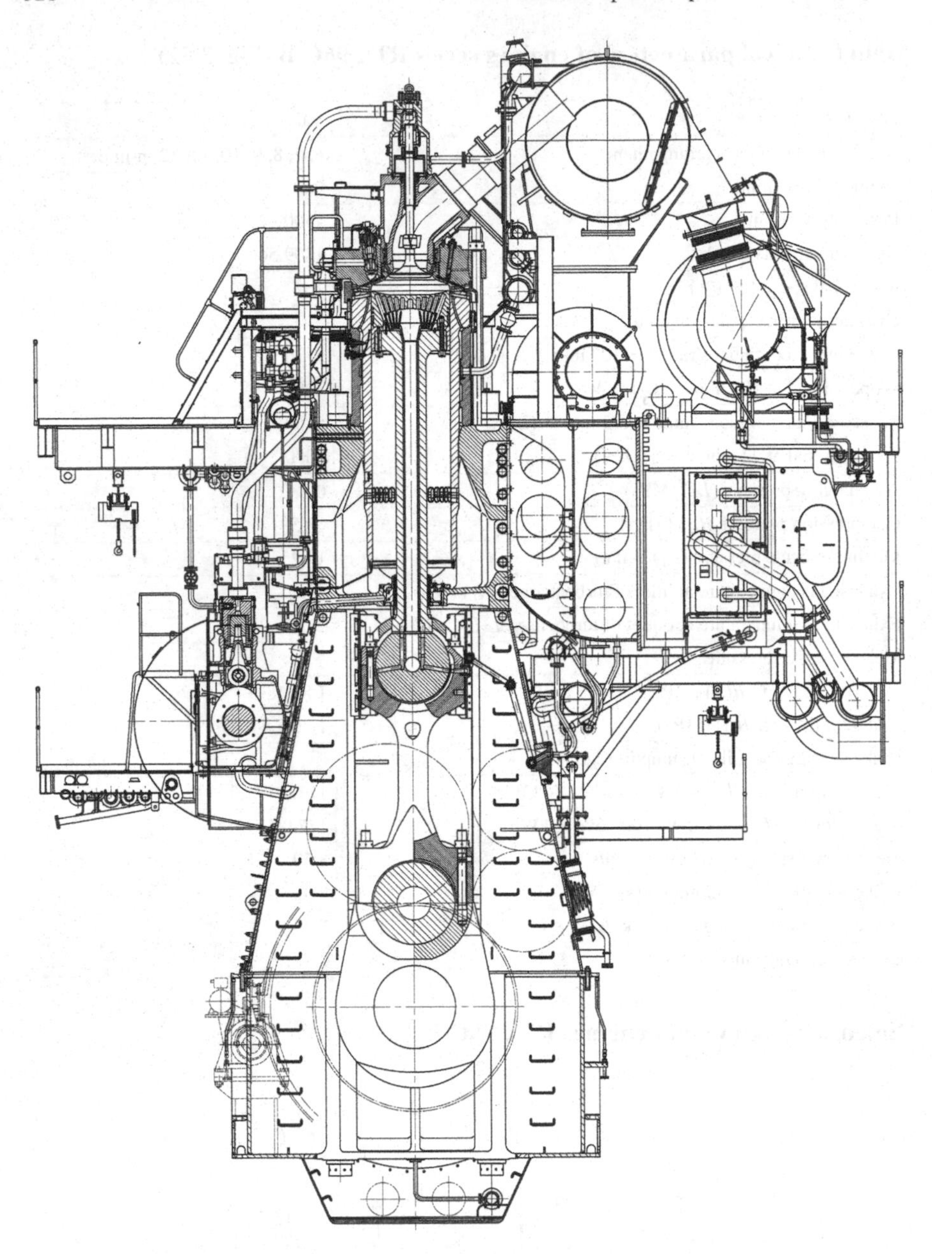

Fig. 3.52 Cross-section of the engine RTA 96C-B by Sulzer firm [51]

Number of cylinders	6	7	8	9	10	11	12
L_{min}	11,564	13,244	15,834	17,514	19,194	20,874	22,554
Weight (kg)	1,160,000	1,290,000	1,470,000	1,620,000	1,760,000	1,910,000	2,050,000

B (mm)	C (mm)	D (mm)	E (mm)	F (mm)	G (mm)	I (mm)	K (mm)
4480	1800	10925	5232	12950	2594	723	676

3.3.2 *Engines of the RT-Flex Series*

In the 90s of the twentieth century, it became clear, that the trend of tightening environmental standards for ship engines in the future will only increase, and the rise in prices for oil fuels will require a significant increase in their efficiency. In response to these trends, Sulzer began to develop a new generation of low-speed engines with electronic control systems for the main elements, that influence on the nature of the workflow in the engine. The new series was named RT-flex, since it was created on the basis of the well-proven series RTA. For the new series of engines was chosen the concept of using the accumulatory fuel injection system, called Common-Rail.

The first series RT-flex engine equipped with the Common-Rail system was a full-size laboratory engine with a cylinder diameter of 580 mm, installed at the Wärtsilä Research Center in Winterthur in 1998. The first serial engine was the six-cylinder diesel RT-flex58T-B with a capacity of 11275 kW, installed on the "Gypsum Centennial" bulk carrier, put into operation in 2001.

Next years, the RT-flex system was extended to 11 sizes of low-speed engines, produced by Wartsila, with cylinder diameters from 350 to 960 mm and a power range from 3475 to 80,080 kW. The largest of them are 14-cylinder engines RT-flex96C, developing a power of 80,080 kW at a speed of 102 min^{-1}, designed for installation on large container ships. The RT-flex series engines have become a significant step in the development of low-speed two-stroke diesel engines.

Structurally, the RT-flex engines largely inherited technical solutions, developed on RTA engines. At the same time, a fundamentally new fuel injection system was developed for them, designed to work on heavy residual HFO fuel in accordance with the ISO DIN 8217 specification (viscosity up to 700 cSt at 50 °C) at a temperature of up to 150 °C, the exhaust valve hydraulic drive system, systems of start-up and reverse and lubrication of cylinders. All of these systems are electronically controlled by the WECS-9500 microprocessor module (Wärtsilä Engine Control System), which can change their operating modes depending on the current conditions of engine operation.

The Common-Rail system includes two pressure accumulators: for control oil, which is pumped into the accumulator by axial-plunger pumps under a pressure of 20 MPa, and for fuel, supplied to the accumulator under pressure up to 100 MPa by plunger pumps.

Prepared and heated fuel at a pressure of 100 MPa through a non-return valve enters a large-capacity accumulator, which is a thick-walled pipe, stretched along the entire engine. The fuel feed control is carried out by the injection control units, mounted on the accumulator, which are individual for each cylinder. They receive control signals from the WECS 9500 microprocessor unit. Under the influence of these signals, high-speed control valves open or close fuel access to the fuel injectors, that inject fuel into the engine's combustion chamber. In this case, the injectors can supply fuel to the combustion chamber of the engine, both simultaneously and with a shift relative to each other, and when operating at partial loads, one or two injectors can be turned off to ensure better atomization of fuel while reducing the cycle supply.

Fuel and oil are pumped into the corresponding accumulators by a pump unit, which is driven by the engine's crankshaft through an intermediate gear system. Axial-type plunger pumps are used to supply oil to the hydraulic system, and plunger-type high-pressure sections with spool-throttle capacity control are used to supply fuel to the high-pressure section.

Main technical parameters of engines series RT-flex 35 (Fig. 3.53)

Parameter	Value
Number and cylinders arrangement	5, 6, 7, 8 in-line
Cylinder bore (mm)	350
Piston stroke (mm)	1550
Cylinder capacity (dm^3)	149.13
Rotational speed (min^{-1})	167
Cylinder power, N_e (layout area)	
– maximum continuous rating at 167 min^{-1} (L_1) (kW)	870
– operational at 167 min^{-1} (L_2) (kW)	695
– maximum continuous rating at 118 min^{-1} (L_3) (kW)	615
– operational at 118 min^{-1} (L_4) (kW)	450
Air charging pressure (L_1) (MPa)	0.370
Compression pressure (L_1) (MPa)	15.2
Maximum cycle pressure (L_1) (MPa)	17.7
Exhaust gas temperature at inlet of turbocharger (L_1) (°C)	430.0
Exhaust gas temperature at outlet of turbocharger (L_1) (°C)	280.0
Mean effective pressure	
– at 167 min^{-1} (L_1/L_2) (MPa)	2.10/1.68
– at 118 min^{-1} (L_3/L_4) (MPa)	2.10/1.68
Brake specific fuel oil consumption (g/kWh)	

(continued)

(continued)

Parameter	Value
– at 167 min^{-1} (L_1/L_2) for N_e 100/75% (g/kWh)	174.8/168.8
– at 118 min^{-1} (L_3/L_4) for N_e 100/75% (g/kWh)	174.8/168.8
Mean piston speed at 167 min^{-1} (m/s)	8.63
Brake specific air consumption (kg/kWh)	7.49
Brake specific exhaust gas flow (kg/kWh)	7.31
Cylinder oil consumption (g/kWh)	0.6–1.30

Dimensions and weight engines RT-flex 35

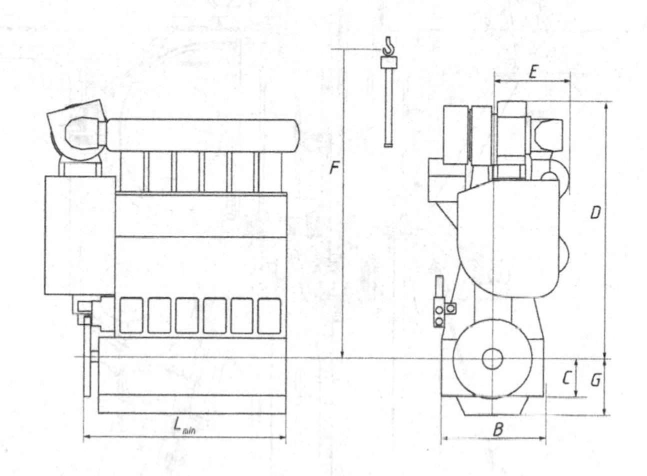

Number of cylinders	5	6	7	8
L_{min}	4398	5010	5622	6234
Weight (kg)	74,000	84,000	95,000	105,000

B (mm)	C (mm)	D (mm)	E (mm)	F (mm)	G (mm)
2284	830	5556	1605	6736	1326

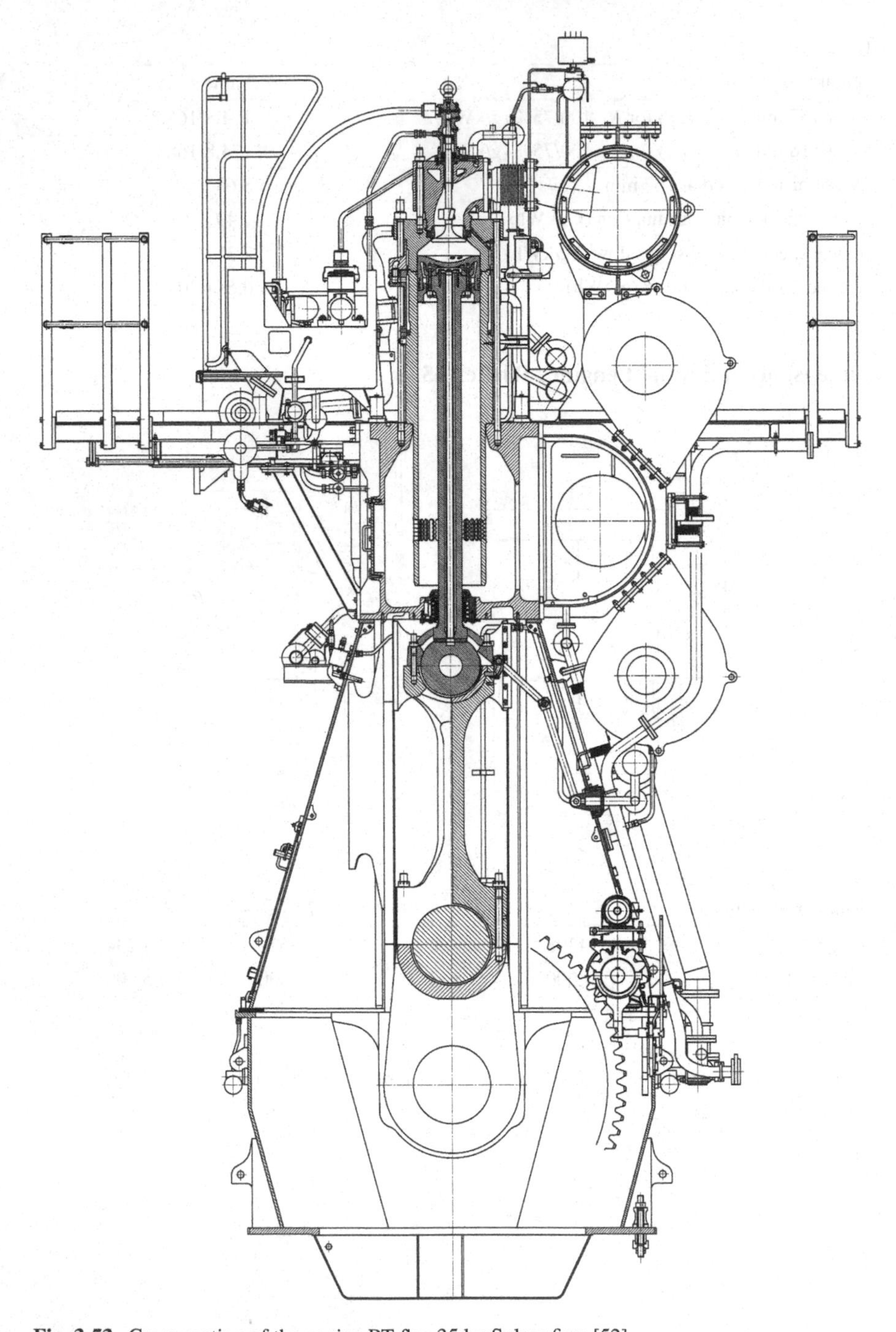

Fig. 3.53 Cross-section of the engine RT-flex 35 by Sulzer firm [52]

Main technical parameters of engines series RT-flex 48-TD (Fig. 3.54)

Parameter	Value
Number and cylinders arrangement	5, 6, 7, 8 in-line
Cylinder bore (mm)	480
Piston stroke (mm)	2000
Cylinder capacity (dm^3)	361.91
Rotational speed (min^{-1})	127
Cylinder power, N_e (layout area)	
– maximum continuous rating at 127 min^{-1} (L_1) (kW)	1455
– operational at 127 min^{-1} (L_2) (kW)	1020
– maximum continuous rating at 102 min^{-1} (L_3) (kW)	1165
– operational at 102 min^{-1} (L_4) (kW)	1020
Air charging pressure (L_1) (MPa)	0.250
Compression pressure (L_1) (MPa)	15.70
Maximum cycle pressure (L_1) (MPa)	17.50
Exhaust gas temperature at inlet of turbocharger (L_1) (°C)	417.0
Exhaust gas temperature at outlet of turbocharger (L_1) (°C)	252.0
Mean effective pressure	
– at 127 min^{-1} (L_1/L_2) (MPa)	1.90/1.33
– at 102 min^{-1} (L_3/L_4) (MPa)	1.89/1.66
Brake specific fuel oil consumption (g/kWh)	
– at 127 min^{-1} (L_1/L_2) for N_e 100/75% (g/kWh)	170.0/164.0
– at 102 min^{-1} (L_3/L_4) for N_e 100/75% (g/kWh)	170.0/166.0
Mean piston speed at 127 min^{-1} (m/s)	8.47
Brake specific air consumption (kg/kWh)	7.76
Brake specific exhaust gas flow (kg/kWh)	7.92
Cylinder oil consumption (g/kWh)	0.60 1.3

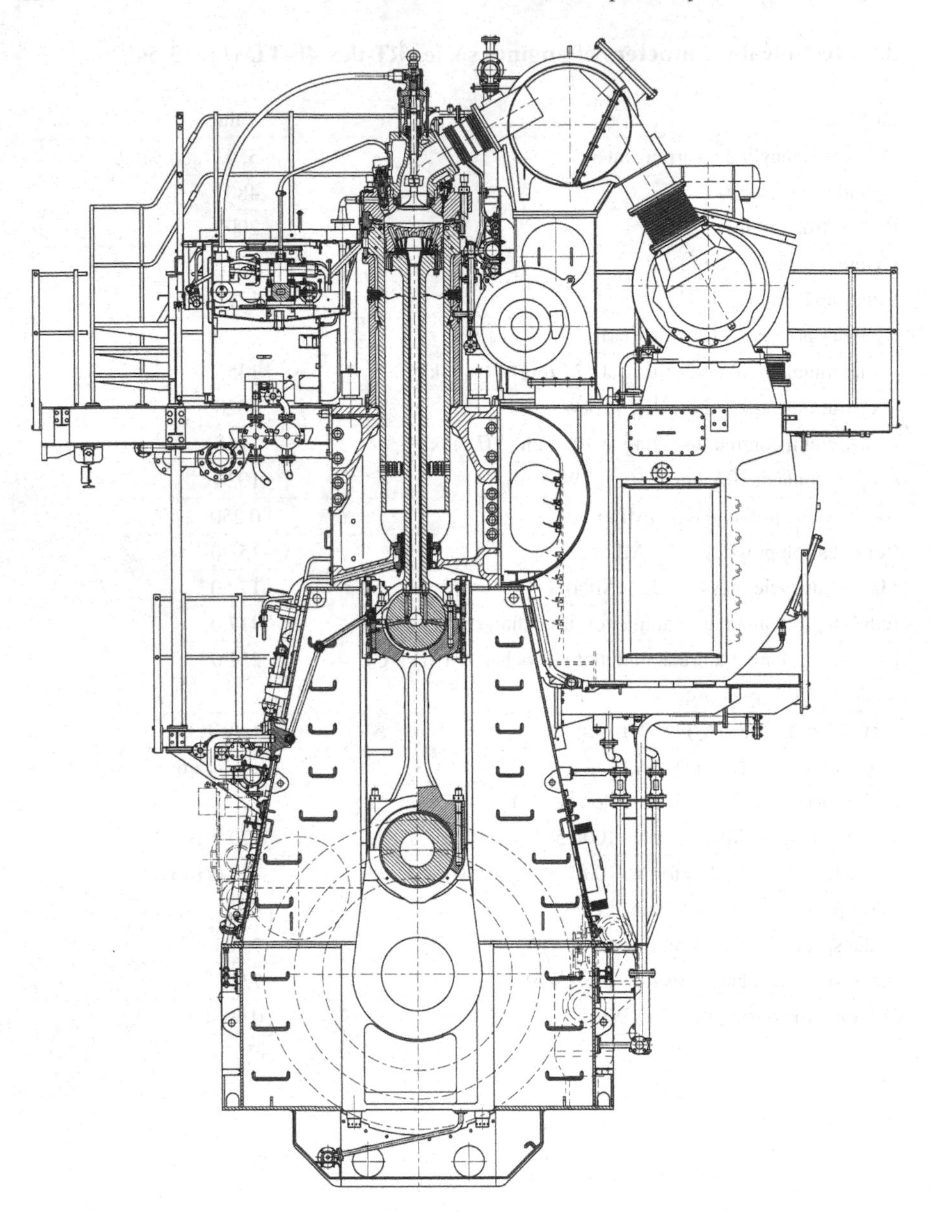

Fig. 3.54 Cross-section of the engine RT-flex 48-TD by Sulzer firm [53]

Dimensions and weight engines RT-flex 48-TD

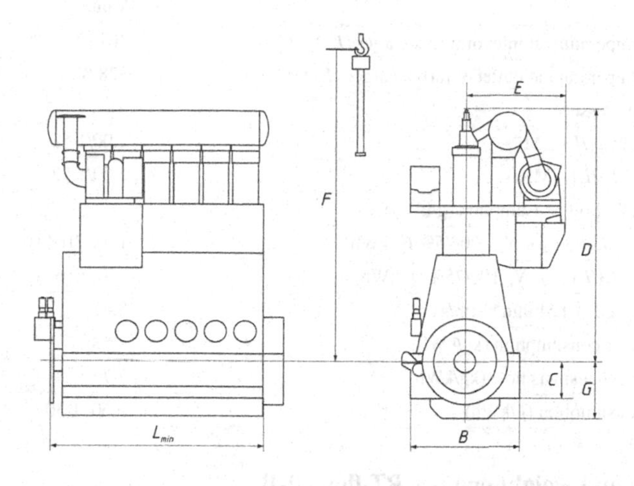

Number of cylinders	5	6	7	8
L_{min}	5314	6148	6982	7816
Weight (kg)	171,000	205,000	225,000	250,000

B (mm)	C (mm)	D (mm)	E (mm)	F (mm)	G (mm)
3170	1085	7334	3253	9030	1700

Main technical parameters of engines series RT-flex 50-B (Fig. 3.55)

Parameter	Value
Number and cylinders arrangement	5, 6, 7, 8 in-line
Cylinder bore (mm)	500
Piston stroke (mm)	2050
Cylinder capacity (dm^3)	402.52
Rotational speed (min^{-1})	124
Cylinder power, N_e (layout area)	
– maximum continuous rating at 124 min^{-1} (L_1) (kW)	1660
– operational at 124 min^{-1} (L_2) (kW)	1265
– maximum continuous rating at 95 min^{-1} (L_3) (kW)	1275
– operational at 95 min^{-1} (L_4) (kW)	970
Air charging pressure (L_1) (MPa)	0.36
Compression pressure (L_1) (MPa)	13.8
Maximum cycle pressure (L_1) (MPa)	15.3

(continued)

(continued)

Parameter	Value
Exhaust gas temperature at inlet of turbocharger (L_1) (°C)	407.0
Exhaust gas temperature at outlet of turbocharger (L_1) (°C)	328.0
Mean effective pressure	
– at 124 min^{-1} (L_1/L_2) (MPa)	2.00/1.52
– at 95 min^{-1} (L_3/L_4) (MPa)	2.00/1.52
Brake specific fuel oil consumption (g/kWh)	
– at 124 min^{-1} (L_1/L_2) for N_e 100/75% (g/kWh)	170.0/164.0
– at 95 min^{-1} (L_3/L_4) for N_e 100/75% (g/kWh)	170.0/164.0
Mean piston speed at 124 min^{-1} (m/s)	8.47
Brake specific air consumption (kg/kWh)	7.58
Brake specific exhaust gas flow (kg/kWh)	7.74
Cylinder oil consumption (g/kWh)	0.60–1.30

Dimensions and weight engines RT-flex 50-B

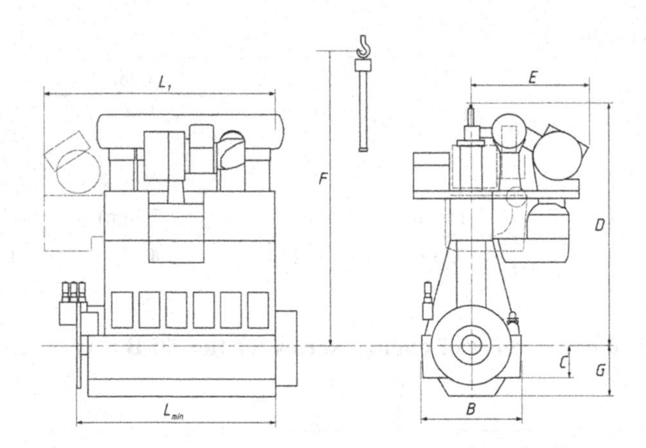

Number of cylinders	5	6	7	8
L_{min}	5576	6456	7336	8216
L_1	6793	7670	–	–
Weight (kg)	200,000	225,000	255000	280,000

B (mm)	C (mm)	D (mm)	E (mm)	F (mm)	G (mm)
3150	1088	7646	3570	9270	1636

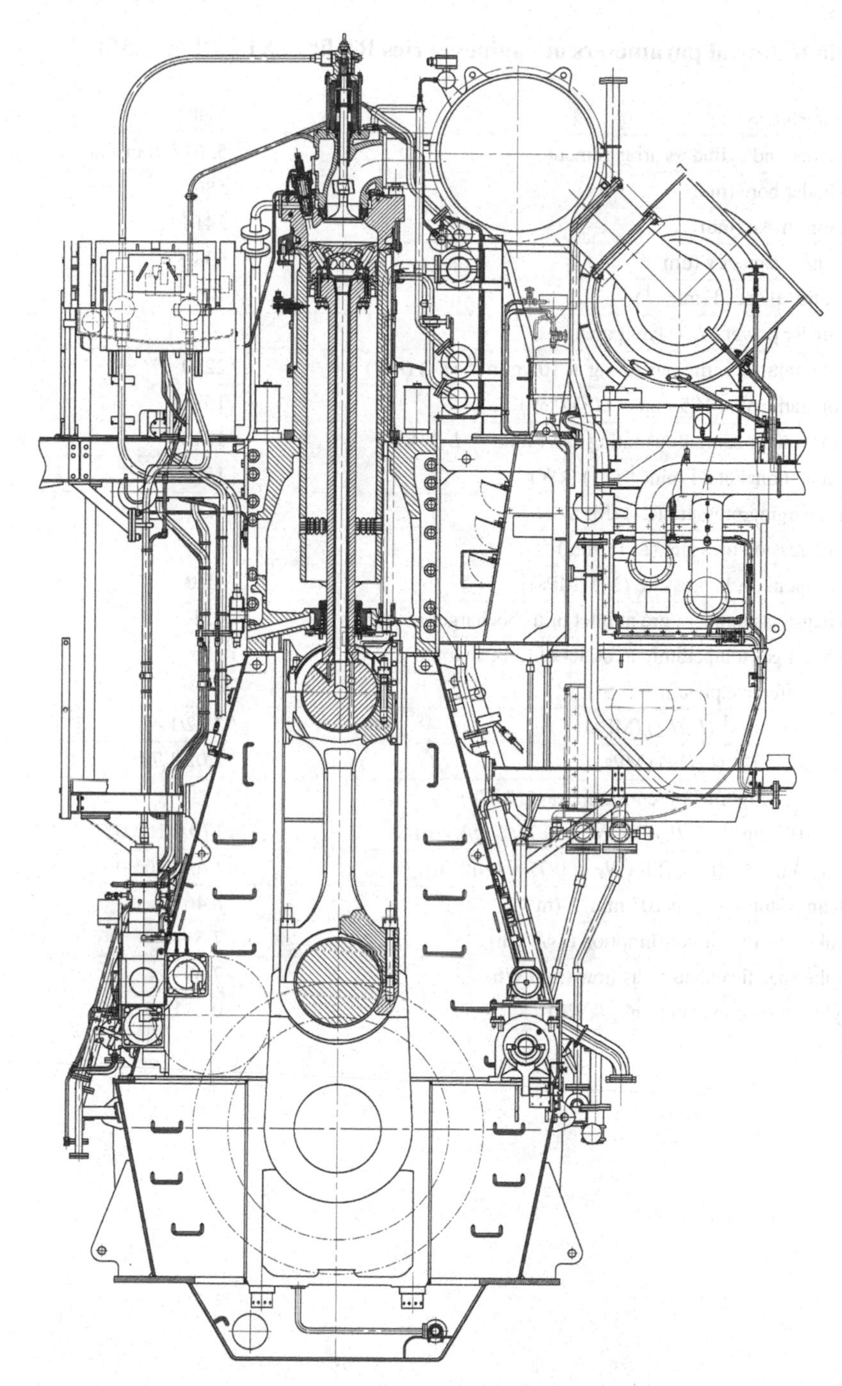

Fig. 3.55 Cross-section of the engine RT-flex 50-B by Sulzer firm [54]

Main technical parameters of engines series RT-flex 58T-D (Fig. 3.56)

Parameter	Value
Number and cylinders arrangement	5, 6, 7, 8 in-line
Cylinder bore (mm)	580
Piston stroke (mm)	2416
Cylinder capacity (dm^3)	638.33
Rotational speed (min^{-1})	105
Cylinder power, N_e (layout area)	
– maximum continuous rating at 105 min^{-1} (L_1) (kW)	2260
– operational at 105 min^{-1} (L_2) (kW)	1580
– maximum continuous rating at 84 min^{-1} (L_3) (kW)	1810
– operational at 84 min^{-1} (L_4) (kW)	1580
Air charging pressure (L_1) (MPa)	0.34
Compression pressure (L_1) (MPa)	13.5
Maximum cycle pressure (L_1) (MPa)	15.0
Exhaust gas temperature at inlet of turbocharger (L_1) (°C)	431
Exhaust gas temperature at outlet of turbocharger (L_1) (°C)	255
Mean effective pressure	
– at 105 min^{-1} (L_1/L_2) (MPa)	2.02/1.41
– at 84 min^{-1} (L_3/L_4) (MPa)	2.02/1.77
Brake specific fuel oil consumption (g/kWh)	
– at 105 min^{-1} (L_1/L_2) for N_e 100/75% (g/kWh)	170.0/162.0
– at 84 min^{-1} (L_3/L_4) for N_e 100/75% (g/kWh)	170.0/166.0
Mean piston speed at 105 min^{-1} (m/s)	8.46
Brake specific air consumption (kg/kWh)	7.51
Brake specific exhaust gas flow (kg/kWh)	7.69
Cylinder oil consumption (g/kWh)	0.6–1.13

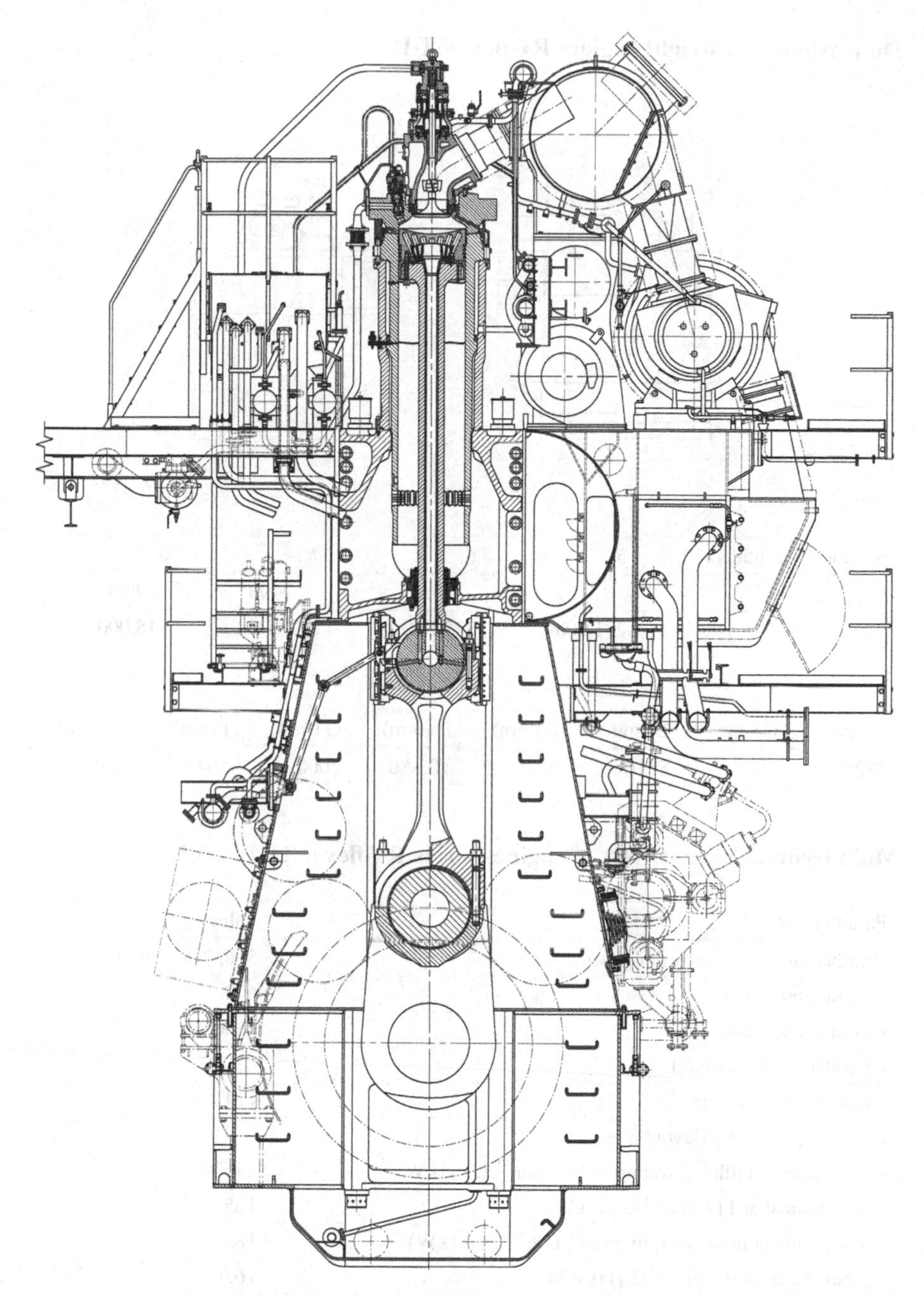

Fig. 3.56 Cross-section of the engine RT-flex 58T-D by Sulzer firm [55]

Dimensions and weight engines RT-flex 58T-D

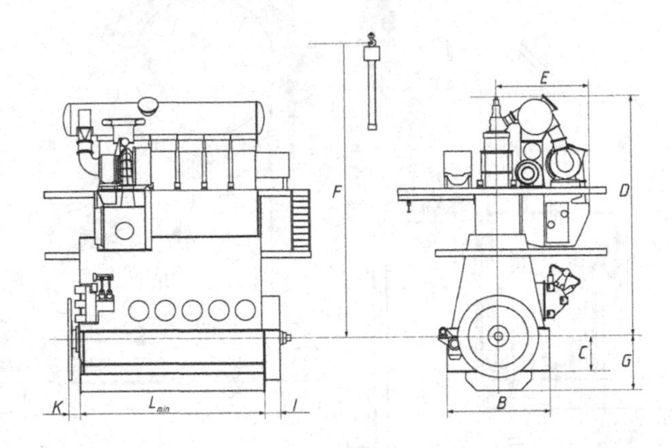

Number of cylinders	5	6	7	8
L_{min}	5981	6987	7993	8999
Weight (kg)	281,000	322,000	377,000	418,000

B (mm)	C (mm)	D (mm)	E (mm)	F (mm)	G (mm)	I (mm)	K (mm)
3820	1300	8810	3475	10,880	2000	604	400

Main technical parameters of engines series RT-flex 60C (Fig. 3.57)

Parameter	Value
Number and cylinders arrangement	5, 6, 7, 8 in-line
Cylinder bore (mm)	600
Piston stroke (mm)	2250
Cylinder capacity (dm^3)	636.17
Rotational speed (min^{-1})	114
Cylinder power, N_e (layout area)	
– maximum continuous rating at 114 min^{-1} (L_1) (kW)	2360
– operational at 114 min^{-1} (L_2) (kW)	1650
– maximum continuous rating at 91 min^{-1} (L_3) (kW)	1880
– operational at 91 min^{-1} (L_4) (kW)	1650
Air charging pressure (L_1) (MPa)	0.262
Compression pressure (L_1) (MPa)	13.70
Maximum cycle pressure (L_1) (MPa)	15.50
Exhaust gas temperature at inlet of turbocharger (L_1) (°C)	400.0

(continued)

(continued)

Parameter	Value
Exhaust gas temperature at outlet of turbocharger (L_1) (°C)	284.0
Mean effective pressure	
– at 114 min^{-1} (L_1/L_2) (MPa)	1.95/1.37
– at 91 min^{-1} (L_3/L_4) (MPa)	1.95/1.70
Brake specific fuel oil consumption (g/kWh)	
– at 114 min^{-1} (L_1/L_2) for N_e 100/75% (g/kWh)	170.0/164.0
– at 91 min^{-1} (L_3/L_4) for N_e 100/75% (g/kWh)	170.0/166.0
Mean piston speed at 114 min^{-1} (m/s)	8.55
Brake specific air consumption (kg/kWh)	8.03
Brake specific exhaust gas flow (kg/kWh)	7.86
Cylinder oil consumption (g/kWh)	0.9–1.3

Dimensions and weight engines RT-flex 60C

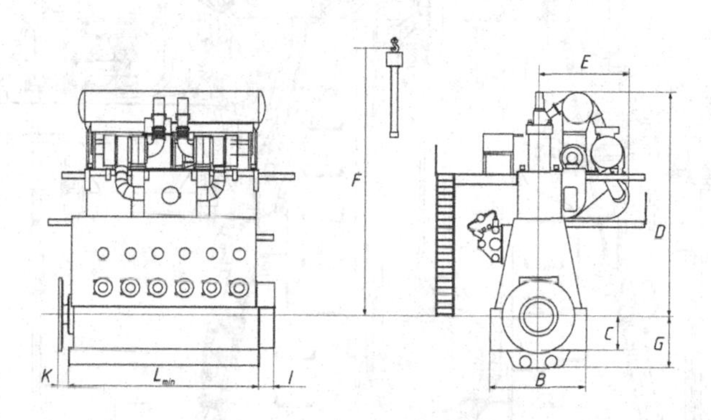

Number of cylinders	5	6	7	8
L_{min}	6213	7253	8293	9333
Weight (kg)	290,000	330,000	375,000	415,000

B (mm)	C (mm)	D (mm)	E (mm)	F (mm)	G (mm)	I (mm)	K (mm)
3700	1300	8520	3960	10400	1955	650	405

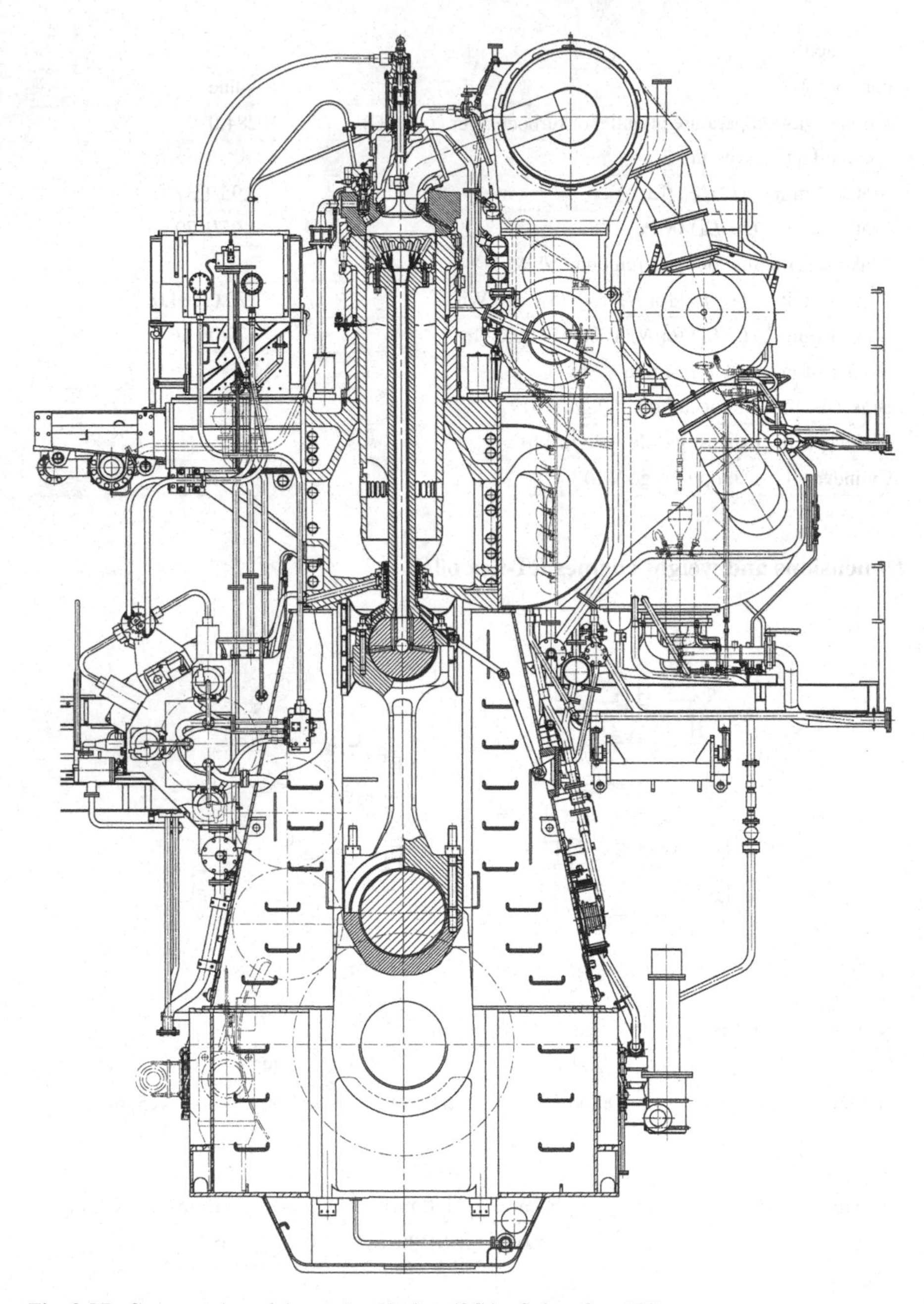

Fig. 3.57 Cross-section of the engine RT-flex 60C by Sulzer firm [56]

Main technical parameters of engines series RT-flex 68D (Fig. 3.58)

Parameter	Value
Number and cylinders arrangement	5, 6, 7, 8 in-line
Cylinder bore (mm)	680
Piston stroke (mm)	2720
Cylinder capacity (dm^3)	987.82
Rotational speed (min^{-1})	95
Cylinder power, N_e (layout area)	
– maximum continuous rating at 95 min^{-1} (L_1) (kW)	3130
– operational at 95 min^{-1} (L_2) (kW)	2190
– maximum continuous rating at 76 min^{-1} (L_3) (kW)	2500
– operational at 76 min^{-1} (L_4) (kW)	2190
Air charging pressure (L_1) (MPa)	0.26
Compression pressure (L_1) (MPa)	13.5
Maximum cycle pressure (L_1) (MPa)	15.0
Exhaust gas temperature at inlet of turbocharger (L_1) (°C)	400.0
Exhaust gas temperature at outlet of turbocharger (L_1) (°C)	310.0
Mean effective pressure	
– at 95 min^{-1} (L_1/L_2) (MPa)	2.00/1.40
– at 76 min^{-1} (L_3/L_4) (MPa)	2.00/1.75
Brake specific fuel oil consumption (g/kWh)	
– at 95 min^{-1} (L_1/L_2) for N_e 100/85% (g/kWh)	169.0/161.0
– at 76 min^{-1} (L_3/L_4) for N_e 100/85% (g/kWh)	169.0/165.0
Mean piston speed at 95 min^{-1} (m/s)	8.61
Brake specific air consumption (kg/kWh)	7.40
Brake specific exhaust gas flow (kg/kWh)	7.57
Cylinder oil consumption (g/kWh)	0.6–1.3

Dimensions and weight engines RT-flex 68D

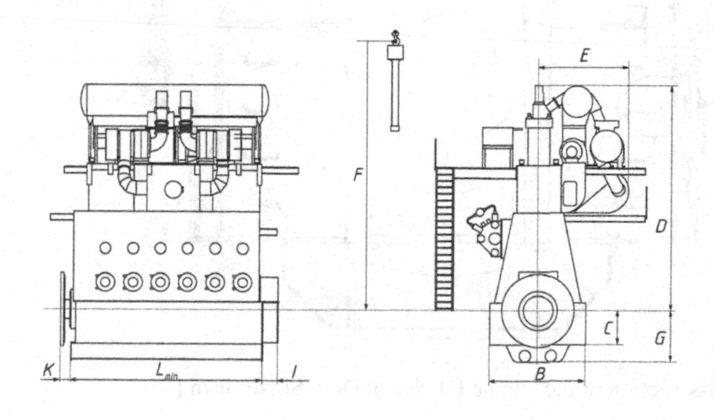

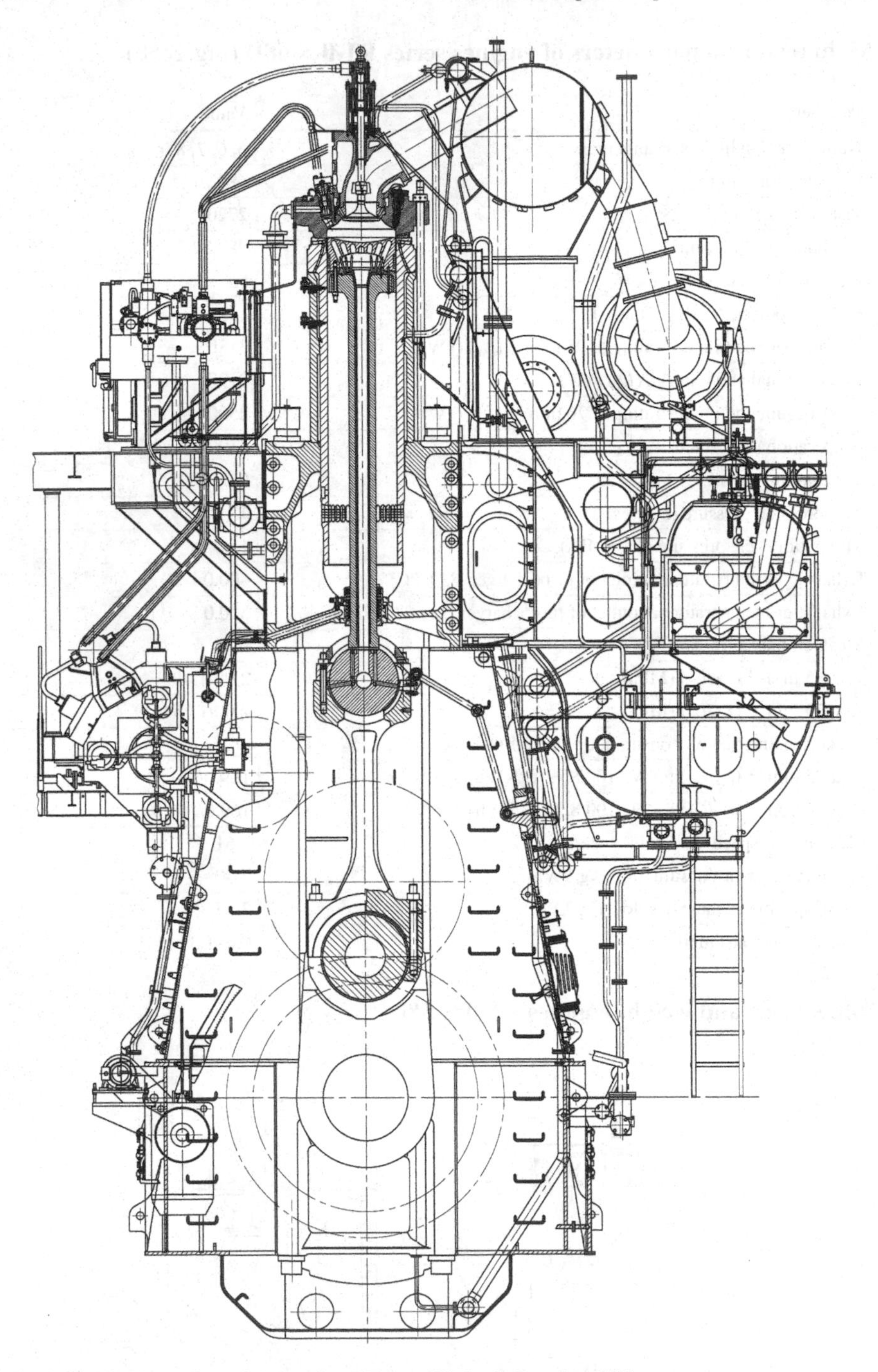

Fig. 3.58 Cross-section of the engine RT-flex 68D by Sulzer firm [57]

Number of cylinders	5	6	7	8
L_{min}	7005	8185	9365	10,545
Weight (kg)	412,000	472,000	533,000	593,000

B (mm)	C (mm)	D (mm)	E (mm)	F (mm)	G (mm)	I (mm)	K (mm)
4300	1520	10,400	4490	12,545	2340	789	525

Main technical parameters of engines series RT-flex 82C (Fig. 3.59)

Parameter	Value
Number and cylinders arrangement	6, 7, 8, 9, 10, 11, 12 in-line
Cylinder bore (mm)	820
Piston stroke (mm)	2646
Cylinder capacity (dm^3)	1397.36
Rotational speed (min^{-1})	102
Cylinder power, N_e (layout area)	
– maximum continuous rating at 102 min^{-1} (L_1) (kW)	4520
– operational at 102 min^{-1} (L_2) (kW)	3620
– maximum continuous rating at 87 min^{-1} (L_3) (kW)	4050
– operational at 87 min^{-1} (L_4) (kW)	3620
Air charging pressure (L_1) (MPa)	0.255
Compression pressure (L_1) (MPa)	12.7
Maximum cycle pressure (L_1) (MPa)	14.5
Exhaust gas temperature at inlet of turbocharger (L_1) (°C)	450.0
Exhaust gas temperature at outlet of turbocharger (L_1) (°C)	300.0
Mean effective pressure	
– at 102 min^{-1} (L_1/L_2) (MPa)	2.00/1.60
– at 87 min^{-1} (L_3/L_4) (MPa)	2.00/1.79
Brake specific fuel oil consumption (g/kWh)	
– at 102 min^{-1} (L_1/L_2) for N_e 100/80% (g/kWh)	171.0/165.0
– at 87 min^{-1} (L_3/L_4) for N_e 100/80% (g/kWh)	171.0/167.0
Mean piston speed at 102 min^{-1} (m/s)	9.00
Brake specific air consumption (кg/kWh)	7.60
Brake specific exhaust gas flow (кg/kWh)	7.77
Cylinder oil consumption (g/kWh)	0.6–1.3

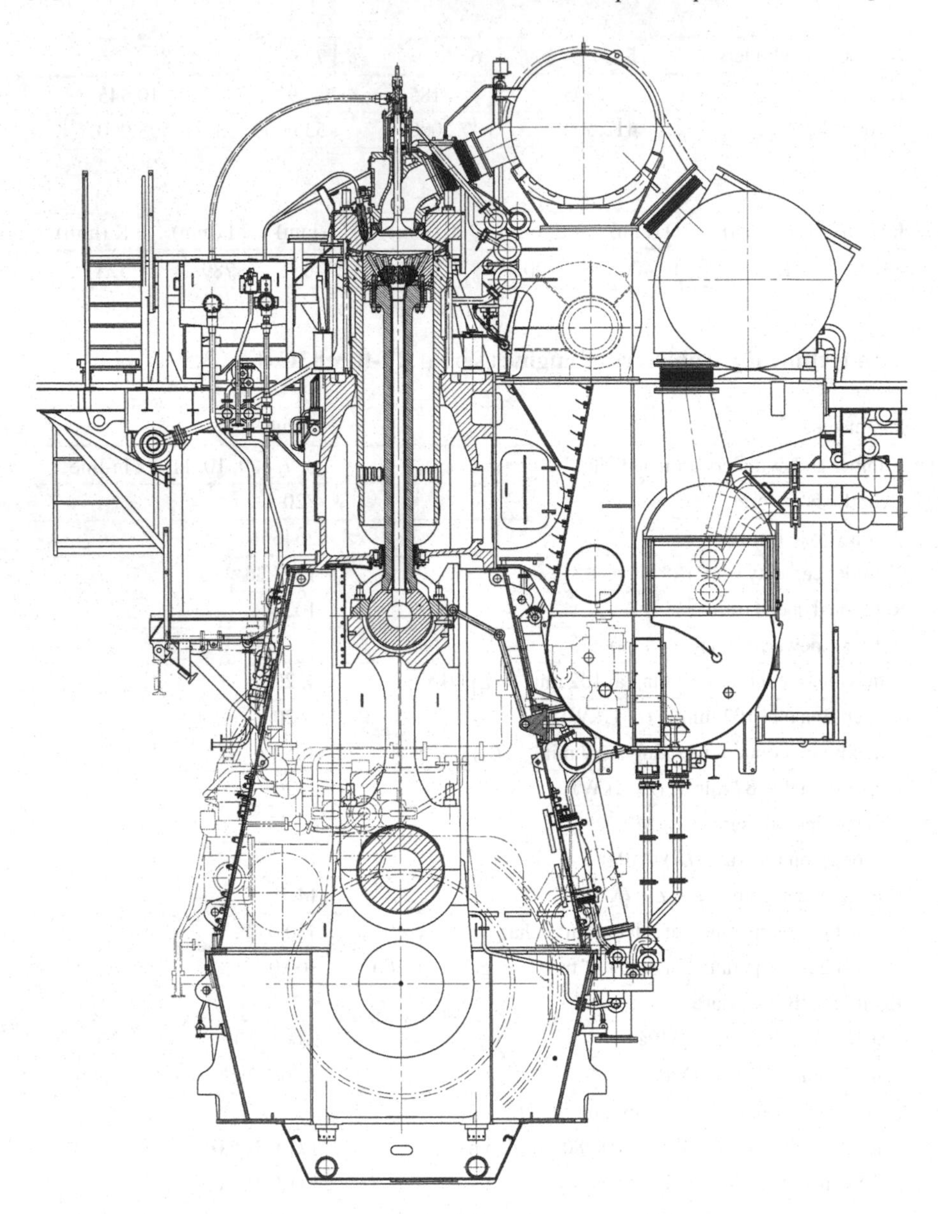

Fig. 3.59 Cross-section of the engine RT-flex 82C by Sulzer firm [58]

Dimensions and weight engines RT-flex 82C

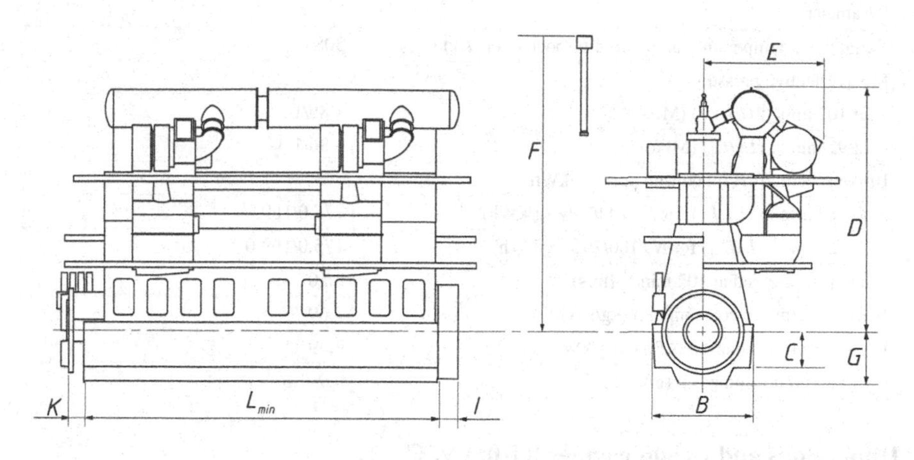

Number of cylinders	6	7	8	9	10	11	12
L_{min}	10,415	11,920	14,425	15,930	17,435	18,940	20,445
Weight (kg)	800,000	910,000	1,020,000	1,140,000	1,250,000	1,360,000	1,470,000

B (mm)	C (mm)	D (mm)	E (mm)	F (mm)	G (mm)	I (mm)	K (mm)
4570	1600	10930	5100	12700	2310	940	690

Main technical parameters of engines series RT-flex 96C (Fig. 3.60)

Parameter	Valuc
Number and cylinders arrangement	6, 7, 8, 9, 10, 11, 12, 13, 14 in-line
Cylinder bore (mm)	960
Piston stroke (mm)	2500
Cylinder capacity (dm^3)	1809.56
Rotational speed (min^{-1})	102
Cylinder power, N_e (layout area)	
– maximum continuous rating at 102 min^{-1} (L_1) (kW)	5720
– operational at 102 min^{-1} (L_2) (kW)	4000
– maximum continuous rating at 92 min^{-1} (L_3) (kW)	5160
– operational at 92 min^{-1} (L_4) (kW)	4000
Air charging pressure (L_1) (MPa)	0.280
Compression pressure (L_1) (MPa)	13.80
Maximum cycle pressure (L_1) (MPa)	14.60
Exhaust gas temperature at inlet of turbocharger (L_1) (°C)	400.0

(continued)

(continued)

Parameter	Value
Exhaust gas temperature at outlet of turbocharger (L_1) (°C)	308.0
Mean effective pressure	
– at 102 min^{-1} (L_1/L_2) (MPa)	1.86/1.30
– at 92 min^{-1} (L_3/L_4) (MPa)	1.86/1.44
Brake specific fuel oil consumption (g/kWh)	
– at 102 min^{-1} (L_1/L_2) for N_e 100/75% (g/kWh)	175.0/169.0
– at 92 min^{-1} (L_3/L_4) for N_e 100/75% (g/kWh)	175.0/169.0
Mean piston speed at 102 min^{-1} (m/s)	8.50
Brake specific air consumption (кg/kWh)	7.69
Brake specific exhaust gas flow (кg/kWh)	7.79
Cylinder oil consumption (g/kWh)	0.9–1.3

Dimensions and weight engines RT-flex 96C

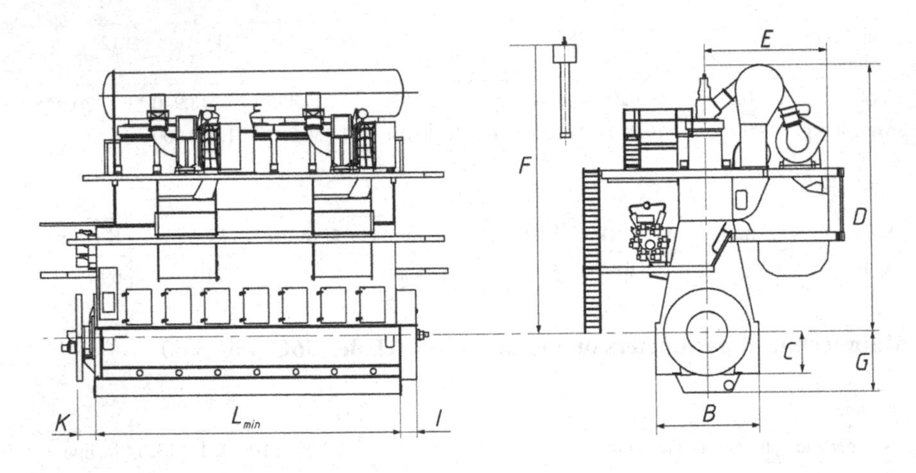

Number of cyl.	6	7	8	9	10
L_{min}	11,564	13,244	15,834	17,514	19,194
Weight (kg)	1,160,000	1,290,000	1,470,000	1,620,000	1,760,000

Number of cyl.	11	12	13	14
L_{min}	20,874	22,554	24,234	25,914
Weight (kg)	1,910,000	2,050,000	2,160,000	2,300,000

B (mm)	C (mm)	D (mm)	E (mm)	F (mm)	G (mm)	I (mm)	K (mm)
4480	1800	10925	6020	12950	2594	723	676

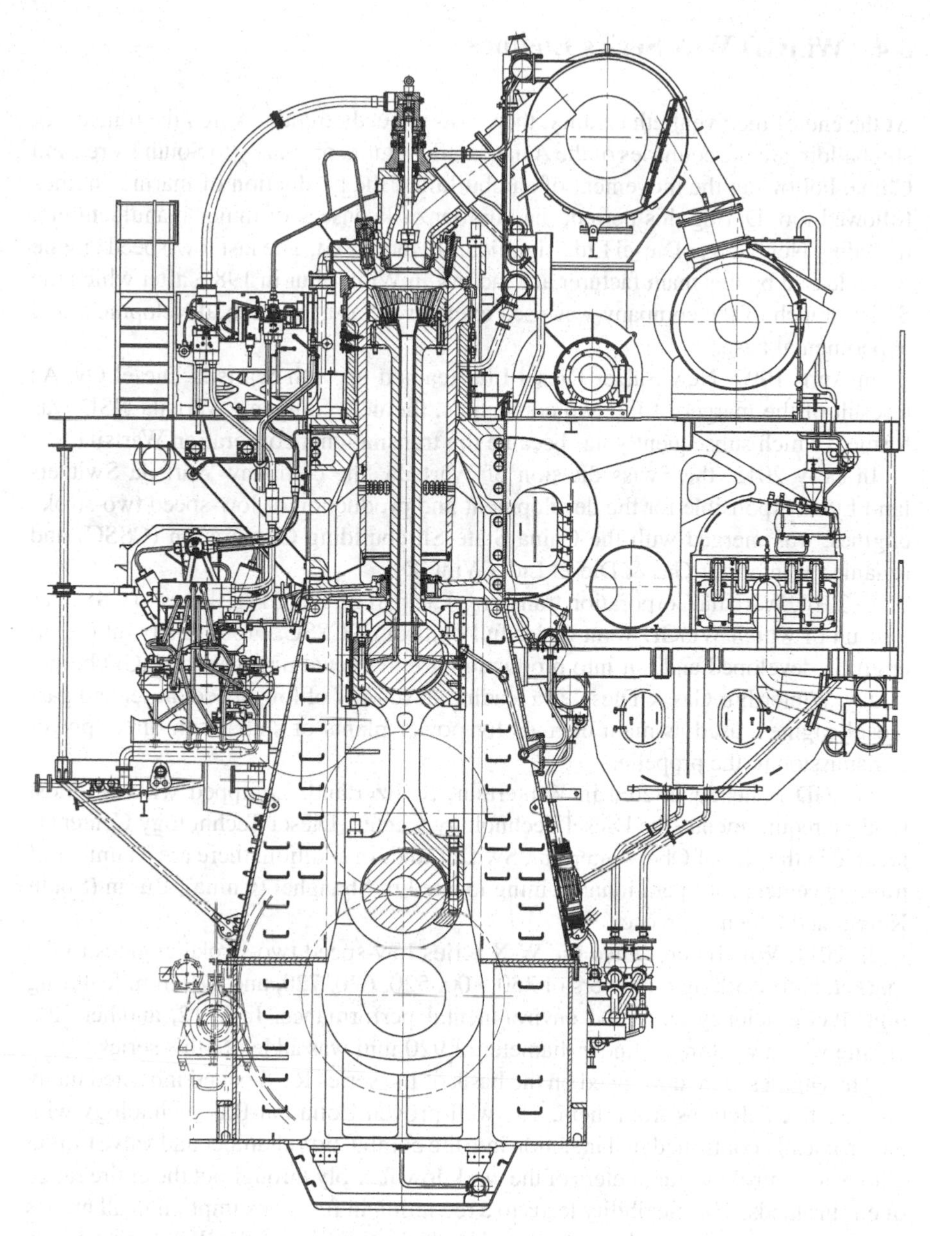

Fig. 3.60 Cross-section of the engine RT-flex 96C by Sulzer firm [59]

3.4 WinGD W-X Series Engines

At the end of the twentieth century, there was a steady trend towards the transfer of shipbuilding to the countries of the Asia-Pacific region, primarily to South Korea and China. Following the movement of shipbuilding, the production of marine engines followed too. During this period, the number of licensees of major manufacturers, including New Sulzer Diesel Ltd., significantly increased. The last low-speed engine was released by this manufacturer at a factory in Winterthur in 1986, after which the Swiss branch of the company was used exclusively as a research, development and experimental base.

In April 1997, New Sulzer Diesel Ltd. teamed up with Wärtsilä Diesel Oy. As a result of the merger of the two companies, the new company Wärtsilä NSD was formed, which subsequently has became the transnational corporation Wärtsilä.

In early 2015, the Swiss division of Wärtsilä, the company Wärtsilä Switzerland Ltd., responsible for the development and production of low-speed two-stroke engines, was merged with the China State Shipbuilding Corporation (CSSC) and renamed Winterthur Gas & Diesel Ltd. (WinGD).

In 2016, Wärtsilä Corporation transferred its part of WinGD's shares to CSSC, as a result of which WinGD went under full control of CSSC. From this point on, all engines, developed and put into production are marketed under the WinGD brand. Today, Winterthur Gas & Diesel is a leading developer of low-speed diesel and gas-diesel engines, used as main engines for power plants of ships with direct power transmission to the propeller.

WinGD is headquartered in Winterthur, Switzerland. Equipped with the most modern requirements, the Diesel Technology Center (Diesel Technology Center) is located in the city of Oberwinterthur, Switzerland. In addition, there are a number of training centers for operational training located in Shanghai (China), Busan (South Korea) and Athens (Greece).

In 2011, WinGD presented the W-X series low-speed two-stroke engines to the market, with working diameters of 350, 400, 520, 620, 720, and 820 mm, featuring high fuel efficiency and good environmental performance. In 2012, another X92 engine with a working cylinder diameter of 920 mm was added to this series.

The engines were developed on the basis of the series RT-flex and inherited many constructive solutions from them. The well-proven Common-Rail technology with electronically controlled fuel injection into the combustion chamber and valve timing allows to control the parameters of the workflow flexibly throughout the entire range of engine loads. This flexibility leads to a reduction in fuel consumption in all modes of operation, especially at low and partial loads. For engines of the W-X type, various settings for the working process (Standard, Delta, Delta By-Pass and Low Load) are available, which are carried out for specific customer requirements. The Intelligent Combustion Control (ICC) system provides additional fuel savings and balanced operation of each cylinder. Engines are fully compliant with NO_x Tier III NO_x emissions when equipped with a catalytic system (SCR).

The introduction of VI by the application of the MARPOL 73/78 Convention on the Energy Efficiency Index of Vessels (EEDI) aimed at reducing CO_2 emissions and the overall efficiency of the ship was taken into account when building W-X engines. To increase the energy efficiency of the engines, it is possible to use different power takeoffs to produce electricity on board the ship, as well as waste heat recovery regeneration (WHR), which provides additional opportunities to increase maximum energy efficiency and reduce emissions.

Main technical parameters of engines series X 35-B (Fig. 3.61)

Parameter	Value
Number and cylinders arrangement	5, 6, 7, 8 in-line
Cylinder bore (mm)	350
Piston stroke (mm)	1550
Cylinder capacity (dm^3)	149.13
Rotational speed (min^{-1})	167
Cylinder power, N_e (layout area)	
– maximum continuous rating at 167 min^{-1} (L_1) (kW)	870
– operational at 167 min^{-1} (L_2) (kW)	695
– maximum continuous rating at 118 min^{-1} (L_3) (kW)	615
– operational at 118 min^{-1} (L_4) (kW)	450
Air charging pressure (L_1) (MPa)	0.370
Compression pressure (L_1) (MPa)	16.2
Maximum cycle pressure (L_1) (MPa)	17.7
Exhaust gas temperature at inlet of turbocharger (L_1) (°C)	430.0
Exhaust gas temperature at outlet of turbocharger (L_1) (°C)	251.0
Mean effective pressure	
– at 167 min^{-1} (L_1/L_2) (MPa)	2.10/1.68
– at 118 min^{-1} (L_3/L_4) (MPa)	2.10/1.68
Brake specific fuel oil consumption (g/kWh)	
– at 167 min^{-1} (L_1/L_2) for N_e 100/75% (g/kWh)	174.8/168.8
– at 118 min^{-1} (L_3/L_4) for N_e 100/75% (g/kWh)	174.8/168.8
Mean piston speed at 167 min^{-1} (m/s)	8.63
Brake specific air consumption (kg/kWh)	7.55
Brake specific exhaust gas flow (kg/kWh)	7.71
Cylinder oil consumption (g/kWh)	0.6–1.30
System oil consumption per cylinder and per day (kg)	2.5

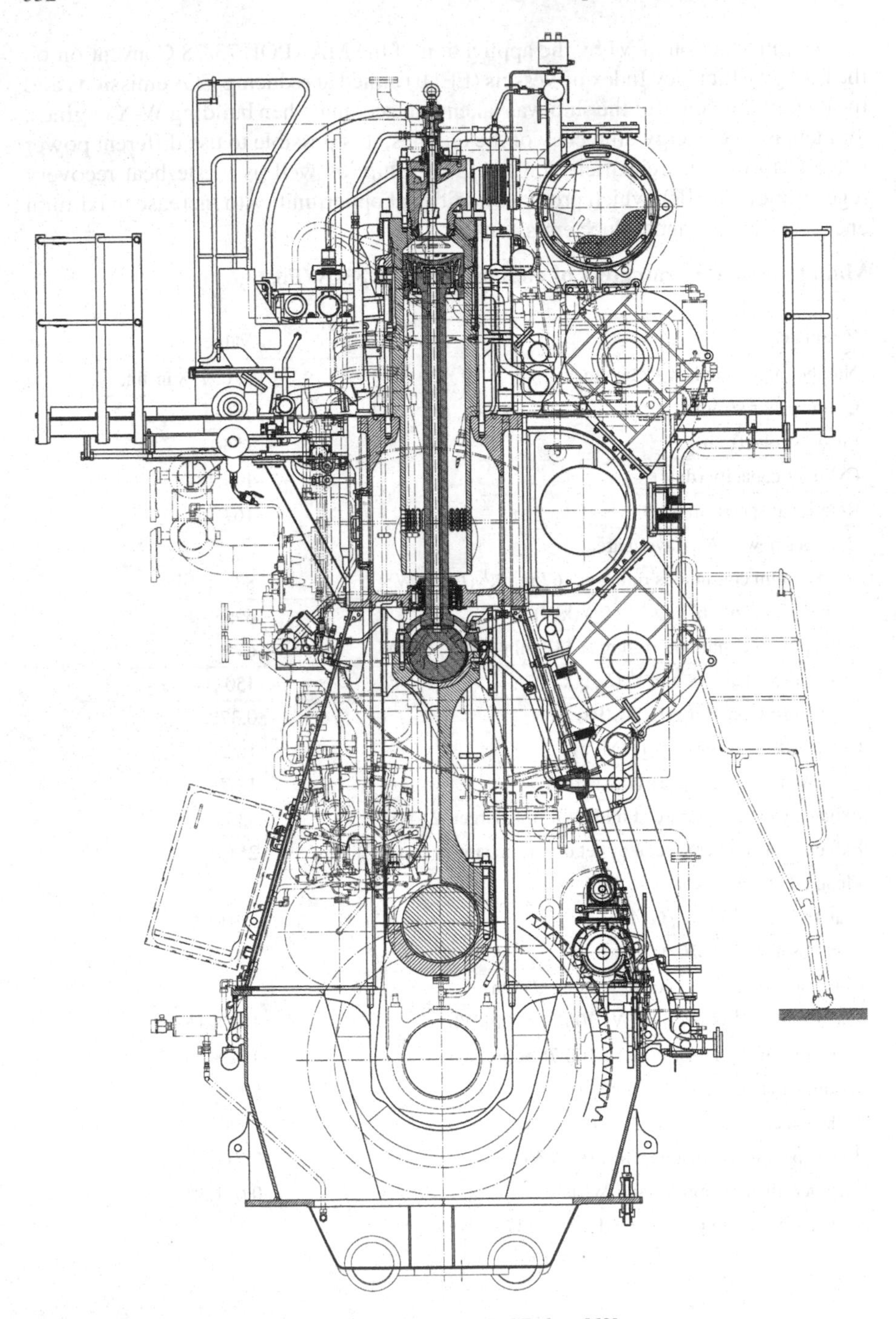

Fig. 3.61 Cross-section of the engine X-35-B by WinGD firm [60]

Dimensions and weight engines X 35-B

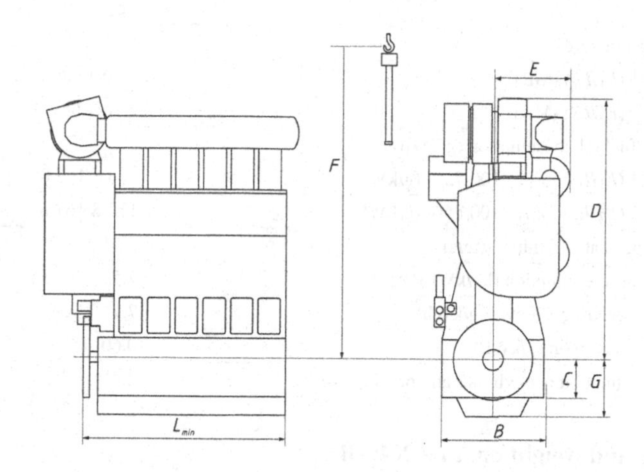

Number of cylinders	5	6	7	8
L_{min}	3838	4450	5062	5674
Weight (kg)	74,000	84,000	95,000	105,000

B (mm)	C (mm)	D (mm)	E (mm)	F (mm)	G (mm)
2284	830	5556	1605	6850	1326

Main technical parameters of engines series X40-B (Fig. 3.62)

Parameter	Valuc
Number and cylinders arrangement	5, 6, 7, 8 in-line
Cylinder bore (mm)	400
Piston stroke (mm)	1770
Cylinder capacity (dm^3)	222.42
Rotational speed (min^{-1})	146
Cylinder power, N_e (layout area)	
– maximum continuous rating at 146 min^{-1} (L_1) (kW)	1135
– operational at 146 min^{-1} (L_2) (kW)	910
– maximum continuous rating at 104 min^{-1} (L_3) (kW)	810
– operational at 104 min^{-1} (L_4) (kW)	650
Air charging pressure (L_1) (MPa)	0.354
Compression pressure (L_1) (MPa)	15.20
Maximum cycle pressure (L_1) (MPa)	17.20
Exhaust gas temperature at inlet of turbocharger (L_1) (°C)	428.0
Exhaust gas temperature at outlet of turbocharger (L_1) (°C)	247.0

(continued)

(continued)

Parameter	Value
Mean effective pressure	
– at 146 min^{-1} (L_1/L_2) (MPa)	2.10/1.68
– at 104 min^{-1} (L_3/L_4) (MPa)	2.10/1.68
Brake specific fuel oil consumption (g/kWh)	
– at 146 min^{-1} (L_1/L_2) for N_e 100/75% (g/kWh)	173.8/167.8
– at 104 min^{-1} (L_3/L_4) for N_e 100/75% (g/kWh)	173.8/167.8
Mean piston speed at 146 min^{-1} (m/s)	8.61
Brake specific air consumption (kg/kWh)	7.53
Brake specific exhaust gas flow (kg/kWh)	7.73
Cylinder oil consumption (g/kWh)	0.60–1.3
System oil consumption per cylinder and per day (kg)	2.8

Dimensions and weight engines X40-B

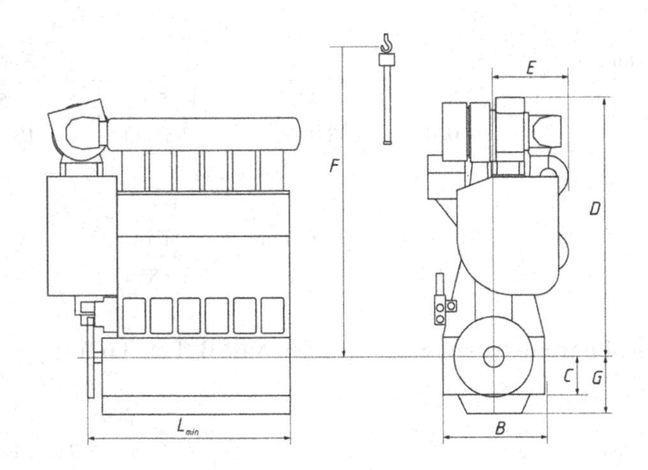

Number of cylinders	5	6	7	8
L_{min}	4390	5090	5790	6490
Weight (kg)	109,000	125,000	140,000	153,000

B (mm)	C (mm)	D (mm)	E (mm)	F (mm)	G (mm)
2610	950	6344	1647	7750	1411

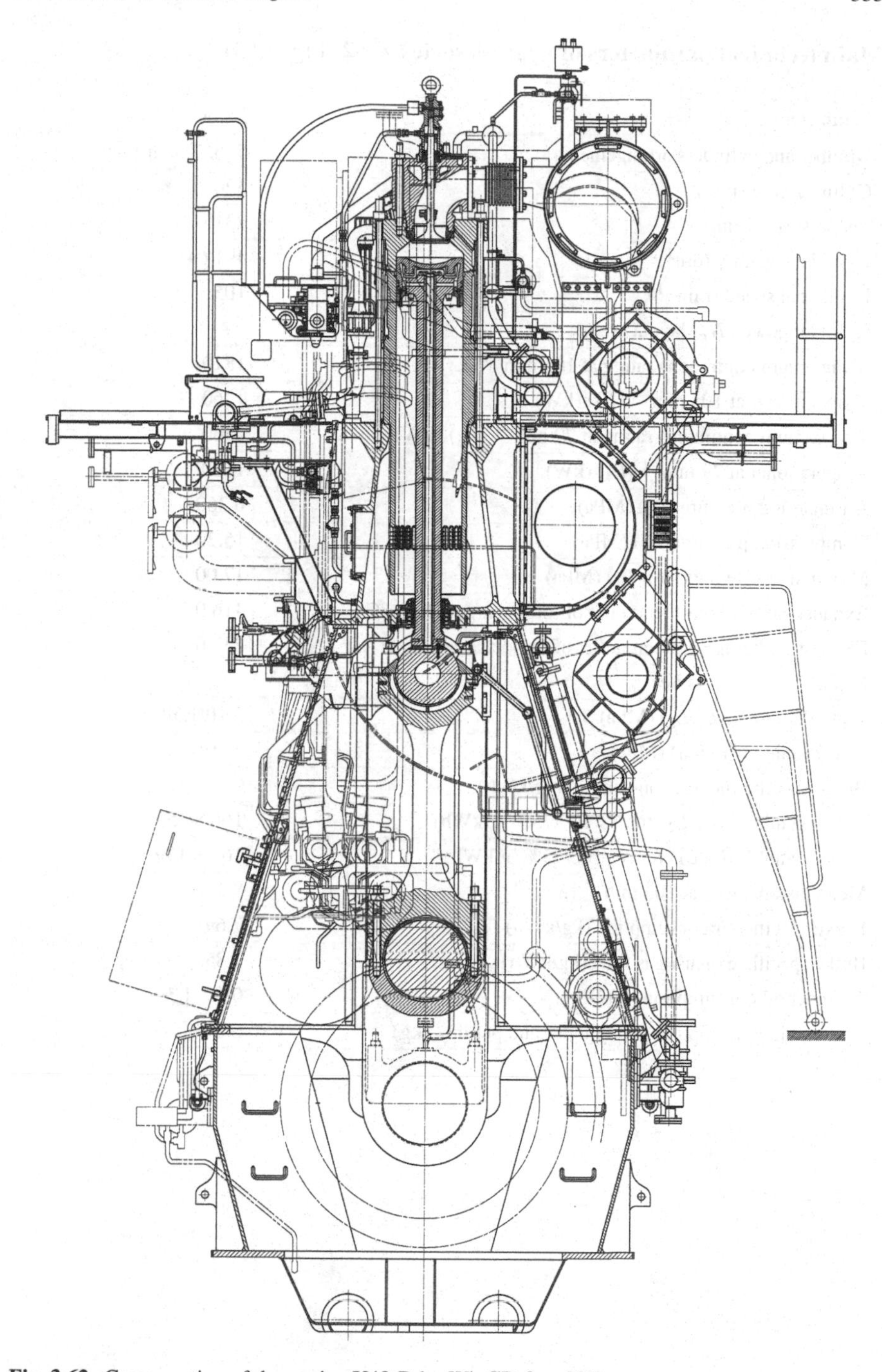

Fig. 3.62 Cross-section of the engine X40-B by WinGD firm [61]

Main technical parameters of engines series X 52 (Fig. 3.63)

Parameter	Value
Number and cylinders arrangement	5, 6, 7, 8 in-line
Cylinder bore (mm)	520
Piston stroke (mm)	2315
Cylinder capacity (dm^3)	491.64
Rotational speed (min^{-1})	105
Cylinder power, N_e (layout area)	
– maximum continuous rating at 105 min^{-1} (L_1) (kW)	1810
– operational at 105 min^{-1} (L_2) (kW)	1360
– maximum continuous rating at 79 min^{-1} (L_3) (kW)	1360
– operational at 79 min^{-1} (L_4) (kW)	1020
Air charging pressure (L_1) (MPa)	0.360
Compression pressure (L_1) (MPa)	15.20
Maximum cycle pressure (L_1) (MPa)	17.00
Exhaust gas temperature at inlet of turbocharger (L_1) (°C)	416.0
Exhaust gas temperature at outlet of turbocharger (L_1) (°C)	237.0
Mean effective pressure	
– at 105 min^{-1} (L_1/L_2) (MPa)	2.10/1.58
– at 79 min^{-1} (L_3/L_4) (MPa)	2.10/1.58
Brake specific fuel oil consumption (g/kWh)	
– at 105 min^{-1} (L_1/L_2) for N_e 100/75% (g/kWh)	166.8/159.8
– at 79 min^{-1} (L_3/L_4) for N_e 100/75% (g/kWh)	166.8/159.8
Mean piston speed at 105 min^{-1} (m/s)	8.1
Brake specific air consumption (kg/kWh)	7.69
Brake specific exhaust gas flow (kg/kWh)	7.85
Cylinder oil consumption (g/kWh)	0.60–1.3
System oil consumption per cylinder and per day (kg)	6.0

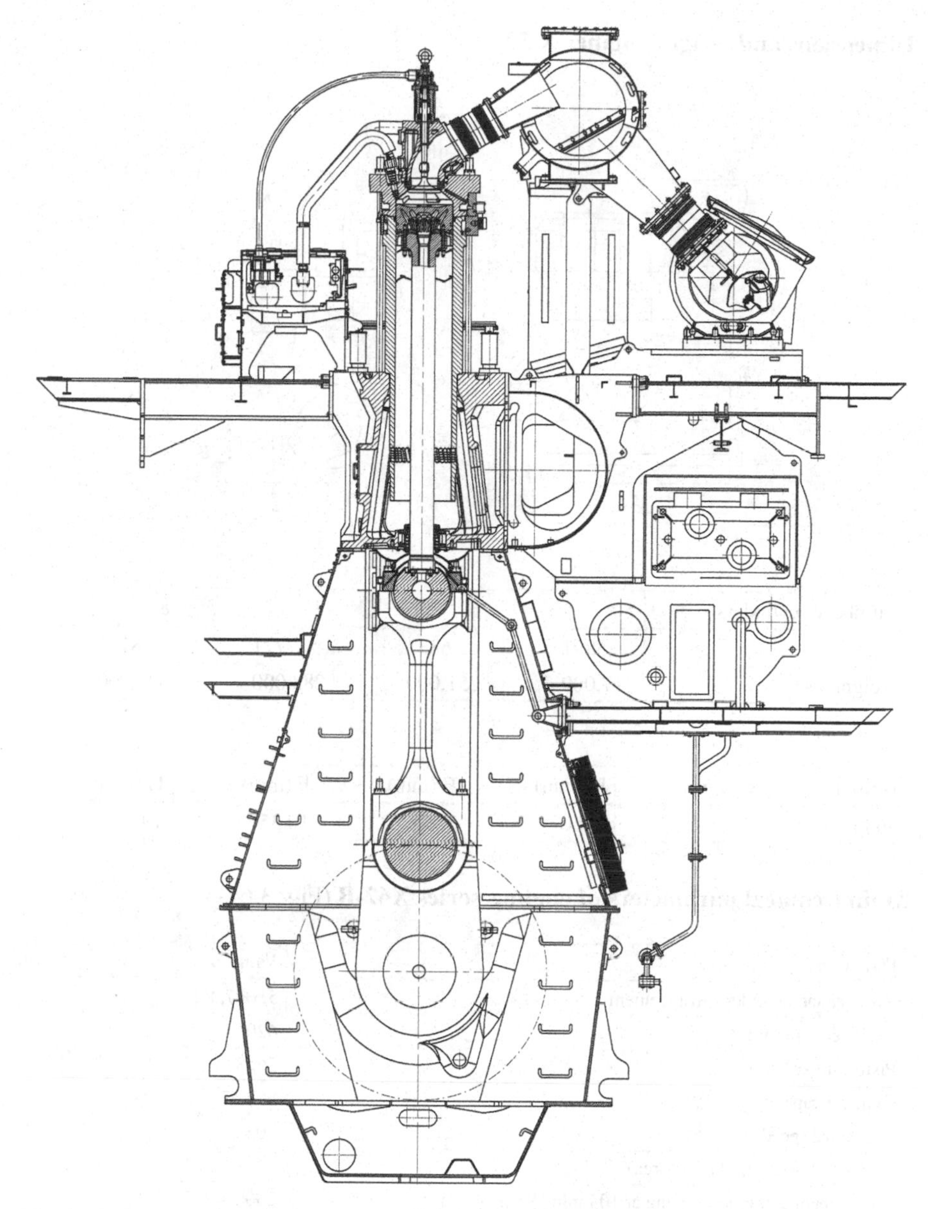

Fig. 3.63 Cross-section of the engine X 52 by WinGD firm [62]

Dimensions and weight engines X 52

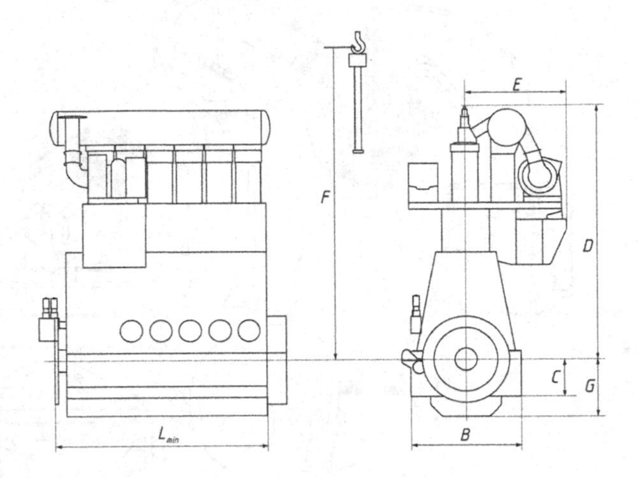

Number of cylinders	5	6	7	8
L_{min}	5891	6831	7771	8711
Weight (kg)	217,000	251,000	288,000	323,000

B (mm)	C (mm)	D (mm)	E (mm)	F (mm)	G (mm)
3630	1205	8550	3555	10,350	1910

Main technical parameters of engines series X62-B (Fig. 3.64)

Parameter	Value
Number and cylinders arrangement	5, 6, 7, 8 in-line
Cylinder bore (mm)	620
Piston stroke (mm)	2658
Cylinder capacity (dm^3)	802.47
Rotational speed (min^{-1})	103
Cylinder power, N_e (layout area)	
– maximum continuous rating at 103 min^{-1} (L_1) (kW)	2900
– operational at 103 min^{-1} (L_2) (kW)	2130
– maximum continuous rating at 77 min^{-1} (L_3) (kW)	2160
– operational at 77 min^{-1} (L_4) (kW)	1590
Air charging pressure (L_1) (MPa)	0.360
Compression pressure (L_1) (MPa)	14.2
Maximum cycle pressure (L_1) (MPa)	16.5
Exhaust gas temperature at inlet of turbocharger (L_1) (°C)	424.0
Exhaust gas temperature at outlet of turbocharger (L_1) (°C)	244.0

(continued)

(continued)

Parameter	Value
Mean effective pressure	
– at 103 min^{-1} (L_1/L_2) (MPa)	2.10/1.54
– at 77 min^{-1} (L_3/L_4) (MPa)	2.10/1.54
Brake specific fuel oil consumption (g/kWh)	
– at 103 min^{-1} (L_1/L_2) for N_e 100/75% (g/kWh)	166.8/159.3
– at 77 min^{-1} (L_3/L_4) for N_e 100/75% (g/kWh)	166.8/159.3
Mean piston speed at 103 min^{-1} (m/s)	9.13
Brake specific air consumption (kg/kWh)	7.31
Brake specific exhaust gas flow (kg/kWh)	7.53
Cylinder oil consumption (g/kWh)	0.60–1.30
System oil consumption per cylinder and per day (kg)	6.0

Dimensions and weight engines X62-B

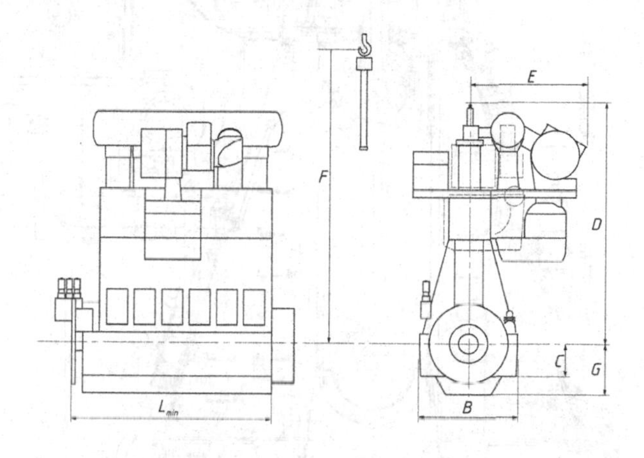

Number of cylinders	5	6	7	8
L_{min}	7000	8110	9215	10,320
Weight (kg)	325,000	377,000	435,000	482,000

B (mm)	C (mm)	D (mm)	E (mm)	F (mm)	G (mm)
4200	1360	9580	3915	11,775	2110

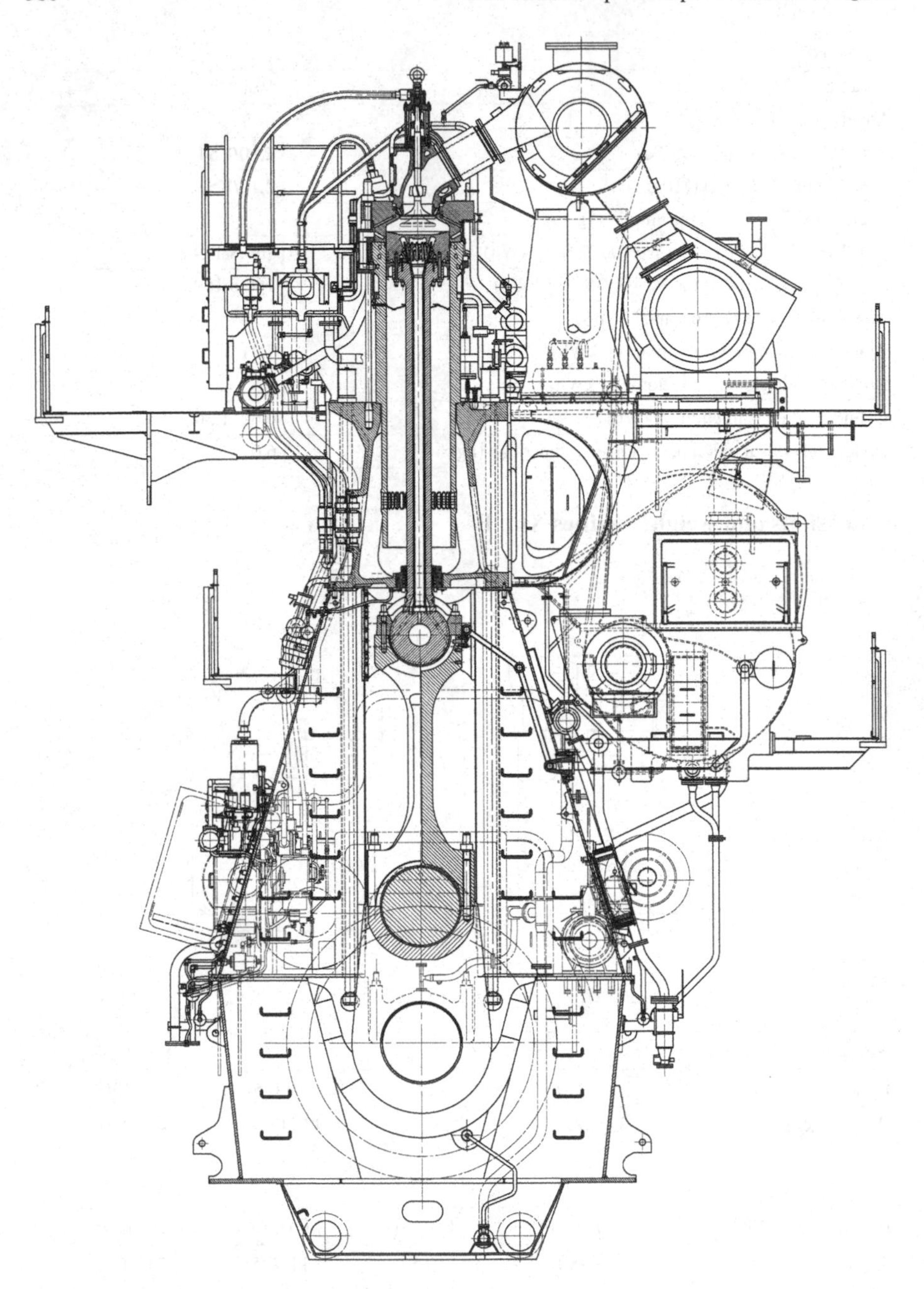

Fig. 3.64 Cross-section of the engine X62 by WinGD firm [63]

Main technical parameters of engines series X72-B (Fig. 3.65)

Parameter	Value
Number and cylinders arrangement	5, 6, 7, 8 in-line
Cylinder bore (mm)	720
Piston stroke (mm)	3086
Cylinder capacity (dm^3)	1256.47
Rotational speed (min^{-1})	89
Cylinder power, N_e (layout area)	
– maximum continuous rating at 89 min^{-1} (L_1) (kW)	3920
– operational at 89 min^{-1} (L_2) (kW)	2860
– maximum continuous rating at 66 min^{-1} (L_3) (kW)	2910
– operational at 66 min^{-1} (L_4) (kW)	2120
Air charging pressure (L_1) (MPa)	0.360
Compression pressure (L_1) (MPa)	12.20
Maximum cycle pressure (L_1) (MPa)	16.20
Exhaust gas temperature at inlet of turbocharger (L_1) (°C)	427
Exhaust gas temperature at outlet of turbocharger (L_1) (°C)	246
Mean effective pressure	
– at 89 min^{-1} (L_1/L_2) (MPa)	2.10/1.54
– at 66 min^{-1} (L_3/L_4) (MPa)	2.10/1.54
Brake specific fuel oil consumption (g/kWh)	
– at 89 min^{-1} (L_1/L_2) for N_e 100/75% (g/kWh)	166.8/159.3
– at 66 min^{-1} (L_3/L_4) for N_e 100/75% (g/kWh)	166.8/159.3
Mean piston speed at 89 min^{-1} (m/s)	9.16
Brake specific air consumption (kg/kWh)	7.47
Brake specific exhaust gas flow (kg/kWh)	7.63
Cylinder oil consumption (g/kWh)	0.6–1.14
System oil consumption per cylinder and per day (kg)	8.0

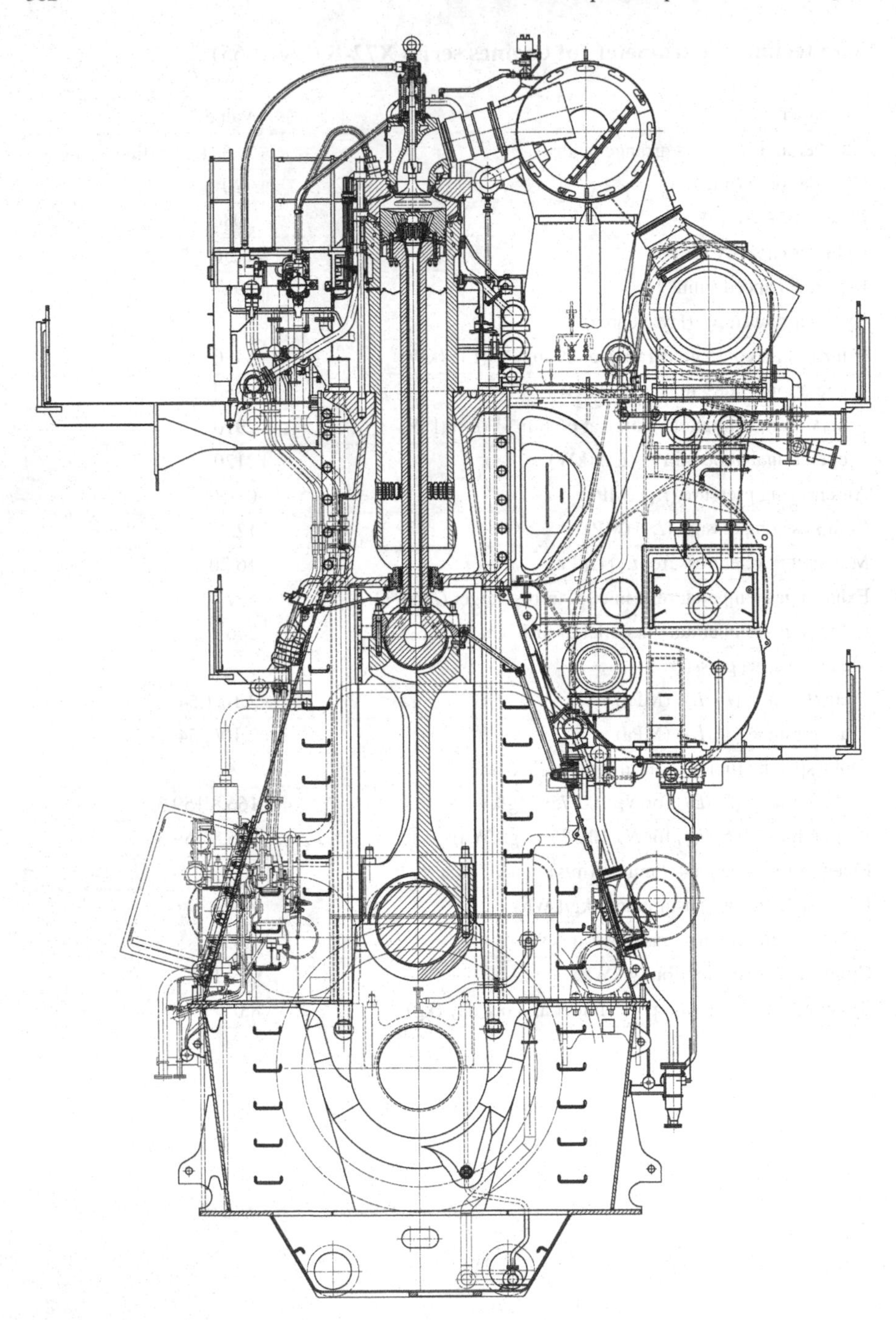

Fig. 3.65 Cross-section of the engine X 72-B by WinGD firm [64]

Dimensions and weight engines X72-B

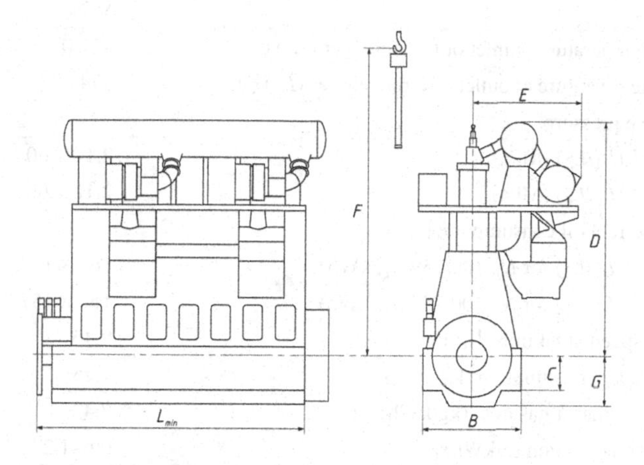

Number of cylinders	5	6	7	8
L_{min}	8085	9375	10,665	11,960
Weight (kg)	481,000	561,000	642,000	716,000

B (mm)	C (mm)	D (mm)	E (mm)	F (mm)	G (mm)
4780	1575	10,790	4710	13,655	2455

Main technical parameters of engines series X 82-B (Fig. 3.66)

Parameter	Valuc
Number and cylinders arrangement	6, 7, 8, 9 in-line
Cylinder bore (mm)	820
Piston stroke (mm)	3375
Cylinder capacity (dm^3)	1782.34
Rotational speed (min^{-1})	84
Cylinder power, N_e (layout area)	
– maximum continuous rating at 84 min^{-1} (L_1) (kW)	4750
– operational at 84 min^{-1} (L_2) (kW)	3620
– maximum continuous rating at 58 min^{-1} (L_3) (kW)	3625
– operational at 58 min^{-1} (L_4) (kW)	2765
Air charging pressure (L_1) (MPa)	0.342
Compression pressure (L_1) (MPa)	12.60
Maximum cycle pressure (L_1) (MPa)	15.80

(continued)

(continued)

Parameter	Value
Exhaust gas temperature at inlet of turbocharger (L_1) (°C)	422.0
Exhaust gas temperature at outlet of turbocharger (L_1) (°C)	274.0
Mean effective pressure	
– at 84 min^{-1} (L_1/L_2) (MPa)	2.10/1.60
– at 58 min^{-1} (L_3/L_4) (MPa)	2.10/1.60
Brake specific fuel oil consumption (g/kWh)	
– at 84 min^{-1} (L_1/L_2) for N_e 100/75% (g/kWh)	164.8/157.8
– at 58 min^{-1} (L_3/L_4) for N_e 100/75% (g/kWh)	164.8/157.8
Mean piston speed at 84 min^{-1} (m/s)	9.45
Brake specific air consumption (kg/kWh)	7.25
Brake specific exhaust gas flow (kg/kWh)	7.42
Cylinder oil consumption (g/kWh)	0.6–1.2
System oil consumption per cylinder and per day (kg)	9.0

Dimensions and weight engines X 82-B

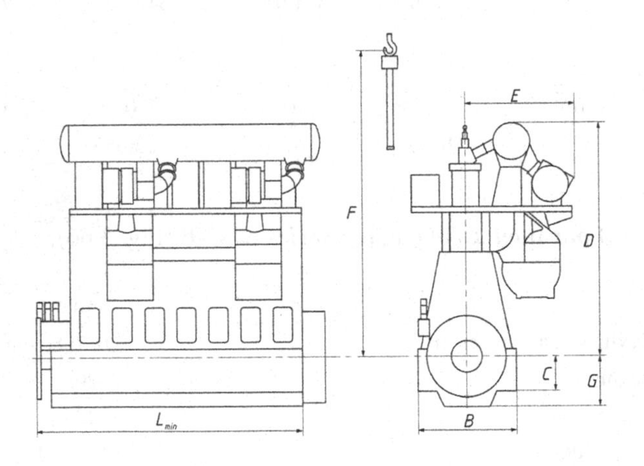

Number of cylinders	6	7	8	9
L_{min}	11,045	12,550	14,055	16,500
Weight (kg)	805,000	910,000	1,020,000	1,160,000

B (mm)	C (mm)	D (mm)	E (mm)	F (mm)	G (mm)
5320	1800	12,250	5400	14,820	27,005

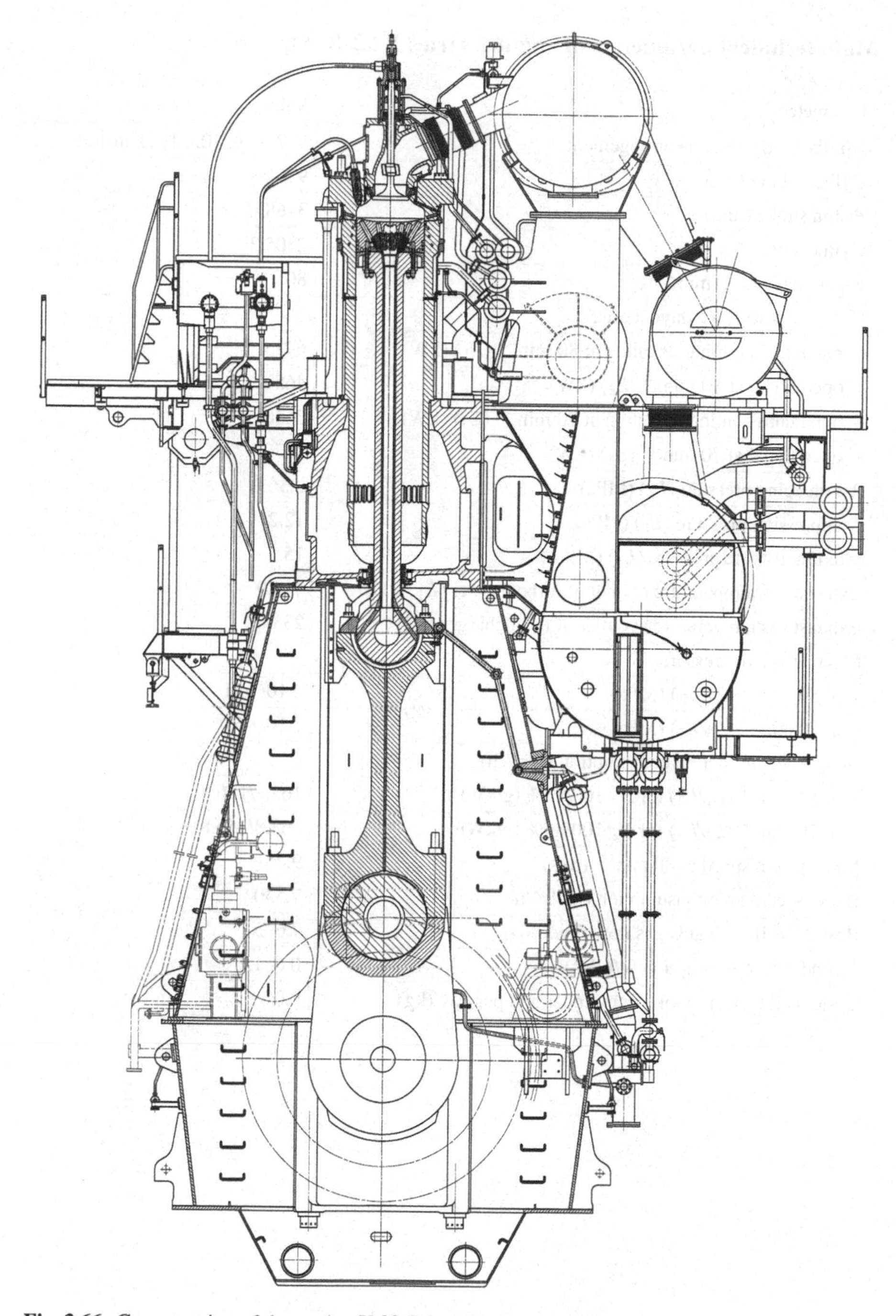

Fig. 3.66 Cross-section of the engine X 82-B by WinGD firm [65]

Main technical parameters of engines series X 92-B (Fig. 3.67)

Parameter	Value
Number and cylinders arrangement	6, 7, 8, 9, 10, 11, 12 in-line
Cylinder bore (mm)	920
Piston stroke (mm)	3468
Cylinder capacity (dm^3)	2305.39
Rotational speed (min^{-1})	80
Cylinder power, N_e (layout area)	
– maximum continuous rating at 80 min^{-1} (L_1) (kW)	6450
– operational at 80 min^{-1} (L_2) (kW)	4650
– maximum continuous rating at 70 min^{-1} (L_3) (kW)	5650
– operational at 70 min^{-1} (L_4) (kW)	4070
Air charging pressure (L_1) (MPa)	0.363
Compression pressure (L_1) (MPa)	12.2
Maximum cycle pressure (L_1) (MPa)	15.5
Exhaust gas temperature at inlet of turbocharger (L_1) (°C)	413.0
Exhaust gas temperature at outlet of turbocharger (L_1) (°C)	233.0
Mean effective pressure	
– at 80 min^{-1} (L_1/L_2) (MPa)	2.10/1.51
– at 70 min^{-1} (L_3/L_4) (MPa)	2.10/1.51
Brake specific fuel oil consumption (g/kWh)	
– at 80 min^{-1} (L_1/L_2) for N_e 100/85% (g/kWh)	163.8/156.8
– at 70 min^{-1} (L_3/L_4) for N_e 100/85% (g/kWh)	163.8/156.8
Mean piston speed at 80 min^{-1} (m/s)	9.25
Brake specific air consumption (kg/kWh)	7.530
Brake specific exhaust gas flow (kg/kWh)	7.695
Cylinder oil consumption (g/kWh)	0.6–1.3
System oil consumption per cylinder and per day (kg)	9.0

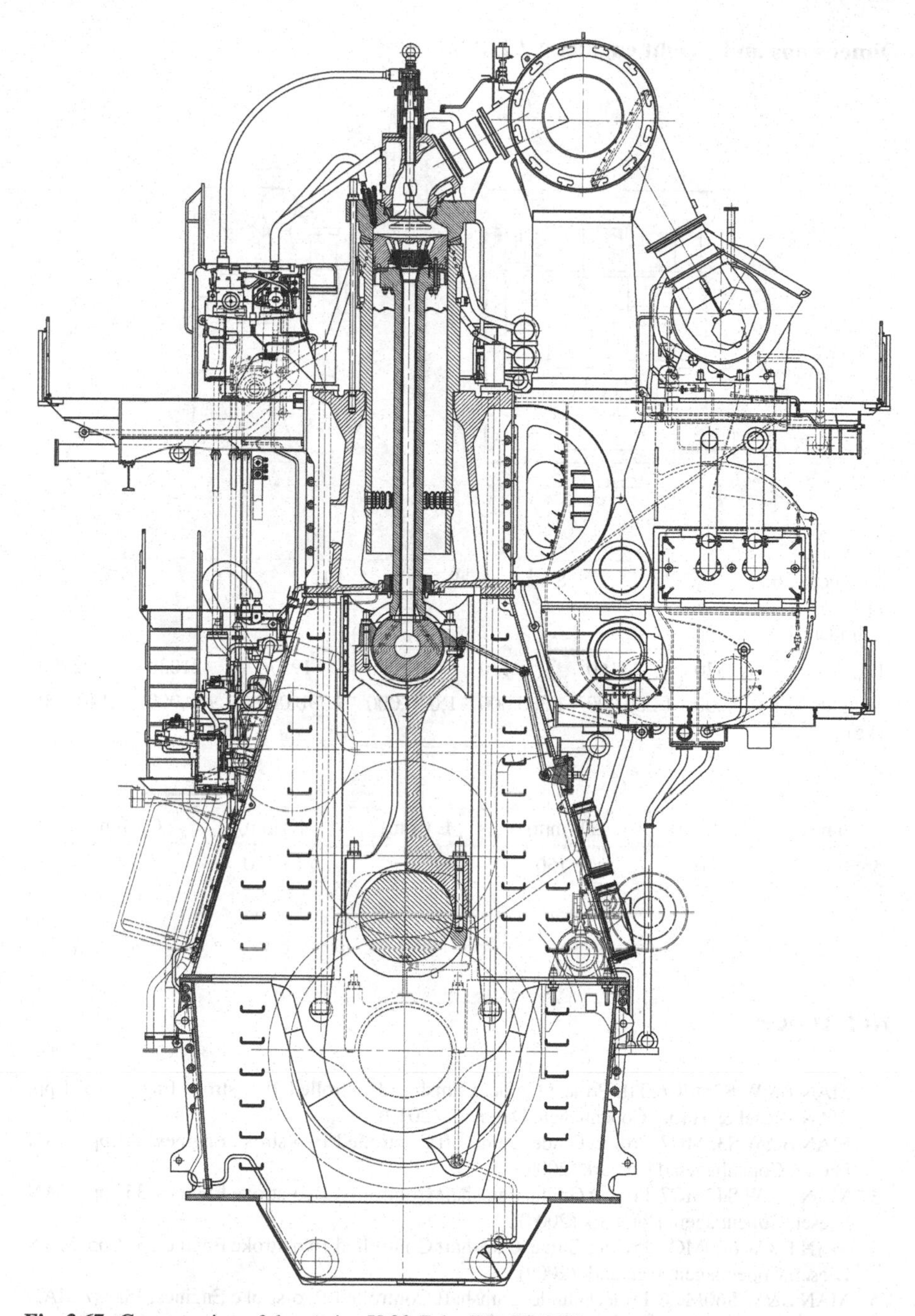

Fig. 3.67 Cross-section of the engine X 92-B by WinGD [66]

Dimensions and weight engines X 92-B

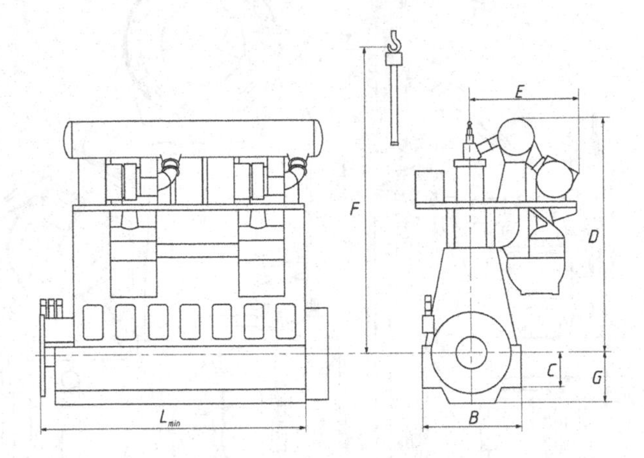

Number of cylinders	6	7	8	9	10	11	12
L_{min}	11570	13160	14750	17780	19370	21030	22700
Weight (kg)	1,120,000	1,260,000	1,380,000	1,630,000	1,790,000	1,960,000	2,140,000

B (mm)	C (mm)	D (mm)	E (mm)	F (mm)	G (mm)
5550	1900	13,150	6050	15,420	2970

References

1. MAN B&W S26MC6-TII. Project Guide. Camshaft Controlled Two-stroke Engines, 371 pp. MAN Diesel & Turbo, Copenhagen, Denmark (2010)
2. MAN B&W S35MC7. Project Guide. Camshaft Controlled Two-stroke Engines, 351 pp. MAN Diesel, Copenhagen, Denmark (2009)
3. MAN B&W S42MC7. Project Guide. Camshaft Controlled Two-stroke Engines, 353 pp. MAN Diesel, Copenhagen, Denmark (2009)
4. MAN B&W S50MC6. Project Guide. Camshaft Controlled Two-stroke Engines, 371 pp. MAN Diesel, Copenhagen, Denmark (2009)
5. MAN B&W S60MC6. Project Guide. Camshaft Controlled Two-stroke Engines, 383 pp. MAN Diesel, Copenhagen, Denmark (2009)
6. MAN B&W S70MC6. Project Guide. Camshaft Controlled Two-stroke Engines, 363 pp. MAN Diesel, Copenhagen. Denmark (2009)
7. MAN B&W S80MC6. Project Guide. Camshaft Controlled Two-stroke Engines, 361 pp. MAN Diesel, Copenhagen, Denmark (2009)

8. MAN B&W L35MC6. Project Guide. Camshaft Controlled Two-stroke Engines, 351 pp. MAN Diesel, Copenhagen, Denmark (2009)
9. L42MC Mk 6. Project Guide. Two-stroke Engines, 240 pp. MAN B&W Diesel A/S, Copenhagen, Denmark (1999)
10. L80MC Mk 6. Project Guide. Two-stroke Engines, 246 pp. MAN B&W Diesel A/S, Copenhagen, Denmark (1998)
11. K98MC Project Guide. Project Guide. Two-stroke Engines, 243 pp. MAN B&W Diesel A/S, Copenhagen, Denmark (1999)
12. MAN B&W S46MC-C7. Project Guide. Camshaft Controlled Two-stroke Engines, 369 pp. MAN B&W Diesel, Copenhagen, Denmark (2009)
13. MAN B&W S50MC-C7. Project Guide. Camshaft Controlled Two-stroke Engines, 393 pp. MAN B&W Diesel, Copenhagen, Denmark (2009)
14. MAN B&W S60MC-C7. Project Guide.Camshaft Controlled Two-stroke Engines, 399 pp. MAN B&W Diesel, Copenhagen, Denmark (2009)
15. MAN B&W S70MC-C7. Project Guide. Camshaft Controlled Two-stroke Engines, 389 pp. MAN B&W Diesel, Copenhagen, Denmark (2009)
16. MAN B&W S80MC-C7. Project Guide. Camshaft Controlled Two-stroke Engines, 375 pp. MAN B&W Diesel, Copenhagen, Denmark (2009)
17. MAN B&W S90MC-C7. Project Guide. Camshaft Controlled Two-stroke Engines, 361 pp. MAN B&W Diesel, Copenhagen, Denmark (2009)
18. MAN B&W L60MC-C7. Project Guide. Camshaft Controlled Two-stroke Engines, 371 pp. MAN B&W Diesel, Copenhagen, Denmark (2009)
19. MAN B&W L70MC-C7. Project Guide. Camshaft Controlled Two-stroke Engines, 383 pp. MAN B&W Diesel, Copenhagen, Denmark (2009)
20. L90MC-C MC Project Guide. Two-stroke Engines, 245 pp. MAN B&W Diesel A/S, Copenhagen, Denmark (1999)
21. MAN B&W K80MC-C6-TII. Project Guide Camshaft Controlled Two-stroke Engines, 391 pp. MAN Diesel & Turbo, Copenhagen, Denmark (2010)
22. MAN B&W K90MC-C6. Project Guide. Camshaft Controlled Two-stroke Engines, 375 pp. MAN Diesel, Copenhagen. Denmark (2009)
23. MAN B&W K98MC-C6. Project Guide. Camshaft Controlled Two-stroke Engines, 383 pp.MAN Diesel, Copenhagen, Denmark (2009)
24. MAN B&W S30ME-B9.5-TII. Project Guide. Electronically Controlled Two-stroke Engines with Camshaft Controlled Exhaust Valves, 327 pp. MAN Diesel & Turbo, Copenhagen, Denmark (2014)
25. MAN B&W S46ME-B8.5-TII. Project Guide. Electronically Controlled Two-stroke Engines with Camshaft Controlled Exhaust Valve, 289 pp. MAN Diesel & Turbo, Copenhagen, Denmark (2014)
26. MAN B&W S50ME-B9.5-TII. Project Guide. Electronically Controlled Two-stroke Engines with Camshaft Controlled Exhaust Valves, 304 pp. MAN Diesel & Turbo, Copenhagen, Denmark (2014)
27. MAN B&W S50ME-C7. Project Guide. Electronically Controlled Two-stroke Engines, 363 pp. MAN Diesel, Copenhagen, Denmark (2009)
28. MAN B&W S60MC-C8.2-TII. Project Guide. Electronically Controlled Two-stroke Engines, 351 pp. MAN Diesel & Turbo, Copenhagen, Denmark (2014)
29. MAN B&W S65ME-C8.5-TII. Project Guide. Electronically Controlled Two-stroke Engines, 343 pp.MAN Diesel & Turbo, Copenhagen, Denmark (2014)
30. MAN B&W S70ME-C8.5-TII. Project Guide. Electronically Controlled Two-stroke Engines, 365 pp. MAN Diesel & Turbo, Copenhagen, Denmark (2014)
31. MAN B&W S80ME-C9.5-TII. Project Guide. Electronically Controlled Two-stroke Engines, 343 pp. MAN Diesel & Turbo. Copenhagen, Denmark (2014)
32. MAN B&W S90ME-C10.5-TII. Project Guide. Electronically Controlled Two-stroke Engines, 345 pp. MAN Diesel & Turbo, Copenhagen, Denmark (2014)

33. MAN B&W L70ME-C8.5-TII. Project Guide. Electronically Controlled Two-stroke Engines, 367 pp. MAN Diesel & Turbo, Copenhagen, Denmark (2014)
34. MAN B&W K98ME-C6. Project Guide. Electronically Controlled Two-stroke Engines, 345 pp. MAN Diesel, Copenhagen, Denmark (2009)
35. MAN B&W G50ME-B9.3-TII. Project Guide. Electronically Controlled Two-stroke Engines with Camshaft Controlled Exhaust Valves, 347 pp. MAN Diesel & Turbo, Copenhagen, Denmark (2014)
36. MAN B&W G60ME-C9.2-TII. Project Guide. Electronically Controlled Two-stroke Engines, 375 pp. MAN Diesel & Turbo, Copenhagen, Denmark (2013)
37. MAN B&W G70ME-C9.2-TII. Project Guide. Electronically Controlled Two-stroke Engines, 377 pp. MAN Diesel & Turbo, Copenhagen, Denmark (2013)
38. MAN B&W G80ME-C9.5-TII. Project Guide. Electronically Controlled Two-stroke Engines, 396 pp. MAN Diesel & Turbo, Copenhagen, Denmark (2018)
39. UEC35LSE-Eco-B2. Technical Data, 14 pp. Mitsubishi Heavy Industries Marine Machinery & Engine Co., Ltd.
40. UEC45LSE-Eco-B2. Technical Data, 226 pp. Mitsubishi Heavy Industries Marine Machinery & Engine Co., Ltd.
41. UEC50LSH-Eco-C2. Technical Data, 228 pp. Mitsubishi Heavy Industries Marine Machinery & Engine Co., Ltd.
42. UEC60LSE-Eco-B1. Technical Data, 34 pp. Mitsubishi Heavy Industries Marine Machinery & Engine Co., Ltd.
43. UEC80LSE-Eco-B1. Technical Data, 18 pp. Mitsubishi Heavy Industries Marine Machinery & Engine Co., Ltd.
44. Sulzer RTA 48T Spare Parts Code Book "Marine", 1005 pp. Wärtsilä NSD, Winterthur, Switzerland (1999)
45. Operating Instructions for Sulzer Diesel Engine RTA52U "Marine", 282 pp. New Sulzer Diesel Limited, Winterthur, Switzerland
46. Sulzer RTA58T Operating Manual "Marine", 999 pp. Wärtsilä NSD Switzerland, Winterthur, Switzerland (1998)
47. Sulzer RTA68T-B Spare Parts Code Book "Marine", 2004 pp. Wärtsilä Switzerland Ltd., Winterthur, Switzerlan (2005)
48. Sulzer RTA72U-B Maintenance Manual "Marine", 396 pp. Wärtsilä Switzerland Ltd., Winterthur, Switzerland (2000)
49. Sulzer RTA84C Maintenance Manual "Marine", 388 pp. Wärtsilä Switzerland Ltd., Winterthur, Switzerland (2001)
50. Sulzer RTA84T-B Maintenance Manual "Marine", 373 pp. Wärtsilä Switzerland Ltd., Winterthur, Switzerland (2004)
51. Sulzer RTA96C Maintenance Manual "Marine", 354 pp. Wärtsilä Switzerland Ltd., Winterthur, Switzerland (2004)
52. Wärtsilä RT-flex 35, 4 pp. Wärtsilä Corporation (2009)
53. Wärtsilä RT-flex 48 TD. Operation Manual "Marine". Pulse Lubrication, 444 pp. Wärtsilä Switzerland Ltd., Winterthur, Switzerland (2012)
54. Wärtsilä RT-flex 50-B. Operation Manual "Marine" (Pulse Lubrication), 412 pp. Wärtsilä Switzerland Ltd., Winterthur, Switzerland (2010)
55. Wärtsilä RT-flex 58T-B. Operation Manual "Marine", 484 pp. Wärtsilä Switzerland Ltd., Winterthur, Switzerland (2006)
56. Wärtsilä RT-flex 60C. Maintenance Manual "Marine" (WECS-9520/Mk II), 538 pp. Wärtsilä Switzerland Ltd., Winterthur, Switzerland (2011)
57. Wärtsilä RT-flex 68-B. Maintenance Manual "Marine" (with CLU-3 Lubrication), 418 pp. Wärtsilä Switzerland Ltd., Winterthur, Switzerland (2010)
58. Wärtsilä RT-flex82C. Maintenance Manual "Marine", 437 pp., Wärtsilä Switzerland Ltd., Winterthur, Switzerland (2008)
59. Wärtsilä RT-flex96C. Maintenance Manual "Marine" (with Dynex Servo Oil Pumps), 512 pp. Wärtsilä Switzerland Ltd., Winterthur, Switzerland (2006)

60. Wärtsilä W-X35. Maintenance Manual « Marine » Document ID: BAC352248. Winterthur Gas & Diesel Ltd. Winterthur, Switzerland. 2015. – 488 p
61. WinGD X40-B. Maintenance Manual "Marine" Document ID: DBAD802920, 494 pp. Winterthur Gas & Diesel Ltd., Winterthur, Switzerland (2016)
62. WinGD X52. Maintenance Manual, 1120 pp. Winterthur Gas & Diesel Ltd., Winterthur, Switzerland. Issue 001 (2017)
63. WinGD X62. Maintenance Manual "Marine", 518 pp. Winterthur Gas & Diesel Ltd., Winterthur, Switzerland (2018)
64. WinGD X72. Maintenance Manual "Marine". Document ID: DBAC848633, 560 pp. Winterthur Gas & Diesel Ltd., Winterthur, Switzerland (2018)
65. WinGD X82-B. Maintenance Manual "Marine", 606 pp. Winterthur Gas & Diesel Ltd., Winterthur, Switzerland (2017)
66. WinGD X92. Maintenance Manual "Marine". Document ID: DBAC873183, 566 pp. Winterthur Gas & Diesel Ltd., Winterthur, Switzerland (2017)

Chapter 4
Gas-Diesel Two-Stroke Ship Low-Speed Engines

The constant rise in prices for petroleum fuels and the tightening of environmental requirements for marine vessels forced manufacturers of ship low-speed engines to look for alternative solutions, related to finding new types of fuels and ways to use them. The most promising fuels that simultaneously reduce the cost of transportation and the amount of harmful emissions are gas fuels of various origins, and, first of all, natural gas.

The use of gas fuels can significantly reduce the amount of harmful emissions in comparison with fuels of petroleum origin and to completely eliminate sulfur emissions, drastically (by 90%) reduce emissions of nitrogen oxides (NO_x) and significantly (by 30%) reduce emissions of particulate matter and carbon dioxide (CO_2).

Other advantages of gaseous fuels include the absence of liquid fractions, which eliminates the dilution of oil in the working area of piston rings, and the almost complete absence of ash leads to improved lubrication conditions and increased service life of the main friction pairs. As a result, the resource of gas-powered engines can be increased by 1.3–1.5 times, while maintenance and repair costs are reduced as much.

Initially, the question of the use of gas fuel in the fleet has come for gas carriers, where the gas, evaporated in cargo tanks, had to be disposed of. The most effective way of such disposal is burning it in main and auxiliary engines. The accumulation of operating experience of the main gas-diesel engines of gas carriers allowed to expand the scope of gas fuel to the engines of other types of ships. Today, the use of natural gas is seen as a promising direction for container ships, passenger ships, car carriers, ferries, etc.

I. Bilousov et al., *Modern Marine Internal Combustion Engines*, Springer Series on Naval Architecture, Marine Engineering, Shipbuilding and Shipping 8,
https://doi.org/10.1007/978-3-030-49749-1_4

The specific operating conditions of the main engines of the vessels required the search for new original solutions for low-speed two-stroke gas-powered engines. This is primarily due to the need to maintain the possibility of operation on liquid fuels, which occurs whenever the vessel moves in ballast or the gas fuel on board is running out. For gas carriers and oil tankers, the composition of the gases, used by the power plant, may vary significantly depending on the type of cargo, navigation conditions and time. The control system of the main engine must adequately respond to such changes and ensure its operation throughout the entire range of operating modes.

On this basis, all low-speed engines are produced dual-fuel (DF), and are able to run on gas fuel with forced ignition, on liquid fuel, or on both fuels at once in different proportions. In two-stroke engines, as opposed to four-stroke engines, it is impossible to organize a relatively simple external mixture formation, since before it enters the working cylinder, the air fills a sub-piston cavity, which has a sufficiently large volume. The presence of a large amount of gas-air mixture increases the danger of an explosion and the serious implications. Therefore, in low-speed two-stroke engines, internal mixture formation is used, when gas fuel is supplied to the working cylinder after the gas distribution mechanism are closed.

There are two main approaches to internal mixing:

- gas is supplied to the working cylinder immediately after the exhaust valve is closed at the initial stage of the compression stroke under relatively low pressure, due to which such systems are called low pressure supply systems;
- gas is supplied into the combustion chamber together with the ignition fuel at the end of a high-pressure compression stroke, therefore such systems are called high-pressure supplied systems or direct gas injection (GD).

4.1 Gas-Diesel Engines of Low Pressure Series X-DF

After analyzing the various concepts of creating a low-speed gas-diesel engine conducted by WinGD in early 2011, it was decided to develop a technology for supplying gas fuel to the engine under low pressure. The first studies were performed on a test base in the Italian city of Trieste in 2013. The diesel engine RT-flex50DF was re-equipped to operate on gas fuel. Liquid fuel was used as a reserve. In this case, the gas fuel engine should work on a cycle, close to the theoretical Otto cycle.

Since 2011, WinGD has conducted a full-scale gas-diesel engine test. These developments were a continuation of the work of the firm Sulzer to create in the 70s of the last century gas-diesel low-pressure engines, based on diesel engines of the RD and RNMD series. In addition, the company also had the experience of creating gas-diesel engines with high-pressure gas supply. So in 1986, Sulzer, together with the Japanese company Diesel United, designed and tested a high-pressure gas supply system for the RTA84 engine. At that time, these technologies were unclaimed, and it was only with the rising prices for petroleum products and the tightening of environmental requirements for marine diesel engines that interest in using gas fuel in them again aroused.

In 2013, WinGD introduced the X-DF series of engines. The engines of this series were developed on the basis of diesel engines of the W-X type and are dual-fuel gas-diesel engines with a low-pressure gas supply system and an flame ignition of the gas-air mixture.

The low-pressure gas technology, developed for the dual-fuel low-speed engine, is designed to burn lean fuel mixture. Gas is supplied to the working cylinder immediately after the exhaust valve is closed, while the pressure is still relatively low. In practice, the gas supply valves are installed at a certain height from the purge ports to provide the necessary time to fill the cylinder with gas. The intake of gas fuel during the compression stroke allows it to be supplied to the cylinders under a relatively low pressure of 1.6 MPa.

In the process of compression, the gas mixes well with air and ignites with the help of an ignition fuel portion. At the same time, its value in the whole range of loads does not exceed 1% of the total injection rate at the nominal mode. To ensure dual fuel, the engine is equipped with three independent fuel supply systems, controlled by an electronic microprocessor module for individual programs, depending on the fuel, used and the operating mode.

For supplying reserved liquid fuel, a standard accumulatory fuel system is used, which is typical of all W-X engines. In this case, the engine retains the ability to work on heavy grades of fuel in the entire range of load-speed modes. Supply of pilot fuel is carried out by a separate low-capacity accumulator system. This solution allows to obtain a stable supply of small portions of fuel with the possibility of flexible regulation of the mode of operation of the flame ignition system.

To improve the ignition conditions of the lean gas-air mixture, two ignition modules are installed on each cylinder, consisting of a pre-chamber with an injector installed in it, to supply the pilot diesel fuel. The chamber is made of heat-resistant steel in the form of two separate liners and is cooled on the outer surface with water, supplied from the cooling circuit of the cylinder cover.

The internal cavity of the prechamber is connected to the combustion chamber via a tangential channel. The presence of the prechamber contributes to good mixing of air with fuel and effective self-ignition. In this case, plasma jets are ejected into the combustion chamber, effectively igniting the lean gas-air mixture.

Strictly directed distribution of plasma jets over the entire volume of the combustion chamber contributes to effective ignition and uniform combustion of the gas-air mixture without the formation of local hot spots, which reduces the level of NO_x in the combustion products by 90% compared to liquid fuel. This, in turn, makes it possible to fulfill the requirements of IMO Tier-III for NO_x emissions without exhaust gas treatment after the engine.

When the engine is running on liquid fuel, to prevent the ignition nozzles from coking, they continue to operate in the minimum steady feed mode. The pressure in the ignition fuel injection system accumulator is maintained at 120 MPa.

Gas fuel is supplied to the working space of the engine, using two gas valves, which are diametrically opposed to each other at a height of about 1/3 of the piston stroke. Valves are fastened to the jacket of the cylinder block and through the holes in the liner gas is supplied to the working cylinder (Fig. 4.2). The valve is opened by means of a hydraulic piston, and closing and retention due to a cylindrical spring. The gas valve drive oil is drawn from the exhaust valve drive system, which greatly simplifies the system design. The oil flow control in the gas valve drive system is controlled by a bistable solenoid valve, which are used in the W-X engines to control the fuel supply.

The signal to the control valve comes from a microprocessor-based engine control module. The valve is equipped with a displacement sensor, through which feedback is provided to the module. If the valve is not closing, the control module automatically stops the flow of gas and converts the engine to liquid fuel.

Gas is supplied to the valves through double-walled bellows from gas pipelines, laid on both sides of the engine. All highways are double-walled, and the space between the walls is constantly ventilated. Gas sensors are installed at the outlet of the ventilation system, which, in case of a gas leak, transmit a signal to the control unit, on arrival of which the engine is automatically transferred to liquid fuel. For safety reasons, all gas pipelines of the engine are made of stainless steel.

The engine gas operation allows to obtain power at the output flange at the level of 80% of the nominal power of the base diesel. If it is necessary to obtain higher power, the engine is converted to liquid fuel. Starting and stopping the engine for the sake of safety is also carried out on liquid fuel.

The main factor, limiting power, is the occurrence of a detonation phenomenon in the cylinder's working space. In order to prevent the engine from detonating, each cylinder has detonation sensors, which, if it occurs, transmit a signal to the electronic control unit, which reduces the load on the engine or transfers it to liquid fuel. In addition to the knock sensors, the engines are equipped with pressure sensors in the working cylinder, the main task of which is to monitor misfire. If it happens, unburned gas-air mixes can accumulate in the exhaust receiver, which can cause an explosion and damage the engine. In addition, it is possible to control the ingress of gas fuel into the sub-piston space.

Currently, WinGD has launched production of low-speed two-stroke gas-diesel engines, based on previously developed W-X diesel engines. These engines include the W-X52DF W-X62DF, W-X72DF, W-X82DF and W-X92DF models. In general, the developed technology of converting engines to gas fuel with maintaining the possibility of their operation on liquid fuel, is focused both on new engines and on upgrading engines, already in operation.

As an example, in Fig. 4.1 shows a cross section of such an engine in a gas-diesel model W-X62DF.

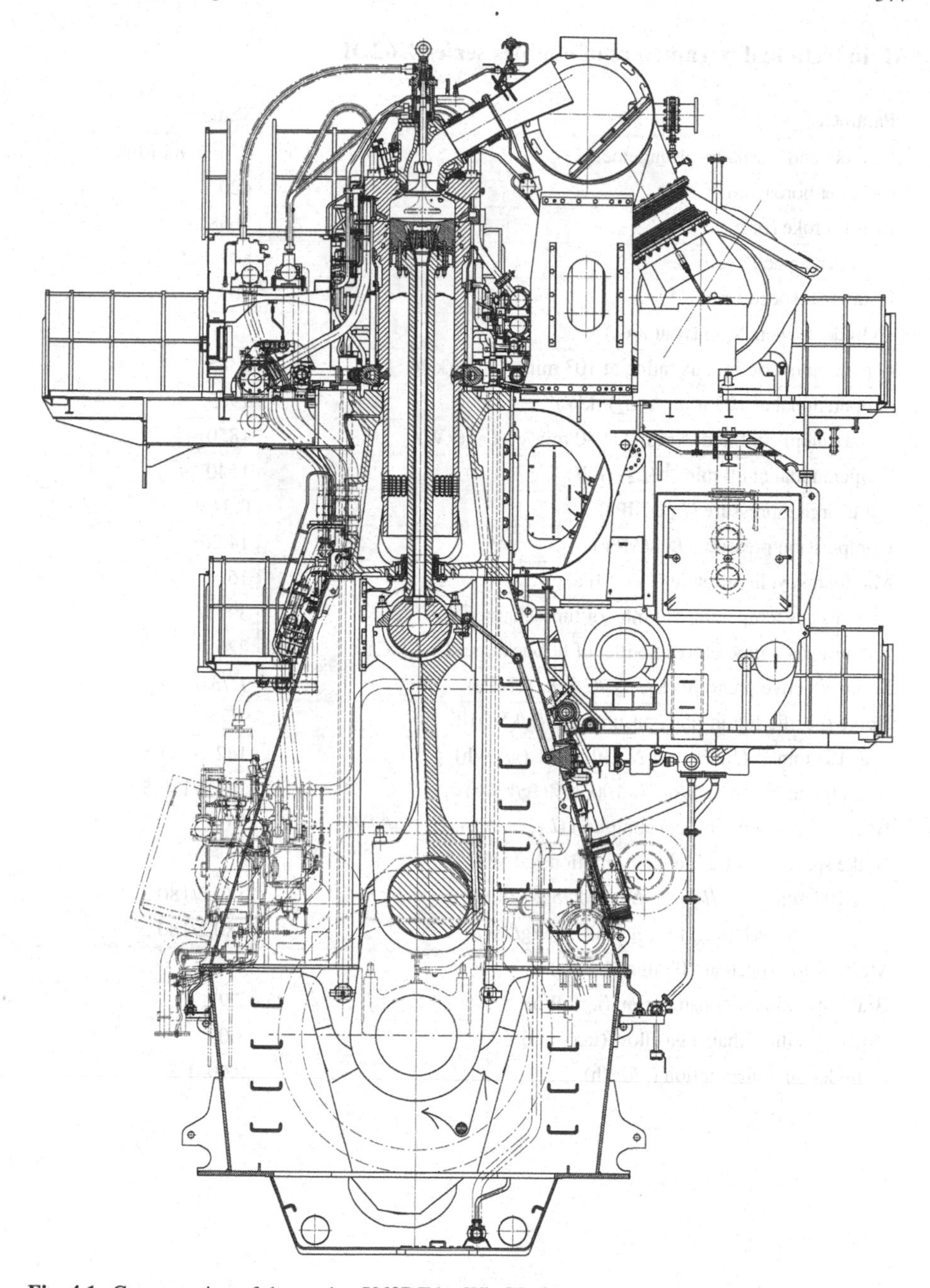

Fig. 4.1 Cross-section of the engine X62DF by WinGD firm [1]

Main technical parameters of engines series X62DF

Parameter	Value
Number and cylinders arrangement	5, 6, 7, 8 in-line
Cylinder bore (mm)	620
Piston stroke (mm)	2658
Cylinder capacity (dm^3)	802.47
Rotational speed (min^{-1})	103
Cylinder power, N_e (layout area)	
– maximum continuous rating at 103 min^{-1} (L_1) (kW)	2385
– operational at 103 min^{-1} (L_2) (kW)	1985
– maximum continuous rating at 80 min^{-1} (L_3) (kW)	1850
– operational at 80 min^{-1} (L_4) (kW)	1540
Air charging pressure (L_1) (MPa)	0.345
Compression pressure (L_1) (MPa)	14.20
Maximum cycle pressure (L_1) (MPa)	16.50
Exhaust gas temperature at inlet of turbocharger (L_1) (°C)	398.0
Exhaust gas temperature at outlet of turbocharger (L_1) (°C)	225.0
Mean effective pressure (L_1, L_3)/(L_2, L_4) (MPa)	1.731/1.44
Brake specific liquid fuel consumption (g/kWh)	
– at 103 min^{-1} (L_1/L_2) for N_e 100/75% (g/kWh)	142.5/137.5
– at 80 min^{-1} (L_3/L_4) for N_e 100/75% (g/kWh)	144.5/139.5
Brake specific pilot oil consumption (L_1, L_3)/(L_2, L_4) (g/kWh)	1.0/1.2
Brake specific liquid fuel consumption (g/kWh)	
– at 103 min^{-1} (L_1/L_2) for N_e 100/75% (g/kWh)	180.2/180.2
– at 80 min^{-1} (L_3/L_4) for N_e 100/75% (g/kWh)	180.2/180.2
Mean piston speed at 103 min^{-1} (m/s)	9.13
Brake specific air consumption (кg/kWh)	7.92
Brake specific exhaust gas flow (кg/kWh)	8.09
Cylinder oil consumption (g/kWh)	0.60–1.3

Dimensions and weight engines X62DF

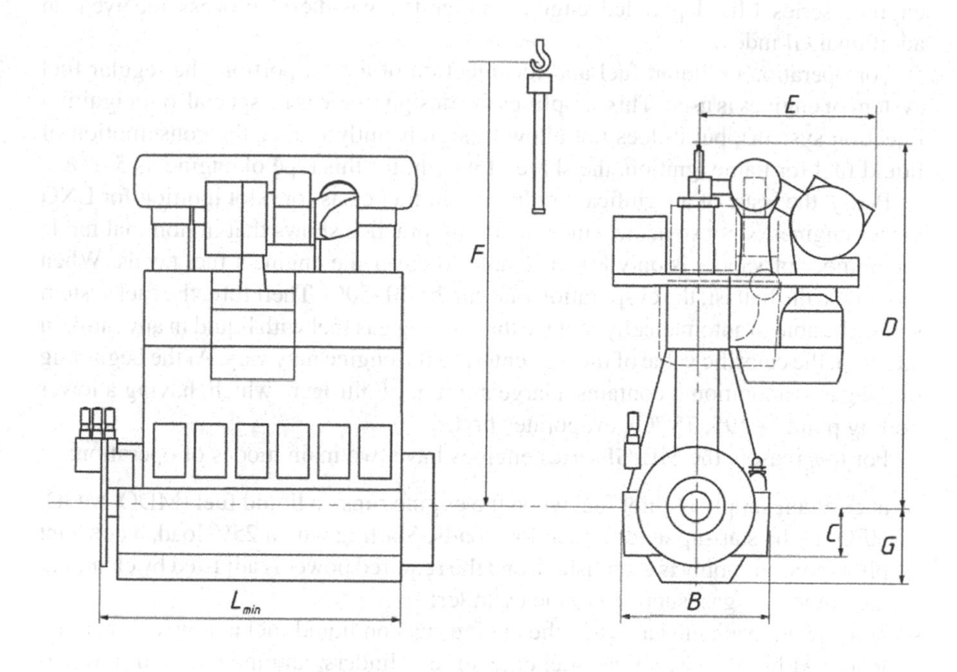

Number of cylinders		5	6	7	8
L_{min}		7000	8110	9215	10,320
Weight (kg)		325,000	377,000	435,000	482,000

B (mm)	C (mm)	D (mm)	E (mm)	F (mm)	G (mm)
4200	1360	9580	3915	11,775	2110

4.2 Gas Diesel Engines High Pressure Series ME-GI

The development and implementation of high-pressure systems has been carried out by MAN, the largest manufacturer of ship low-speed engines, since the 1990s. The first low-speed gas-diesel engine 12K80MC-GI with a capacity of 40 MW was built by Mitsui and put into operation at a power station in Chiba, not far from Tokyo in 1994. In May 2011, the Copenhagen Diesel Engine Research Center (Copenhagen Diesel Research Center) hosted a presentation of a new series of gas-diesel engines ME-GI, which became the culmination of many years of work on their creation. Initially, the engines were developed for LNG tankers, but later their use was extended to almost all types of ships.

The basic models for ME-GI gas-diesel engines are two-cycle low-speed diesel engines series ME. Upgraded engines under the gas-diesel process received an additional GI index.

For operation on liquid fuel and for injection of a pilot portion, the regular fuel system of engines is used. This simplifies the design (there is no special pilot ignition injection system), but it does not allow to significantly reduce the consumption of liquid fuel for flame ignition, the share of which, for this type of engine, is 5–8%.

Partly, the issue of a significant reduction in fuel costs for pilot ignition for LNG tanker engines is not so acute, since operating practice shows that in nominal mode the number of vapors is only 80–90% able to cover the engine's fuel needs. When moving in the ballast, the evaporation rate can be 40–50%. Therefore, the fuel system should be able to automatically replace the missing gas fuel with liquid in any ratio. In addition, the calorific value of the gas, entering the engine may vary. At the beginning of the gas evaporation it contains a large amount of nitrogen, which, having a lower boiling point (−195, 75 °C), evaporates first.

For this reason, the ME-GI series engines have two main modes of operation:

- at constant supply of pilot fuel when the engine runs on liquid fuel (MDO, MGO, HFO) at the start-up mode and at low loads. Starting with a 25% load, a constant pilot ignition supply is established, and the required power is adjusted by changing the amount of gas, supplied to the cylinder;
- at using all available gas when the engine runs on liquid fuel at low and medium loads. At high loads, all gas fuel enters the cylinders, and the required power is adjusted by changing the cycle supply of liquid fuel.

The transition from one type of fuel to another, as well as the switching from mode to mode, is carried out automatically without reducing the shaft power over the entire range of possible engine loads.

Equipment for supplying gas fuel under high pressure includes compressors, heat exchangers, a system for supplying gas fuel to the working cylinders, gas supply control modules and gas injectors.

All gas pipelines on the engine are all-welded and only in the joints of the pipes, that divert gas fuel to the flow control units, are flange connections necessary for servicing the elements of the gas system. The piping design is engineered to compensate for thermal expansion during engine heating. All pipes of the gas system are designed for pressures, exceeding the working one by 50%, and during factory tests they are tested with pressure 150% higher, than the working one. All gas pipes are placed in protective shells, that can withstand the pressure, that can occur when the main line is broken. The inner space between the shell and the pipeline is connected to a forced ventilation system, which provides about 30 times the air change for an hour. Vented also includes cavities, adjacent to the main elements of the fuel system, where gas leakage may occur.

To improve the safety of operation of the engines, the inert gas system is provided as part of the power plant, which allows purging both the entire gas supply system and its individual elements under a pressure of 0.4–0.8 MPa. Such cleaning is a mandatory procedure when switching to a diesel cycle operation or if any part of the gas supply system is damaged.

The gas supply to the combustion chamber is carried out immediately after the ignition portion of the liquid fuel is supplied into the cylinder and ignited. Thus, a high degree of burnout of the fuel is achieved and the danger of missing the ignition is prevented, as well as the ingress of gas through the gaps of the piston rings into the sub-piston space. All gas supply control elements are arranged in a single module, which includes: a gas accumulator, a main shut-off valve with a hydraulic drive, an inert gas system purge valve, an injectors hydraulic control valve. The module itself is attached to the cylinder head (Fig. 4.2), which has internal drilling to supply gas from the control module to the gas injectors, installed in the cylinder head next to the injectors for the injection of liquid fuel.

Gas fuel from the supply line through the back-pressure valve enters the pressure accumulator, designed as a cavity in the module case and calculated on approximately 20 cycle gas supplies at the nominal load. The gas fuel pressure in the system is maintained at 30 MPa. The presence of an accumulator in the control unit serves to reduce the pressure drop in the fuel injection process. Stable pressure is necessary for the control system, that could be correctly determine the valve opening time, which determines the amount of cyclic supply.

In the absence of a control signal to the control unit of the main shut-off valve, the valve closes and the gas does not flow to the gas injectors. When an electric signal is received from the engine control unit to the control unit of the main shut-off valve, its valve moves and supplies control oil to the hydraulic drive mechanism of the main shut-off valve. The valve opens and gas flows to the injectors, the needle valves of which remain closed at this time. Filling the channels between the module and the injectors, the gas acts on the pressure sensor. Information about the actual pressure, obtained from the sensor, goes to the engine control unit and is used, when calculating the required injector opening time. Based on the received information, the control unit generates a signal, supplied to the control unit of the injectors hydraulic actuator. Under the action of a signal, the block spool moves and delivers control oil to the injector drive. Opening, the injectors make gas supply to the engine's combustion chamber.

After removing the control signal from the control unit of the injector's hydraulic actuator, the valve, moving, switches the oil from the hydraulic circuit to the drain, the valve of the injectors is closed. Removing the signal from the control unit of main shut-off valve, leads to its closure, and the system returns to its original state.

There are two gas injectors installed cylinder in special wells, made in the cylinder head and located near the liquid fuel nozzles. In the closed state, the needle valve of the gas injector is held by the force of the spring. In the lower part of the needle valve there is a collar, precision fitted to the body, acting as a hydraulic piston. The channels in the guide and in the body of the needle valve oil from the supply control module enters the annular cavity under the collar, forcing the needle valve to open. To prevent gas leakage between the needle valve and the atomizer body, a special sealing oil under pressure of 0.2–0.3 MPa higher, than the gas pressure before the atomizer, is constantly comes into the gap between them.

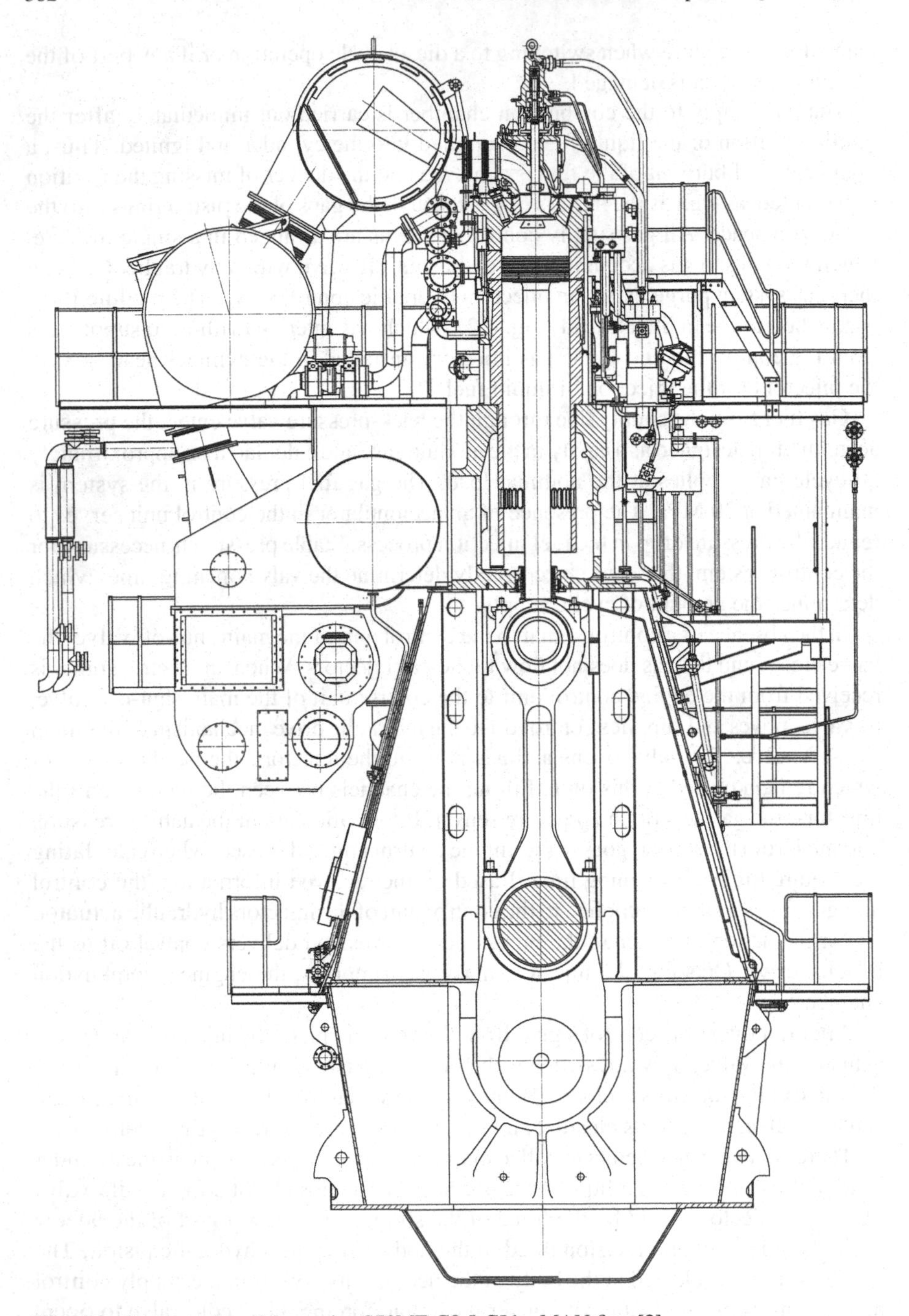

Fig. 4.2 Cross-section of the engine S70ME-C8.5-GI by MAN firm [2]

Main technical parameters of engines series S70ME-C8.5-GI

Parameter	Value
Number and cylinders arrangement	5, 6, 7, 8 in-line
Cylinder bore (mm)	700
Piston stroke (mm)	2800
Cylinder capacity (dm^3)	1077.57
Rotational speed (min^{-1})	91
Cylinder power, N_e (layout area)	
– maximum continuous rating at 91 min^{-1} (L_1) (kW)	3270
– operational at 91 min^{-1} (L_2) (kW)	2610
– maximum continuous rating at 73 min^{-1} (L_3) (kW)	2620
– operational at 73 min^{-1} (L_4) (kW)	2100
Air charging pressure (L_1) (MPa)	0.365
Compression pressure (L_1) (MPa)	13.50
Maximum cycle pressure (L_1) (MPa)	15.20
Exhaust gas temperature at inlet of turbocharger (L_1) (°C)	395.0
Exhaust gas temperature at outlet of turbocharger (L_1) (°C)	231.0
Brake specific liquid fuel consumption (g/kWh)	
– at 91 min^{-1} (L_1/L_2) for N_e 100/75% (g/kWh)	139.2/133.0
– at 73 min^{-1} (L_3/L_4) for N_e 100/75% (g/kWh)	139.2/133.0
Brake specific pilot oil consumption (L_1, L_3)/(L_2, L_4) (g/kWh)	5.0/6.3
Brake specific liquid fuel consumption (g/kWh)	
– at 91 min^{-1} (L_1/L_2) for N_e 100/75% (g/kWh)	169.0/163.0
– at 73 min^{-1} (L_3/L_4) for N_e 100/75% (g/kWh)	169.0/163.0
Mean piston speed at 91 min^{-1} (m/s)	8.49
Brake specific air consumption (кg/kWh)	8.07
Brake specific exhaust gas flow (кg/kWh)	8.32
Cylinder oil consumption (g/kWh)	0.70

Dimensions and weight engines S70ME-C8.5-GI

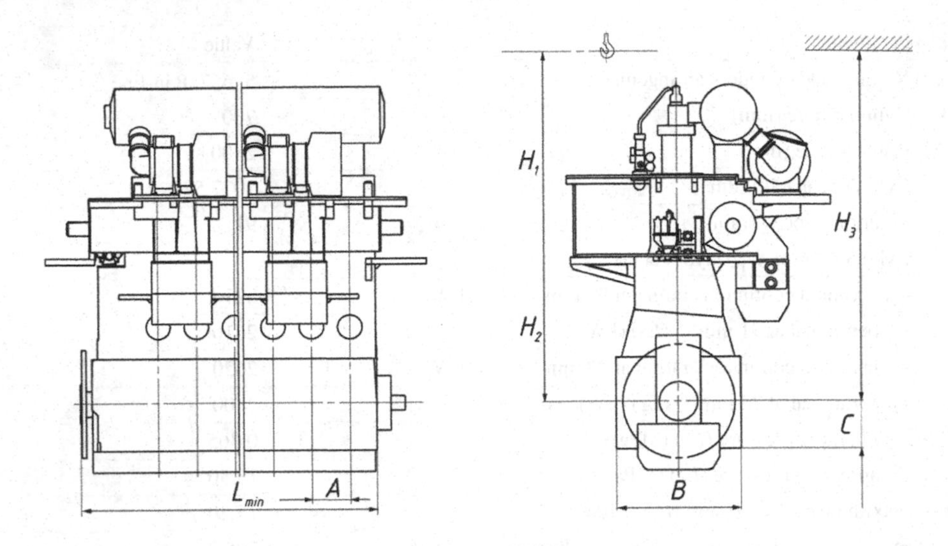

Number of cylinders		5	6	7	8
L_{min}		7514	8704	9894	11,084
Weight (kg)		451,000	534,000	605,000	681,000
A (mm)	B (mm)	C (mm)	H_1 (mm)	H (mm)	H_3 (mm)
1190	4390	1520	12,550	11,725	11,500

References

1. WinGD X62DF.: Maintenance Manual "Marine". Document ID: DBAD220106, 560 pp. Winterthur Gas & Diesel Ltd. Winterthur, Switzerland (2018)
2. MAN B&W L70ME-C8.5-GI-TII. Project Guide. Electronically Controlled Dual Fuel Two-stroke Engines, 417 pp. MAN Diesel & Turbo. Copenhagen, Denmark (2014)

Bibliography

1. Техническая спецификация двигателя ABC тип (V)DZC. Anglo Belgian Corporation, Gent (Belgium), 4 pp. DATASHEET 12/16DZC-RU-03/2013
2. Техническая спецификация двигателя ABC тип DL36. Anglo Belgian Corporation, Gent (Belgium), 4 pp. DATASHEET 6/8DL36-RU-10/2012
3. Marine Engine Application and Installation. Guide. Introduction. General Information. LEKM8460, 536 pp. Caterpillar Inc. Printed in U.S.A. (1998)

4. Gesamt-Motor-Betriebsanleitung Typ M601C, 1982 pp. MaK Motoren GmbH & Co. KG, Kiel (1982)
5. INSTRUCTION BOOK VOLUME I Engine type H32/40 Hyundai Heavy Industries Co., Ltd. Engine & Machinery Division 1, 376 pp. Cheonha-Dong, Dong-Gu, Ulsan, Korea
6. MAN B&W Diesel Ltd. Selection Guide 2001, 52 pp. MAN B&W Diesel Ltd Paxman Hythe Hill, Colchester, Essex, C01 2HW England
7. M 46 DF Project guide/Propulsion, 198 pp. Caterpillar Motoren GmbH & Co. Kiel, Germany (2015)
8. MAN Group History 1758–2006, 29 pp. MAN Aktiengesellschaft Corporate Communications, Munich (2008)
9. The Sulzer diesel engine centenary, pp. 57–60. Schip & Werf de ZEE (1998)
10. Brown, D.T., Sulzer, A.: History of the Sulzer low-speed marine diesel engine. Published in celebration of the 150-th anniversary of Sulzer Brothers Ltd., 47 pp. Winterthur, Switzerland (1984)
11. Smil, V.: Two Prime Movers of Globalization. The History and Impact of Diesel Engines and Gas Turbines, England, London, 261 pp. The MIT Press Cambridge, Massachusetts (2010)

Printed in the United States
by Baker & Taylor Publisher Services